VOLUME FORTY SIX

ADVANCES IN ECOLOGICAL RESEARCH

ADVANCES IN ECOLOGICAL RESEARCH

Series Editor

GUY WOODWARD
School of Biological and Chemical Sciences
Queen Mary University of London
London, UK

VOLUME FORTY SIX

Advances in ECOLOGICAL RESEARCH

Global Change in Multispecies Systems Part 1

Edited by

UTE JACOB
Institute for Hydrobiology and Fisheries Science
University of Hamburg
Hamburg, Germany

GUY WOODWARD
School of Biological and Chemical Sciences
Queen Mary University of London
London, UK

ELSEVIER
AMSTERDAM • BOSTON • HEIDELBERG • LONDON
NEW YORK • OXFORD • PARIS • SAN DIEGO
SAN FRANCISCO • SINGAPORE • SYDNEY • TOKYO
Academic Press is an imprint of Elsevier

Academic Press is an imprint of Elsevier

32 Jamestown Road, London NW1 7BY, UK
The Boulevard, Langford Lane, Kidlington, Oxford, OX51GB, UK
Radarweg 29, PO Box 211, 1000 AE Amsterdam, The Netherlands
225 Wyman Street, Waltham, MA 02451, USA
525 B Street, Suite 1900, San Diego, CA 92101-4495, USA

First edition 2012

ISBN: 978-0-12-396992-7
ISSN: 0065-2504

For information on all Academic Press publications
visit our website at store.elsevier.com

Printed and bound in UK
12 13 14 15 11 10 9 8 7 6 5 4 3 2 1

CONTENTS

CONTRIBUTORS

Cecilia Alonso
Ecología Funcional de Sistemas Acuáticos, Centro Universitario Regional Este (CURE), Universidad de la República, Ruta, Rocha, Uruguay

Isabel Alves-Dos-Santos
Departamento de Ecologia, Instituto de Biociências, Universidade de São Paulo, São Paulo, São Paulo, Brazil

Matías Arim
Departamento de Ecología y Evolución, Facultad de Ciencias, Centro Universitario Regional Este (CURE), Universidad de la República, Burnett s/n; Facultad de Ciencias, Universidad de la República, Iguá, CP 11400, Montevideo, Uruguay, and Center for Advanced Studies in Ecology and Biodiversity (CASEB), Depto. de Ecología, Facultad de Ciencias Biológicas, Pontificia Universidad Católica, CP 6513677, Santiago, Chile

Patrick D. Armitage
Freshwater Biological Association River Laboratory, East Stoke, Wareham, Dorset, United Kingdom

Wolf E. Arntz
Alfred Wegener Institute for Polar and Marine Research, Bremerhaven, Germany

Esteban R. Barrera-Oro
Instituto Antártico Argentino and CONICET, and Museo Argentino de Ciencias Naturales 'Bernardino Rivadavia', Buenos Aires, Argentina

Meryem Beklioğlu
Department of Biology, Limnology Laboratory, Middle East Technical University, Üniversiteliler Mahallesi, Dumlupınar Bulvarı, Çankaya, Ankara, Turkey

Alice Boit
Department of Ecology and Ecosystem Modeling, Institute of Biochemistry and Biology, University of Potsdam, Potsdam, Germany

Thomas Brey
Alfred Wegener Institute for Polar and Marine Research, Bremerhaven, Germany

Lee E. Brown
School of Geography, University of Leeds, Leeds, United Kingdom

Sandra Brucet
European Commission, Joint Research Centre, Institute for Environment and Sustainability, Ispra, Italy

Daniel W. Carstensen
Department of Bioscience, Aarhus University, Aarhus, Denmark

Marcus A.M. De Aguiar
Instituto de Física Gleb Wataghin, Universidade Estadual de Campinas, Campinas, São Paulo, Brazil

Yoko L. Dupont
Department of Bioscience, Aarhus University, Aarhus, Denmark

Scott D. Dyer
Procter & Gamble, Cincinnati, Ohio, USA

Francois K. Edwards
Centre for Ecology and Hydrology, Wallingford, United Kingdom

Leslie Faggiano
Institute of Aquatic Ecology, Universitat de Girona, Girona, Spain, and Laboratoire Evolution et Diversité Biologique, Université Paul Sabatier, CNRS, Toulouse Cedex 9, France

Stefan Geisen
Department of Terrestrial Ecology, Institute of Zoology, University of Köln, Biozentrum Köln, Köln, Germany

Julieta Genini
Departamento de Botânica, Laboratório de Fenologia, UNESP Univ Estadual Paulista, Rio Claro, São Paulo, Brazil

Iván González-Bergonzoni
Departamento de Ecología y Evolución, Facultad de Ciencias, Centro Universitario Regional Este (CURE), Universidad de la República, Burnett s/n, Maldonado, Uruguay, and Department of Bioscience, Aarhus University, Vejlsøvej, Silkeborg, Denmark

Angélica L. González
Department of Zoology, University of British Columbia, Vancouver, British Columbia, Canada

Guillermo Goyenola
Departamento de Ecología y Evolución, Facultad de Ciencias, Centro Universitario Regional Este (CURE), Universidad de la República, Burnett s/n, Maldonado, Uruguay

Paulo R. Guimarães Jr.
Departamento de Ecologia, Instituto de Biociências, Universidade de São Paulo, São Paulo, São Paulo, Brazil

Melanie Hagen
Department of Bioscience, Aarhus University, Aarhus, Denmark

Rebecca M.L. Harris
School of Geography, Earth and Environmental Sciences, University of Birmingham, Edgbaston, Birmingham, United Kingdom

Carlos Iglesias
Departamento de Ecología y Evolución, Facultad de Ciencias, Centro Universitario Regional Este (CURE), Universidad de la República, Burnett s/n, Maldonado, Uruguay

Ute Jacob
Institute for Hydrobiology and Fisheries Science, University of Hamburg, Hamburg, Germany

Gareth B. Jenkins
School of Biological and Chemical Sciences, Queen Mary University of London, London, United Kingdom

Erik Jeppesen
Department of Bioscience, Aarhus University, Vejlsøvej, Silkeborg, Denmark; Greenland Climate Research Centre (GCRC), Greenland Institute of Natural Resources, Kivioq, P.O. Box 570 3900, Nuuk, Greenland, and Sino-Danish Centre for Education and Research (SDC), Beijing, China

Pedro Jordano
Integrative Ecology Group, Estación Biológica de Doñana, CSIC, Sevilla, Spain

Christopher N. Kaiser-Bunbury
Department of Bioscience, Aarhus University, Aarhus, Denmark

Michael Kaspari
Department of Zoology, University of Oklahoma, Norman, Oklahoma, USA, and Smithsonian Tropical Research Institute, Balboa, Panama

W. Daniel Kissling
Department of Bioscience, Aarhus University, Aarhus, Denmark

Rainer Knust
Alfred Wegener Institute for Polar and Marine Research, Bremerhaven, Germany

Sarian Kosten
Department of Aquatic Ecology and Water Quality Management, Wageningen University, Wageningen, The Netherlands, and Leibniz-Institute of Freshwater Ecology and Inland Fisheries (IGB), Berlin/Neuglobsow, Germany

Carla Kruk
Laboratory of Ethology, Ecology and Evolution, Instituto de Investigaciones Biológicas Clemente Estable, Italia, CP 11600, and Ecología Funcional de Sistemas Acuáticos, Limnología, IECA, Facultad de Ciencias, Universidad de la República, Iguá, CP 11400, Montevideo, Uruguay

Gissell Lacerot
Ecología Funcional de Sistemas Acuáticos, Centro Universitario Regional Este (CURE), Universidad de la República, Ruta, Rocha, Uruguay

Sandra Lavorel
Laboratoire d'Ecologie Alpine, CNRS UMR 5553, Grenoble Cedex 9, France

Mark E. Ledger
School of Geography, Earth and Environmental Sciences, University of Birmingham, Edgbaston, Birmingham, United Kingdom

Kate P. Maia
Departamento de Ecologia, Instituto de Biociências, Universidade de São Paulo, São Paulo, São Paulo, Brazil

Felix C. Mark
Alfred Wegener Institute for Polar and Marine Research, Bremerhaven, Germany

Pablo A. Marquet
Center for Advanced Studies in Ecology and Biodiversity, Institute of Ecology and Biodiversity, Pontificia Universidad Catolica de Chile, Santiago, Chile, and The Santa Fe Institute, Santa Fe, New Mexico, USA

Flavia M. Darcie Marquitti
Departamento de Ecologia, Instituto de Biociências, Universidade de São Paulo, São Paulo, São Paulo, Brazil

Néstor Mazzeo
Departamento de Ecología y Evolución, Facultad de Ciencias, Centro Universitario Regional Este (CURE), Universidad de la República, Burnett s/n, and South American Institute for Resilience and Sustainability Studies (SARAS), Maldonado, Uruguay

Órla Mclaughlin
Environmental Research Institute, and School of Biological, Earth, and Environmental Sciences, University College Cork, Cork, Ireland

Mariana Meerhoff
Departamento de Ecología y Evolución, Facultad de Ciencias, Centro Universitario Regional Este (CURE), Universidad de la República, Burnett s/n; Department of Bioscience, Aarhus University, Vejlsøvej, Silkeborg, Denmark, and South American Institute for Resilience and Sustainability Studies (SARAS), Maldonado, Uruguay

Alexander M. Milner
School of Geography, Earth and Environmental Sciences, University of Birmingham, Edgbaston, Birmingham, United Kingdom, and Institute of Arctic Biology, University of Alaska, Fairbanks, Alaska, USA

Katja Mintenbeck
Alfred Wegener Institute for Polar and Marine Research, Bremerhaven, Germany

Eugenia Moreira
Instituto Antártico Argentino and CONICET, Buenos Aires, Argentina

L. Patricia C. Morellato
Departamento de Botânica, Laboratório de Fenologia, UNESP Univ Estadual Paulista, Rio Claro, São Paulo, Brazil

Shigeta Mori
Forestry and Forest Products Research Institute, Tsukuba, Japan

Christian Mulder
National Institute for Public Health and the Environment (RIVM), Bilthoven, The Netherlands

Eoin J. O'Gorman
School of Biological and Chemical Sciences, Queen Mary University of London, London, United Kingdom

Jens M. Olesen
Department of Bioscience, Aarhus University, Aarhus, Denmark

Juan Pablo Pacheco
Departamento de Ecología y Evolución, Facultad de Ciencias, Centro Universitario Regional Este (CURE), Universidad de la República, Burnett s/n, Maldonado, Uruguay

Claus Rasmussen
Department of Bioscience, Aarhus University, Aarhus, Denmark

Axel G. Rossberg
Centre for Environment, Fisheries and Aquaculture Science (Cefas), Lowestoft Laboratory, Suffolk, United Kingdom, and Medical Biology Centre, School of Biological Sciences, Queen's University Belfast, Belfast, United Kingdom

Robert W. Sterner
Ecology, Evolution & Behavior, University of Minnesota, St. Paul, Minnesota, USA

Anneli Strobel
Alfred Wegener Institute for Polar and Marine Research, Bremerhaven, Germany

Franco Teixeira-de Mello
Departamento de Ecología y Evolución, Facultad de Ciencias, Centro Universitario Regional Este (CURE), Universidad de la República, Burnett s/n, Maldonado, Uruguay

Kristian Trøjelsgaard
Department of Bioscience, Aarhus University, Aarhus, Denmark

Jason M. Tylianakis
School of Biological Sciences, University of Canterbury, Christchurch, New Zealand

Mariana Morais Vidal
Departamento de Ecologia, Instituto de Biociências, Universidade de São Paulo, São Paulo, São Paulo, Brazil

Winfried Voigt
Community Ecology Group, Institute of Ecology, Friedrich Schiller University, Jena, Germany

J. Arie Vonk
Institute for Biodiversity and Ecosystem Dynamics, University of Amsterdam, Amsterdam, and Institute for Water and Wetland Research, Radboud University Nijmegen, Nijmegen, The Netherlands

Diana H. Wall
Department of Biology & Natural Resource Ecology Laboratory (NREL), School of Global Environmental Sustainability, Colorado State University, Fort Collins, Colorado, USA

Guy Woodward
School of Biological and Chemical Sciences, Queen Mary University of London, London, United Kingdom

PREFACE

Editorial Commentary: Global Change in Multispecies Systems Part 1

Ute Jacob*, Guy Woodward†
*Institute for Hydrobiology and Fisheries Science, University of Hamburg, Hamburg, Germany
†School of Biological and Chemical Sciences, Queen Mary University of London, London, United Kingdom

Evaluating the consequences of global change is a major challenge in current research, especially in the light of rising concerns over anthropogenic impacts in general and climate forcing and its potential impacts on ecosystems in particular (Daufresne et al., 2009; Parmesan, 2006; Parmesan and Yohe, 2003; Raffaelli, 2004; Sarmiento et al., 2004; Walther et al., 2002; Woodward et al., 2010a,b). To address how multispecies systems might change over the next century, ecologists need to focus on how their properties will be affected by the main drivers behind global change that have already been identified (Sala et al., 2000; Thomas et al., 2004), as well as dealing with emerging and potentially synergistic multiple stressors (e.g. Feuchtmayr et al., 2009; McKee et al., 2003; Memmott et al., 2007; Moss et al., 2003). This collection of papers in Volume 46 of *Advances in Ecological Research* represents the first of a set of three under the theme of *Global Change in Multispecies Systems*, each of which contains a different blend of papers from both empirical and theoretical approaches: this commentary is concerned primarily with the papers appearing in this and the two subsequent volumes, while briefly placing them in the wider context of recent advances in the field, rather than giving an in-depth review.

Global ecological change is driven by a suite of environmental parameters that control the structure and dynamics of multispecies systems (Thomas et al., 2004; Tylianakis, 2009; Woodward et al., 2010a). The main drivers investigated in Volumes 46–48 include land use change and habitat fragmentation (Hagen et al., 2012), drought (Ledger et al., 2012a,b) and other components of climate change (Meerhoff et al., 2012; Mintenbeck et al., 2012; Möllmann and Diekmann, 2012; O'Gorman et al., 2012; Peck et al., 2012), eutrophication (Jeppesen et al., 2012), and resource overexploitation (Peck et al., 2012; Rossberg, 2012). Other aspects of global change that are touched on in the papers published here and in other recent volumes in the series include invasive species (Jacob et al., 2011; Woodward et al., 2010a,b), marine and freshwater acidification (Layer et al., 2010; Mintenbeck et al., 2012; Peck et al., 2012), and agricultural intensification (Feld et al., 2011; Hladyz et al., 2011a,b; Mulder et al., 2011, 2012). There is also considerable potential for synergies to arise from different combinations of these stressors, as it is rare that a single driver will be operating in isolation, but these interactive effects are still poorly understood (Feld et al., 2011; Friberg et al., 2011; Woodward et al., 2010a,b).

In terms of response variables, the six papers in this volume are concerned primarily with populations, communities and ecosystems over ecological timescales, whereas some of those in the two subsequent volumes encompass an even wider range of organisational levels (e.g. genes, individuals: Moya-Laraño et al., 2012; O'Gorman et al., 2012) and spatiotemporal scales (e.g. eco-evolutionary dynamics of food webs: Moya-Laraño et al., 2012).

Habitat fragmentation is one of the key threats to both global and local biodiversity, as species and higher-level responses will be determined by their spatial context and source-sink dynamics within the landscape (Raffaelli, 2004). Surprisingly, though, very little attention has been paid to gauging how ecological networks (e.g. food webs, mutualistic networks) might be affected by habitat size or fragmentation, despite the fact that the configuration and strength of interactions between species may be just as important as the identity and number of species themselves (Hagen et al., 2012; McLaughlin et al., 2010). Even less attention has been paid to how potential synergistic effects of additional stressors, such as climate change, may amplify or mitigate the effects of habitat fragmentation. Here, Hagen et al. (2012) make a first attempt at considering how best to integrate spatial and ecological networks, to explore the effects of dispersal, colonisation, extinction and habitat fragmentation on network structure and dynamics. They also make the first steps towards embedding network approaches more explicitly within applied and landscape ecology and ideas arising from metacommunity theory, highlighting the great potential for improving on the current species-based or habitat-centric approaches to management and conservation of biodiversity in the face of global change. Hagen et al.'s paper reflects a growing realisation within the applied ecology fraternity that effective monitoring, management and conservation of multispecies systems must move beyond the spatial and temporal boundaries that are typically used to delimit them at present. The potential for eco-evolutionary dynamics to create feedbacks in fragmented food webs are also touched on, with species both responding to and shaping their biotic and abiotic environment over a range of timescales. These points are covered in greater depth in the preceding and subsequent volumes by Melian et al. (2011) and Moya-Laraño et al. (2012), which represent pioneering attempts to bridge the gap between ecology and evolution in multispecies systems.

Understanding the likely cause-and-effect relationships behind global change in natural systems requires an understanding of how they change through both time and space. These themes recur throughout these three volumes, especially in those papers focused on fresh waters (Jeppesen et al., 2012; Ledger et al., 2012a,b; Meerhoff et al., 2012; O'Gorman et al., 2012). These seemingly fragile ecosystems cover only a tiny percentage of Earth's surface, yet they are disproportionately important, especially in terms of the biodiversity they hold and the ecosystem goods and services they provide. They are also fragmented islands of water in a predominantly terrestrial landscape (Hynes, 1975), and as such, they are vulnerable to an array of anthropogenic stressors (Hladyz et al., 2011a,b), especially as much of the world's human population lives clustered close to their shores or on their floodplains. In the context of climate change, the analysis of flood and drought risks is critically important for preserving the structure and functioning of fresh waters (Milly et al., 2006). Unfortunately, our ability to make accurate and predictive assessments is still severely constrained by the current lack of data and understanding, with most examples to date being limited to correlational studies (Lake, 2003). Here an attempt to help redress this balance is made by the Ledger et al. (2012a,b) paper, which presents a comprehensive set of results building from earlier studies (Ledger and Hildrew, 2001; Ledger et al., 2008, 2009) in one of the first mesocosm field experiments to measure the effects of drought on replicate macroinvertebrate communities, providing a rare experimental test of community resilience in aquatic ecosystems at intergenerational scales. The communities were resilient to relatively low-frequency disturbance (quarterly droughts), but the capacity for recovery was soon exceeded as disturbance frequency increased (monthly droughts), skewing community structure and functioning. This highlights how multispecies responses to climate change may be highly non-linear, with marked thresholds and tipping points, rather than a simple progressive and gradual erosion of ecological integrity.

The Ledger et al. (2012a,b) study isolates a single component of climate change (drought), as does the O'Gorman et al. (2012) study in Volume 47, although the latter focuses on the effects of warming on stream ecosystems. Climate change is among the most important but also most complex drivers of global change, and quantifying and anticipating its effects on the structure and functioning of multispecies systems are daunting. This is especially challenging because climate change represents an amalgam of changes in both the abiotic (atmospheric and hydrological) and the biotic (invasions, extinctions and evolution) environments, in addition to the enormous scope for unexpected feedbacks to arise between these components.

Identifying correlations between climate variables and the biota is therefore only a beginning, and we need to move rapidly beyond the species-centric bioclimatic envelope approach to address responses in multispecies systems (Pearson and Dawson, 2003). A deeper mechanistic understanding is needed urgently, and considerable effort is being devoted to achieving this goal via the use of increasingly sophisticated experiments and models. While this gap in our knowledge is being bridged, correlational data continue to provide invaluable insights and help guide further studies by providing the wider context within which to search for likely mechanisms (e.g. Konig et al., 2002; Milner et al., 2000, 2008, 2009; Rawcliffe et al., 2010), as shown here in a new synthesis by Meerhoff et al. (2012), even though they cannot demonstrate causality unequivocally. Community or ecosystem responses to climate change are likely to be complex, because species populations interact within their respective ecological networks at both ecological and evolutionary timescales (Melian et al., 2011; Moya-Laraño et al., 2012; Olesen et al., 2010; Woodward et al., 2010a). Competition, natural enemies, and physiological constraints (among many others) influence species distributions, and these will change with both the climate and shifts in local communities on a global scale (Möllmann and Diekmann, 2012; Walther, 2010; Woodward et al., 2010a,b). Here, Meerhoff et al. (2012) provide an overview of multispecies responses to climatic change, as detected using space-for-time substitutions in lake ecosystems. They consider not only structural properties (species richness, biomass, density, body size) but also processes (e.g. reproduction, the intensity of trophic interactions and potential for top-down control of resources by consumers) for the major taxonomic and functional groups within food webs across a broad latitudinal range. This approach is complemented in the subsequent volume by O'Gorman et al. (2012), who also employ space-for-time substitution, but within a single catchment, to assess the effects of warming in fresh waters.

Mintenbeck et al. (2012) also use a model system approach, this time focusing on the seas of the high Antarctic, where fishes play a central role in the food web, as highlighted previously by Jacob et al. (2011). These key taxa are important conduits of energy and also often agents of top-down control, yet they are affected by climate change in different ways: directly, via increasing water temperatures and/or CO_2 concentrations and decreasing salinity; indirectly, via changes in food web structure and dynamics, and by habitat change due to sea-ice retreat and scouring of the sea floor. They identified potential bottlenecks arising from the loss of key species, which could ripple through the entire food web, altering the ecosystem as a whole. Highly connected or "strategically positioned" nodes within food webs can exert powerful effects and may even trigger trophic cascades or regime shifts in extreme cases, which could not be predicted without considering how they are connected to other species in the system. These ideas, which are well-supported by both data and theory (e.g. Dunne et al., 2002; Jacob et al., 2011; Jeppesen et al., 2003; Kishi et al., 2005; Montoya et al., 2009), resurface several times in Volumes 46–48 (e.g. Mintenbeck et al., 2012; O'Gorman et al., 2012). In contrast, other systems seem relatively robust to perturbations (e.g. Twomey et al., 2012): the challenge is to be able to identify what attributes of multispecies systems make them especially vulnerable, or stable, in the face of the drivers of global change. It is also increasingly clear that not all ecological response variables are equally sensitive and that huge change at one level of biological organisation might have little effect at another: for example, if different species have similar ecological roles, ecosystem functioning may be maintained in the early stages of species loss (Petchey et al., 2004; Reiss et al., 2009). For instance, in this volume, Ledger et al. (2012a) demonstrate dramatic shifts in population abundances, community composition and biomass production in response to drought, yet many structural food web attributes remain highly conserved (Ledger et al., 2012b; Woodward et al., 2012).

Overfishing is an unfortunately familiar example of resource overexploitation that threatens marine biodiversity on a global scale, and its consequences across multiple levels of organisation are also considered by Rossberg (2012), who presents a novel, elegant and yet comprehensive analytic approach to modelling the dynamics of marine ecosystems. This offers an important means of improving our understanding of the mechanisms that control community and population size structure and likely responses to perturbations. Although focused primarily on marine fisheries, many of these principles can clearly be extended to many other aquatic and terrestrial systems, where, for example, size-spectra and other allometric scaling approaches are rapidly gaining favour (e.g. Mulder et al., 2012) as cross-fertilisation between these previously disparate disciplines gathers apace (Jennings and Brander, 2010; Woodward et al., 2011).

ON THE WISHLIST: BETTER DATA AND PREDICTIVE FRAMEWORKS

Global change will continue to challenge theoretical ecologists, empirical scientists, and conservation biologists over the coming decades as both natural and managed multispecies systems are exposed to ever-increasing levels of environmental stress from a growing range of sources (Feld et al., 2011; Friberg et al., 2011; Woodward et al., 2010a). The six papers presented here, and those that will follow in Volumes 47 and 48, provide just a glimpse of the scale of the task ahead, but they also offer some glimmers of hope as to how we might take the necessary first steps into this huge and largely uncharted territory.

Monitoring and assessment of medium- to long-term impacts of global change on multispecies systems is needed across a wide range of organisational levels: from molecules, to individuals, to entire ecosystems (Purdy et al., 2010), and we need to understand the role of both ecological and evolutionary drivers at appropriate timescales (Melian et al., 2011; Moya-Laraño et al., 2012). Ideally, studies of entire multispecies communities should span a wide range of organism sizes and trophic levels (e.g. Hagen et al., 2012; Meerhoff et al., 2007, 2012; Mulder et al., 2012; O'Gorman and Emmerson, 2010), as well as helping to align more strategic multidisciplinary research to aid the longer term development of the field as a whole (Woodward et al., 2011). The various topics addressed here and in related volumes bridge several disciplinary gaps, including (1) the effects of nutrients on biomass and production (Jeppesen et al., 2012), (2) energy and material flows through ecosystems (Mulder et al., 2011, 2012; Yvon-Durocher et al., 2010), (3) the ecological effects of species richness and the roles of microbial organisms (Mulder et al., 2012; Perkins et al., 2010; Ptacnik et al., 2010), (4) the role of feeding behaviour in system dynamics and trophic controls (Jacob et al., 2011; Layer et al., 2010, 2011; Mintenbeck et al., 2012), (5) the dynamics of communities and links between different ecosystem types (Hagen et al., 2012; Hladyz et al., 2011a,b; Layer et al., 2010), (6) the combined effects of body size and behaviour in determining the structure and dynamics of food webs (Jacob et al., 2011; Mintenbeck et al., 2012; O'Gorman and Emmerson, 2010), and (7) the impacts of external drivers on food web dynamics and functioning (Mulder et al., 2012; Woodward et al., 2010a). Several common threads link the different conceptual frameworks that span these various topics, emphasising the importance of tackling multispecies system studies with a variety of theoretical and empirical approaches.

A MATTER OF TIME AND SPACE: TEMPORAL AND SPATIAL SCALE AND LEVELS OF ORGANISATION

The impact of global change in multispecies systems is intimately connected to the spatiotemporal scales of system variability. Climate change is already affecting natural systems, as is clear from the ample data on shifts in the seasonal timing of reproduction and migration, in body size and species' distribution ranges (Peck et al., 2012; Visser, 2010). The spatial distribution and impact of a species in an environment are a consequence of a combination of both intrinsic and extrinsic factors that govern its population dynamics and how it interacts with others (Layer et al., 2010; Nakazawa et al., 2011; Olesen et al., 2010; Riede et al., 2010). Intrinsic factors include dispersal, growth, survival and reproduction, constrained by physiological and morphological capabilities, whereas extrinsic factors include the spatial and temporal variation in physicochemical conditions required for populations to be viable (Layer et al., 2010, 2011; Mintenbeck et al., 2012; Mulder et al., 2012), in addition to sufficient habitat availability (Hagen et al., 2012; Ledger et al., 2012a,b).

Prey availability and predation risk are important determinants of habitat use, but both vary across spatial (and temporal) scales. In multispecies systems, consumer and resource distributions may covary at large spatial scales but do not necessarily coincide at small spatial scales: in extreme cases, opposite trends may even be evident (e.g. positive aggregative responses of predators to prey in local patches, but negative correlations in their abundance across habitats) (Hagen et al., 2012; Olesen et al., 2010). Rates of growth, survival and reproductive success of species in multispecies systems are intimately linked to spatial and/or temporal variability in the composition and abundance of potential prey items. Within multispecies systems, foraging theory can be developed as a useful framework for describing changes in species behaviour in response to global stressors, and contributions here (e.g. Hagen et al., 2012) and in recent or forthcoming volumes (e.g. Moya-Laraño et al., 2012; Olesen et al., 2010) provide a foundation for pursuing these challenges in future research and for coupling them more explicitly with metacommunity theory (Hagen et al., 2012).

A QUESTION OF TRAITS

Functional traits of species and systems are important (Litchman et al., 2007; Mulder et al., 2012; Perkins et al., 2010; Ptacnik et al., 2010; Reiss et al., 2009) controls of many ecosystem processes—perhaps even more so than taxonomic identity—yet our mechanistic understanding is still far from complete (Jacob et al., 2011).

Here, Mulder et al. (2012) provide a collection of empirical examples from aquatic and terrestrial ecosystems examining how biodiversity supports ecosystem functioning, both within and across trophic levels. They also provide an in-depth evaluation of B–EF relationships that includes aspects of taxonomic diversity, functional categorisation and metabolic scaling, as well as suggesting rules for their appropriate use. This study complements several others that have appeared in recent volumes (e.g. Perkins et al., 2010; Ptacnik et al., 2010) and draws the conclusion that, while community complexity and the abundance of organisms reflect functional diversity, the influence of biodiversity (in terms of species richness) on community stability and ecosystem functioning is less clear. This highlights the need for a more mechanistic framework, based on functional traits that drive community structure and ecosystem processes, for understanding how altering species composition will affect community stability and the provisioning of ecosystem services in response to global change (Palmer and Febria, 2012).

To achieve this, a clear understanding of what the functional characteristics really are is needed to make the appropriate choice of the relevant traits: several papers in recent volumes have focused on body size (Arim et al., 2011; Gilljam et al., 2011; Jacob et al., 2011; Nakazawa et al., 2011; Perkins et al., 2010; Woodward et al., 2010a,b), whereas others have revealed how other traits that may be correlated with size (dispersal ability in fragmented habitats) or orthogonal to it (sociality, taxonomy) also need to be taken into account (Hagen et al., 2012; Henri and van Veen, 2011; Jacob et al., 2011; Ledger et al., 2012a,b; Meerhoff et al., 2012; Moya-Laraño et al., 2012; Mulder et al., 2012). Developing a trait-based framework that can incorporate these different functional properties could provide a more unified platform for understanding and predicting how multispecies systems will respond to the multiple drivers of global change (Spooner and Vaughn, 2008). In essence, this re-emphasises the need to view biodiversity as more than just species richness, and the use of traits also forges an appealing intuitive link to how species interact within ecological networks, providing a useful bridge between the structural and functional attributes of multispecies systems.

Multispecies systems are complex but that does not necessarily mean they are unpredictable. We know, for instance, that large species are especially vulnerable to many drivers of global change (Jennings and Brander, 2010; Perry et al., 2010; Raffaelli, 2004) and that these species have many functional attributes linked to their autecology and synecology that should also respond predictably (Ings et al., 2009; Woodward et al., 2012). For instances, as the largest fishes are removed selectively by overharvesting, the trophic height of the food web as a whole should decline, just as is seen throughout the world's fisheries (Ings et al., 2009; Rossberg, 2012).

Detecting significant impacts and potential synergies among global change drivers in complex multispecies systems is an important, but challenging, task. If the community structure is (relatively) simple, we may be able to analyse the roles of different species relatively easily simply by determining how environmental impacts translate into species loss. However, as most communities contain complex food webs in reality, it is far harder to envisage the full range of responses. One approach to dealing with such complexity is to focus on the functional traits of species, in order to identify their main defining role within the system, rather than becoming too beholden to the Latin binomial, especially as many species are cryptic and/or incompletely described in many systems, particularly in the Tropics (Hagen et al., 2012; Jacob et al., 2011; Meerhoff et al., 2012; Mulder et al., 2011; Purdy et al., 2010).

THE NEXT STEPS FORWARD . . .

Global change affects biodiversity and ecosystem functioning and will ultimately lead to the emergence of novel ecological communities. While the separate effects of the main drivers, such as climate change, habitat loss and pollution, as well as species loss and gain, are becoming increasingly well documented, still very little is known about their consequences when acting in concert (Walther et al., 2002; Woodward et al., 2010a). Predicting the combined effects of multiple and interacting drivers represents a significant challenge and highlights where the next major advances need to be made as a matter of urgency. We need to be able to identify the critical gaps in our current knowledge and to gauge the extent to which they might be addressed with existing data, or whether new primary research is required. As time is short, we do not have the luxury of always being able to wait for the empirical data to catch up with the theoretical explorations of what the future might hold, so modelling is certain to become increasingly important.

It is also essential to develop studies that can integrate applied and basic research and empirical and theoretical approaches so that management regimes in the "real world" can cope with the anticipated changes in future biodiversity (Feld et al., 2011).

An essential prerequisite for sustaining our own species in the lifestyle to which we have become accustomed in the twenty-first century into the future is to ensure that natural systems continue to provide the services required for human well-being. A mix of observation, experimentation, modelling and synthesis that can be applied across a wide array of ecological communities is needed to develop the necessary predictive frameworks (e.g. Rossberg, 2012). From the empirical perspective, we will need to employ a mixed approach that combines model systems with more extensive comparative approaches, and there are encouraging signs that such studies are becoming more commonplace. Much of this is largely fortuitous, as previously established research programmes that were set up decades ago in a few model systems start to bear often unexpected fruit, as long-term trends start to emerge (e.g. Hildrew, 2009; Kratz et al., 2003; Lane, 1997; Layer et al., 2011). There is also an active movement in contemporary ecology towards integrating experiments, surveys and models within the same system, especially as interdisciplinary, collaborative research becomes evermore common practice (e.g. Layer et al., 2010, 2011; O'Gorman et al., 2012; Twomey et al., 2012).

In conclusion, it is clear that impacts of environmental stressors and global change often cannot be predicted reliably from single-factor and single-species studies because different drivers affect community structure and functioning in different and potentially synergistic ways. To attain a more complete understanding, we need to adopt a more integrated approach that spans multiple organisational levels, disciplines and spatiotemporal scales of study. The six papers compiled in this thematic volume, and also in the two subsequent volumes, should help us move closer to achieving this goal.

REFERENCES

Arim, M., Berazategui, M., Barreneche, J.M., Ziegler, L., Zarucki, M., Abades, S.R., 2011. Determinants of density-body size scaling within food webs and tools for their detection. Adv. Ecol. Res. 45, 1–39.

Daufresne, M., Lengfellner, K., Sommer, U., 2009. Global warming benefits the small in aquatic ecosystems. Proc. Natl. Acad. Sci. USA. 106, 12788–12793.

Dunne, J.A., Williams, R.J., Martinez, N.D., 2002. Network structure and biodiversity loss in food webs, robustness increases with connectance. Ecol. Lett. 5, 558–567.

Feld, C.K., Birk, S., Bradley, D.C., Hering, D., Kail, J., Marzin, A., Melcher, A., Nemitz, D., Pedersen, M.L., Pletterbauer, F., Pont, D., Verdonschot, P.F.M., et al., 2011. From natural to degraded rivers and back again: a test of restoration ecology theory and practice. Adv. Ecol. Res. 44, 119–210.

Feuchtmayr, H., Moran, R., Hatton, K., Connor, L., Heyes, T., Moss, B., Harvey, I., Atkinson, D., 2009. Global warming and eutrophication: effects on water chemistry and autotrophic communities in experimental hypertrophic shallow lake mesocosms. J. Appl. Ecol. 46, 713–723.

Friberg, N., Bonada, N., Bradley, D.C., Dunbar, M.J., Edwards, F.K., Grey, J., Hayes, R.B., Hildrew, A.G., Lamouroux, N., Trimmer, M., Woodward, G., 2011. Biomonitoring of human impacts in freshwater ecosystems: the good, the bad, and the ugly. Adv. Ecol. Res. 44, 1–68.

Gilljam, D., Thierry, A., Figueroa, D., Jones, I., Lauridsen, R., Petchey, O., Woodward, G., Ebenman, B., Edwards, F.K., Ibbotson, A.T.J., 2011. Seeing double: size-based versus taxonomic views of food web structure. Adv. Ecol. Res. 45, 67–133.

Hagen, M., Kissling, W.D., Rasmussen, C., de Aguiar, M.A.M., Brown, L.E., Carstensen, D.W., Alves-dos-Santos, I., Dupont, Y.L., Edwards, F.K., Genini, J., Guimarães, P.R., Jenkins, G.B., Jordano, P., Kaiser-Bunbury, C.N., Ledger, M., Maia, K.M., Marquitti, F.M.D., McLaughlin, O., Morellato, L.P.C., O'Gorman, E., Trøjelsgaard, K., Tylianakis, J.M., Vidal, M.M., Woodward, G., Olesen, J.M., 2012. Biodiversity, species interactions and ecological networks in a fragmented world. Adv. Ecol. Res. 46, 89–210.

Henri, D.C., van Veen, F.J.F., 2011. Body size, life history and the structure of host-parasitoid networks. Adv. Ecol. Res. 45, 135–180.

Hladyz, S., Abjornsson, K., Cariss, H., Chauvet, E., Dobson, M., Elosegi, A., Ferreira, V., Fleituch, T., Gessner, M.O., Giller, P.S., Grac¸a, M.A.S., Gulis, V., et al., 2011a. Stream ecosystem functioning in an agricultural landscape: the importance of terrestrial-aquatic linkages. Adv. Ecol. Res. 44, 211–276.

Hladyz, S., Åbjörnsson, K., Giller, P.S., Woodward, G., 2011b. Impacts of an aggressive riparian invader on community structure and ecosystem functioning in stream food webs. J. Appl. Ecol. 48, 443–452.

Hildrew, A.G., 2009. Sustained Research on Stream Communities: A Model System and The Caomparative Approach. ADV ECOL RES 41, 175–312.

Hynes, H.B.N., 1975. The stream and its valley. Verhandlungen der Internationalen Vereinigung für Theoretische und Angewandte Limnologie. Proc. Int. Assoc. Theor. Appl. Limnol. 19, 1–15.

Ings, T.C., Montoya, J.M., Bascompte, J., Bluthgen, N., Brown, L., Dormann, C.F., Edwards, F., Figueroa, D., Jacob, U., Jones, J.I., Lauridsen, R.B., Ledger, M.E., Lewis, H.M., Olesen, J.M., Van Veen, F.J.F., Warren, P.H., Woodward, G., 2009. Ecological networks—beyond food webs. J. Anim. Ecol. 78, 253–269.

Jacob, U., Thierry, A., Brose, U., Arntz, W.E., Berg, S., Brey, T., Fetzer, I., Jonsson, T., Mintenbeck, K., Mollmann, C., Petchey, O., Riede, J.O., et al., 2011. The role of body size in complex food webs: a cold case. Adv. Ecol. Res. 45, 181–223.

Jennings, S.J., Brander, K., 2010. Predicting the effects of climate change on marine communities and the consequences for fisheries. J. Mar. Syst. 79, 418–426.
Jeppesen, E., Søndergaard, M., Jensen, J.P., 2003. Climatic warming and regime shifts in lake food webs: some comments. Limnol. Oceanogr. 48, 1346–1349.
Jeppesen, E., Søndergaard, M., Lauridsen, T.L., Davidson, T.A., Liu, Z., Mazzeo, N., Trochine, C., Özkan, K., Jensen, H.S., Trolle, D., Starling, F., Lazzaro, X., Johansson, L.S., Bjerring, R., Liboriussen, L., Larsen, S.E., Landkildehus, F., Egemose, S., Meerhoff, M., 2012. Biomanipulation as a restoration tool to combat eutrophication—recent advances and future challenges. Adv. Ecol. Res. 47, in press.
Kishi, D., Murakami, M., Nakano, S., Maekawa, K., 2005. Water temperature determines strength of top-down control in a stream food web. Freshw. Biol. 50, 1315–1322.
Konig, K.A., Kamenik, C., Schmidt, R., Agustí-Panareda, A., Appleby, P., Lami, A., Prazakova, M., Rose, N., Schnell, O.A., Tessadri, R., Thompson, R., Psenner, R., 2002. Environmental changes in an alpine lake (Gossenköllesee, Austria) over the last two centuries—the influence of air temperature on biological parameters. J. Paleolimnol. 28, 147–160.
Kratz, T.K., Deegan, L.A., Harmon, M.E., Lauenroth, W.K., 2003. Ecological variability in space and time: insights gained from the US LTER Program. Bioscience 53, 57–67.
Lake, P.S., 2003. Ecological effects of perturbation by drought in flowing water. Freshw. Biol. 48, 1161 1172.
Lane, A.M.J., 1997. The U.K. Environmental Change Network database: an integrated information resource for long-term monitoring and research. J. Environ. Manage. 51, 87–105.
Layer, K., Riede, J.O., Hildrew, A.G., Woodward, G., 2010. Food web structure and stability in 20 streams across a wide pH gradient. Adv. Ecol. Res. 42, 265–301.
Layer, K., Hildrew, A.G., Jenkins, G.B., Riede, J., Rossiter, S.J., Townsend, C.R., Woodward, G., 2011. Long-term dynamics of a well-characterised food web: four decades of acidification and recovery in the Broadstone Stream model system. Adv. Ecol. Res. 44, 69–117.
Ledger, M.E., Hildrew, A.G., 2001. Recolonization by the benthos of an acid stream following a drought. Arch. Hydrobiol. 152, 1–17.
Ledger, M.E., Harris, R.M.L., Armitage, P.D., Milner, A.M., 2008. Disturbance frequency influences patch dynamics in stream benthic algal communities. Oecologia 155, 809–819.
Ledger, M.E., Harris, R.M.L., Armitage, P.D., Milner, A.M., 2009. Realism of model ecosystems: an evaluation of physicochemistry and macroinvertebrate assemblages in artificial streams. Hydrobiologia 617, 91–99.
Ledger, M.E., Harris, R.M.L., Armitage, P.D., Milner, A.M., 2012a. Climate change impacts on community resilience: evidence from a drought disturbance experiment. Adv. Ecol. Res. 46, 211–258.
Ledger, M.E., Brown, L.E., Edwards, F.K., Milner, A.M., Woodward, G., 2012b. Drought alters the structure and functioning of complex food webs. Nat. Clim. Chang. http://dx.doi.org/10.1038/nclimate1684.
Litchman, E., Klausmeier, C.A., Schofield, O.M., Falkowski, P.G., 2007. The role of functional traits and trade-offs in structuring phytoplankton communities: scaling from cellular to ecosystem level. Ecol. Lett. 10, 1170–1181.
McKee, D., Atkinson, D., Collings, S.E., Eaton, J.W., Gill, A.B., Harvey, I., Hatton, K., Heryes, T., Wilson, D., Moss, B., 2003. Response of freshwater microcosm communities to nutrients, fish, and elevated temperature during winter and summer. Limnol. Oceanogr. 48, 707–722.
McLaughlin, O., Jonsson, T., Emmerson, M., 2010. Temporal variability in predator-prey relationships of a forest floor food web. Adv. Ecol. Res. 42, 171–264.

Meerhoff, M., Clemente, J.M., De Mello, F.T., Iglesias, C., Pedersen, A.R., Jeppesen, E., 2007. Can warm climate-related structure of littoral predator assemblies weaken the clear water state in shallow lakes? Glob. Change Biol. 13, 1888–1897.

Meerhoff, M., Teixeira-de Mello, F., Kruk, C., Alonso, C., González-Bergonzoni, I., Pacheco, J.P., Arim, M., Beklioğlu, M., Brucet, S., Goyenola, G., Iglesias, C., Lacerot, G., Mazzeo, N., Kosten, S., Jeppesen, E., 2012. Environmental warming in shallow lakes: a review of effects on community structure as evidenced from space-for-time substitution approaches. Adv. Ecol. Res. 46, 259–350.

Melian, C.J., Vilas, C., Baldo, F., Gonzalez-Ortegon, E., Drake, P., Williams, R.J., 2011. Eco-evolutionary dynamics of individual-based food webs. Adv. Ecol. Res. 45, 225–268.

Memmott, J., Craze, P.G., Waser, N.M., Price, M.V., 2007. Global warming and the disruption of plant–pollinator interactions. Ecol. Lett. 10, 710–717.

Milly, P.C.D., Dunne, K.A., Vecchia, A.V., 2006. Global pattern of trends in streamflow and water availability in a changing climate. Nature 438, 347–350.

Milner, A.M., Knudsen, E.E., Soiseth, C., Robertson, A.L., Schell, D., Phillips, I.T., Magnusson, K., 2000. Colonization and development of stream communities across a 200-year gradient in Glacier Bay National Park, Alaska. Can. J. Fish. Aquat. Sci. 57, 2319–2335.

Milner, A.M., Robertson, A.E., Monaghan, K., Veal, A.J., Flory, E.A., 2008. Colonization and development of a stream community over 28 years; Wolf Point Creek in Glacier Bay, Alaska. Front. Ecol. Environ. 6, 413–419.

Milner, A.M., Brown, L.E., Hannah, D.M., 2009. Hydroecological effects of shrinking glaciers. Hydrol. Process. 23, 62–77.

Mintenbeck, K., Barrera-Oro, E.R., Brey, T., Jacob, U., Knust, R., Mark, F.C., Moreira, E., Strobel, A., Arntz, W.E., 2012. Impact of climate change on fish in complex Antarctic ecosystems. Adv. Ecol. Res. 46, 351–426.

Möllmann, C., Diekmann, R., 2012. Marine ecosystem regime shifts induced by 1387 climate and overfishing—a review for the Northern hemisphere. Adv. Ecol. Res. 47, in press.

Montoya, J.M., Woodward, G., Emmerson, M.C., Sole, R.C., 2009. Press perturbations and indirect effects in real food webs. Ecology 90, 2426–2433.

Moss, B., McKee, D., Atkinson, D., Collings, S.E., Eaton, J.W., Gill, A.B., Harvey, I., Hatton, K., Heyes, T., Wilson, D., 2003. How important is climate? Effects of warming, nutrient addition and fish on phytoplankton in shallow lake microcosms. J. Appl. Ecol. 40, 782–792.

Moya-Laraño, J., Verdeny-Vilalta, O., Rowntree, J., Melguizo-Ruiz, N., Montserrat, M., Laiolo, P., 2012. Climate change and eco-evolutionary dynamics in food webs. Adv. Ecol. Res. 47, in press.

Mulder, C., Boit, A., Bonkowski, M., de Ruiter, P.C., Mancinelli, G., van der Heijden, M.G.A., van Wijnen, H.J., Vonk, J.A., Rutgers, M., 2011. A belowground perspective on Dutch agroecosystems: how soil organisms interact to support ecosystem services. Adv. Ecol. Res. 44, 277–358.

Mulder, C., Boit, A., Mori, S., Vonk, J.A., Dyer, S.D., Faggiano, L., Geisen, S., González, A.L., Kaspari, M., Lavorel, S., Marquet, P.A., Rossberg, A.G., Sterner, R.W., Voigt, W., Wall, D.H., 2012. Distributional (in)congruence of biodiversity-ecosystem functioning. Adv. Ecol. Res. 46, 1–88.

Nakazawa, T., Ushio, M., Kondoh, M., 2011. Scale dependence of predator-prey mass ratio: determinants and applications. Adv. Ecol. Res. 45, 269–302.

O'Gorman, E., Emmerson, M., 2010. Manipulating interaction strengths and the consequences for trivariate patterns in a marine food web. Adv. Ecol. Res. 42, 301–419.

O'Gorman, E., Pichler, D.E., Adams, G., Benstead, J.P., Craig, N., Cross, W.F., Demars, B.O.L., Friberg, N., Gislason, G.M., Gudmundsdottir, R., Hawczak, A., Hood, J.M., Hudson, L.N., Johansson, L., Johansson, M., Junker, J.R., Laurila, A., Manson, J.R., Mavromati, E., Nelson, D., Olafsson, J.S., Perkins, D.M., Petchey, O.L., Plebani, M., Reuman, D.C., Rall, B.C., Stewart, R., Thompson, M.S.A., Woodward, G., 2012. Impacts of warming on the structure and functioning of aquatic communities: individual- to ecosystem-level responses. Adv. Ecol. Res. 47, in press.

Olesen, J.M., Dupont, Y.L., O'Gorman, E., Ings, T.C., Layer, K., Melian, C.J., Troejelsgaard, K., Pichler, D.E., Rasmussen, C., Woodward, G., 2010. From Broadstone to Zackenberg: space, time and hierarchies in ecological networks. Adv. Ecol. Res. 42, 1–71.

Palmer, M.A., Febria, C.M., 2012. The heartbeat of ecosystems. Science 336, 1393–1394.

Parmesan, C., 2006. Ecological and evolutionary responses to recent climate change. Annu. Rev. Ecol. Evol. Syst. 37, 637–669.

Parmesan, C., Yohe, G., 2003. A globally coherent fingerprint of climate change impacts across natural systems. Nature 421, 37–42.

Pearson, R.G., Dawson, T.P., 2003. Predicting the impacts of climate change on the distribution of species: are bioclimate envelope models useful? Glob. Ecol. Biogeogr. 12, 361–371.

Peck, M.A., Huebert, K.B., Llopiz, J.K., 2012. Intrinsic and extrinsic factors driving match-mismatch dynamics 1460 during the early life history of marine fishes. Adv. Ecol. Res. 47, in press.

Perkins, D.M., McKie, B.G., Malmqvist, B., Gilmour, S.G., Reiss, J., Woodward, G., 2010. Environmental warming and biodiversity-ecosystem functioning in freshwater microcosms: partitioning the effects of species identity, richness and metabolism. Adv. Ecol. Res. 43, 177–209.

Perry, R.I., Cury, P., Brander, K., Jennings, S., Möllmann, C., Planque, B., 2010. Sensitivity of marine systems to climate and fishing: concepts, issues and management responses. J. Mar. Syst. 79, 427–435.

Petchey, O.L., Downing, A.L., Mittelbach, G.G., Persson, L., Steiner, C.F., Warren, P.H., Woodward, G., 2004. Species loss and the structure and functioning of multitrophic aquatic systems. Oikos 104, 467–478.

Ptacnik, R., Moorthi, S.D., Hillebrand, H., 2010. Hutchinson reversed, or why there need to be so many species. Adv. Ecol. Res. 43, 1–44.

Purdy, K.J., Hurd, P.J., Moya-Larano, J., Trimmer, M., Woodward, G., 2010. Systems biology for ecology: from molecules to ecosystems. Adv. Ecol. Res. 43, 87–149.

Raffaelli, D., 2004. How extinction patterns affect ecosystems. Science 306, 1141–1142.

Rawcliffe, R., Sayer, C.D., Woodward, G., Grey, J., Davidson, T.A., Jones, J.I., 2010. Back to the future: using paleolimnology to infer long-term changes in shallow lake food web. Freshwater Biology 55, 600–613.

Reiss, J., Bridle, J., Montoya, J.M., Woodward, G., 2009. Emerging horizons in biodiversity and ecosystem functioning research. Trends Ecol. Evol. 24, 505–514.

Riede, J.O., Rall, B.C., Banasek-Richter, C., Navarrete, S.A., Wieters, E.A., Brose, U., 2010. Scaling of food web properties with diversity and complexity across ecosystems. Adv. Ecol. Res. 42, 139–170.

Rossberg, A., 2012. A complete analytic theory for structure and dynamics of populations and communities spanning wide ranges in body size. Adv. Ecol. Res. 46, 427–522.

Sala, O.E., Chapin Iii, F.S., Armesto, J.J., Berlow, E., Bloomfield, J., Dirzo, R., Huber-Sanwald, E., Huenneke, L.F., Jackson, R.B., Kinzig, A., et al., 2000. Global biodiversity scenarios for the year 2100. Science 287, 1770–1774.

Sarmiento, J.L., Slater, R., Barber, R., Bopp, L., Doney, S.C., Hirst, A.C., Kleypas, J., Matear, R., Mikolajewicz, U., Monfray, P., Soldatov, V., Spall, S.A., Stouffer, R., 2004. Response of ocean ecosystems to climate warming. Global Biogeochem. Cycles 18, 1–23.

Spooner, D.E., Vaughn, C.C., 2008. A trait-based approach to species' roles in stream ecosystems: climate change, community structure, and material cycling. Oecologia 158, 307–317.

Thomas, C.D., Cameron, A., Green, R.E., Bakkenes, M., Beaumont, L.J., Collingham, Y.C., Erasmus, B.F.N., De Siqueira, M.F., Grainger, A., Hannah, L., 2004. Extinction risk from climate change. Nature 427, 145–148.

Twomey, M., Jacob, U., Emmerson, M.E., 2012. Perturbing a marine food web - consequences for food web structure and trivariate patterns. Adv. Ecol. Res. 47, in press.

Tylianakis, J., 2009. Warming up food webs. Science 323, 1300–1301.

Visser, M.E., Caro, S.P., van Oers, K., Schaper, S.V., Helm, B., 2010. Phenology, seasonal timing and circannual rhythms: towards a unified framework. Philos. Trans. R. Soc. B 365, 3113–3127.

Walther, G.R., 2010. Community and ecosystem responses to recent climate change. Phil. Trans. R. Soc. B 365, 2019–2024.

Walther, G.-R., Post, E., Convey, P., Menzel, A., Parmesan, C., Beebee, T.J.C., Fromentin, J.-M., Hoegh-Guldberg, O., Bairlein, F., 2002. Ecological responses to recent climate change. Nature 416, 389–395.

Woodward, G., Benstead, J.P., Beveridge, O.S., Blanchard, J., Brey, T., Brown, L., Cross, W.F., Friberg, N., Ings, T.C., Jacob, U., Jennings, S., Ledger, M.E., et al., 2010a. Ecological networks in a changing climate. Adv. Ecol. Res. 42, 72–138.

Woodward, G., Blanchard, J., Lauridsen, R.B., Edwards, F.K., Jones, J.I., Figueroa, D., Warren, P.H., Petchey, O.L., 2010b. Individual-based food webs: species identity, body size and sampling effects. Adv. Ecol. Res. 43, 211–266.

Woodward, G., Andersen, K.H., Belgrano, A., Blanchard, J., Reiss, J., 2011. Preface editorial commentary: body size and the (re)unification of ecology. Adv. Ecol. Res. 45, xv–xxix.

Woodward, G., Brown, L.E., Edwards, F.K., Hudson, L.N., Milner, A.M., Reuman, D.C., Ledger, M.E., 2012. Climate change impacts in multispecies systems: drought alters food web size-structure in a field experiment. Philos. Trans. R. Soc. Lond. B 1, 1–5.

Yvon-Durocher, G., Allen, A.P., Montoya, J.M., Trimmer, M., Woodward, G., 2010. The temperature dependence of the carbon cycle in aquatic systems. Adv. Ecol. Res. 43, 267–313.

Distributional (In)Congruence of Biodiversity–Ecosystem Functioning

Christian Mulder[*,1], Alice Boit[†], Shigeta Mori[‡], J. Arie Vonk[§,¶], Scott D. Dyer[||], Leslie Faggiano[#,], Stefan Geisen[††], Angélica L. González[‡‡], Michael Kaspari[§§,¶¶], Sandra Lavorel[||||], Pablo A. Marquet[##,***], Axel G. Rossberg[†††,‡‡‡], Robert W. Sterner[§§§], Winfried Voigt[¶¶¶], Diana H. Wall[||||||]**

[*]National Institute for Public Health and the Environment (RIVM), Bilthoven, The Netherlands
[†]Department of Ecology and Ecosystem Modeling, Institute of Biochemistry and Biology, University of Potsdam, Potsdam, Germany
[‡]Forestry and Forest Products Research Institute, Tsukuba, Japan
[§]Institute for Biodiversity and Ecosystem Dynamics, University of Amsterdam, Amsterdam, The Netherlands
[¶]Institute for Water and Wetland Research, Radboud University Nijmegen, Nijmegen, The Netherlands
[||]Procter & Gamble, Cincinnati, Ohio, USA
[#]Institute of Aquatic Ecology, Universitat de Girona, Girona, Spain
[**]Laboratoire Evolution et Diversité Biologique, Université Paul Sabatier, CNRS, Toulouse, France
[††]Department of Terrestrial Ecology, Institute of Zoology, University of Köln, Biozentrum Köln, Köln, Germany
[‡‡]Department of Zoology, University of British Columbia, Vancouver, British Columbia, Canada
[§§]Department of Zoology, University of Oklahoma, Norman, Oklahoma, USA
[¶¶]Smithsonian Tropical Research Institute, Balboa, Panama
[||||]Laboratoire d'Ecologie Alpine, CNRS, Grenoble, France
[##]Center for Advanced Studies in Ecology and Biodiversity, Institute of Ecology and Biodiversity, Pontificia Universidad Catolica de Chile, Santiago, Chile
[***]The Santa Fe Institute, Santa Fe, New Mexico, USA
[†††]Centre for Environment, Fisheries and Aquaculture Science (Cefas), Lowestoft Laboratory, Suffolk, United Kingdom
[‡‡‡]Medical Biology Centre, School of Biological Sciences, Queen's University Belfast, Belfast, United Kingdom
[§§§]Ecology, Evolution & Behavior, University of Minnesota, St. Paul, Minnesota, USA
[¶¶¶]Community Ecology Group, Institute of Ecology, Friedrich Schiller University, Jena, Germany
[||||||]Department of Biology & Natural Resource Ecology Laboratory (NREL), School of Global Environmental Sustainability, Colorado State University, Fort Collins, Colorado, USA
[1]Corresponding author: e-mail address: christian.mulder@rivm.nl

Contents

Advances in Ecological Research, Volume 46
ISSN 0065-2504
http://dx.doi.org/10.1016/B978-0-12-396992-7.00001-0

Abstract

The majority of research on biodiversity–ecosystem functioning in laboratories has concentrated on a few traits, but there is increasing evidence from the field that functional diversity controls ecosystem functioning more often than does species number. Given the importance of traits as predictors of niche complementarity and community structures, we (1) examine how the diversity *sensu lato* of forest trees, freshwater fishes and soil invertebrates might support ecosystem functioning and (2) discuss the relevance of productive biota for monophyletic assemblages (taxocenes).

In terrestrial ecosystems, correlating traits to abiotic factors is complicated by the appropriate choice of body-size distributions. Angiosperm and gymnosperm trees, for example, show metabolic incongruences in their respiration rates despite their pronounced macroecological scaling. Scaling heterotrophic organisms within their monophyletic assemblages seems more difficult than scaling autotrophs: in contrast to the generally observed decline of mass-specific metabolic rates with body mass within metazoans, soil organisms such as protozoans show opposite mass-specific trends.

At the community level, the resource demand of metazoans shapes multitrophic interactions. Hence, population densities and their food web relationships reflect functional diversity, but the influence of biodiversity on stability and ecosystem functioning remains less clear. We focused on fishes in 18 riverine food webs, where the ratio of primary versus secondary extinctions (hereafter, 'extinction partitioning') summarizes the responses of fish communities to primary species loss (deletions) and its consequences. Based on extinction partitioning, our high-diversity food webs were just as (or even more) vulnerable to extinctions as low-diversity food webs.

Our analysis allows us to assess consequences of the relocation or removal of fish species and to help with decision-making in sustainable river management. The study highlights that the topology of food webs (and not simply taxonomic diversity) plays a greater role in stabilizing the food web and enhancing ecological services than is currently acknowledged.

ABBREVIATIONS

B biomass
B–EF biodiversity–ecosystem functioning
C carbon content
C connectance of a food web or network
eNPP ecosystem's Net Primary Productivity
FD functional diversity
L trophic links
m mass at individual level
$\bar{M}$ mass average at population level (site-specific)
M species-specific estimate of body-mass average
MIH More Individuals Hypothesis
N numerical abundance at population level
N nitrogen content
P phosphorus content
PD fraction of primary deletions (1 − #SD)
R metabolic rate at individual level
R_{SD} robustness against SD
S number of species within one monophyletic taxocene (taxonomic diversity)
SD fraction of secondary deletions (1 − #PD)

All substances, in so far as they can be perceived in space at the same time, exist in a state of complete reciprocity of action.
Immanuel Kant *(1781) Kritik der reinen Vernunft:* Dritte Analogie.

1. INTRODUCTION

1.1. Vexing drivers and responses

Despite scientific rationalism, too many generalizations and recent extrapolations on the so-called sixth Great Extinction Event are widely supported and spread by modern media (criticism on NGO's statements and current concerns already by Mann, 1991). On the one hand, the growing human impact on Earth is beyond discussion and many scientists even assigned the term Anthropocene to the present epoch (Crutzen, 2002; Estes et al., 2011). On the other hand, forecasting global changes is hampered by the lack of consensus on interactions among the causes (e.g. Sala et al., 2000), the existence of overstating implications (e.g. 'taxa committed to extinction' by Thomas et al. (2004a) are soon claimed as lost in many press releases) and contradictory conclusions (e.g. Samanta et al., 2010 vs. Xu et al., 2011). Too often, in the public opinion, biodiversity seems therefore to sound vague despite of full awareness of resource exploitation and habitat loss. Hence, a dangerous consequence that must be avoided is a possibly growing cynicism and complacency about the current changes at planetary scale as a whole, although the interest with which policy-decision makers and stakeholders look to models is higher than ever.

Given that biodiversity on Earth is only superficially explored, functional groupings used up to date have arisen from a pragmatic approach to categorize biota into ecologically meaningful aggregates (Brussaard, 2012; Kerkhoff et al., 2005; Loreau et al., 2001). For example, body size, among other (frequently related) traits, is ultimately important in determining interaction strengths between consumers and resources. Moreover, organisms of different sizes can have very different effects on ecosystem functioning (EF), both within and among species (Perkins et al., 2010; Reiss et al., 2010, 2011). Size measurements can be carried out at either the individual or the species level, might be used comparatively across species, and have the power to become more directly correlated with properties that influence the performance of organisms and communities (Hodgson et al., 1999; Ledger et al., 2012; McGill et al., 2006).

Life is a matter of scale: faunal dispersal over broad spatial scales favours plasticity (Sultan and Spencer, 2002), in contrast to vascular plants, for which adaptation is limited by seed dispersal mechanisms (Hagen et al., 2012; Olesen et al., 2010). Differently sized plants with variable leaf N and P contents may affect ongoing ecological processes, either actively, due to their direct influence on decomposition efficiency, or passively, through biomass production (Bradford et al., 2002; Fortunel et al., 2009; Garnier et al., 2004; Reich et al., 2010). Further, plants change the amount and composition of root exudates depending on life form (Du Rietz, 1931; Raunkiaer, 1934; Walter, 1964) and nutrient status (Johnson, 2010; Ladygina and Hedlund, 2010; Lipton et al., 1987; Lynch and Ho, 2005; Richardson et al., 2009; Yoneyama et al., 2007).

Still, most researchers have not addressed the role of size as an effect trait at the species level, but have instead preferred to address the response trait as biomass at community level, as for many aboveground ecosystems with different productivity (e.g. Hartnett and Wilson, 1999; Klironomos et al., 2000; Wardle, 2002; Watkinson and Freckleton, 1997). Given the correlation between response and effect traits, that is, the 'response–effect hypothesis' (Lavorel and Garnier, 2002), the complexity and (mutual) importance of such direct and indirect interactions among biodiversity, EF and the environment is challenging (Bradford et al., 2002; Lavorel et al., 2009; Zobel, 1997). A resulting niche complementarity is in fact the product of not only species interactions but also a direct consequence of combinations of traits (Flombaum and Sala, 2012). Hence, many of these phenomena are interwoven and are commonly merged together into 'services', like nutrient availability, soil structure, water regulation, biological pest control and resilience (Millennium Ecosystem Assessment, 2005).

Effects of dominant species at the ecosystem level (whether a certain community composition is necessary to form and support a given ecosystem) and the EF are two 'linchpins' which matter at several levels (Perrings et al., 1992, 2011). Due to closely interrelated mechanisms, B–EF relationships have been described at many operational levels in an attempt to forecast effects of global change: although determinants of structural variability across different operational levels are not fully understood yet, changes in organismal, demographic and abundance responses might be predicted by nutrient availability or disturbance (Caswell and Cohen, 1991; Elser and Urabe, 1999; Lavorel et al., 1997; Sterner and Elser, 2002; Suding et al., 2003; Tilman, 1988).

1.2. Contrasting dichotomies

EF depends on 'dynamic relationships within species, among species and between species and their abiotic environment, as well as the physical and chemical interactions within the environment' (Millennium Ecosystem Assessment, 2005; UNEP/Convention on Biodiversity, 2004; Wall, 2008). Quantifying EF in terms of biomass, productivity and size structure within and among different ecosystems is important in ecology as it can provide clues about the underlying processes that shape communities. But how close to reality are the correlations between EF and biodiversity? And why are so many research papers and their related questions scale-specific? An excellent starting point would be Waide et al. (1999), who performed an authoritative and extensive meta-analysis of the correlations between biomass and/or production (both excellent

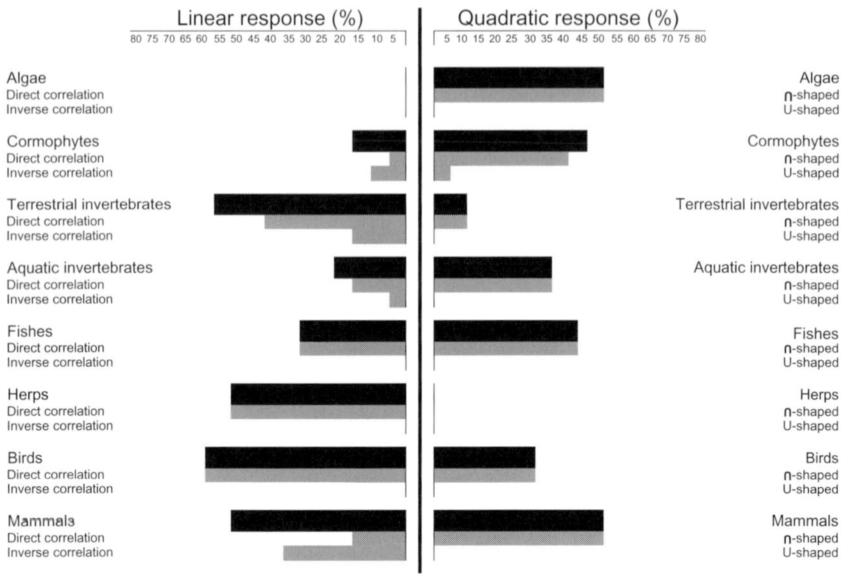

Figure 1 Meta-analysis of the biomass/production and biodiversity relationships published between 1967 and 1996, modified from Waide et al. (1999). For each of these groups of organisms, the amount of biomass–biodiversity studies was set equal to 100%: Trends in biomass as predicted by biodiversity (here, B–EF relationships) can be either linear or quadratic, including unimodal and U-shaped distributions, and can be lumped into different subunits (in grey). The horizontal sum of the black units is always $\leq$100%, because the difference is the percentage of non-significant trends. The vertical sum of the grey subunits is equal to the black unit just above them.

proxies to quantify EF) and biodiversity published between 1967 and 1996, summarized in Fig. 1.

The relationships between biodiversity and primary productivity show the extent to which, under different scales, most controlling processes differ as well, because biodiversity is not merely a simple function of primary productivity, but it may feed back onto it (Adler et al., 2011; Fridley, 2001; Hooper et al., 2005; Loreau et al., 2001). In reality, experimental communities in the field (e.g. Dukes et al., 2005; Menge and Field, 2007) and in micro- or mesocosms (e.g. Benton and Beckerman, 2005; Hunting et al., 2012; Reiss et al., 2011) represent one assemblage of randomly chosen species from a virtually available species pool (Huston, 1997; Naeem, 2008).

According to Lepš (2001), this can be a problem with the design of experiments, as the responses to change determining the success of an experimental community must be viewed with caution, due to either species that avoid competitive exclusion under low productivity, or species adapted to productive environments given the limited number of species able to compete in high-nutrient environments. Regardless of the species pool and experimental design, if vascular plants (cormophytes) share a significant response, they tend to have a bell-shaped (unimodal) biomass–biodiversity relationship (cf. Adler et al., 2011; Fig. 1). In contrast, for algae, both biomass/productivity proxies seem to have no linear response with increasing biodiversity, and for herps, there is only evidence of direct linear correlations with increasing diversity of amphibians and reptiles (Fig. 1).

Fishes are a powerful example for contrasting biomass–biodiversity correlations: along a biodiversity gradient, does function respond linearly or not? Considering that fishes are commonly overexploited (FAO, 2000) despite their intrinsic capacity to respond to environmental changes, fish assemblages must be representative in their abiotic and biotic properties as well as their faunal composition for a range of sites. The proxies for aquatic and terrestrial invertebrates are even more contrasting, with direct correlates with soil biodiversity less than three times more frequent than inverse correlates (Fig. 1). If so, contrasting biomass–biodiversity relationships within and between taxonomic groups might have clear implications for the ecosystem.

During the same period of the extensive review by Waide and others, Hector et al. (1999, 2002) published BIODEPTH, resulting in a flood of B–EF studies showing robust linear trends on one side and strongly debated statistical arguments on the validity of the experimental design on the other (e.g. Cottingham et al., 2001; Huston et al., 2000). Few experimental studies have since measured biodiversity and biomass production: in those cases, discontinuous relations between biodiversity and productivity, rather far from linearity, were shown (Boit et al., 2012; Roscher et al., 2008).

There is compelling evidence that process rates associated with animals that influence ecosystem services vary with body size: small organisms vary more rapidly in population density and behave differently from larger organisms (Fenchel and Finlay, 1983; Huston and Wolverton, 2011; Sterner and Elser, 2002), such as having a very different metabolic capacity per unit biomass if an assemblage is comprised of many small versus a few large individuals (Perkins et al., 2010). At individual and population levels, plants exhibit contrasting responses to the animal framework of Table 1: plants are mostly growing slowly in unfertile ecosystems (i.e. low eNPP), although they may live longer. For animals, tradeoffs in physical, biochemical and ecological constraints related to parental energy

Table 1 Predictions of ecologically and evolutionarily relevant properties for low and high net primary production scenarios across organizational levels as defined by Huston and Wolverton (2011); table modified (Michael Huston, personal communication) and redrawn with permission from ESA

	Low eNPP	**High eNPP**
Culture, socioeconomics	Small stature, short	Large stature, tall
	Low per capita income	High per capita income
	Malnutrition, vitamin deficiencies	Good health, nutrition
	Homes small, crowded	Homes large, spacious
	Low educational attainment	High educational levels
	Small social groups, cooperation	Hierarchy social stratification
Community	Facilitation, mutualism common	Competition, aggression common
	Small species predominate	Large species dominant
	High species evenness	Low species evenness
	High species richness	Low species richness
Species	'K' traits predominate	'*r*' and 'K' traits present
	Sensitive to mortality	Robust to mortality
	Locally rare	Locally common
	Small average size	Large average size
Population	Low emigration rate	High emigration rate
	Low biomass density	High biomass density
	Low population density	High population density
	Low rate of increase	High rate of increase
Individual	Poor health, strength	Good health, strength
	Low longevity	High longevity
	Few or small offspring	Many or large offspring
	Low adult size	High adult size
	Low growth rate	High growth rate
	Low birth mass	High birth mass

Ubiquitous biogeochemical effects of nutrients and proteins on organisms support at ecosystem level an elemental-affected net primary production (eNPP).

investment have been hypothesized to be responsible for many observed body growth patterns and behavioural traits (e.g. Blomberg et al., 2003; Bongers, 1999; Calder, 1984; Carbone et al., 2011; Guénard et al., 2011; Hendriks and Mulder, 2008, 2012; Peters, 1983; Suding et al., 2003). Many of these faunal attributes are mentioned in Table 1, such as offspring number, dispersal rate, maximum lifespan and territory, and vary predictably with the organism's body size with respect to the species' phylogenetic position (Guénard et al., 2011).

Body size remains a fundamental determinant of an organism's ecology, including territory and niche (Jenkins et al., 2007; Werner and Gilliam, 1984), and is one of the most-studied aspects of animal ecology (Blackburn and Gaston, 1994; Isaac and Carbone, 2010). Therefore, we can hypothesize that 'body size' might at least constrain the dispersal rate, population density and 'foraging' of smaller organisms in a different way from those of larger organisms (e.g. Castle et al., 2011; Finlay, 1998, 2002; Foissner, 2006, 2008; Hagen et al., 2012; Mulder and Elser, 2009). And if so, consistent relationships between bio(geo)chemistry and multitrophic interactions will open exciting ways to assess EF (Friberg et al., 2011).

1.3. Aims of our study

While framing our questions on B–EF relationships is relatively straightforward, testing them is not. Biodiversity collectively refers to all aspects of biotic diversity (Naeem et al., 1999) and its effects are believed to differ among ecosystem types (Hooper et al., 2005; Schmid et al., 2009). To avoid possible confusion, biodiversity will be used here for 'biodiversity as a whole', *S* (species diversity) for taxonomic diversity and FD for functional diversity. As Ghilarov addressed (2000: p. 410), any meaning of biodiversity for EF is strictly dependent on the definitions of ecosystem types and EF; like Ghilarov—and Lindeman (1942) before him—we adhere here to a functional definition of the ecosystem, separated from the surrounding 'environment'. For this purposes, we chose monophyletic ecological assemblages ('taxocenes' *sensu* Hutchinson, 1978) as units to investigate numerical abundance and species diversity relationships (Kaspari, 2001). Productivity and species diversity are influenced by resource limitation and nutrient supply (Allen et al., 2005; Brown et al., 2004; Hubbell, 2001; Huston and Wolverton, 2009; Mulder et al., 2005a; Sterner and Elser, 2002). Hence, the total number of coexisting individual monophyletic assemblages is hypothesized to reflect the ability to harvest and divide energy within a single taxocene (Kaspari, 2001).

Previous attempts to compare different monophyletic assemblages with each other show an unrecorded parallel among insect, bird and plant species (Thomas et al., 2004b). In an attempt to foster a new mechanistic debate, which is important for understanding EF (Chapin et al., 2000), we selected three characteristic taxocenes from the plant and animal kingdoms as well: vascular plants (seagrasses and forest trees), terrestrial invertebrates (soil nematodes and social insects) and freshwater fishes. These three taxocenes are representative of two terrestrial systems, the 'green world' (Polis, 1999) and the 'brown world' (Allison, 2006), and one aquatic system, the 'blue world' (Fig. 2). We also focus on two harsh soils systems, the Atacama Desert with its hypolithic communities of underneath living phototrophs and the Antarctic Dry Valleys with their extremely low biodiversity (Wall, 2008), and two temperate, human-disturbed biota, the rivers of Ohio (Burton et al., 2012), and agroecosystems across the Netherlands (Mulder et al., 2011a–c).

We believe that our study has broad implications with respect to developing more effective management of our biotic resources and consequently we shall:

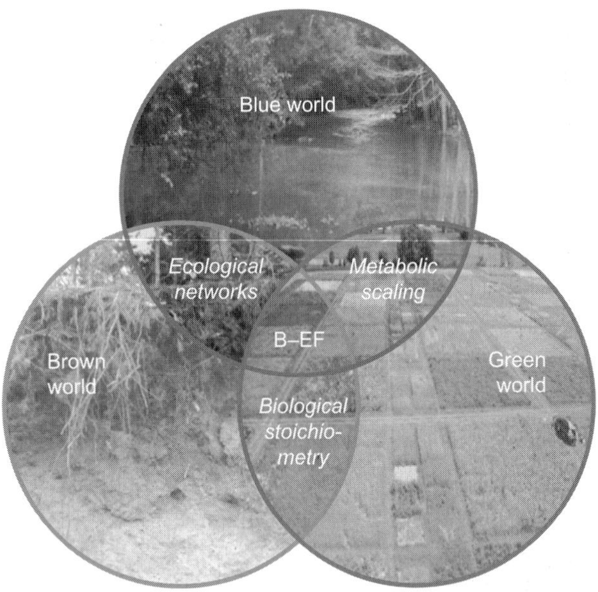

Figure 2 Operational classification of the blue world (water compartment), the brown world (belowground) and the green world (aboveground). The overlapping parts address the kind of B–EF responses measured in this study for three independent taxocenes (freshwater fishes, soil invertebrates and vascular plants). Photo credits: Scott D. Dyer, Shigeta Mori and Winfried Voigt, respectively. (For colour version of this figure, the reader is referred to the online version of this chapter.)

1. analyze biodiversity–productivity relationships in the framework of EF to provide empirical evidence for allometric scaling in reference to ecological stoichiometry,
2. relate the environmental abiotics to presence, mass and abundance of organisms within taxocenes, to assess the importance of species traits that can be seen as stoichiometrically similar and
3. investigate the extent to which aquatic and terrestrial communities composed of species that are stoichiometrically similar may differ from those where species have wide differences in their elemental composition.

2. SCALING B–EF

2.1. Implications of scaling

The scaling of the rates of organismal functions with body size in communities or ecosystems is effectively addressed by allometry, a central—although still somewhat controversial—feature of ecosystems. Allometric scaling has been successfully used, among others, in macroecology (e.g. Arim et al., 2011; Brown et al., 2004; Jacob et al., 2011; Nakazawa et al., 2011; Savage et al., 2004; Storch et al., 2007; West and Brown, 2004), ecological stoichiometry (Mulder and Elser, 2009), the assessment of human-induced biomass exploitation by fishing (Jennings and Blanchard, 2004; Jennings et al., 1999), the impact of global warming on freshwater communities (Dossena et al., 2012; Yvon-Durocher et al., 2010, 2011a) and even for the characterization of fossil food webs (Dunne et al., 2008).

The ecological implications of scaling are great. Figure A1 shows that allometric diversity–yield relationships between species mass and species density (mass–abundance) can be translated into ecological processes transcending discrete boundaries. The metabolic rate, in particular, can be easily estimated by allometric scaling (Enquist et al., 1999; Ernest et al., 2003; Mulder et al., 2005b), with the metabolic respiration rate *per capita*, R, as function of the individual organismal body mass m:

$$R \propto m^{3/4} \qquad [1]$$

A notable example of such research beyond biogeographical boundaries comes from a continental transect across Asia, where the field investigation of tree species of different ages (from saplings up to giant trees) is possibly the best physiological example of macroecological scaling (Fig. 3, recalculated from Mori et al., 2010) in which respiration and fresh weight were determined for

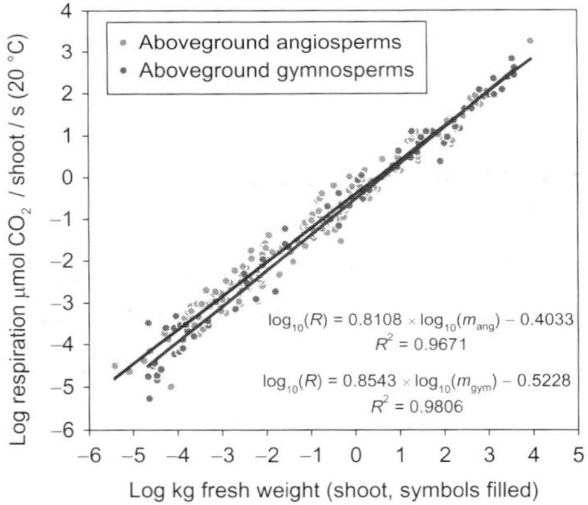

Above ground

Fresh weight (shoot kg)	Gymnosperm (µmol CO_2)	Angiosperm (µmol CO_2)	Respiration rate (adimensional)
0.0001	0.00011	0.00023	0.51
0.001	0.0008	0.0015	0.56
0.01	0.006	0.009	0.62
0.1	0.04	0.06	0.69
1	0.3	0.4	0.76
10	2.1	2.6	0.84
100	15.3	16.5	0.93
1000	109.7	106.9	1.03
10,000	784.15	691.67	1.13

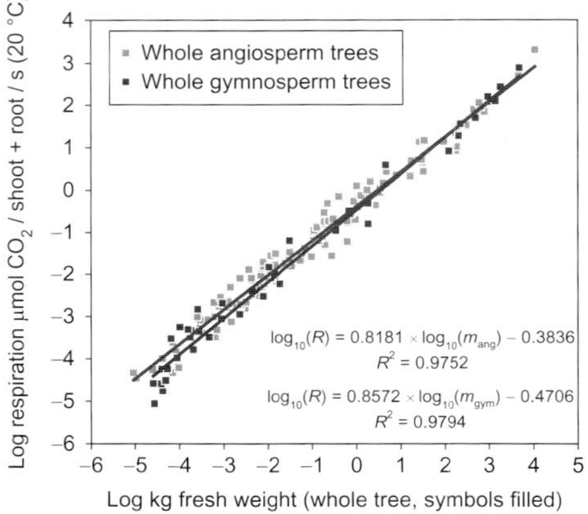

Above and below ground

Fresh weight (whole-tree kg)	Gymnosperm (µmol CO_2)	Angiosperm (µmol CO_2)	Respiration rate (adimensional)
0.0001	0.00013	0.00022	0.57
0.001	0.0009	0.0015	0.62
0.01	0.007	0.010	0.68
0.1	0.05	0.06	0.75
1	0.3	0.4	0.82
10	2.4	2.7	0.90
100	17.5	17.9	0.98
1000	126.2	117.7	1.07
10,000	908.24	774.11	1.17

Figure 4 Plant physiological efficiencies for gymnosperm and angiosperm trees as predicted by the scaling analyses for the aboveground part (upper panel) and the whole-tree weight (lower panel). In the headers from left to right, the fresh weight of the plant (kg shoots), the tree respiration at 20 °C in µmol CO_2 forecasted for gymnosperms and angiosperms, and the physiological efficiency rate between gymnosperm and angiosperm trees of the same weight. The aboveground part of angiosperm saplings is about two times as efficient in the respiration rate as the shoot of gymnosperms of the same weight. For taller trees (> 100 kg fresh shoot weight), the switch in the aboveground respiration for (adult) gymnosperms versus angiosperms is expected to occur around 600 kg fresh shoot weight. For the whole tree, angiosperm saplings remain much more efficient in

and can show contrasting relationships between metabolic rates and body mass (Makarieva et al., 2008; Reiss et al., 2010).

According to the data of Makarieva et al. (2008), the metabolic rate of free-living amoebae scales with their body mass to the $-^{1}/_{6}$ power, opposite to non-amoeboid parasites (i.e. human endopathogenic protozoans) whose metabolic rate scales with body mass to the $^{1}/_{6}$ power. One feature is that we might speculate that non-amoeboid parasites are adapted to constant, high temperatures and downregulate gene expression or even lose genes as organelles, that is, mitochondria, are commonly lost by such parasites (Cavalier-Smith, 1993; Walker et al., 2011). Thus, we may need to take the external conditions, that is, the host, into consideration to evaluate metabolic rates or generally treat obligate parasites separately. Another feature is that in the case of free-living protist groups, this might indicate that locomotion in viscous water is less energy demanding for amoebae which move forward attached to surfaces and do not swim actively like ciliates and flagellated organisms. Amoebae were already regarded by Fenchel and Finlay (1983) to be metabolically different from other protozoans. In summary, currently available data do not enable the recognition of global allometric trends between and within all taxocenes occurring in the green, brown or blue worlds.

2.3. Allometry and management

These allometric approaches to B–EF relationships have been applied increasingly in the real-world setting of assessing human impacts on fisheries and understanding the causes and consequences of the current global collapses in fish stocks. On a local scale, the taxocene that describes the fish assemblage provides not only critical ecosystem processes but also goods and services of huge economic value to humans. The historical correlations between density-dependent stocks, mesh size, fishing efforts and resulting overexploitation are clearly evident (e.g. Cardinale and Svedang, 2004; Jackson et al., 2001; Walters and Maguire, 1996) and have consequently contributed to make allometry an accepted tool in fisheries and marine sciences (Jennings, 2005; Shin et al., 2005; White et al., 2008; Section 4.5).

Despite the abundance of papers unravelling aspects of the blue world, fewer examples are known for the green and the brown world. Although a comparable correlation between canopy density and forest productivity also seems to be a representative example of B–EF congruences, the data by Mori et al. (2010) also show B–EF incongruences between the metabolic scaling of angiosperms and gymnosperms (Fig. 4). Lumping the variability of individual metabolism in a community (e.g. roughly comparing angiosperms with gymnosperms or evergreen trees with deciduous trees) sums over the limits of forecasting (Fig. 3).

Our findings might have implications for different aspects of forest management: in *conservation* management, significant carbon stocks are protected in living biomass, whereas in *sequestration* management, carbon is retained in ecosystems by (increasing) reforestation. The decomposition process of converting the organic carbon in the (surface or root) litter to CO_2, making nitrogen available for plants without rhizobia, is influenced by the chemical nature of carbon compounds (cellulose vs. lignin), by the kind of mycorrhizal symbiosis, by root exudates and by the microbial pools (bacteria vs. fungi) that support plant life and therefore ultimately underpin terrestrial EF (Beerling and Woodward, 2001; Gams, 1992; Lynch and Whipps, 1990; Moore et al., 2004).

Although the capability of plants to sequester carbon and emit CO_2 to the atmosphere varies across species (Bala et al., 2007), allometry has been used scarcely to forecast or manage global changes (Fahey et al., 2010). Size-related allometry provides dynamic tools for wild and domestic population management, such as in the framework of restoration ecology, reducing carbon footprints and implementing activities to minimize deforestation effects. Despite many countries focusing on conservation (e.g. planted trees must belong to native species) or thinning wood, a sustainable agroforestry management should avoid the current large-scale recommendation of gymnosperm trees (such as in United Kingdom, see www.direct.gov.uk/thebigtreeplant, and in the United States, see http://apps.fs.fed.us/fido) because the different capabilities of gymnosperm and angiosperm adults to emit CO_2. Such considerations will surely demand more attention during the planning of afforestation projects in the near future, especially given the increasing socioeconomic momentum behind developing low-carbon-based economies.

3. CONSTRAINING B–EF

3.1. Allometry rules the world

There is a need to investigate B–EF to gain understanding of the biological and ecological factors underpinning sensitivities and traits of species in the context of environmental stressors. In the previous sections, we show the extent to what EF may become recognizable with macroecological approaches such as allometric scaling. Allometry is a suitable method to assess the emergent characteristics of large data sets of organisms (Jonsson et al., 2005; Marquet et al., 2005). According to Brown et al. (2004) and Marquet et al. (2004), the fundamental rules of chemistry, physics and biology provide the means to link individual organisms and their populations to ecosystems and their ecological processes. However, although 'it is clear that scaling relationships hold best when examining patterns across a wide spectrum of body sizes' (Tilman et al. (2004): p. 1798), Brown and Gillooly (2003) show that separate taxocenes derived from small data sets exhibit biomass and mass–abundance scaling relationships that can be opposite from the scaling relationships for all data sets together (Cohen et al., 2003). These divergent relationships raise the question about predictability of species sensitivity to stoichiometrically driven processes, even within comparable size classes. For instance, certain taxa sharing comparable sizes may occupy the most extreme trophic positions not only in a food web but even within a loop, as in the case of viruses as top predators (despite their viral host specificity) and bacteria as basal producers (e.g. Thingstad, 2000). Therefore, it seems difficult to always extrapolate (opposite) results to a wider context, although similarities in the response of phyla and biota become more evident as soon as studies are addressed across scales.

Some, but not all, organisms can be easily identified at species level and a comparable methodology does not *per se* imply equivalent taxonomic resolution: microbial taxa, which drive so many ecosystem processes, remain a particular challenge in this respect (Mulder et al., 2005a,b, 2009; Purdy et al., 2010; Reuman et al., 2009). Regardless of their Latin binomial, all taxa within one community can be modelled using either the unbinned body mass (size) versus numerical abundance scaling or the binned biomass-size spectrum. Successful examples on large-scale investigations come from the blue world (Clarke and Johnston, 1999; Cohen et al., 2003; Killen et al., 2010; Pope et al., 1994, 2006), focusing on traits for behavioural adaptation (like in the case of suspension feeders: Goldbogen et al., 2012; Jeschke et al., 2004; Jørgensen, 1966). Investigations on whales in particular show extreme trophic positions in pelagic ecosystems, because the huge baleens are not only *able* to feed on very small prey (Goldbogen et al., 2012; Jacob et al., 2011), in contrast to toothed whales and teleosts, but are *specialized* to feed on patchy resources. Moreover, trophic levels do not imply a discrete body size (Borgmann, 1987), for although sharing the same trophic level across a wide size distribution, phytoplankton belonging to smaller size classes may achieve faster nitrogen uptake rates than phytoplankton belonging to large size classes (Hein et al., 1995).

This is in contrast to the nitrogen uptake by zooplankton, whose smaller individuals are forced to feed on algae only and whose larger individuals can feed on both phytoplankton and zooplankton (Boit et al., 2012; Fry and Quiñones, 1994; Ptacnik et al., 2010). Body size is thus a fundamental trait for both autotrophs and heterotrophs, as a taxon occupies in a size-based model a much more defined position than it does in a trophic level model' (Cohen, 1994; Cousins, 1980).

Body size can also greatly influence ecological interactions among terrestrial organisms, although perhaps less obviously so than in the blue world, with important consequences at the community and ecosystem level (Yvon-Durocher et al., 2011b). The metabolic scaling of a given organism (Calder, 1984; Damuth, 1981, 1991; Peters, 1983) is one of the best examples of B–EF because functional scaling is species-independent (i.e. unrelated to taxonomic diversity). Across species and within one or more taxocenes, many physiological models may hold. Figure 5 shows that pollinating insects (here, some extremely diverse bees, wasps, butterflies and moths) can cover the entire allometric range of reported measurements. The metabolic rate (R) of *all* insects (data recomputed from the publicly available data of Chown et al., 2007) scales with insect mass to the 0.87 (±0.02 SE) power, but different metabolic scaling exponents are recognizable within finer taxonomic groupings. Scaling with mass to the 0.78 (±0.06) and the 0.72 (±0.03) power, respectively, hymenopterans and coleopterans are the groupings closest to the scaling exponent for the metabolic rate for all insects. In the upper part of the scatter-plot of Fig. 5, the metabolic rates of lepidopterans and orthopterans scale to the 0.67 (±0.06) and the 0.60 (±0.09) power, respectively (much lower than the isopterans, which scale isometrically with mass to the 1.04 (±0.18) power). Some dipterans, hemipterans and coleopterans (two genera of Curculionidae) have the lowest metabolic rate among all insects. Given the ubiquity of insects in freshwater ecosystems, their different allometric scaling must have implications for the blue world as well and should be addressed in the future.

3.2. How local biodiversity determines individual abundances at taxocene level

The number of species is believed to be particularly critical for B–EF in environments with low biodiversity, where there is less scope for redundancy to be manifested (i.e. 1–10 species; Wall, 2007). For protozoa and microorganisms,

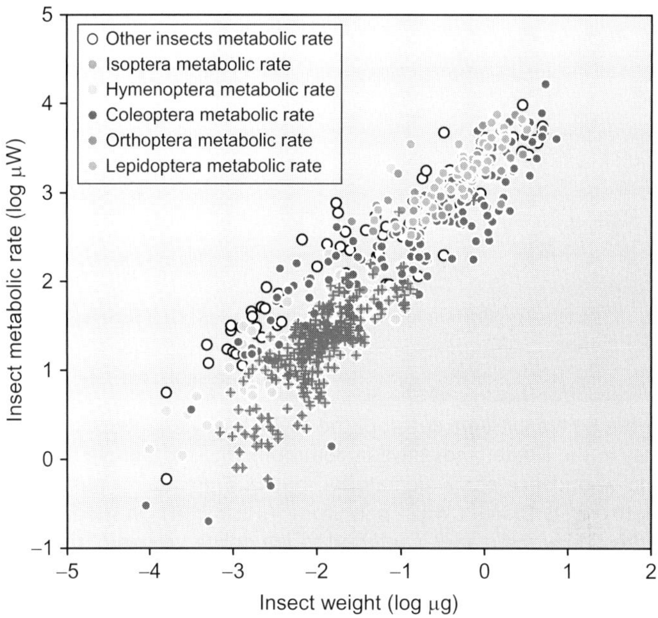

Figure 5 Power laws at synecological scale: allometry is suitable for functional forecasting of brown food web invertebrates such as insects. All eusocial insects like size-polymorphic colony ants (Hymenoptera) and termites (Isoptera) are additionally marked by one cross. Raw data from Chown et al. (2007, their Supplementary Material); these authors converted different metabolic rates from scientific literature to microwatts assuming $Q_{10}=2.0$ at 25 °C. Values for the scaling exponents vary among groups and fall approximately between $^2/_3$ and 1. (For colour version of this figure, the reader is referred to the online version of this chapter.)

rare species are likely to compose the majority of species within a habitat (Dawson and Hagen, 2009; Finlay, 1998). Less abundant microorganisms might have pronounced bottom-up effects, as shown for several bacterial species under lab conditions (Höppener-Ogawa et al., 2009). This holds for soil invertebrates as well: in the McMurdo Dry Valleys of Antarctica, there may be three to five (or even fewer) nematode species (Moorhead et al., 2002; Treonis et al., 1999, 2000). However, the numerical abundances of nematodes in pristine Antarctica can be comparable to those of temperate agroecosystems (Fig. 6), in apparent contradiction to the More Individuals Hypothesis (MIH), as originally defined by Srivastava and Lawton (1998), who related the higher biodiversity of productive locations *to the ability of such sites to support large populations of each species*. However, although the most

Figure 6 What do the extreme desert of the Taylor Valley in Antarctica and one recovered sea clay in the Netherlands have in common? One soil sample might contain an almost equal abundance of soil nematodes, but with greatly contrasting numbers of species. Photo credits: Diana H. Wall/Emily Stone (A) and Christian Mulder (B). (For colour version of this figure, the reader is referred to the online version of this chapter.)

productive sites of that study (physically isolated water-filled microhabitats such as tree holes; see further Hagen et al., 2012) contained more species (Srivastava and Lawton, 1998), this was not a matter of more individuals, as the increase in species richness with productivity occurred only when the energy amount was reduced.

Assuming that the opposite holds as well (the fewer the species, the lower the total abundance), smaller populations under low productivity are likely to be prone to extinction. In that case, the polar deserts are a unique exception, because they do not only support fewer species, but exhibit large populations with far more individuals (Wall, 2007, 2008). The low human-induced disturbance in most deserts makes such environments attractive to assess the ecosystem responses to climate and therefore, other drylands received more attention as well. Recently, Maestre et al. (2012) clearly show that sustainability and multifunctionality (defined, among others, as the ecosystems' ability to maintain productivity, to support carbon storage and to buildup nutrient pools) are positively related to species richness. Deserts like those investigated by Maestre et al. (2012) and the Antarctic Dry Valleys are most valuable to test the MIH because drylands are less affected by sampling effects and patchiness, in contrast to fragmented landscapes such as the moss carpets of Gonzalez et al. (1998). Deserts can therefore reveal the key function (if any) of biodiversity *sensu lato* under environmental stress.

To test the MIH model of Srivastava and Lawton (1998), we analyzed the nematofauna from 200 agroecosystems with different habitat fertility (Mulder and Vonk, 2011), that is, productivity proxies: in contrast to the first study, where the authors used debris, we used primary nutrients to characterize productivity. Being the Shannon-Wiener index function of the (number of) species and given that higher values indicate even species distribution, Fig. A2 shows a hump-shaped relationship for both the exponential Shannon index and the taxonomic diversity of nematodes. Soil fertility (i.e. resource supply) and nematode species richness seem there to be interdependent (a classical B–EF example), indirectly supporting the historical paradigm that productivity drives species richness in contrast to the contemporary view that species richness drives productivity (Cardinale et al., 2009; Gross and Cardinale, 2007). However, such a set of variables might have a predictive power that dynamically changes in space (e.g. Hurlbert and Jetz, 2010; Huston, 1994, 1997; Loreau and Hector, 2001), possibly due to sampling bias or species competition, and in time (e.g. Kaspari (2005) for temperature and Yee and Juliano (2007) for phenology).

Spatial scaling predicts a positive decelerating relationship between abundance and species richness in a way comparable to the MIH: in the sampling hypothesis, for a given species pool, a tropical plot should *per se* contain more species than a low-productivity boreal plot. Kaspari et al. (2003) tested this by randomly sampling simulated m^2 plots with 1, 2,... *n* individuals from the measured species pool for a site and compared that curve with those observed. Where species richness versus total abundance relationships is predictable, it is at such a large aggregation that abundance reflects immigration and/or extinction processes, an intriguing topic when considering that over 100 ant species can be found in 100 m^2 of forest (Kaspari et al., 2001).

To test the extent to which EF increases with biodiversity, we plotted several communities of invertebrates, from ants in pristine rainforests and temperate forests up to nematodes and non-flying arthropods in deserts in the southern hemisphere. All numerical abundances in Fig. 7 were converted to densities per m^2. Assuming the N of the entire population within one taxocene represents a proxy for local resource availability, the taxonomic diversity *within one taxocene* (e.g. the number of arthropod species) scales directly with the abundance of all individuals (e.g. all the arthropods of a given location as in Mulder et al., 2005a). All significant species–density relationships of Table 2 follow power laws with exponents smaller than ½ (i.e. the total species diversity within one taxocene increases monotonically with abundance N), whereas density–species relationships

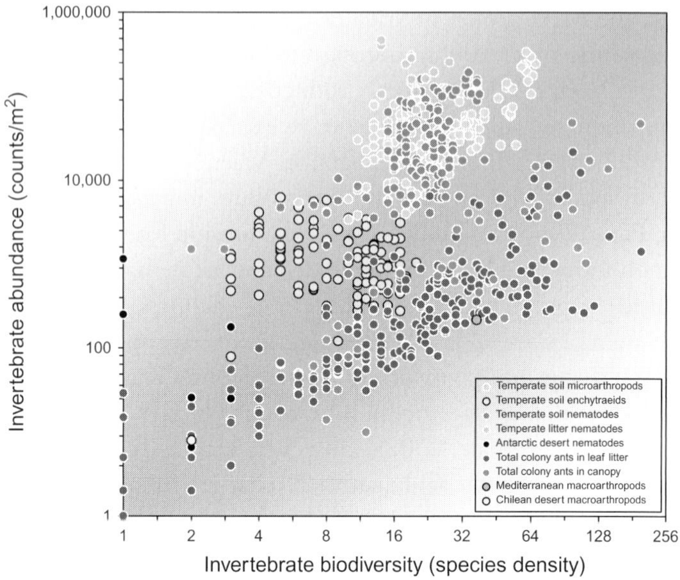

Figure 7 Direct B–EF correlations between species density (biodiversity) and total abundance of individuals are expected to be widespread in comparable plots. Ants data downloadable from Weiser et al. (2011); polar nematofauna from the Taylor Valley, Antarctica, as in Courtright et al. (2001: their Table III) and Barrett et al. (2006: their Table IV); temperate nematofauna downloadable from Mulder and Elser (2009) and Mulder and Vonk (2011). Macroarthropods from the coastal zone of the Atacama Desert, Chile, were described in González et al. (2011); temperate soil microarthropods—mites and collembolans—described in Mulder et al. (2005a); Dutch data on soil enchytraeids from grasslands, heathlands and forests are novel; data on the litter nematofauna from pine forests in the Netherlands and on the litter macroarthropods from a Mediterranean beech forest in Italy are unpublished (C. Mulder and G. Mancinelli, respectively). Please note the two logarithmic scales, being different bases (2 and 10) used. (For colour version of this figure, the reader is referred to the online version of this chapter.)

follow power laws with exponents larger than 1 (i.e. the total abundance within one taxocene increases with species diversity S, cf. Fig. 7).

All macroarthropods in Fig. 7 had $R^2 = 0.66$ ($P < 0.00001$), rejecting the null hypothesis of no correlation between density N and diversity S. With different combinations, including or excluding soil and litter invertebrates, respectively, the results for the taxocenes are given in Table 2, along with the regression lines of density as function of biodiversity and vice versa. Assuming that resource availability within a sampling area is homogeneous, different distributions become recognizable, irrespective of environmental conditions: smaller animals belonging to the micro- and the mesofauna

Table 2 Scaling at different aggregation levels of the total density N as function of biodiversity S and vice versa for the invertebrates shown in Fig. 7

ID	Faunal taxocenes	Environment types	Plots	N scales to S	S scales to N	Pearson's r	Variance explained (%)
A	Microfauna	Soil and litter	142	2.24 ± 0.17	0.25 ± 0.02	0.742***	55.1
A1	Free-living soil nematodes	Soil	120	1.70 ± 0.26	0.16 ± 0.02	0.521***	27.1
A2	Moss-inhabiting nematodes	Litter	22	3.15 ± 0.34	0.26 ± 0.03	0.901***	81.2
B	Mesofauna	Soil	246	1.98 ± 0.12	0.27 ± 0.02	0.733***	53.7
B1	Mites and other microarthropods	Soil	146	1.24 ± 0.13	0.32 ± 0.03	0.635***	40.3
B2	Enchytraeids	Soil	100	-0.21 ± 0.16	-0.09 ± 0.06	−0.136	<2
C	Macrofauna	Litter and canopy	259	1.42 ± 0.06	0.46 ± 0.02	0.809***	65.5
C1	Litter macroarthropods	Litter	225	1.40 ± 0.06	0.49 ± 0.02	0.831***	69.2
C2	Canopy macroarthropods	Canopy	34	1.26 ± 0.24	0.37 ± 0.07	0.680***	46.3
	Partial aggregation						
A + B2	Enchytraeids + nematodes	Soil and litter	242	2.18 ± 0.14	0.23 ± 0.01	0.704***	49.6
A1 + B	Soil micro + mesofauna	Soil	366	2.04 ± 0.10	0.25 ± 0.01	0.721***	52.0
B1 + C	Micro + Macroarthropods	Soil, litter and canopy	405	1.66 ± 0.13	0.17 ± 0.01	0.529***	28.0
	Complete aggregation						
A + B + C	All taxocenes together	Soil, litter and canopy	647	1.48 ± 0.11	0.15 ± 0.01	0.472***	22.3

N and S values were $\log_{10}$-transformed to measure strength and direction of their linear dependence by the Pearson's correlation coefficient and the standard error. Such base-10 log–log linear regressions can be easily transformed in power laws. Besides enchytraeids ($P = 0.179$, implying a random, non-linear relationship), all these linear relationships were significant: ***($P < 0.00001$). Note the steepness increase in abundance–biodiversity relationships from the larger-sized macroarthropods (here: colony ants under pristine conditions) down to the microarthropods (mites and collembolans) and nematodes.

(like nematodes, enchytraeids, mites and collembolans) clearly have a much higher average density per species than is true for larger macroarthropods (ants, beetles, etc.). Although taxocenes showed significant positive correlations between biodiversity and total abundance, enchytraeids showed no significant trend (Table 2).

3.3. The extent to which scaling changes between taxocenes

The way in which the correlation between density N and number of species S changed between taxocenes may largely be interpreted as secondary tradeoffs (demographic responses to abiotic or biotic factors *sensu* Suding et al., 2003). In 2006, Meehan classified the occurring taxa in his brown world study into two broad guilds: 'Grazers', including soil nematodes, oribatid mites, collembolans and enchytraeids, and 'Carnivores', including spiders, ants, chilopods and non-oribatid mites (these in apparent contrast to freshwater literature, where grazers typically mean consumers of algae and carnivores do not *per se* imply only piscivorous fishes). Notwithstanding the difference of more than five orders of magnitude in the total faunal density N (Fig. 7), for differently sized invertebrates the slopes of the density–biodiversity regression lines were 1.42, 1.98 and 2.24 for litter and soil macrofauna, mesofauna and microfauna, respectively (Table 2).

The exclusion of litter macroinvertebrates from our log(N) on log(S) analysis makes the slope of the regression line steeper than lumping all invertebrates together. This was expected, given that invertebrates using different resources (i.e. different fractions of energy supply, as in the case of litter vs. canopy arthropods), collectively deplete energy more effectively (Kaspari and Weiser, 2012). Hence, scaling to higher levels of aggregation may track the ecosystem's productivity (eNPP) more accurately (Kaspari and Weiser, 2012). Scaling varied markedly between taxocenes, from the numerical abundance of Neotropical canopy ants (highest eNPP), which scale to the $^5/_4$ power of (high) species diversity, up to the abundance of Antarctic nematodes (lowest eNPP), which scale to almost the $^7/_2$ power to their (low) species diversity (Table 2).

Conversely, the log(S) on log(N) functions show a sharper increase in steepness, varying from 0.25 to 0.27, and finally 0.46 for macrofauna, mesofauna and microfauna, respectively. The coefficients of variation (CVs $= 100 \times$ SD/degree average) for diversity S, as predicted by density N, were higher than for density N as predicted by diversity

S (32.7% vs. 9.4%, respectively). Again, the power scaling differs from the species diversity of litter nematodes, which scales to the ¼ power to their numerical abundance, up to the species diversity of litter arthropods, which scales to the ½ power to their numerical abundance (Table 2). This reflects an antagonism between belowground 'Grazers' and 'Carnivores', as the biodiversity of the entire 'Grazer' group scales to the ¼ power of the total group and the biodiversity of the entire 'Carnivores' group scales to ~½ power of the total group (Section 3.2).

Our results reflected those from other macroecological studies. Marquet et al. (2005) found linearity of log-transformed total number of species S versus log-transformed mass average $\bar{M}$: for South American mammals, their log(S) on log(M) slope was close to −¾. If we assume an isometric mass–abundance scaling (a log(M) on log(N) regression slope equal to −1, implying a constant biomass across populations), then $N \propto \bar{M}^{-1}$ (Cohen et al., 2003; Mulder et al., 2005b; Woodward et al., 2005a). In Table 2, however, all soil invertebrates belonging to micro- and mesofauna (nematodes, enchytraeids, mites and collembolans) share $S = N^{0.25}$. If we merge these two equations together, we get for our soil invertebrates

$$S = N^{0.25} = \bar{M}^{0.25-1} = \bar{M}^{-0.75} \qquad [2]$$

as expected from general metabolic scaling and macroecological theory (Marquet et al., 2005; Storch et al., 2007; Wardle, 2006) and as empirically supported by randomly chosen assemblages of soil nematodes (Mulder and Vonk, 2011).

There are a few exceptions to this law (Eq. 2) in soil food webs, one of which is the enchytraeids, which occupy only one trophic level (this taxocene comprises strictly peripheral consumers of microbial resources). In high productivity grasslands, soil microbivores like enchytraeids seem to be the most sensitive to density-dependent regulation according to classical theory (Lack, 1954), but the relationship between the species diversity and density of enchytraeids is the only non-significant correlation in Fig. 7. Already in 1982, Standen recognized a certain tendency for sites with few enchytraeid species to have high abundances (Standen, 1982), while sites with many enchytraeid species rarely achieve high abundances (see also Standen, 1980). The possibility that this ecologically important size-scaled taxocene is also stoichiometrically different from others (like soil nematodes) merits further investigation in the future.

4. PREDICTING B–EF

4.1. B–EF and functional redundancy in the blue world: Theoretical background

Generalist feeding strategies and omnivory are well-known in the food web literature (e.g. Gilljam et al., 2011; Woodward et al., 2010a). Therefore, a quantification of the degree of redundancy within food webs, either as interspecific or intraspecific (if at different life stages) differences in trophic position and diet, is crucial for B–EF modelling. In contrast to empirical studies, where complexity provides according to Polis (1998) 'an interwoven matrix that holds ... a community together', few studies have really addressed the role of the redundancy within a predator–prey matrix (Reiss et al., 2009). The assignment to a specific guild (trait) is important because it determines the amount (and vertical direction) of possible links. For the classification of freshwater fish species, we followed Goldstein and Simon (1999) and Goldstein and Meador (2004). Fish species are good indicators of freshwater ecosystems health (e.g. Simon, 1999) and cover many feeding types (Attrill and Depledge, 1997). We assigned fishes to five main feeding guilds: planktivore, detritivore, invertivore, herbivore and carnivore *sensu stricto* (Table 3).

Assuming that no species is isolated and that all species are part of one network with more subunits, then possible trophic links depict consumer–resource interactions within the fish assemblages (Table 4 shows empirically validated trophic links as recorded in current literature). After Winemiller (1989), Martinez (1992) and Dunne et al. (2002, 2004), interaction richness is defined as the trophic links L per species S, also referred to as link density, and by connectance C. If both interspecific and intraspecific effects are considered, C is defined as

$$C = \frac{L}{S^2} \quad [3]$$

and if intraspecific effects are not considered, that is, the realized fraction of all pair-wise interactions, besides cannibalism, as

$$C = \frac{L}{[S \times (S-1)]} \quad [4]$$

C as defined in Eq. (3) is used more frequently, and termed 'directed connectance' (Beckerman et al., 2006; Ebenman and Jonsson, 2005; Martinez,

Table 3 Lessons from stress ecology. Abiotic predictors are known to affect the occurrence of fish species (e.g. Hawkins, 2006; Hendrixson et al., 2007; Posthuma and De Zwart, 2006), albeit fish traits are influenced as well, as expected by the relevant properties across taxonomical and organizational levels from Table 1 (Huston and Wolverton, 2011; but see Sterner and Elser, 2002). The fish traits at population level are based on species assignment according to Goldstein and Meador (2004); traits at individual level are based on empirical data across the State of Ohio

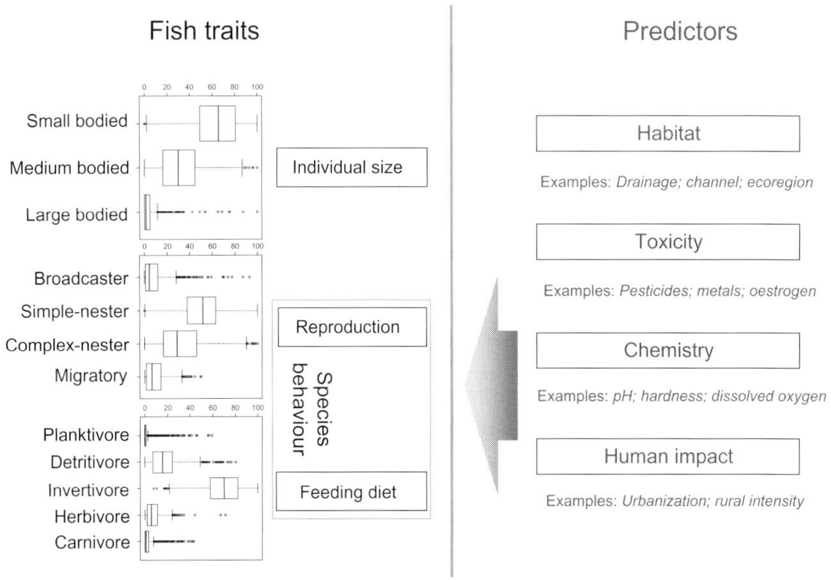

1991). The 'link density' is also known as linkage density (Pimm et al., 1991; Winemiller 1989, but see the original definition in Briand, 1985).

To avoid confounding C of different food webs by differences in sampling methods, we focused in the next section on a consistent methodology (Havens, 1992; Martinez, 1993; Romanuk et al., 2009). Rather than computing the link density as the number of realized trophic interactions per locally occurring species, we chose an adapted taxocene-specific web connectance (hereafter, C_t as in Fig. A3), where the proportion of all trophic links between fish species (i.e. who eats whom but not who eats what) that are realized in one fish assemblage as derived from the matrix in Table 4. C_t reflects either a dominance of generalists (high C_t values:

Table 4 Predator–prey matrix showing the dominant freshwater fish species occurring in 18 rivers of Ohio, horizontally and vertically ranked according their average fresh weight. Rows as resources, columns as consumers, black cells the realized trophic links the largest consumer (predator) is plotted upper left of the table, the smallest consumer (*Etheostoma nigrum* feeds on eggs of other fishes) is plotted bottom right. More details in Table A1

Preys \ Predators	*Pylodictis olivaris*	*Aplodinotus grunniens*	*Ictalurus punctatus*	*Carpiodes cyprinus*	*Micropterus dolomieux*	*Pomoxis nigromaculatus*	*Pomoxis annularis*	*Ameiurus natalis*	*Micropterus punctulatus*	*Micropterus salmoides*	*Ambloplites rupestris*	*Perca flavescens*	*Esox americanus*	*Lepomis gulosus*	*Nocomis micropogon*	*Lepomis gibbosus*	*Noturus flavus*	*Lepomis cyanellus*	*Semotilus atromaculatus*	*Percina caprodes*	*Percopsis omiscomaycus*	*Umbra limi*	*Cottus bairdii*	*Percina maculata*	*Cyprinella spiloptera*	*Pimephales notatus*	*Pimephales promelas*	*Fundulus notatus*	*Notropis stramineus*	*Etheostoma flabellare*	*Etheostoma caeruleum*	*Etheostoma nigrum*
Pylodictis olivaris	0	0	0	0	0	0	0	0	0	0	0	0	0	0	0	0	0	0	0	0	0	0	0	0	0	0	0	0	0	0	0	0
Moxostoma anisurum	1	0	0	0	0	0	0	0	0	0	0	0	0	0	0	0	0	0	0	0	0	0	0	0	0	0	0	0	0	0	0	0
Aplodinotus grunniens	1	0	0	0	0	0	0	0	0	0	0	0	0	0	0	0	0	0	0	0	0	0	0	0	0	0	0	0	0	0	0	0
Ictalurus punctatus	1	1	0	0	0	0	0	0	0	0	0	0	0	0	0	0	0	0	0	0	0	0	0	0	0	0	0	0	0	0	0	0
Carpiodes cyprinus	1	1	1	0	0	0	0	0	0	0	0	0	0	0	0	0	0	0	0	0	0	0	0	0	0	0	0	0	0	0	0	0
Moxostoma breviceps	1	1	1	1	0	0	0	0	0	0	0	0	0	0	0	0	0	0	0	0	0	0	0	0	0	0	0	0	0	0	0	0
Moxostoma duquesnei	1	1	1	1	0	0	0	0	0	0	0	0	0	0	0	0	0	0	0	0	0	0	0	0	0	0	0	0	0	0	0	0
Moxostoma erythrurum	1	1	1	1	0	0	0	0	0	0	0	0	0	0	0	0	0	0	0	0	0	0	0	0	0	0	0	0	0	0	0	0
Minytrema melanops	1	1	1	1	0	0	0	0	0	0	0	0	0	0	0	0	0	0	0	0	0	0	0	0	0	0	0	0	0	0	0	0
Micropterus dolomieux	1	1	1	1	0	0	0	0	0	0	0	0	0	0	0	0	0	0	0	0	0	0	0	0	0	0	0	0	0	0	0	0
Pomoxis nigromaculatus	1	1	1	1	1	0	0	0	0	0	0	0	0	0	0	0	0	0	0	0	0	0	0	0	0	0	0	0	0	0	0	0
Pomoxis annularis	1	1	1	1	1	1	0	0	0	0	0	0	0	0	0	0	0	0	0	0	0	0	0	0	0	0	0	0	0	0	0	0
Dorosoma cepedianum	1	1	1	1	1	1	1	0	0	0	0	0	0	0	0	0	0	0	0	0	0	0	0	0	0	0	0	0	0	0	0	0
Hypentelium nigricans	1	1	1	1	1	1	1	0	0	0	0	0	0	0	0	0	0	0	0	0	0	0	0	0	0	0	0	0	0	0	0	0
Ameiurus natalis	1	1	1	1	1	1	1	0	0	0	0	0	0	0	0	0	0	0	0	0	0	0	0	0	0	0	0	0	0	0	0	0
Micropterus punctulatus	1	1	1	1	1	1	1	1	0	0	0	0	0	0	0	0	0	0	0	0	0	0	0	0	0	0	0	0	0	0	0	0
Catostomus commersonii	1	1	1	1	1	1	1	1	1	0	0	0	0	0	0	0	0	0	0	0	0	0	0	0	0	0	0	0	0	0	0	0
Micropterus salmoides	1	1	1	1	1	1	1	1	1	0	0	0	0	0	0	0	0	0	0	0	0	0	0	0	0	0	0	0	0	0	0	0
Ambloplites rupestris	1	1	1	1	1	1	1	1	1	1	0	0	0	0	0	0	0	0	0	0	0	0	0	0	0	0	0	0	0	0	0	0
Perca flavescens	1	1	1	1	1	1	1	1	1	1	1	0	0	0	0	0	0	0	0	0	0	0	0	0	0	0	0	0	0	0	0	0
Esox americanus	1	1	1	1	1	1	1	1	1	1	1	1	0	0	0	0	0	0	0	0	0	0	0	0	0	0	0	0	0	0	0	0
Lepomis gulosus	1	1	1	1	1	1	1	1	1	1	1	1	1	0	0	0	0	0	0	0	0	0	0	0	0	0	0	0	0	0	0	0
Nocomis micropogon	1	1	1	1	1	1	1	1	1	1	1	1	1	1	0	0	0	0	0	0	0	0	0	0	0	0	0	0	0	0	0	0
Lepomis gibbosus	1	1	1	1	1	1	1	1	1	1	1	1	1	1	1	0	0	0	0	0	0	0	0	0	0	0	0	0	0	0	0	0
Noturus flavus	1	1	1	1	1	1	1	1	1	1	1	1	1	1	1	1	0	0	0	0	0	0	0	0	0	0	0	0	0	0	0	0
Lepomis macrochirus	1	1	1	1	1	1	1	1	1	1	1	1	1	1	1	1	1	0	0	0	0	0	0	0	0	0	0	0	0	0	0	0
Lepomis cyanellus	1	1	1	1	1	1	1	1	1	1	1	1	1	1	1	1	1	0	0	0	0	0	0	0	0	0	0	0	0	0	0	0
Lepomis megalotis	1	1	1	1	1	1	1	1	1	1	1	1	1	1	1	1	1	1	0	0	0	0	0	0	0	0	0	0	0	0	0	0
Semotilus atromaculatus	1	1	1	1	1	1	1	1	1	1	1	1	1	1	1	1	1	1	0	0	0	0	0	0	0	0	0	0	0	0	0	0
Percina caprodes	1	1	1	1	1	1	1	1	1	1	1	1	1	1	1	1	1	1	1	0	0	0	0	0	0	0	0	0	0	0	0	0
Notemigonus crysoleucas	1	1	1	1	1	1	1	1	1	1	1	1	1	1	1	1	1	1	1	1	0	0	0	0	0	0	0	0	0	0	0	0
Luxilus cornutus	1	1	1	1	1	1	1	1	1	1	1	1	1	1	1	1	1	1	1	1	0	0	0	0	0	0	0	0	0	0	0	0
Luxilus chrysocephalus	1	1	1	1	1	1	1	1	1	1	1	1	1	1	1	1	1	1	1	1	0	0	0	0	0	0	0	0	0	0	0	0
Lepomis humilis	1	1	1	1	1	1	1	1	1	1	1	1	1	1	1	1	1	1	1	1	0	0	0	0	0	0	0	0	0	0	0	0
Lampetra aepyptera	1	1	1	1	1	1	1	1	1	1	1	1	1	1	1	1	1	1	1	1	0	0	0	0	0	0	0	0	0	0	0	0
Percopsis omiscomaycus	1	1	1	1	1	1	1	1	1	1	1	1	1	1	1	1	1	1	1	1	0	0	0	0	0	0	0	0	0	0	0	0
Umbra limi	1	1	1	1	1	1	1	1	1	1	1	1	1	1	1	1	1	1	1	1	1	0	0	0	0	0	0	0	0	0	0	0
Campostoma anomalum	1	1	1	1	1	1	1	1	1	1	1	1	1	1	1	1	1	1	1	1	1	1	0	0	0	0	0	0	0	0	0	0
Cottus bairdii	1	1	1	1	1	1	1	1	1	1	1	1	1	1	1	1	1	1	1	1	1	1	0	0	0	0	0	0	0	0	0	0
Clinostomus elongatus	1	1	1	1	1	1	1	1	1	1	1	1	1	1	1	1	1	1	1	1	1	1	1	0	0	0	0	0	0	0	0	0
Phenacobius mirabilis	1	1	1	1	1	1	1	1	1	1	1	1	1	1	1	1	1	1	1	1	1	1	1	0	0	0	0	0	0	0	0	0
Notropis photogenis	1	1	1	1	1	1	1	1	1	1	1	1	1	1	1	1	1	1	1	1	1	1	1	0	0	0	0	0	0	0	0	0
Percina maculata	1	1	1	1	1	1	1	1	1	1	1	1	1	1	1	1	1	1	1	1	1	1	1	0	0	0	0	0	0	0	0	0
Cyprinella spiloptera	1	1	1	1	1	1	1	1	1	1	1	1	1	1	1	1	1	1	1	1	1	1	1	1	0	0	0	0	0	0	0	0
Etheostoma blennioides	1	1	1	1	1	1	1	1	1	1	1	1	1	1	1	1	1	1	1	1	1	1	1	1	1	0	0	0	0	0	0	0
Hybopsis amblops	1	1	1	1	1	1	1	1	1	1	1	1	1	1	1	1	1	1	1	1	1	1	1	1	1	0	0	0	0	0	0	0
Pimephales notatus	1	1	1	1	1	1	1	1	1	1	1	1	1	1	1	1	1	1	1	1	1	1	1	1	1	0	0	0	0	0	0	0
Notropis atherinoides	1	1	1	1	1	1	1	1	1	1	1	1	1	1	1	1	1	1	1	1	1	1	1	1	1	1	0	0	0	0	0	0
Lythrurus umbratilis	1	1	1	1	1	1	1	1	1	1	1	1	1	1	1	1	1	1	1	1	1	1	1	1	1	1	0	0	0	0	0	0
Notropis buccatus	1	1	1	1	1	1	1	1	1	1	1	1	1	1	1	1	1	1	1	1	1	1	1	1	1	1	0	0	0	0	0	0
Lythrurus fasciolaris	1	1	1	1	1	1	1	1	1	1	1	1	1	1	1	1	1	1	1	1	1	1	1	1	1	1	0	0	0	0	0	0
Pimephales promelas	1	1	1	1	1	1	1	1	1	1	1	1	1	1	1	1	1	1	1	1	1	1	1	1	1	1	0	0	0	0	0	0
Phoxinus erythrogaster	1	1	1	1	1	1	1	1	1	1	1	1	1	1	1	1	1	1	1	1	1	1	1	1	1	1	1	0	0	0	0	0
Fundulus notatus	1	1	1	1	1	1	1	1	1	1	1	1	1	1	1	1	1	1	1	1	1	1	1	1	1	1	1	0	0	0	0	0
Notropis stramineus	1	1	1	1	1	1	1	1	1	1	1	1	1	1	1	1	1	1	1	1	1	1	1	1	1	1	1	1	0	0	0	0
Labidesthes sicculus	1	1	1	1	1	1	1	1	1	1	1	1	1	1	1	1	1	1	1	1	1	1	1	1	1	1	1	1	1	0	0	0
Etheostoma flabellare	1	1	1	1	1	1	1	1	1	1	1	1	1	1	1	1	1	1	1	1	1	1	1	1	1	1	1	1	1	0	0	0
Notropis volucellus	1	1	1	1	1	1	1	1	1	1	1	1	1	1	1	1	1	1	1	1	1	1	1	1	1	1	1	1	1	1	0	0
Etheostoma caeruleum	1	1	1	1	1	1	1	1	1	1	1	1	1	1	1	1	1	1	1	1	1	1	1	1	1	1	1	1	1	1	0	0
Etheostoma zonale	1	1	1	1	1	1	1	1	1	1	1	1	1	1	1	1	1	1	1	1	1	1	1	1	1	1	1	1	1	1	1	0
Etheostoma spectabile	1	1	1	1	1	1	1	1	1	1	1	1	1	1	1	1	1	1	1	1	1	1	1	1	1	1	1	1	1	1	1	0
Etheostoma nigrum	1	1	1	1	1	1	1	1	1	1	1	1	1	1	1	1	1	1	1	1	1	1	1	1	1	1	1	1	1	1	1	0

omnivory and piscivory dominate) or a skew towards specialists (low C_t values: high sensitivity to environmental stress and low resilience).

The definition of species categories in a food web determines the interactions (these two issues are mutually dependent for Cousins, 1996: p. 244) and static food webs use aggregate attributes like connectance to predict other aggregate attributes, such as the proportion of omnivores (Loreau, 2010: p. 56). However, based on our empirical knowledge of both species-related traits and size-related guilds within the Pisces monophyletic taxocene, the C_t proportion of all links is suggested to reflect the dominance of piscivory without requiring a more systematic data structure. Hence, being unrelated to other compartments or taxonomical groups, C_t values indirectly represent the mutual fish behaviour *within one taxocene* (Fig. A3), analogous to intraguild predation, which is a common feature of many food webs, especially in the blue world (Woodward et al., 2010b).

4.2. Inland water biodiversity: Effects of landscape complexity on B–EF

4.2.1 Streams and ecoregions

Fishes are mobile and useful to assess large-scale (and long-term) effects because many species cover a wide area during their lifespan. The taxonomy of fish is well defined, reflecting their importance as a food resource. Three American Midwestern ecoregions presenting a comparable number of sampling sites were selected for a pilot study on Ohio fish assemblages: Eastern Corn Belt Plains, Erie Drift Plain and Western Allegheny Plateau (Fig. A4). These North-American regional communities were used to quantify and compare the variation in fish body mass and the *properties* of individuals, populations and assemblages across ecoregions.

An ecoregion is defined here as a unit of land that is homogenous with respect to multiple landscape characteristics such as geology, soil characteristics, natural vegetation and climate (after Wagner et al., 2007). The underlying assumption behind the use of ecoregions and watersheds is that classification of surface waters will reduce natural within-class variation of biological and ecological data (Gerritsen et al., 2000). Streams within ecoregions generally respond in a broadly similar manner to comparable management practices (Lyons, 1989), although heterogeneity in physical habitat and water quality conditions may confound the measurement of their biotic responses (cf. Feld et al., 2011; Friberg et al., 2011; Hawkins, 2006; Larrañaga et al., 2010; Toepfer et al., 1998).

4.2.2 Computational methods

Fish species and their assemblages (numerical abundance, body size and interactions) are hypothesis to reflect environmental conditions (e.g. Feld et al., 2011; Layer et al., 2010, 2011). Our analyses of the freshwater fish assemblages were performed at different levels: firstly, via the construction of networks for sites sampled in a comparable way and secondly via the calculation of size spectra and power laws to describe network topology.

At the first level, we used the US-EPA raw counts and body mass of individual fishes in the State of Ohio (www.epa.gov). After data mining, we had 2656 fish taxocenes sampled in different ways. Electrofishing (often in conjunction with seines or nets) was the principal sampling methods for fish individuals in wadeable and boatable streams (Flotemersch, 2001): in wadeable streams, block nets are placed downstream and upstream of the sampling pool, in contrast to boatable streams, where boat-based electrofishing is done throughout.

At the second level, an allometric assessment of the fish size distribution was performed. We conducted analyses for the 2656 locations on possible deviations from linearity of the upper tail of the binned biomass-size spectra: reflecting the use of different sampling methods, the 534 boatable streams exhibited three times lower standard deviations than the 2132 wadeable streams (0.0787 SD vs. 0.2440 SD, respectively). Based on the allometric uncertainties in biomass estimates as derived from the size spectra for all the locations (summarized in Fig. A5), we chose a subset of 302 locations unaffected by either habitat heterogeneity, sampling bias or survey differences, aiming for a consistent sampling methodology (in our case, boatable streams).

Some of these fish assemblages will be discussed in the next sections. In particular, we investigated whether the food web structure within one taxocene might be relatively vulnerable to different static deletion sequences to diagnose the magnitude of changes in biodiversity, for example, through potential fish extinctions caused by ecological impacts in rivers, and allowing for a more effective environmental management. We shall use three different deletion scenarios:

1. the 'connectivity descending' scenario (we will erode from the top of the food web where the most connected nodes are, thus well-connected hubs will get removed as first, so it is likely that we remove a top predator and only that node is gone and neighbouring nodes get isolated),
2. the 'connectivity ascending' scenario (we will erode from the basis of the food web where the less connected nodes are, thus we will get inevitably a high probability of removing a resource which will starve its consumers and the web will collapse with several secondary extinctions) and
3. the 'random' scenario (intermediate simulation, removing nodes randomly from the top, the middle and the bottom of the food web).

4.2.3 Can a web be robust?

Most studies in the last decades have focused on the values of biodiversity. When considering wide ranges in body size, species richness cannot have one value *per se*, even if consistent resolution is used (Leaper and Raffaelli, 1999). These authors showed that different taxonomic resolutions affect allometric scaling and they advocated the use of evenly resolved taxa (for instance, all nodes at either species level, genus level or as life stages). The horizontal axis of Fig. 8 shows a clear increase in the fish species diversity (CV = 36%, from 13 species up to 40 species) and the scatter suggests a (weak) inverse correlation between the species diversity of the ecological networks S and connectance C_t (Pearson's product–moment correlation coefficient of −0.41, $P = 0.090$), in partial agreement with theory (Havens, 1992; May, 1972; Winemiller, 1989). Connectance typically shows greater variability for low-diversity aquatic webs, but $^2/_3$ of our aquatic webs fall into a narrow range between $L/S^2 = 0.2$ and 0.3. For comparison, only $^1/_5$ of the soil food webs addressed in Section 6.1 fall into the same 0.2–0.3 range ($^4/_5$ share $L/S^2 < 0.2$). These results are notable because such differences in connectance values imply that fishes attack up to three times more often other individuals within their own taxocene than soil nematodes do.

A second feature that the food webs have in common is that although the linkage density shows a strong direct correlation with biodiversity both in the blue and brown worlds (Pearson's r = correlation coefficient of 0.90, $P < 0.00001$ for the 18 freshwater food webs of Fig. 8 and Pearson's $r = 0.63$, $P = 0.0016$ for the 22 soil food webs in Mulder and Elser, 2009), the most significant relationships between primary versus secondary extinctions and food web attributes are provided by number of nodes and connectance (Fig. 9). Although strong correlations between S and L/S as well as S and L/S^2 are acknowledged and may seem trivial here, the ecological implications are great. The correlation between S and L/S is more than a statistical artefact, because it indirectly shows the degree to which the possibility to encounter an(other) omnivore species in the food web *increases* with a larger species pool. Hence, it also implies that the average linkage density for a food web *must increase* with the total of investigated species.

According to Williams and Martinez (2004), with no information on link-strengths, the short-weighted trophic level is the most accurate approximation for quantifying trophic levels within (real) food webs that include omnivory, cannibalism and mutual predation. In freshwater ecosystems, like those investigated here (Table 4), fishes consume resources from many

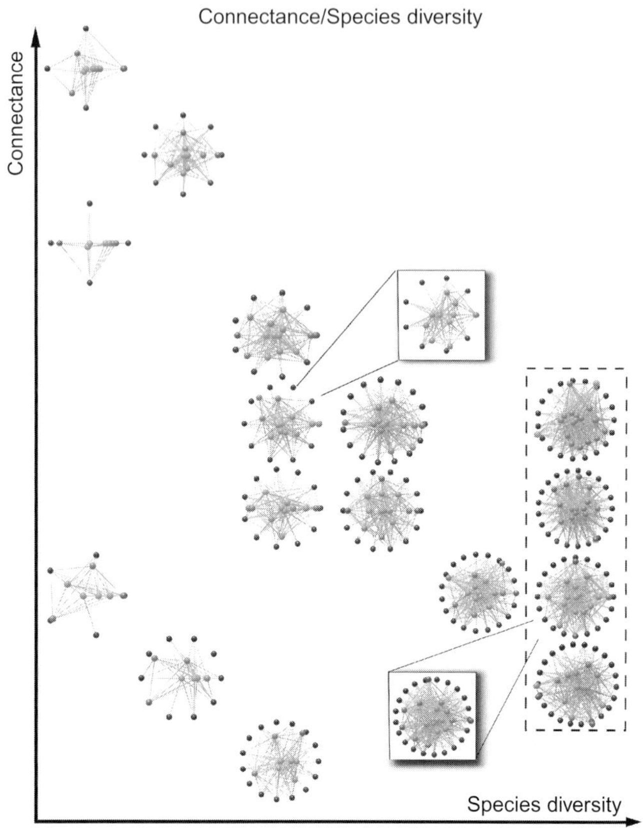

Figure 8 Connectance in relationship to fish species diversity of 18 river food webs visualized with Network3D (Yoon et al., 2004). Two food webs are depicted as overlays (solid boxes) because they fall into the same range in the connectance–diversity space (cf. Table 5). The variability and maximum value of connectance are highest in low-diversity webs, while high-diversity food webs (dashed frame) show a more consistent connectance pattern according to the constant connectance hypothesis (please see the text). The columns of this matrix show that food webs with similar (or even statistically undistinguishable) biodiversity can have different linkage patterns. (For colour version of this figure, the reader is referred to the online version of this chapter.)

trophic levels from hatching to death, including other fishes that were feeding on them during early life stages (Froese and Pauly, 2005; Gilljam et al., 2011; Layer et al., 2010; Montoya et al., 2006; O'Gorman and Emmerson, 2010; Woodward et al., 2010a). This structural complexity is reflected by certain food web properties such as the mean of the short-weighted trophic level, the characteristic path length and the probability that two nodes linked to the same resource are clustered (Dunne, 2009).

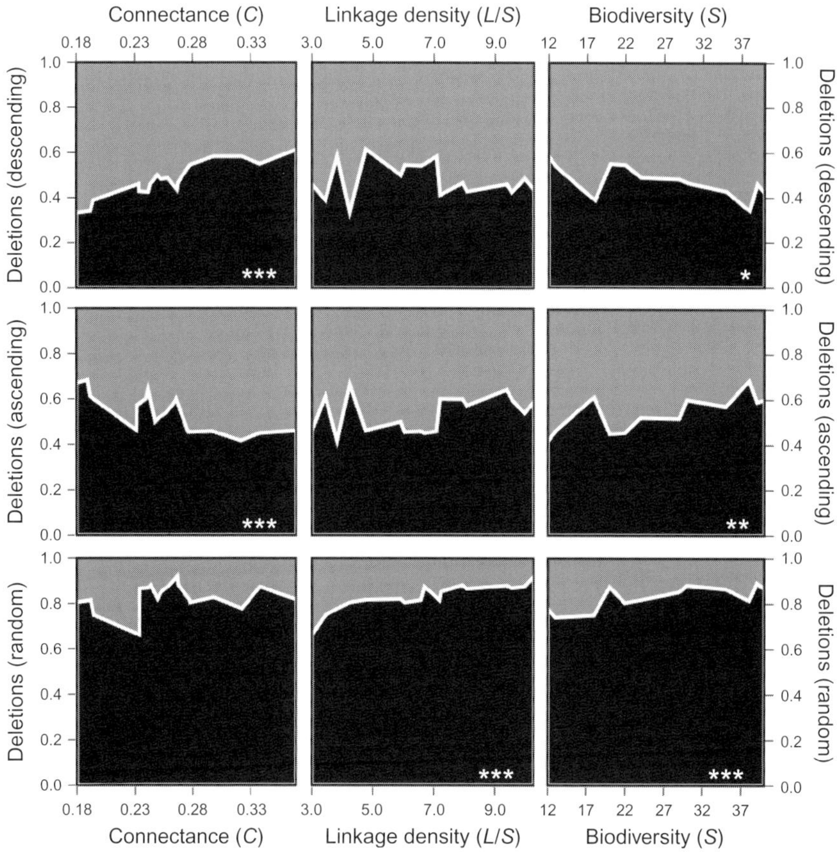

Figure 9 Extinction partitioning for the 18 river food webs according to the 'connectivity descending' (upper three plots), the 'connectivity ascending' (middle three plots) and 'random' (bottom three plots) scenarios along connectance (left part), linkage density (central part) and biodiversity (right part) gradient. The black part is the fraction of species lost by primary deletions and the red part is the fraction of species lost by secondary deletions (all trends as moving averages). The mean primary deletion fraction in the 'descending', 'ascending' and 'random' scenarios is 0.48, 0.56 and 0.83, respectively, and the mean secondary deletion fraction until breakdown of the web is 0.52, 0.44 and 0.17, respectively. If statistically significant, significance is as follows: $^{*}P = 0.035$, $^{**}P = 0.003$ and $^{***}P < 0.001$. Along our linkage density gradient, the fraction of secondary deletions in the descending scenario is very similar to the fraction of primary deletions in the ascending scenario. See text for details. (For interpretation of the references to colour in this figure legend, the reader is referred to the online version of this chapter.)

One common measure of a food web's vulnerability to extinctions is the fraction of primary and secondary deletions until final collapse of the web. For our fish assemblages (Table 5), the two fractions #PD and #SD are roughly similar in the 'connectivity descending' scenario. In contrast, the other scenarios ('connectivity ascending' and 'random') show considerably higher fractions of primary deletions and less secondary extinctions than 'connectivity descending' does. In Fig. 9, we show the extinction partitioning for the 18 food webs according to the 'connectivity descending', 'connectivity ascending' and 'random' scenarios. Food webs are most sensitive to the 'connectivity descending' scenario because their well-connected hubs get removed as first, so secondary deletions occur rapidly, as neighbouring nodes get isolated (Fig. 9, upper left plot). Most connected hubs are at the top in our food web structures. In the 'connectivity descending' scenario, the threshold for web collapse is below 50% of primary deletions, that is, primary species loss. On the other hand, in the 'connectivity ascending' scenario, we removed the less connected (basal) node. As they are resources to others (Table 4), it is inevitable that as soon all of them have gone extinct, any web will collapse quickly, with secondary extinctions becoming prevalent (Fig. 9, middle left plot). Considering the average fraction of basal nodes in the 18 food webs, the value of 51% (Table 5) is close to the average of 54% for the threshold for web collapse in the 'connectivity ascending' scenario. Therefore, the web must collapse quickly close to this point because all basal nodes were already removed. This is the explanation why these descending and ascending scenarios have similar thresholds, but behave differently in reaching them (Fig. 9).

The random scenario differs from the previous scenarios because species were removed from the middle, bottom and top of the web regardless of their connectivity. Therefore, the web is not eroded systematically (either removing all hubs, as in the upper part of Fig. 9, or all basal nodes, as in the middle part of Fig. 9), but merely by chance. Hence, the probability for each species occupying either a hub or a basal node to persist is, on average, higher in the random scenario (Fig. 9, bottom plots). Random changes in the resulting connectivity might generate large extinction events (Solé et al., 1999: p. 159), but not in the freshwater assemblages investigated here.

The average biodiversity of our food webs is 27 fish species: food webs with a lower than average biodiversity have a higher than average connectance, and vice versa (Table 5). Low-connected and more diverse webs

Table 5 Food web properties and vulnerability to simulated species deletion of 18 fish assemblages (location in Fig. A4) ranked according to their initial biodiversity. Summary of properties (red) and vulnerability to three deletion sequences: connectivity descending (green), connectivity ascending (yellow) and random (blue, 1000 simulations). *S*, Number of species; L/S^2, connectance; *L/S*, linkage density; Top, fraction of piscivorous species; Inter, fraction of intermediate species; Basal, fraction of zooplanktivorous species; Omni, fraction of species eating on more trophic levels; GenSD, standard deviation of mean consumer generality; VulSD, standard deviation of mean consumer vulnerability; ConnSD, standard deviation of connectivity for a consumer; SWTL, mean short-weighted trophic level; Char, characteristic path length; Clu, mean cluster coefficient; # PD, fraction of primary deletions (=species deleted/initial species diversity); # SD, fraction of secondary deletions; R_{SD}, robustness against secondary deletions computed as $R_{SD} = PD_1/S$ with PD_1 as number of primary deletions that caused the first secondary deletion(s). If R_{SD} equals # PD, all the secondary deletions occurred at the last primary deletion with web collapse. (For interpretation of the references to colour in this Table legend, the reader is referred to the online version of this chapter.)

	Food web properties													Descending			Ascending			Random		
River name	*S*	L/S^2	*L/S*	Top	Inter	Basal	Omni	GenSD	VulSD	ConnSD	SWTL	Char	Clu	#PD	#SD	R_{SD}	#PD	#SD	R_{SD}	# PD	# SD	R_{SD}
1 Cuyahoga River	12	0.32	3.83	0.08	0.50	0.42	0.50	1.05	0.66	0.39	1.73	1.30	0.36	0.58	0.42	0.17	0.42	0.58	0.42	0.78	0.22	0.66
2 Sandusky River	13	0.23	3.00	0.08	0.46	0.46	0.46	1.31	0.75	0.59	1.71	1.50	0.31	0.46	0.54	0.08	0.46	0.54	0.46	0.66	0.34	0.37
3 Mahoning River	13	0.37	4.77	0.08	0.54	0.38	0.54	1.01	0.54	0.27	1.76	1.21	0.43	0.62	0.38	0.46	0.46	0.54	0.46	0.82	0.18	0.60
4 Buck Creek	18	0.19	3.44	0.06	0.33	0.61	0.33	1.58	0.65	0.70	1.45	1.59	0.31	0.39	0.61	0.06	0.61	0.39	0.61	0.75	0.25	0.38
5 West Mahoning River	20	0.34	6.70	0.05	0.50	0.45	0.50	1.08	0.54	0.36	1.67	1.29	0.43	0.55	0.45	0.20	0.45	0.55	0.45	0.87	0.13	0.77
6 Duck Creek	22	0.27	6.05	0.05	0.50	0.45	0.50	1.23	0.59	0.49	1.72	1.42	0.38	0.55	0.45	0.09	0.45	0.55	0.45	0.80	0.20	0.44
7 Paint Creek East	24	0.18	4.25	0.04	0.29	0.67	0.29	1.69	0.61	0.76	1.38	1.63	0.33	0.33	0.67	0.04	0.67	0.33	0.67	0.81	0.20	0.50
8 Hocking River	24	0.25	5.92	0.04	0.46	0.50	0.46	1.31	0.62	0.56	1.64	1.49	0.39	0.50	0.50	0.04	0.50	0.50	0.50	0.82	0.18	0.48
9 Blanchard River	24	0.27	6.58	0.04	0.50	0.46	0.50	1.23	0.58	0.50	1.72	1.43	0.39	0.54	0.46	0.04	0.46	0.54	0.46	0.82	0.18	0.39
10 West Mahoning River	24	0.30	7.08	0.04	0.54	0.42	0.54	1.12	0.62	0.46	1.79	1.39	0.40	0.58	0.42	0.04	0.46	0.54	0.42	0.83	0.17	0.49
11 Hocking River	29	0.25	7.24	0.03	0.45	0.52	0.45	1.27	0.64	0.56	1.61	1.48	0.38	0.48	0.52	0.07	0.52	0.48	0.52	0.85	0.15	0.65
12 Scioto Brush Creek	30	0.26	7.93	0.03	0.43	0.53	0.47	1.35	0.49	0.53	1.59	1.46	0.42	0.47	0.53	0.17	0.60	0.40	0.60	0.88	0.12	0.76
13 Big Darby Creek	35	0.23	8.09	0.03	0.40	0.57	0.40	1.42	0.57	0.61	1.52	1.52	0.39	0.43	0.57	0.03	0.57	0.43	0.57	0.87	0.13	0.46
14 Paint Creek West	38	0.19	7.18	0.03	0.37	0.61	0.32	1.60	0.61	0.73	0.62	1.61	0.36	0.34	0.66	0.03	0.68	0.32	0.66	0.82	0.18	0.26
15 Little Miami River	39	0.24	9.38	0.03	0.44	0.54	0.44	1.35	0.58	0.59	1.58	1.51	0.41	0.46	0.54	0.05	0.64	0.36	0.54	0.88	0.12	0.59
16 Little Miami River	39	0.26	9.97	0.03	0.46	0.51	0.49	1.29	0.59	0.55	1.67	1.48	0.41	0.49	0.51	0.03	0.54	0.46	0.54	0.88	0.12	0.42
17 Walhonding River	39	0.26	10.28	0.03	0.41	0.56	0.41	1.34	0.48	0.54	1.52	1.46	0.43	0.44	0.56	0.13	0.59	0.41	0.56	0.92	0.08	0.78
18 Little Miami River	40	0.24	9.55	0.03	0.43	0.55	0.40	1.37	0.57	0.59	0.68	1.51	0.40	0.43	0.58	0.03	0.60	0.40	0.58	0.87	0.13	0.26
Average	27	0.26	6.74	0.04	0.44	0.51	0.44	1.31	0.59	0.54	1.52	1.46	0.39	0.48	0.52	0.10	0.54	0.46	0.53	0.83	0.17	0.51

behaved differently from highly connected and less diverse webs (Fig. 9). In the 'connectivity descending' scenario, food webs with high connectance were more robust (higher fraction of primary deletions until collapse) because the probability to remove a 'resource node' for other nodes was low: it was far more likely that a top predator was removed. Webs with higher connectance have a higher probability that there are still enough links left so no species becomes isolated, limiting species loss to the primary deletion in each event. In contrast, in a 'connectivity ascending' scenario, the food web is eroded from its base, starving predators of prey. So, low connectance webs are more robust in this scenario because the resource node has fewer predators that will also disappear when the resource node is gone. In contrast, those webs with higher connectance collapse faster because more secondary extinctions occur when one of the resource nodes is removed.

The kind of deletion sequence clearly affects the vulnerability of the network to species loss (Dunne et al., 2002; Fig. 9). To further evaluate the importance of diversity and connectance on the vulnerability to species deletions, we compared two creeks with intermediate biodiversity (Scioto Brush Creek: $S=30$, Eastern Paint Creek: $S=24$), but different connectance (Scioto Brush Creek: $C=0.26$, Eastern Paint Creek: $C=0.18$) in the 'connectivity descending' scenario (Table 5). Figure A6 shows how rapidly the total deletions equal S and the entire fish network disappears. The shape of the blue curves reflects an increasing number of secondary extinctions and the number of trophic links per fish species descends approximately linearly (Fig. A6B and D), with highly connected species being the most vulnerable. The 'threshold period' until secondary extinctions occur is 17% higher in Scioto Brush Creek than in Eastern Paint Creek (Fig. A6A and C). Despite substantial biological improvement in the environmental health of the Scioto Brush Creek (Burton et al., 2012; OHIO-EPA, 2008), simulated secondary deletions with connectivity descending happen quickly after only 20% of total species removed on average, and all species are gone with on average about 47% of primary deletions. Eastern Paint Creek's web appears even more vulnerable, with the first secondary extinction occurring at 17% of total species loss and the food web collapses at only 33% of primary species loss. Low linkage density implies that the removal of a highly connected node (here, one fish species) results in a loss of fewer links than for webs with higher linkage density (but lower connectance), limiting species loss to primary deletion (Figs. 8 and 9).

4.3. Inland water biodiversity: Vulnerability of B–EF across ecoregions

Another, less frequently discussed aspect of food web vulnerability to species loss is the robustness R_{SD} against secondary deletions, as measured by the fraction of primary deletions that occurred in the web without triggering a secondary extinction. The higher the R_{SD}, the later the secondary extinctions occur as a consequence of a primary one. Secondary species extinctions mostly occur because a primary extinction creates an unfeasible interaction between the predator and the lost prey (Table 4). The different topologies of the webs (cf. Fig. 8) translated into greatly varying vulnerabilities to secondary extinctions in the three species deletion scenarios of Table 5 (Fig. 9). Secondary deletions occur much earlier in the 'connectivity descending' (R_{SD}: mean = 0.10) than in the 'connectivity ascending' scenario (R_{SD}: mean = 0.53). This again implies that, although the total number of primary and secondary extinctions is similar in these two scenarios, the extinction process to final web collapse is very different. Most webs in the 'connectivity descending' start disintegrating early and continuously with one or a few secondary extinctions following each primary one. In contrast, most webs in the 'connectivity ascending' scenario only suffer primary extinctions for a long time, but then break down suddenly with a high number of secondary extinctions. The extinction processes in the 'random' scenario fall between the two extremes (i.e. high and low vulnerability to secondary extinctions for the descending and ascending scenarios) and are more similar to the 'connectivity ascending' scenario (R_{SD}: mean = 0.51). The standard deviations for the 1000 random scenarios are low for the primary deletions (~10% of the mean) and high for the secondary deletions (~50% of the mean).

The differences between the three deletion scenarios demonstrate that web attributes as well as the order of species deletions have a considerable impact on the (species) vulnerability to extinction events: that is, it is not just biodiversity of nodes but also of the interactions within a food web that will affect B–EF trajectories of species loss. On the other hand, even sites from a polluted watershed such as the Hocking River (Burton et al., 2012) seem to have intermediate connectances (Table 5), implying that sensitive fish species that may have previously stabilized the web have already been lost. Although higher connectance led to higher robustness against secondary extinctions in earlier studies (Dunne and Williams, 2009; Dunne et al., 2002), our results only show such an effect ($R_{SD} > 0.1$) for two low-diversity webs (Cuyahoga and Mahoning rivers) and three webs of intermediate to high species richness (West branch of the Mahoning

River, Scioto Brush Creek and Walhonding River). An explanation of this effect for the low-diversity webs may be that they have higher connectance (scales with S^2), but also lower linkage density (scales with S) than other webs (Table 5). For intermediate to high-diversity webs, higher connectance may convey high robustness in some cases, but our results also demonstrate the very opposite, namely that highly vulnerable webs exist despite of high connectance and high linkage density (i.e. Big Darby Creek, Blanchard River, Duck Creek, Hocking River, Paint Creek and all three locations at the Little Miami River shown in Fig. 10).

Food webs (at least those in Fig. 9) are clearly more sensitive to deletion sequences with 'connectivity descending' than to those with 'connectivity ascending'. Node connectivity plays a critical role for the vulnerability of the food web regardless of species diversity and is largely independent of connectance. Our results indicate that high-diversity webs are just as (or even more than) vulnerable to static extinctions as low-diversity webs, a non-intuitive result which may have important implications for ecosystem management. Since higher trophic levels are strongly interlinked with feeding relationships, in contrast to the producer and herbivorous community, this pattern implies that the loss of well-connected intermediate and top

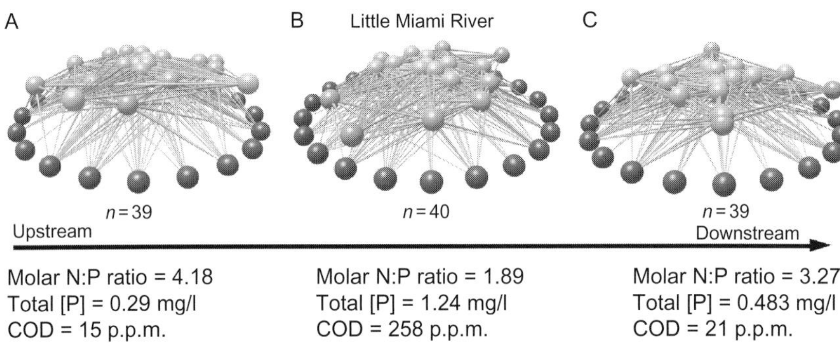

Figure 10 Comparison of species diversity (n = number of fish species) between three locations along the Little Miami River, Ohio, United States (Fig. A4). Although biodiversity is maintained, structural changes in food webs reflect subtle changes in water chemistry, being the central assemblage—collected in a slowly streaming and particulate-rich creek with most organic compounds and high P concentration—the most vulnerable despite of its high connectance and high linkage density (Table 5). The mean of all the coefficients of variation (CVs) for each environmental parameter described by Dyer et al. (1998) for this watershed equals 33%. Below the arrow indicating stream direction, the molar N:P ratio, the total phosphorus concentration and the chemical oxygen demand (COD) in water are provided for each location. (For colour version of this figure, the reader is referred to the online version of this chapter.)

predators is critical to the persistence of species. It could also increase the risk of extinction cascades in ecosystems undergoing environmental and/or anthropogenic change, as is the case in many freshwaters on a global scale (Friberg et al., 2011; Woodward et al., 2012). Such a conclusion has been partly corroborated by previous studies on both model food webs (De Visser et al., 2011; Srinivasan et al., 2007) and empirical food webs (Estes et al., 2011; O'Gorman et al., 2008). Given that the Ohio fish data set comprises, besides predators and consumers, herbivore species only as prey (Table 4), our statistics might have severe implications for extinction risk in a broader context. In fact, even in consumer-free ecosystems, like those modelled by Solé and Montoya (2006), species richness can drop and taxocenes will collapse as soon colonization is no longer sufficient to compensate for habitat fragmentation and habitat destruction.

Apart from connectivity, other species properties such as body size (De Visser et al., 2011), relative abundances (Lyons and Schwartz, 2001) and interaction strength (Allesina and Pascual, 2009) also play a decisive role for a food web's vulnerability to extinction events, especially in dynamic species deletion simulations (Layer et al., 2010; Pimm, 1980; Quince et al., 2005) and long-term empirical studies (Stachowicz et al., 2008). This long-standing, complex issue of the diversity–stability relationship is still controversial (Banašek-Richter et al., 2009; Montoya et al., 2006; Rossberg et al., 2011) and many of its implications for B–EF relationships remain open (as reviewed by Cardinale et al., 2006; Hooper et al., 2005). When species diversity is maintained despite (increased) nutrient loading, biodiversity may act as a kind of buffer against environmental disturbance (Cardinale, 2011) and if this evidence holds for metazoans as well, management or restoration of native fish species becomes desirable (Feld et al., 2011). We provided here an overview that helps to quickly yet coarsely assess the risk of species loss without time-consuming sampling or modelling.

4.4. Population fluctuations at standardized taxonomical resolution: A virtual case study

In both aquatic and terrestrial ecosystems, Srinivasan et al. (2007) and De Visser et al. (2011) showed high sensitivity of (relatively pristine) food webs to the loss of large, dominant or even common species. Moreover, such rare species can inflate allometric relationships, depending on their occurrence and distribution within size classes and areas with different spatial resolution (Valcu et al., 2012; http://cran.r-project.org/package=rangeMapper).

Since size spectra do not distinguish between species, they are easily measured and more robust to inclusion/exclusion of rare species than is the case for species-based community measures of allometric scaling, such as the trivariate food webs that have gained increasing prominence in recent years (Woodward et al., 2010b).

Mass–abundance scatter-plots have the advantage that they can more consistently combine information and, in contrast to size spectra, can be plotted as functions of either endogenous traits (body mass, mostly weight, or body size, mostly length) or exogenous traits (typically numerical abundance). Brown and Gillooly (2003) argued that only traits like endogenous body mass can be used to predict numerical abundance. Unfortunately, the inversion of the *M* and *N* axes in some papers published after Brown and Gillolly's plea contributed to a recent generation of terms which slows down the research itself, as the resulting overlap in terminology may confound many readers. Still, the predictive power of exogenous traits such as *N* for *M* and/or *B* is often surprisingly high. If size-dependent physiology of individuals within one taxocene is extended to entire communities, the allometric scaling of the latter should converge on a biomass-constant isometric line (among others, Cohen et al., 2003; Hildrew, 2009; Mulder et al., 2005b; Rossberg et al., 2008; White et al., 2007; Woodward et al., 2005a).

Previous analyses demonstrated that log(*N*), log(*M*) and log(*B*) are strongly correlated, as theoretically expected (Brown and Gillooly, 2003; Damuth, 1981; Mohr, 1940) and empirically shown (Cohen and Carpenter, 2005; Cohen et al., 2003; Mulder et al., 2008; Reuman et al., 2008). When the classical log–log mass–abundance linear regression model

$$\log(N) = a_1 \times \log(M) + b_1 \quad [5]$$

is merged into the log-transformed biomass (originally weight times abundance)

$$\log(B) = \log(M) + \log(N) \quad [6]$$

we can rewrite Eq. (6) as

$$\log(B) = \log(M) + a_1 \times \log(M) + b_1 = (1 + a_1) \times \log(M) + b_1 \quad [7]$$

which is now in the form of a typical biomass-size spectrum

$$\log(B) = a_2 \times \log(M) + b_2 \quad [8]$$

although the two intercepts b_1 and b_2 are not comparable to each other, in contrast to both slopes which are correlated and are supposed to differ by one unit from each other (Jennings and Mackinson, 2003; Mulder et al., 2008; Schneider et al., 2012).

The linear allometric model of Eq. (8) was fitted to the locations in Ohio separately (confidence interval 99%), and the lumped log(B) for all sampled fishes was plotted at the middle of the respective size class along the binned log(M) gradient. Binned and lumped log(B) with zero observations were excluded, because log(0) is undefined. Size bins can influence the resulting power functions: our fish size spectra tend to show a fluctuating increase in biomass with body size up to a peak near the largest mass-bins comparable to those of Duplisea and Drgas (1999) in the blue world and to those of Mulder et al. (2008, 2009) in the brown world. The huge influence of larger (predatory and omnivorous) fishes is reflected by the regressions that fit the dome before the site-specific modal size bin: the linear regressions fitted to size spectra of the (boat-sampled) fish networks have positive slopes ranging from 0.72 ± 0.074 SE (min) up to 1.24 ± 0.216 SE (max) and the (from Eqs. 5–7 derived) mass–abundance linear regression slopes are rather shallow (their power laws fluctuate between $-1/4$ and $+1/4$, with an average very close to 0). Mass–abundance positive slopes are known as possible within a taxocene (e.g. Ulrich et al., 2005).

For fish assemblages with $1/4$ power scaling, if population density had a body mass scaling exponent of 0.25, a 10-fold increase in weight would increase the fish population by $10^{0.25}$, equal to a 1.78-fold increase in density. Conversely, if population density had a body mass scaling exponent of -0.25, a 10-fold increase in weight would decrease the population as a function of $10^{-0.25}$, which is equal to a 0.56-fold decrease in density of the smaller individuals. To illustrate these opposite trends for further interpretation of freshwater biodiversity, some brief examples may be useful. Let us imagine a very simple freshwater food web consisting of only four fish species, namely Emerald Shiner (*Notropis atherinoides*), Yellow Perch (*Perca flavescens*), Walleye (*Stizostedion vitreum*) and Muskellunge (*Esox masquinongy*). Let their respective wet weights be 4, 40, 400 and 4000 g on a site-specific average. After log-transformation, their log(M) will become 0.6, 1.6, 2.6 and 3.6. Given that with abscissa log(M) and with ordinate log(N) populations fall approximately along a straight line with a negative slope (e.g. Brown and Gillooly, 2003; Cohen et al., 2003; Damuth, 1981, 1987, 1991; Hildrew, 2009; Mulder et al., 2005b; Woodward et al., 2005a), we assume for simplicity that the population densities of these four fish species are equal to 100, 10, 1 and 0.1 individuals, respectively. After log-transformation, their log(N) will become 2, 1, 0 and −1. Their specific log(B) equals log(M)+log(N)=0.6+2=1.6+1=2.6+0=3.6−1=2.6. Hence, these four fish populations will keep a biomass of $10^{2.6} \approx 400$ g and, if plotted on log–log axes, the theoretically resulting linear regression slope should be isometric. In the case of a −¼ power scaling, keeping the aforementioned weights and a comparable number of fishes, the population densities of these species could be 60, 40, 30 and 10, and in the case of a ¼ power scaling, the respective densities should be 10, 30, 40 and 60. In the first case, the resulting specific fish biomass is negatively correlated with the increase in fish body mass, whereas in the second case the opposite occurs. In the case of the 534 boat-sampled sites (Fig. A5), 57.7% showed the negative mass–abundance scaling, albeit on average only $-^1/_8$, but 42.3% showed a positive scaling for the bin approach.

4.5. Superimposed disruption of fish biodiversity on cascading interactions

Cascade effects on other species and trophic levels, for instance due to either invasive or extinct species, can potentially be quantified by allometric analyses and characterization of multitrophic interactions. Sterner and Elser (2002) and Hall (2009) formalized the complexity of elemental constraints and thresholds from stoichiometrically explicit perspectives, reviewing several studies from microbiology to aquatic ecology, emphasizing how important the modulation of chemical imbalances between trophic levels can be for understanding B–EF relationships.

In our study, we modelled the site-specific changes in total biomass (Fig. 11, upper panel) and average weight (Fig. 11, lower panel) along gradients of fish diversity (number of species, left plots) and molar N:P ratio (right plots). The marked decrease in fish size with biodiversity (presumably a consequence of decreasing energy at higher trophic levels) provides a measure for assessing the sensitivity of these species-poor networks (less than 10 species) to predation, even though their environmental conditions (here, the molar N:P ratio) can be considered optimal (Fig. 11D). The average mass of individual fishes does not show a linear correlation with the molar N:P ratio of the water column, in contrast with previous studies where body size (both as fish length and as fish weight) increased with the molar N:P ratio of cyprinids (Sterner and George, 2000) but in line with more recent research, which shows marked stoichiometric imbalances between the environmental availability and tissue content of consumers in freshwaters (Lauridsen et al., 2012).

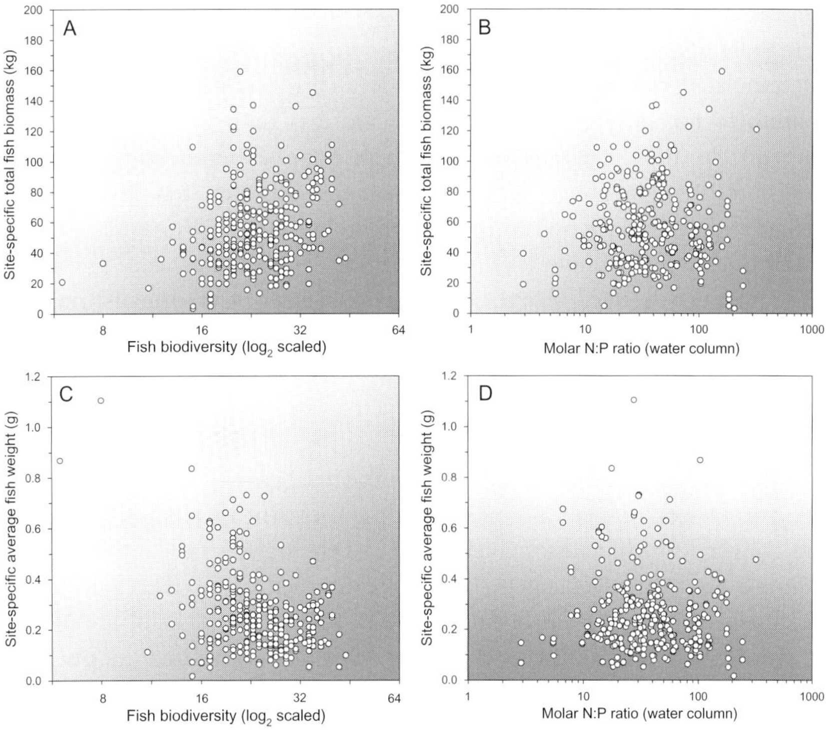

Figure 11 Site-specific fish biomass (A, B) and average fish weight (C, D) related to fish biodiversity (A, C) and water column N:P ratio (B, D) for freshwater fishes from boat-sampled Ohio rivers (Fig. A5). Although American freshwater ecosystems are well-known for their remarkably high N:P ratios (Cotner et al., 2010), the clump in the fish biomass distribution reflects a certain (positive) bias in the amount of Ohio rivers with 'lower' N:P ratios, a log-normal distribution that is known to occur in large datasets (Kattge et al., 2011). According to Pfisterer and Schmid (2002), the species-poor experimental systems achieved under unperturbed conditions show a lower biomass production than the species-rich experimental systems. The left panel clearly resembles their grassland model, where in this case species-poor fish communities not only reduced biomass production under unperturbed conditions (A), but also achieved the highest individual body-size averages (C). The darkness of the grey effect suggests increasing environmental perturbation as derived from abiotic data.

Since regularities might be expected in biodiversity and/or biomass distributions, to what extent can a possible introduction of specifically sized organisms be necessary to preserve ecosystems, for instance counteracting negative effects of overfishing or habitat destruction? As early as 1955, MacArthur pointed out that abundance of species can vary greatly, and that if one species has an abnormal abundance, a community may be unstable if the abundances of other species become inflated (MacArthur, 1955). In Section 4.2.3, we have shown 'cascade effects' on occurring species after simulated removal of fish species in 18 rivers (primary and secondary deletions).

Simulated deletion sequences provide a clear picture of food webs responses to the possible removal of specific taxa (either consumers or resources). In the field, small-scale manipulation experiments (enclosure/exclosure of larger predators) often reveal that relatively few resources are strongly depleted (Woodward and Hildrew, 2002; Woodward et al., 2005b), although whole-lake manipulation experiments can provide a different perspective. Large-scale manipulation experiments include recruitment of Rainbow Trout (*Oncorhynchus mykiss*), that altered the planktivory regime and the water quality (Elser et al., 1995), the addition of Northern Pike (*Esox lucius*), which led to crashes in cyprinid minnow populations (Carpenter et al., 2011; Elser et al., 1998, 2000), and the replacement of planktivorous minnows with a comparable mass of piscivorous bass (Ives et al., 1999), with consequent long-term changes in the zooplankton biomasses (Ives et al., 1999; Jonsson et al., 2005). One example for controlled alteration of abiotic factors are whole-lake N:P treatments to prevent nitrogen limitation with consequences at different trophic levels, as planktivore biomass was inversely related to piscivore biomass (Carpenter et al., 2001) and changes in the fish-driven phosphorus cycle (Carpenter et al., 1992). Given that low linkage density implies that the removal or replacement of a highly connected species results in a loss of fewer links than for webs with higher linkage density, the traits of endangered (or recruited) fish species must be taken into greatest account.

5. CONCEPTUAL UNIFICATION

5.1. Articulating B–EF in terrestrial ecosystems

Different functional responses and effects within and across adjacent trophic levels (x and $x+1$) can be articulated within a conceptual framework to predict EF under future scenarios (Lavorel et al., 2009). To analyze B–EF relationships, the application of such a framework requires that each compartment (box) is specified for individual species in terms of functional diversity (FD) and/or trait attributes and can be applied to most ecosystems stepwise. We will use nutrient cycling and trophic interactions as examples and will define the functional effect trait(s) contributing to the ecosystem function at the trophic level most related to the function itself (x in Fig. 12) and, if relevant, at the adjacent ($x+1$) trophic level.

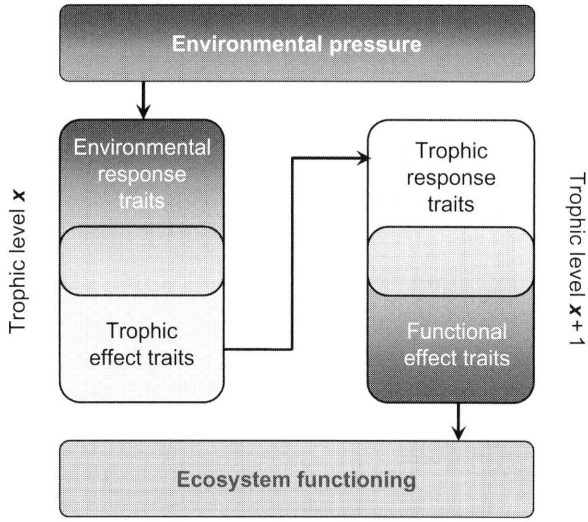

Figure 12 Framework articulating functional responses and effects within and across two adjacent trophic levels to forecast ecosystem functioning. The analysis of the effects of functional diversity on ecosystem functioning will then integrate the quantitative effects of diversity for the environmental response traits, the trophic effect traits, the trophic response traits and the functional effect traits. This can be done by partitioning the variance of the ecosystem functioning according to these trait types (Díaz et al., 2007). According to us, such a framework can be applied to run specific scenarios of environmental change in a predictive approach. (For colour version of this figure, the reader is referred to the online version of this chapter.)

Step 1. Given that the environmental response traits are often taxon-specific, if more than one functional effect trait is involved within the considered (sub)food web, then association patterns between different traits need to be taken into account. The outcome is a trade-off among positive and negative effects at adjacent trophic levels.

Step 2. Trophic effect traits and trophic response traits and associated processes must be identified. In the case of ecological stoichiometry, for example, the chemical quality of soil systems (C:N:P and $[H^+]$ either as pH or pOH) enhances the numerical abundance (and hence the biomass) of soil mesofauna much more than the soil microfauna (Mulder and Elser, 2009). Moreover, soil mesofauna incorporates most fungivores and microfauna incorporates most bacterivores (Mulder et al., 2005a; Wu et al., 2011) and bacteria and fungi respond to chemical resources in different ways according to their ability to break down carbon- versus nitrogen-rich compounds (De Vries et al., 2006; Hunt and Wall, 2002; Krivtsov et al., 2011; Wardle, 2002; Wardle et al., 2004).

Step 3. This step identifies the response traits for each of the trophic levels (starting from the lowest, x in Fig. 12) to the environmental predictor of interest. In the case of total soil phosphorus, a lower C:P ratio directly favours larger arthropods (Mulder and Elser, 2009) and the proliferation of bacteria with an *r*-like strategy (Makino et al., 2003). Given that most microarthropods are predators or fungivores (Mulder et al., 2005a; Wardle, 2002), a shift in the fungi-to-bacteria ratio is expected in soil systems (De Vries et al., 2006; Mulder et al., 2009).

Step 4. Having established the relationships between functional effect traits for a given environmental condition (or a predictor), the responses of different trophic levels to pressure and multiple functional relationships involved in a selected ecosystem service, the final analyses will allow the translation of effects at individual or species level into actual ecological processes at community (or even biome) level.

5.2. Articulating B–EF in aquatic ecosystems

The framework of Fig. 12 may be applied not only to a given ecosystem process influenced by a range of contrasting conditions (and a comparative analysis conducted in order to identify generic vs. contingent relationships), but it can also identify (in)congruences in B–EF. Therefore, it is also potentially suitable to assess the ecological risks of environmental pressure.

Eutrophication, for example, is a widespread kind of environmental pressure which affects key ecosystem services. Global increase in use of urea in both agriculture and manufacturing has resulted in increased run-off to sensitive coastal systems and is important in the nitrogenous nutrition of some harmful algal bloom species (Glibert et al., 2006). Reduction in water quality directly influences important coastal ecosystems like seagrass meadows (Waycott et al., 2009), which trap sediments and nutrients and have a large net primary productivity (Orth et al., 2006). Seagrass meadows are also sensitive to changes in turbidity and nutrient enrichment and provide ecosystem services such as supporting commercial fisheries through habitat provisioning and globally significant sequestration of carbon (Duarte et al., 2005). Besides indirect effects of eutrophication due to reduction of light penetration in the water column (changes in turbidity) and enhance coverage by epiphytes (biological disturbance), direct effects include shifts in nutrient ratios of seagrass leaves (Antón et al., 2011) which influence grazing patterns and cause selective abrasion and even removal of the plants. Leaves of persistent species such as *Thalassia hemprichii* have a lower C:N ratio than leaves of the ephemeral *Halodule uninervis* or the intermediate *Cymodocea rotundata* (Fig. 13). Seagrasses with higher C:N ratios have higher palatability for sea urchins *in situ* (tropical seagrasses, Vonk et al., 2008) or for

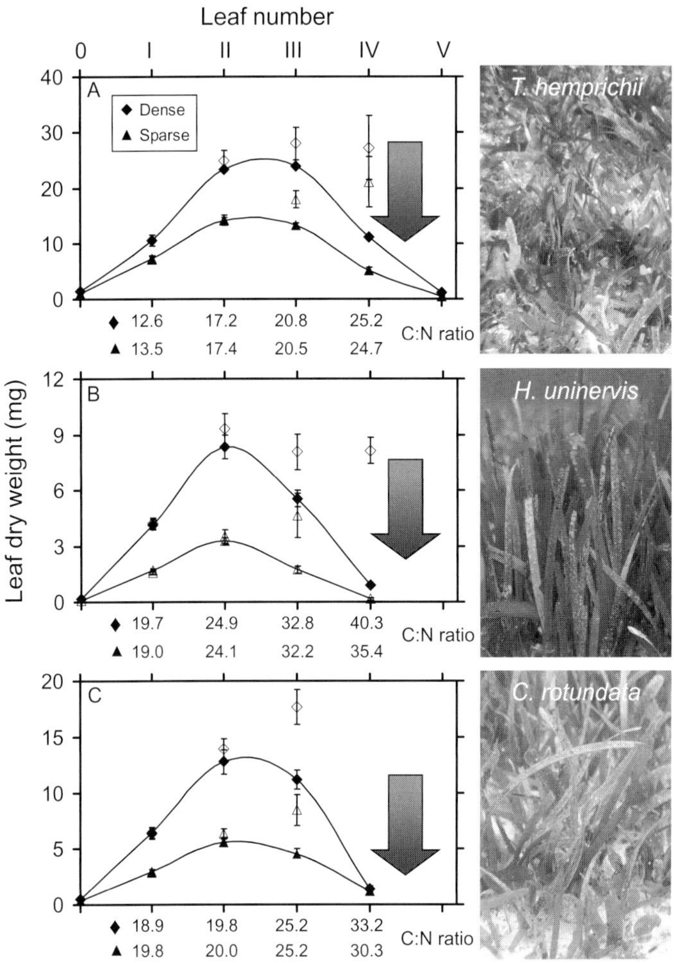

Figure 13 Example of *trophic effect* and *trophic response* traits (*sensu* Fig. 12) from the green world: grazing pressure (vertical arrow) and elemental quality in three species from seagrass meadows of the Spermonde Archipelago (Indonesia). Mean leaf weight development ± SE of seagrasses in dense (♦) and sparse (▲) canopies (Vonk and Stapel, 2008) with comparable PO_4^{3-} concentrations (in water columns: 0.24 and 0.23 μM and in pore water: 0.39 and 0.40 μM, respectively). From top to bottom: (A) *Thalassia hemprichii*, (B) *Halodule uninervis* and (C) *Cymodocea rotundata* (photo credits: Jan Arie Vonk). All three seagrass species continuously produce leaf tissues at a fixed basal meristem (Short and Duarte, 2001); the Roman number refers to the leaf layout during destructive counting (I is the youngest leaf, V the oldest leaf collected). The closed symbols and lines represent the natural mixture of leaves and open symbols represent intact leaves only. Leaf molar C:N ratios for dense and sparse canopies provided below each plot; all samples ($n=9$) of 10–20 shoots per species, depending on size. Despite increased palatability, *T. hemprichii* (A) appears to be the most resistant to grazing pressure. (For colour version of this figure, the reader is referred to the online version of this chapter.)

isopods in mesocosm (temperate seagrasses, Tomas et al., 2011). Contrasting trophic effect and trophic response traits not only affect the composition of seagrass meadows, but also enhance the nitrogen pool, with both the N concentration in aboveground seagrass biomass as the dominance of species with lower C:N ratio higher in grazed treatments (Vonk et al., 2008). Similarly, C:N of detrital resources in freshwaters, which is a key determinant of decomposition rates (Hladyz et al., 2009), is also a function of both the surrounding riparian vegetation and the nutrient status of the waterbody itself (Hladyz et al., 2011).

6. SYSTEM-DRIVEN B–EF

6.1. Elemental changes within one taxocene: Less is more

Abiotics (pH and C:N:P) play a key role in determining the abundance of diversity of the soil nematofauna and nutrients in particular enhance the productivity (here, their total biomass) of free-living nematodes (Fig. 14) as well as the unevenness of the soil nematofauna (Fig. A2). In P-enriched, intensively managed soils (low N:P molar ratios), nematode total biomasses are much greater than in sites with a (relatively) higher N:P molar ratio. In other words, a lack of soil P in agroecosystems kills off the predatory nematodes or strongly diminishes the abundance of all nematodes (Mulder and Vonk, 2011), and there is increasing evidence of similar patterns even in extreme environments (Barrett et al., 2007).

Although comparable patterns have been detected among taxocenes (Mulder and Elser, 2009, more details in Section 6.2), Fig. 14 shows that the distribution of the free-living nematode biomass may overlap constrained bottom-up responses to microbial producers. Under higher grazing pressure (low soil N:P), either the microbial activity is diminished or the density of bacterial cells is low. In contrast, under lower grazing pressure (high soil N:P), microbial activity can become stimulated and the density of bacterial cells is high (Mulder et al., 2009; Reuman et al., 2009). If so, given that most free-living nematodes are bacterial feeders, Fig. 14 resembles the classical 'energy enrichment paradox', which shows here at both tails of the nematode distribution the exacerbated incongruences between the bacterial autotrophs and the bacterivore nematodes under either low or high N (cf. Hall, 2009 and references therein). Nematode patterns seem comparable to those predicted by the 'hump-backed model' (Grime, 1973, 1979), which used an arbitrary scale from 0 to 1 (Grime, 1977). Biomass may in fact increase with respect to the limiting

STxET	pH (H_2O)	Molar ratios C:N	N:P	C:P	N (# / kg soil)	$\overline{M}$ (μg dry wgt)	B (μg / kg soil)
HI×MG	5.51 (0.06)	17.23 (0.41)	13.31 (0.56)	233.3 (14.2)	79,718 (7274)	0.046 (0.003)	3626 (391)
FC×MG	6.40 (0.08)	13.86 (0.38)	8.97 (0.32)	123.7 (5.0)	58,740 (4121)	0.044 (0.003)	2606 (248)
FC×AF	7.76 (0.03)	15.81 (0.60)	4.17 (0.56)	64.9 (8.1)	11,657 (1312)	0.035 (0.002)	391 (40)
LU×AF	6.76 (0.05)	18.61 (0.26)	3.26 (0.09)	60.5 (1.3)	21,299 (3461)	0.058 (0.004)	1220 (208)
PO×MG	5.95 (0.05)	18.76 (0.37)	7.41 (0.28)	138.3 (5.6)	41,620 (1925)	0.059 (0.002)	2391 (133)
PO×AF	5.85 (0.06)	20.73 (1.16)	7.71 (0.96)	172.3 (27.3)	36,258 (3210)	0.052 (0.004)	1868 (275)
PO×SW	3.98 (0.08)	31.71 (2.27)	32.10 (2.96)	998.9 (105.0)	17,609 (1704)	0.027 (0.002)	486 (66)

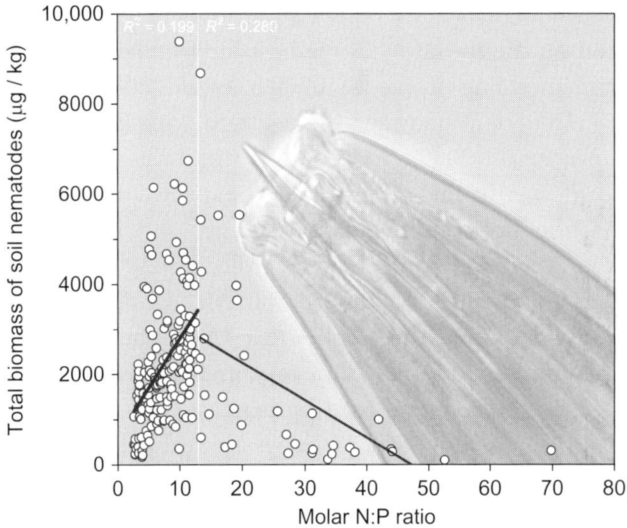

Figure 14 Occurrence of *environmental response* traits in the brown world: synergetic processes of land history and abiotics are reflected in the soil nematodes (upper panel: nematode density (N), mean weight ($\overline{M}$), biomass (B) and (SE) in brackets). From left to right (upper photos) and from top to bottom (synoptic table), managed grasslands on peat (HI × MG), managed grasslands on clay (FC × MG), arable fields on clay (FC × AF), arable fields on Loess (LU × AF), arable fields on sand (PO × AF), managed grasslands on sand (PO × MG) and shrublands on sand (PO × SW). Lower molar N:P ratios seem to enhance the productivity (biomass) of nematodes (lower panel), but also the unevenness within the nematofauna (Fig. A2). *Raw data from Mulder and Vonk (2011).* (For colour version of this figure, the reader is referred to the online version of this chapter.)

nutrient (here, phosphorus) via a saturating, non-linear function, whereas diversity may increase, decrease or exhibit a hump shape (Sterner and Elser, 2002 and Sterner, 2004, respectively). Our threshold of 13 is supported by the two regression trends for high and low fertility (Fig. 14) and is similar to the atomic N:P ratio by Cleveland and Liptzin (2007) for soils and roughly comparable with the historical atomic N:P ratio of

16 by Redfield (1958) for the blue world. The left and right regression lines of Fig. 14 show, in fact, a direct correlation between the total biomass of soil nematodes and the soil N:P ratio until 13 (higher eNPP sites, see Table 1) and an inverse correlation between biomass and N:P afterwards (lower eNPP sites).

We found no consistent relationships between the average mass $\bar{M}$ and the average predator–prey body-mass ratios across ecosystem types and soil types (Mulder et al., 2011a). Widely distributed horizontal distributions of $\bar{M}$ across environmental C:N:P transects (as those in Fig. 15) might revitalize the discussion on the use of $\bar{M}$ as the best independent sole predictor for mass–abundance scaling (compare Cohen et al., 2003; Hildrew, 2009; Mulder et al., 2005b; Woodward et al., 2005a with Brown and Gillooly, 2003; Reuman et al., 2009). Indeed, it is the numerical abundance that changes the most, not the mean mass, as expected from a well-known direct correlation between population density and resource availability (e.g. Kaspari, 2004; Meehan, 2006; Wardle, 2002) and Kaspari (2004) focuses on the variable N instead of $\bar{M}$. Under relatively stable environmental conditions, this implies that EF might be driven more by the total numerical abundances N of organisms than by their body-mass average $\bar{M}$ or by the resulting total biomass ($N \times \bar{M}$ as in Fig. 14). Our productivity gradients show that $\bar{M}$ values are real and vary from place to place less than previously suspected (Kaspari, 2004).

The rather comparable $\bar{M}$ values are surprising and could make terrestrial 'stable states' questionable: as large-scale fluctuations of $\bar{M}$ values were not observed along the C:N:P gradients (Mulder and Vonk, 2011), we might wonder under which kind of environmental conditions (to be held constant for a certain time span) such 'stable states' might actually occur in the brown world. Moreover, an investigation by Gilljam et al. (2011) consistently shows either underestimations or overestimations of predator–prey systems as soon as the (derived) species-specific $\bar{M}$ averages were used instead of the (original) site-specific weights at individual level m.

6.2. Elemental changes across taxocenes: Community mismatches

Assuming that, at least in the brown world, numerical abundance N matters more for EF than individual mass, we might continue to neglect the below-ground variation of individual body-mass values within one taxon or, possibly, even within the same taxocene—as most soil ecologists currently do (overview in Mulder et al., 2011b). If all taxa absorb energy at constant rate, and the metabolic rate of an individual approximately follows a power law in

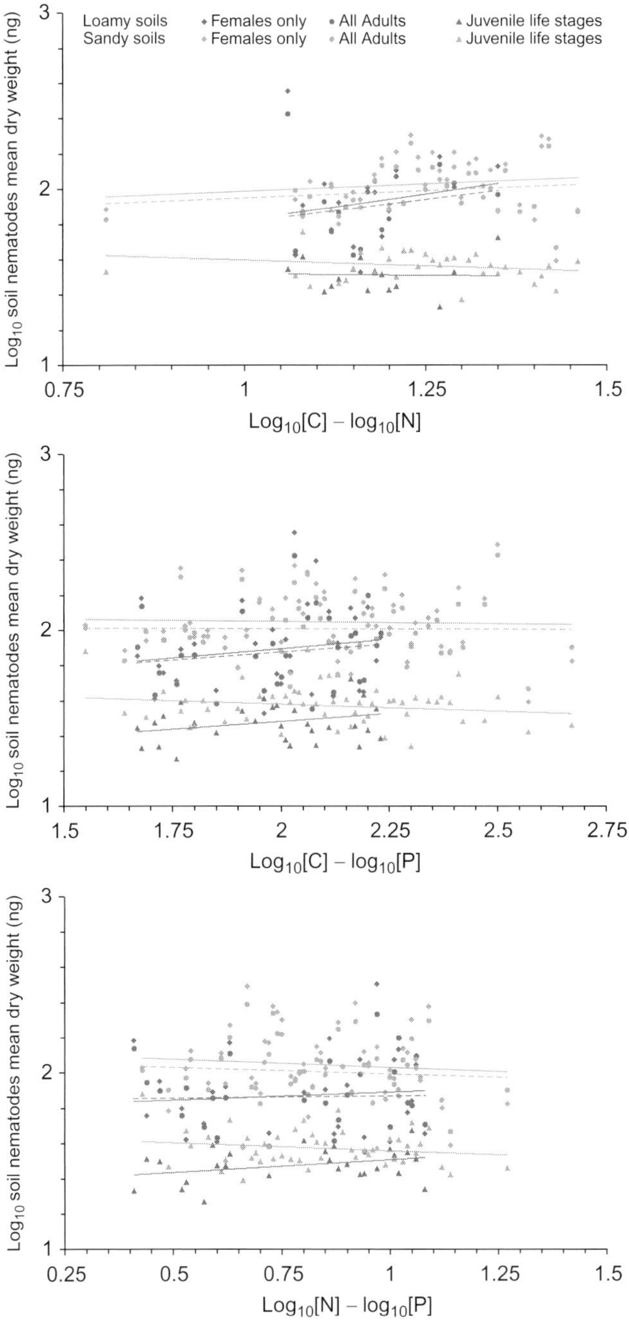

Figure 15 Occurrence of *functional effect* traits in the brown world. The soil types influence the size of individual nematodes more than nutrient ratios do: from top to bottom, the weighted differences in the body mass between the sandy soils (in green)

$\bar{M}$ (Brown et al., 2004; Peters, 1983; Savage et al., 2004), the energetic equivalence hypothesis predicts a mass–abundance slope of $-^3\!/\!_4$. Observing a mass–abundance slope less negative than (respectively, more negative than) $-^3\!/\!_4$ suggests that larger invertebrates absorb more (respectively, less) energy from the environment than smaller invertebrates. However, rather few soil communities scale to the $-^3\!/\!_4$ power (Mulder et al., 2005b, 2009, 2011c), in contrast to species–density scaling (Eq. 2). It should be noted, though, that the scaling power strongly varies between $-⅔$ and -1 for mass–abundance relationships (Mulder, 2010) and between ⅔ and 1 for metabolic rates (Glazier, 2010) because the scaling exponents are sensitive to which taxa are included (Boit et al., 2012; Glazier, 2005; Mulder et al., 2005b, 2009; Prothero, 1986; Reuman et al., 2008). Still, diverse patterns and clear trends remain recognizable under comparable methodologies.

Few studies on food web manipulation have been performed in the brown world, but Wardle et al. (2011) recently showed that ant exclusion enhanced the first and third trophic level of soil food webs, increasing active microbial biomass and predatory soil nematodes but not bacterial-feeding nematodes (second trophic level). Assuming that larger mites (soil mesofauna) are often predatory, one mesofaunal individual and (at least) one microfaunal individual have to come together. The probability that this happens increases approximately as the product of both population abundances. Thus, if in a nutrient-richer soil both microfauna and mesofauna would be two times more abundant, then the probability of encounter would approximately increase by a factor 4, leading to over-proportionally more feeding opportunities for the predatory mesofauna. But again, directions of the responses need to be specified as in the conceptual framework

and the loamy soils (in brown) of all the occurring nematodes per soil type together fluctuate between 7.9% and 15.0%, implying that nematodes in loamy soils are smaller than those in sandy soils. Averages were consolidated separately for C:N, C:P and N:P ratios and log-transformed. Log–log linear regressions are just plotted for clarity, although they are not significant: from top to bottom, log–log linear regressions for all females (upper solid lines), all adults (dotted lines) and all juveniles (lower solid lines). The cross-product soil type (ST) versus ecosystem type (ET) determines the total abundance of individuals (and hence, the total biomass). This is rather surprising, given that the Atom% Excess (APE) ^{13}C and ^{15}N for nematodes is known to be most sensitive to enrichment (Crotty et al., 2011). Only arable fields and grasslands are shown: loamy soils (41 sites, 1094 adults and 4936 juveniles) versus sandy soils (96 sites, 3504 adults and 10,819 juveniles); *raw data from Mulder and Vonk (2011)*. (For interpretation of the references to colour in this figure legend, the reader is referred to the online version of this chapter.)

(Fig. 12): here, a log-linear decrease of mites and collembolans with decreasing soil fertility (higher nutrient ratios), in contrast to a curvilinear increase of bacterivore nematodes and fungi (cf. Santos et al., 1981). Therefore, regardless of the kind of environmental adversity, soil mesofauna might increase over-proportionally in enriched systems such as the real food web plotted on the background of the four scatter-plots (a reference site marked by a cross) in the composite (Fig. 16).

The results support the stoichiometric theory (Elser, 2006; Mulder and Elser, 2009; Sterner and Elser, 2002) which predicts that animals with higher P demands would suffer a competitive disadvantage due to poor stoichiometric food quality. In Fig. 16, omnivorous species with lower P demands are favoured. This seems to be the case within our soil

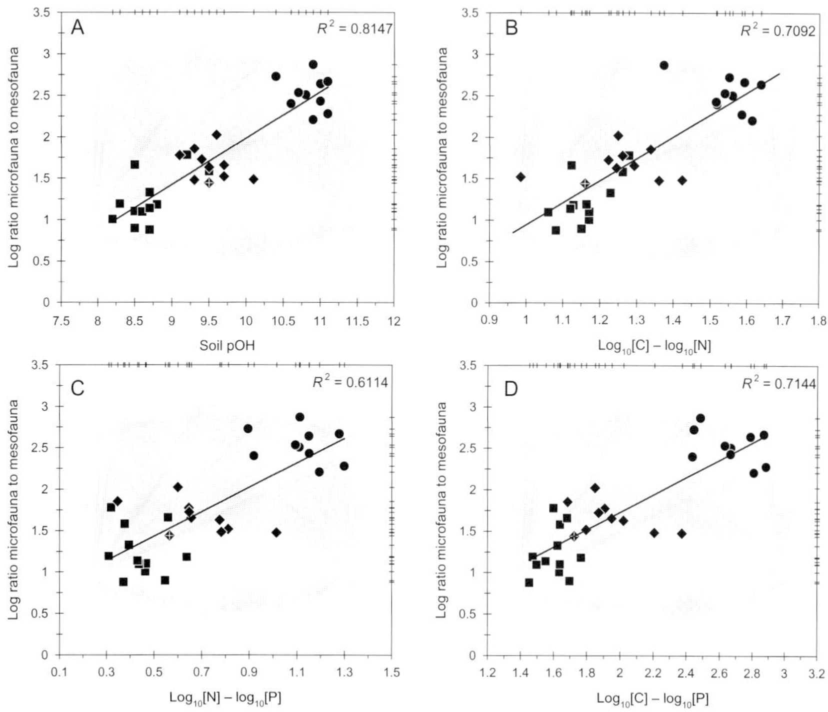

Figure 16 Another example of *trophic effect* and *trophic response* traits (cf. Fig. 13). Logarithmic fits of differently body-size-scaled soil invertebrates across four quantitative gradients of increasing environmental adversity as described by pOH (A), log-transformed C:N ratio (B), N:P ratio (C) and C:P ratio (D) for Dutch dry heathlands (●), abandoned grasslands (◆) and bio-organic farms (■). Lower soil fertility as in the heathlands plotted at the right of the scatters enhances the steepness of the microfauna to mesofauna ratio. Springtails and enchytraeids get eliminated, relative to the overwhelming increase of nematodes, by decreasing [N] and [P].

mesofauna, given that the P contents in the bodies of (predatory) mites are up to three times lower than in detritivorous collembolans (~0.5 body % P-in-Acari vs. 1.4 body % P-in-Collembola; Martinson et al., 2008 and Schneider et al., 2010, respectively). This means that invertebrates at higher trophic levels have a higher P demand than those at lower trophic levels, at least in the brown world. However, during the lifetime of metazoans, the P demand is not necessarily inversely related to the P content as adult (like the aforementioned mites with lower P content than other groupings within the same taxocene); the P demand can, in fact, be needed during growth for structural components like bony skeleton. In the case of fishes, for instance, P demand and P content remain directly correlated with each other (Lauridsen et al., 2012; Sterner and George, 2000).

Therefore, ecological stoichiometry and classical prey–predator chains coexist and contribute to explain apparent difficulties in the application of the Lotka–Volterra model in reality. Still, the possibility of a kind of top-down control has to be taken into account (Wardle, 1999) and causal relationships must be directional and quantitative, such as the pathway analyses performed by Perner and Voigt (2007) and Voigt et al. (2007). Comparing this stoichiometric perspective across soil systems with the large number of terrestrial B–EF studies, it remains surprising that the plea of Chase (2000) to address phosphorus in terrestrial ecosystem types has remained largely ignored by so many soil ecologists (but see Lynch and Ho, 2005).

7. CODA

Macroecology and ecological stoichiometry encompass a wide variety of large-scale phenomena (cf. Gardner et al., 2001; Hall, 2009; Sterner and Elser, 2002), and allometric scaling can link large-scale macroecology to either species- or community levels (Yodzis and Innes, 1992). Together, allometry and ecological stoichiometry are suitable measures to catch the emergent characteristics of large data sets distributed in time and space and offer a reliable tool to outflank difficulties in the environmental assessment of disturbed ecosystems. Like Yodzis and Innes (1992), we argue that allometry and food web theory can be successfully integrated, even if the coupling between biodiversity and EF is less stringent than commonly assumed.

In 2004, the plea for the conservation of ecosystem structure and functioning as priority target came from the United Nations (UNEP/CBD, 2004). Alas, biological findings are often not strengthened for stakeholders and policy-decision makers (Mann, 1991). Many of the studies of the predominance of quarter-power scaling in biology have remained purely descriptive so far, although there are exciting hints of a mechanistic explanation for observed patterns in EF. The evenness in the structure of a food web can be captured by allometric scaling and when the resulting mass–abundance relationship is close to isometry, its equitability roughly implies biomass evenness across trophic levels. Such 'allometric metrics' can take into account traits like the average body size of given species that may often be independent from environmental predictors, indirectly supporting the historical critique on biodiversity by Hurlbert (1971).

Our survey on the understanding of B–EF relationships includes aspects of taxonomic diversity, functional categorization and metabolic scaling as well as rules for their appropriate use. Given the well-known role of traits as predictors of niche complementarity and community structures, we have considered empirical examples examining how biodiversity supports EF. As model organisms, we chose fishes, plants and invertebrates and highlighted distributional (in)congruences of these organisms, the current state of the field and future challenges. Our review of independent case studies from the blue, the brown and the green worlds shows that biodiversity (at least the taxonomic diversity) as the key predictor for EF and multitrophic interactions like those described in food web theory may be overemphasized. There is an urgent need to galvanize ecologists from different subdisciplines, bringing them together for so many existing questions (Carpenter et al., 2009). A greater synergy between theoretical and empirical disciplines during the construction of null hypotheses is necessary to allow a careful differentiation between experimental design and EF. In the past, research on EF was often diluted by a dichotomy between empirical reports (often as grey literature), novel biology journals and theoretical journals, complicating data mining of disparate data sets. Hence, generalization without oversimplification becomes an important objective in its own right, with the ability to identify traits that underlie species responses and ecological processes (Grime, 1997). Such responses, especially when aggregated up to the biome level, can provide critically important ways to predict ecosystem responses to environmental changes at a global scale (Wall et al., 2008).

ACKNOWLEDGEMENTS

Anton M. Breure, Chris Holmes, Michael A. Huston, Katherine Kapo, Owen Lewis, Giorgio Mancinelli, Shahid Naeem, Loreto Rossi, Torbjörn Säterberg, Paulo Sousa, Guy Woodward and one anonymous referee with helpful comments on the earlier version of this paper are gratefully acknowledged. We thank Microsoft Research for supporting the development of Network3D, and Dennis Mischne, of the Ohio Environmental Protection Agency, for making available biological monitoring data. S. M. thanks for the financial support by KAKENHI Grants-in-Aid for Scientific Research (MEXT No. 22658051 and 23380094); J. A. V. acknowledges the WOTRO co-funding by the Netherlands Foundation for the Advancement of Tropical Research (W86-168); M. K. acknowledges the NSF funding No. 090221 (Collaborative Research RUI); A. G. R. acknowledges a Beaufort Marine Research Award, funded under the Marine Research Sub-Programme of the Irish National Development Plan; D. H. W. acknowledges the NSF funding to the McMurdo Dry Valley LTER.

APPENDIX

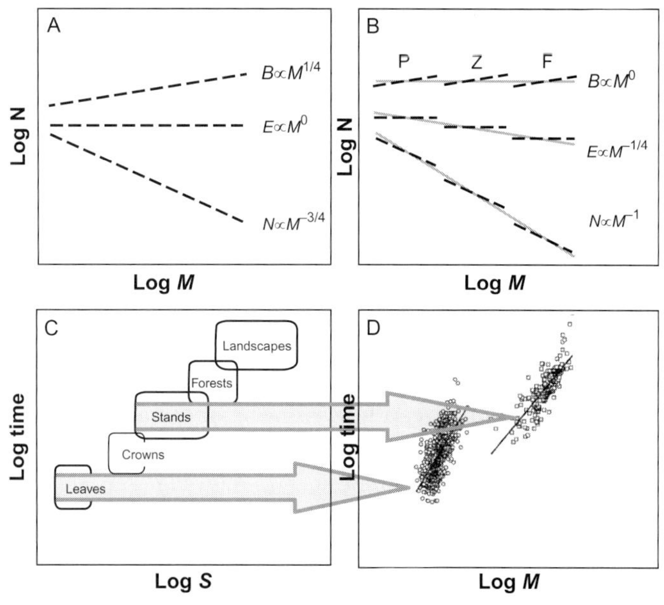

Figure A1 Allometric scaling and diversity–yield relationships. Upper panel: dashed lines describe three log–log relationships between species average weight $\bar{M}$ and species density N for total biomass B, upper line, energy rate E, middle line, and mass–abundance scaling, lower line, as function of $\bar{M}$ for pelagic food webs across trophic levels (A) and within three taxocenes, namely the phytoplankton P, the zooplankton Z and the fishes F (B). Lower panel: dynamic domains of scale (S) in time occupied by different entities enable to address the variation in ecological processes across the discrete boundaries of the investigated domains (C) and allow taking into consideration relationships into a similar space-time domain (D). Adapted from Brown and Gillooly (2003), the upper panels (A, B), and from Kerkhoff and Enquist (2007), the lower panels (C, D).

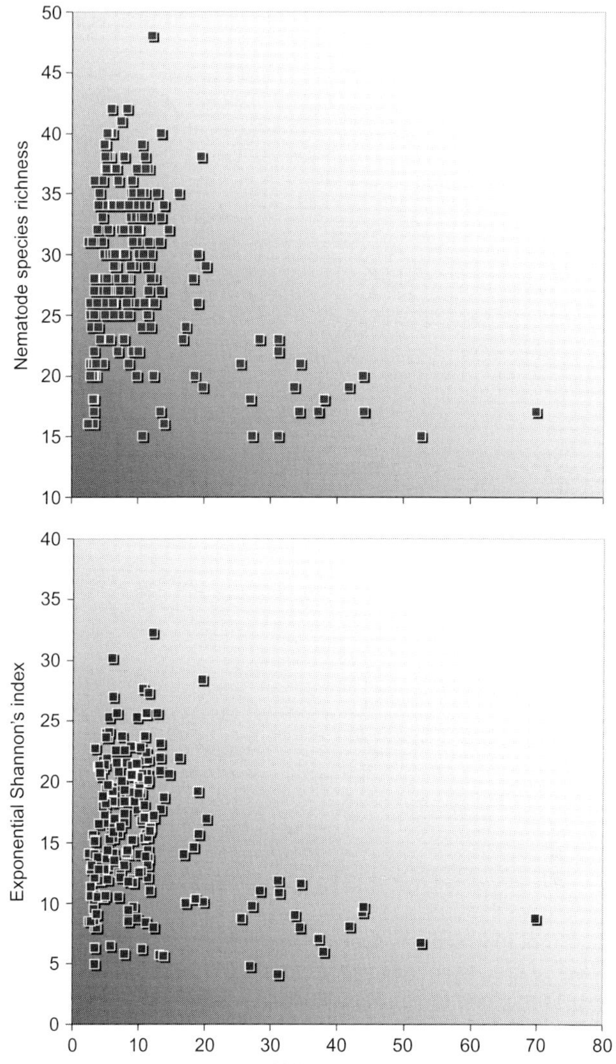

Figure A2 Soil abiotics and diversity–yield relationships for free-living nematodes. Diversity metrics that combine species richness with relative abundance, like the so-called Hill numbers (here the exponential Shannon-Wiener index as in Hill, 1973), are not independent from the (number of) species themselves. We have chosen the molar N:P ratio as proxy for the productivity of agroecosystems (Mulder and Vonk, 2011). If so, productive sites show a higher value of the exponential Shannon-Wiener index (bottom) and a higher amount of nematode species (top). Hence, in more productive sites, species are more even in their spatial distribution than in less productive sites. (For colour version of this figure, the reader is referred to the online version of this chapter.)

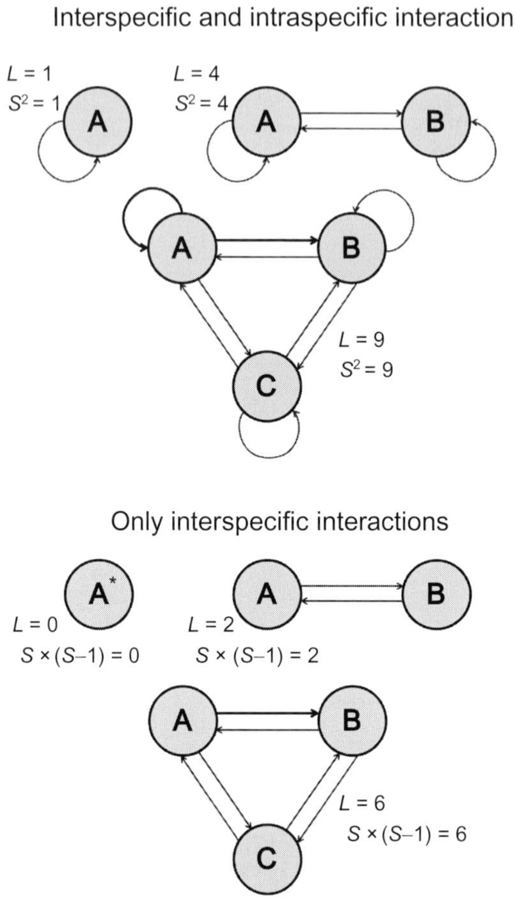

Figure A3 The extent to which one species (A) will feed on other species (here, B and C) can be quantified by species connectivity according to metrical computations (explanations in the text). Within one taxocene, C_t fluctuates between 0% (no trophic links at all) and 100% (maximal aggressive behaviour) for both interspecific and intraspecific interactions (directed connectance, upper panel) as for interspecific interactions only (interactive connectance, no cannibalistic links; lower panel). Here we show the maximal number of possible trophic interactions within the same taxocene ($C_t = 100\%$), besides the unique case of the polar nematode *Scottnema* (this most extreme condition—marked by an asterisk—exhibits $L = 0$, $S \times (S - 1) = 0$ and therefore $C_t = 0\%$ in A*). Many realized trophic links are suggested to reflect a dominance of generalists (high species connectivity implies high omnivory and aggressive feeding behaviour, therefore high resilience at taxocene level), in contrast to a low proportion of realized trophic links, which reflects a skew towards specialists and immature life stages (low species connectivity: low resilience).

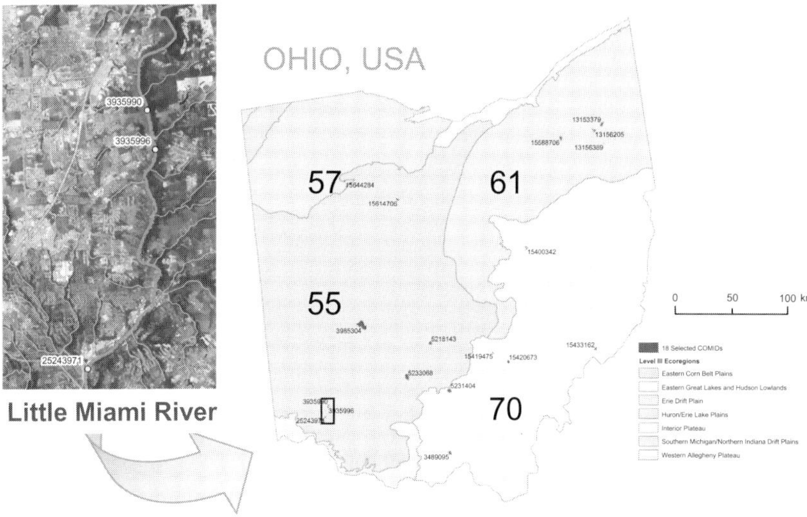

Figure A4 Geographical location of the 18 freshwater assemblages randomly selected after allometric screening (Fig. A5). The rivers Scioto Brush Creek, Hocking (two times), Big Darby Creek, Paint Creek (two times), Blanchard, Duck Creek, Little Miami (three times), Buck Creek, Sandusky, Walhonding, Mahoning (three times) and Cuyahoga were sampled between 2000 and 2007. Many locations (like those in Figs. 10 and A6) are in the Eastern Corn Belt Plains (Ecoregion 55), the most variable in total phosphorus (CV = 88.6%), biochemical oxygen demand (85.5), chemical oxygen demand (118.6), nitrite (125.0), ammonia (231.4) and total suspended solids (153.8), and the least variable in hardness (CV = 22.6%), nitrate (44.4), conductivity (24.6) and total dissolved solids (25.4). Ecoregions as defined in: http://www.eoearth.org/article/Ecoregions_of_Indiana_and_Ohio_%28EPA%29. GIS credit: Katherine Kapo. (For colour version of this figure, the reader is referred to the online version of this chapter.)

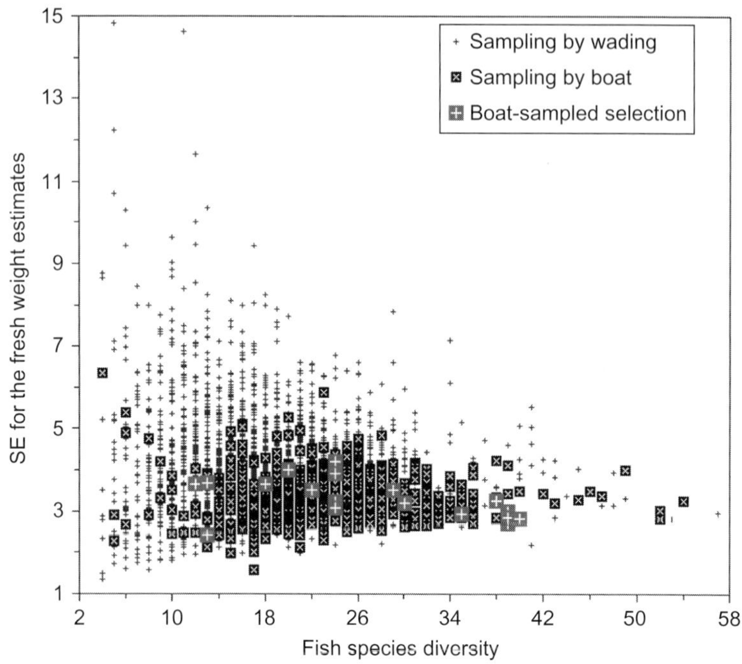

Figure A5 Possible effects of sampling methods associated with the allometric model on uncertainty were investigated by selecting 2656 locations in Ohio sampled either by boat or by wading. The uncertainty in the allometric estimates of the mean fresh biomass (|SE| in grams) of the smallest fish populations in each fish assemblage shows that the kind of sampling (boat or wading) and, indirectly, the river type (large, tributary, etc.), inflates biomass estimates in low-diversity communities, supporting that boat sampling provides the best estimates. As differences in sampling efforts are important for appropriate data mining and computations, we confined our further analyses in the aforementioned boat-sampled locations (Figs. 8 and A6; Table 5). (For colour version of this figure, the reader is referred to the online version of this chapter.)

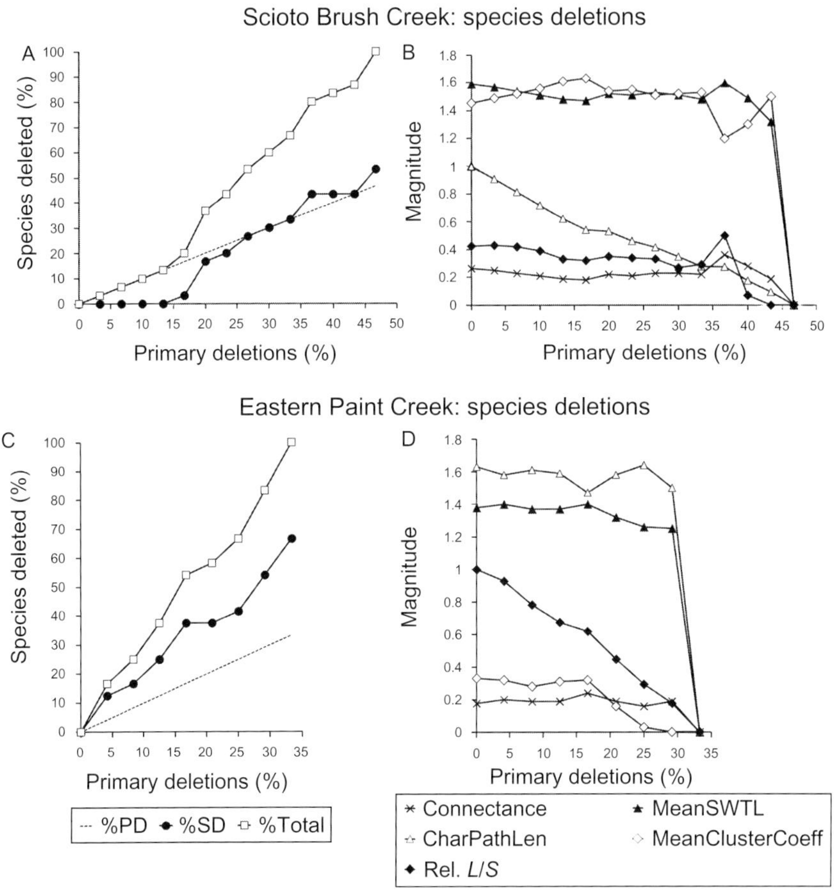

Figure A6 Two case studies in central Ohio on static species deletion scenario 'connectivity descending'. The creeks are both tributaries of the Scioto River. The fish assemblages have different vulnerabilities to secondary deletions (Table 5; Fig 9). Data on the left have been normalized for comparison. (A) The Scioto Brush Creek web with $n=30$ only shows secondary deletions (% SD) from 15–50% of primary deletions (% PD) until its final collapse. At the beginning, secondary extinctions are less than primary extinctions, becoming equal to (and later more than) primary extinctions (dashed line indicates $x=y$). (B) The Scioto Brush Creek's web properties during the species deletion process, with the relative linkage density L/S quickly decreasing as highly connected nodes disappear from the web. (C) In contrast to the previous river, the Eastern Paint Creek with $n=24$ immediately shows more secondary extinctions than primary extinctions (line above the 1:1 dashed line). The food web collapses after only 33% of primary extinctions. (D) Most Paint Creek's web properties behave similarly to those in (B).

Table A1 Occurrence and site-specific body mass (wet weight) of freshwater fishes in the 18 investigated rivers across Ohio

Weight (average of individual measurements in grams) versus COMID (7- and 8-digit code, GIS location in Fig. A4)

Species (Latin Please check whether the entries are aligned properly in this table. binomial, qualitative) Behavioural Level (**1** or **2**, numeric)																	
3489095	3935990	3935996	3985304	5218143	5231404	5233068	13153379	13156205	13156389	15400342	15419475	15420673	15433162	15588706	15614706	15644284	25243971
Ambloplites rupestris **2**																	
36.0	113.3	115.0	46.7	48.3	32.0	0	25.0	0	0	120.0	90.0	0	62.0	110.0	61.8	93.7	120.0
Ameiurus natalis **2**																	
0	0	0	0	0	0	0	0	0	0	54.0	0	0	0	177.0	0	202.0	0
Amia calva **2**																	
0	0	0	0	0	0	0	0	0	0	1040.0	0	0	0	0	0	0	0
Ammocrypta pellucida **2**																	
0	0	0	0	0	0	0	2.0	0	0	0	0	0	0	0	0	0	0
Aplodinotus grunniens **2**																	
2382.1	865.0	1692.8	0	1100.0	677.2	0	0	0	0	2100.0	0	0	254.5	0	0	545.5	492.1
Campostoma anomalum **1**																	
0	6.3	8.2	20.3	9.2	8.3	0	0	0	0	21.4	0	5.0	0	0	0	0	10.0
Carpiodes carpio **1**																	
0	905.0	1028.0	0	1350.0	601.3	750.0	0	0	0	0	583.3	900.0	0	0	0	0	961.4
Carpiodes cyprinus **2**																	
1400.0	991.7	0	0	1425.0	4.5	0	0	0	0	1100.0	384.2	400.0	140.0	0	0	0	0

Table A1 Occurrence and site-specific body mass (wet weight) of freshwater fishes in the 18 investigated rivers across Ohio—cont'd

Carpiodes velifer **1**																	
700.0	631.3	0	0	0	261.7	0	0	0	0	650.0	0	0	0	0	0	0	703.0
Catostomus commersonii **1**																	
0	0	0	178.1	0	0	245.0	202.4	242.4	4.0	0	3.0	0	160.0	907.8	154.5	571.4	0
Cottus bairdii **2**																	
0	0	0	9.0	0	0	0	0	0	0	0	2.5	1.0	0	0	0	0	0
Cyprinella spiloptera **2**																	
2.0	4.0	4.6	0	3.0	4.1	2.5	5.0	0	3.3	3.4	3.7	3.5	2.4	0	5.1	4.7	3.5
Cyprinella whipplei **2**																	
0	8.8	4.6	0	0	2.0	0	0	0	0	0	0	6.0	0	0	0	0	3.6
Cyprinus carpio **2**																	
0	2083.3	2511.0	1681.3	2994.4	0	1460.0	0	4911.8	1787.5	1495.7	2317.9	2236.1	3400.0	3012.5	2150.0	2338.5	2366.7
Dorosoma cepedianum **1**																	
135.5	66.0	176.5	0	347.1	247.9	89.4	46.0	5.0	51.7	114.2	422.0	60.9	104.5	0	0	9.0	234.3
Erimystax dissimilis **1**																	
0	0	0	0	8.9	2.0	0	0	0	0	8.6	0	0	0	0	0	0	0
Erimystax x-punctata **1**																	
0	0	8.0	0	0	5.8	0	0	0	0	4.0	0	0	0	0	0	0	0

Continued

Table A1 Occurrence and site-specific body mass (wet weight) of freshwater fishes in the 18 investigated rivers across Ohio—cont'd

Esox americanus **2**																	
0	0	0	0	0	0	0	52.5	38.6	0	0	12.0	5.0	0	15.0	0	0	0
Esox lucius **2**																	
0	0	0	0	0	0	0	0	0	0	0	0	0	0	2400.0	0	0	0
Esox masquinongy **2**																	
0	0	0	0	0	0	0	0	3991.7	0	0	0	0	0	0	0	0	0
Etheostoma blennioides **1**																	
2.0	8.0	3.9	6.8	3.1	3.3	1.0	0	4.0	0	3.3	4.0	1.0	0	0	0	5.0	5.0
Etheostoma caeruleum **2**																	
0	3.0	1.0	2.8	1.4	1.0	0	0	0	0	0	0	0	0	0	0	0	2.0
Etheostoma camurum **1**																	
0	0	0	0	2.1	1.8	0	0	0	0	4.0	0	0	0	0	0	0	0
Etheostoma flabellare **2**																	
0	0	0	0	0	0	0	7.3	0	0	0	0	0	2.0	0	0	0	0
Etheostoma nigrum **2**																	
2.0	0	0	0	0	0	0	1.4	2.0	0	0	0	0	0	0	0	0	0
Etheostoma tippecanoe **1**																	
0	0	0	0	0.8	1.4	0	0	0	0	0	0	0	0	0	0	0	0
Etheostoma variatum **1**																	
0	4.0	4.9	0	7.5	2.5	0	0	0	0	3.6	0	0	0	0	0	0	5.0

Table A1 Occurrence and site-specific body mass (wet weight) of freshwater fishes in the 18 investigated rivers across Ohio—cont'd

Etheostoma zonale **1**																	
0	4.0	1.8	2.1	0.8	1.7	0	2.0	0	0	1.0	1.5	1.3	0	0	0	0	1.9
Fundulus notatus **2**																	
2.5	0	0	0	0	0	0	0	0	0	0	0	0	0	0	0	2.0	0
Hiodon tergisus **2**																	
0	0	0	0	0	0	0	0	0	0	0	0	0	0	0	0	0	217.0
Hypentelium nigricans **1**																	
130.0	203.3	160.9	156.4	178.1	118.7	95.5	94.7	243.3	0	93.6	172.2	108.2	45.7	0	150.0	200.5	88.7
Ichthyomyzon fossor **1**																	
0	0	0	0	0	0	0	12.5	0	0	0	0	0	0	0	0	0	0
Ictalurus punctatus **2**																	
675.0	1512.5	1960.0	0	808.0	1.0	516.7	0	520.0	882.5	1757.1	1450.0	832.0	1115.0	0	0	250.0	1363.5
Ictiobus bubalus **1**																	
0	2252.5	2215.7	0	0	1500.0	0	0	0	0	0	0	0	0	0	0	0	2500.0
Ictiobus cyprinellus **1**																	
0	0	0	0	0	1900.0	0	0	0	0	0	0	0	0	0	0	0	1500.0
Ictiobus niger **1**																	
0	0	0	0	450.0	0	0	0	0	0	0	0	0	0	0	0	0	0

Continued

Table A1 Occurrence and site-specific body mass (wet weight) of freshwater fishes in the 18 investigated rivers across Ohio—cont'd

Labidesthes sicculus **1**																	
1.3	0	0	0	0	0	0	0	0	2.8	0	0	0	0	0	0	0	0
Lampetra aepyptera **1**																	
0	0	0	0	0	0	0	0	0	0	0	11.0	0	0	0	0	0	0
Lepisosteus osseus **2**																	
1000.0	0	725.5	0	410.0	0	0	0	0	0	0	0	0	0	0	0	0	120.0
Lepomis cyanellus **2**																	
90.0	22.5	0	0	0	20.0	52.5	9.3	9.6	0	0	18.3	0	0	0	6.7	23.9	0
Lepomis gibbosus **2**																	
0	0	0	0	0	0	0	25.0	90.0	0	0	0	0	0	37.4	0	0	0
Lepomis gulosus **2**																	
0	0	0	0	0	0	0	60.0	0	0	0	0	0	0	0	0	0	0
Lepomis macrochirus **1**																	
46.7	21.0	0	9.3	15.4	45.0	11.7	42.2	33.3	135.0	40.0	22.5	36.2	0	25.5	80.0	80.0	76.7
Lepomis megalotis **1**																	
10.2	63.5	32.3	0	20.5	21.8	27.5	0	0	0	0	32.0	0	13.3	0	0	25.9	14.8
Lepomis microlophus **1**																	
0	0	0	0	0	25.0	0	0	0	0	60.0	0	0	0	0	0	0	0
Luxilus chrysocephalus **1**																	
4.3	2.0	2.0	28.0	0	3.0	2.0	0	0	0	2.0	0	0	0	0	0	0	0

Table A1 Occurrence and site-specific body mass (wet weight) of freshwater fishes in the 18 investigated rivers across Ohio—cont'd

Luxilus cornutus **1**																	
0	0	0	0	0	0	0	0	0	0	0	0	0	0	0	21.1	0	0
Lythrurus fasciolaris **1**																	
0	0	0	0	0	0	2.5	0	0	0	0	0	0	0	0	0	0	0
Lythrurus umbratilis **1**																	
0	0	0	0	0	0	0	0	0	0	0	0	0	0	0	0	2.0	0
Micropterus dolomieux **2**																	
712.5	314.3	105.3	124.3	202.7	183.3	326.2	525.0	340.0	168.3	110.3	100.0	159.4	44.0	0	106.7	200.0	124.7
Micropterus punctulatus **2**																	
46.8	140.0	0	0	192.5	49.4	0	0	0	0	187.0	475.0	96.0	170.0	0	0	0	10.0
Micropterus salmoides **2**																	
57.5	90.0	0	0	15.0	0	85.0	70.0	134.3	621.6	233.3	11.3	2.3	169.5	171.8	0	85.7	0
Minytrema melanops **1**																	
96.3	0	0	0	0	0	90.7	60.0	0	0	0	0	0	60.0	697.5	0	167.8	0
Morone chrysops **2**																	
0	190.0	0	0	250.0	0	0	0	0	0	0	0	0	100.0	0	0	0	0
Morone saxatilis **2**																	
0	0	0	0	0	0	0	0	0	0	0	0	0	0	0	0	0	220.0

Continued

Table A1 Occurrence and site-specific body mass (wet weight) of freshwater fishes in the 18 investigated rivers across Ohio—cont'd

Moxostoma anisurum **1**																	
379.4	2075.0	2050.0	0	2150.0	1050.0	1240.0	0	0	0	855.3	560.0	728.8	413.9	0	0	0	1866.7
Moxostoma breviceps **1**																	
51.7	296.0	416.3	0	0	285.0	295.0	0	0	0	229.1	0	0	0	0	0	0	215.2
Moxostoma carinatum **1**																	
0	1950.0	0	0	2065.0	0	0	0	0	0	2245.5	0	0	550.0	0	0	0	0
Moxostoma duquesnei **1**																	
78.3	714.5	814.8	476.9	153.8	700.0	650.0	0	0	0	330.0	750.0	0	0	0	0	0	379.3
Moxostoma erythrurum **1**																	
48.6	526.0	559.6	450.0	398.6	7.3	384.5	51.3	0	0	171.5	391.3	331.0	176.8	0	166.2	386.5	372.8
Moxostoma macrolepidotum **1**																	
0	0	0	0	447.5	0	0	0	0	0	0	0	0	100.0	0	0	0	0
Nocomis biguttatus **1**																	
0	0	0	0	0	0	0	0	7.9	0	0	0	0	0	0	0	0	0
Nocomis micropogon **2**																	
0	0	20.0	0	0	0	0	43.2	0	0	6.0	0	0	0	0	0	0	0
Notemigonus crysoleucas **1**																	
0	0	0	0	0	0	0	0	15.0	0	0	0	0	0	17.4	0	0	0

Table A1 Occurrence and site-specific body mass (wet weight) of freshwater fishes in the 18 investigated rivers across Ohio—cont'd

Notropis amblops **1**																	
0	0	0	0	0	0	0	0	0	0	2.4	0	0	0	0	0	0	0
Notropis atherinoides **1**																	
3.0	2.3	1.7	0	0	2.0	0	0	0	0	0	0	0	1.9	0	0	18.7	1.1
Notropis buccatus **1**																	
0	0	0	0	0	0	0	0	0	0	0	1.5	0	0	0	0	0	0
Notropis photogenis **1**																	
2.0	0	2.0	9.2	5.8	2.0	5.9	0	0	0	8.7	0	0	0	0	0	0	0
Notropis rubellus **2**																	
0	0	2.0	0	0	0	0	0	0	0	2.9	2.0	0	0	0	0	0	0
Notropis stramineus **2**																	
0	0	2.3	0	1.9	1.2	0	0	0	0	2.0	2.0	2.5	0	0	0	3.0	1.5
Notropis volucellus **1**																	
0	3.0	2.3	0	0	0	0	0	0	0	2.0	0	0	0	0	0	0	1.5
Noturus eleutherus **2**																	
0	2.0	3.0	0	0	0	0	0	0	0	0	0	0	0	0	0	0	0
Noturus flavus **2**																	
0	30.0	8.2	0	0	7.1	0	0	0	0	6.0	0	0	0	0	40.0	0	10.0
Noturus miurus **2**																	
0	0	0	0	0	5.0	0	0	0	0	0	0	0	0	0	0	0	0

Continued

Table A1 Occurrence and site-specific body mass (wet weight) of freshwater fishes in the 18 investigated rivers across Ohio—cont'd

Noturus stigmosus **2**																	
0	0	6.0	0	0	5.0	0	0	0	0	0	0	0	0	0	0	0	1.0
Perca flavescens **2**																	
0	0	0	0	0	0	0	70.9	46.4	13.0	0	26.0	0	0	0	0	0	0
Percina caprodes **2**																	
7.6	16.0	8.6	17.0	15.5	12.5	14.9	18.5	0	0	11.0	13.0	0	0	0	0	0	6.3
Percina maculata **2**																	
0	0	0	0	0	0	0	0	0	0	4.0	0	0	0	0	0	0	0
Percina phoxocephala **1**																	
0	4.0	0	0	0	0	0	0	0	0	0	0	0	0	0	0	0	6.0
Percina sciera **1**																	
0	0	0	0	0	0	0	0	0	0	0	0	12.0	0	0	0	0	0
Percopsis omiscomaycus **2**																	
0	0	0	0	0	0	0	0	0	0	0	0	1.3	0	0	0	0	0
Phenacobius mirabilis **1**																	
0	0	4.5	0	0	3.0	0	0	0	0	0	0	0	0	0	0	0	7.0
Pimephales notatus **1**																	
2.6	2.9	2.7	8.0	3.0	1.4	2.0	3.0	4.0	3.8	2.3	0.8	2.5	0	0	3.0	3.1	1.6
Pimephales vigilax **2**																	
0	2.0	1.0	0	0	2.0	0	0	0	0	0	0	0	0	0	0	0	2.4

Table A1 Occurrence and site-specific body mass (wet weight) of freshwater fishes in the 18 investigated rivers across Ohio—cont'd

Pomoxis annularis **2**																	
0	0	0	0	0	0	0	0	0	0	0	0	0	0	0	80.0	116.0	0
Pomoxis nigromaculatus **2**																	
0	0	200.0	0	0	0	215.0	0	51.8	250.0	280.0	0	0	0	139.7	0	0	0
Pylodictis olivaris **2**																	
2700.0	1747.5	338.7	0	0	0	0	0	0	0	0	0	0	0	0	0	0	2020.0
Semotilus atromaculatus **2**																	
0	0	0	29.0	0	0	0	0	0	0	0	0	0.9	0	0	0	5.0	0
Stizostedion canadense **2**																	
0	290.0	0	0	388.0	0	0	0	0	0	0	0	0	137.5	0	0	0	379.0
Stizostedion vitreum **2**																	
0	0	484.0	0	0	0	0	150.0	0	455.6	0	0	0	0	0	0	0	0
Umbra limi **2**																	
0	0	0	0	0	0	0	0	4.0	0	0	0	0	0	0	0	0	0

Please note: Occurrence (if present, weight is provided) and behavioural level: 0 = fish species absent in a given site; 1 = all fish species feeding upon preys from other taxocenes; 2 = piscivorous fishes predating within their own taxocene but not within their own population (cannibalism excluded). All feeding traits and fish taxonomy according to www.FishBase.org version June 2011 accessed August 2011.

REFERENCES

Adl, S.M., Simpson, A.G.B., Farmer, M.A., Andersen, R.A., Anderson, O.R., Barta, J.R., Bowser, S.S., Brugerolle, G., Fensome, R.A., Fredericq, S., James, T.Y., Karpov, S., et al., 2005. The new higher level classification of eukaryotes with emphasis on the taxonomy of protists. J. Eukaryot. Microbiol. 52, 399–451.

Adler, P.B., Seabloom, E.W., Borer, E.T., Hillebrand, H., Hautier, Y., Hector, A., Harpole, W.S., O'Halloran, L.R., Grace, J.B., Anderson, T.M., Bakker, J.D., Biederman, L.A., et al., 2011. Productivity is a poor predictor of plant species richness. Science 333, 1750–1753.

Allen, A.P., Gillooly, J.F., Brown, J.H., 2005. Linking the global carbon cycle to individual metabolism. Funct. Ecol. 19, 202–213.

Allesina, S., Pascual, M., 2009. Googling food webs: can an Eigenvector measure species' importance for coextinctions? PLoS Comput. Biol. 5, e1000494.

Allison, S.D., 2006. Brown ground: a soil carbon analogue for the green world hypothesis? Am. Nat. 167, 619–627.

Amthor, J.S., 2000. The McCree–de Wit–Penning de Vries–Thornley respiration paradigms: 30 years later. Ann. Bot. 86, 1–20.

Antón, A., Cebrian, J., Heck, K.L., Duarte, C.M., Shhehan, K.L., Miller, M.-E.C., Foster, C.D., 2011. Decoupled effects (positive to negative) of nutrient enrichment on ecosystem services. Ecol. Appl. 21, 991–1009.

Arim, M., Berazategui, M., Barreneche, J.M., Ziegler, L., Zarucki, M., Abades, S.R., 2011. Determinants of density-body size scaling within food webs and tools for their detection. Adv. Ecol. Res. 45, 1–39.

Attrill, M.J., Depledge, M.H., 1997. Community and population indicators of ecosystem health: targeting links between levels of biological organisation. Aquat. Toxicol. 38, 183–197.

Bala, G., Caldeira, K., Wickett, M., Phillips, T.J., Lobell, D.B., Delire, C., Mirin, A., 2007. Combined climate and carbon-cycle effects of large-scale deforestation. Proc. Natl. Acad. Sci. U.S.A. 104, 6550–6555.

Banašek-Richter, C., Bersier, L.F., Cattin, M.F., Baltensperger, R., Gabriel, J.P., Merz, Y., Ulanowicz, R.E., Tavares, A.F., Williams, D.D., De Ruiter, P.C., Winemiller, K.O., Naisbit, R.E., 2009. Complexity in quantitative food webs. Ecology 90, 1470–1477.

Barrett, J.E., Virginia, R.A., Wall, D.H., Cary, S.C., Adams, B.J., Hacker, A.L., Aislabie, J.M., 2006. Co-variation in soil biodiversity and biogeochemistry in northern and southern Victoria Land, Antarctica. Antarct. Sci. 18, 535–548.

Barrett, J.E., Virginia, R.A., Lyons, W.B., McKnight, D.M., Priscu, J.C., Doran, P.T., Fountain, A.G., Wall, D.H., Moorhead, D.L., 2007. Biogeochemical stoichiometry of Antarctic Dry Valley ecosystems. J. Geophys. Res. 112, G01010.

Beckerman, A.P., Petchey, O.L., Warren, P.H., 2006. Foraging biology predicts food web complexity. Proc. Natl. Acad. Sci. U.S.A. 103, 13745–13749.

Beerling, D.J., Woodward, F.I., 2001. Vegetation and the Terrestrial Carbon Cycle: Modelling the First 400 Million Years. Cambridge University Press, Cambridge, UK.

Benton, T.G., Beckerman, A.P., 2005. Population dynamics in a noisy world: lessons from a mite experimental system. Adv. Ecol. Res. 37, 143–181.

Blackburn, T.M., Gaston, K.J., 1994. Animal body-size distributions—patterns, mechanisms and implications. Trends Ecol. Evol. 9, 471–474.

Blomberg, S., Garland, T., Ives, A., 2003. Testing for phylogenetic signal in comparative data: behavioral traits are more labile. Evolution 57, 717–745.

Boit, A., Martinez, N.D., Williams, R.J., Gaedke, U., 2012. Mechanistic theory and modelling of complex food-web dynamics in Lake Constance. Ecol. Lett. 15, 594–602.

Bongers, T., 1999. The maturity index, the evolution of nematode life history traits, adaptive radiation and cp-scaling. Plant Soil 212, 13–22.

Borgmann, U., 1987. Model on the slope of, and biomass flow up, the biomass size spectrum. Can. J. Fish. Aquat. Sci. 44, 136–140.
Bradford, M.A., Jones, T.H., Bardgett, R.D., Black, H.I.J., Boag, B., Bonkowski, M., Cook, R., Eggers, T., Gange, A.C., Grayston, S.J., Kandeler, E., McCaig, A.E., et al., 2002. Impacts of soil faunal community composition on model grassland ecosystems. Science 298, 615–618.
Briand, F., 1985. Structural singularities of freshwater food webs. Verh. Int. Ver. Theor. Angew. Limnol. 22, 3356–3364.
Brown, J.H., Gillooly, J.F., 2003. Ecological food webs: high-quality data facilitate theoretical unification. Proc. Natl. Acad. Sci. U.S.A. 100, 1467–1468.
Brown, J.H., Gillooly, J.F., Allen, A.P., Savage, V.M., West, G.B., 2004. Toward a metabolic theory of ecology. Ecology 85, 1771–1789.
Brussaard, L., 2012. Ecosystem services provided by the soil biota. In: Wall, D.H., Bardgett, R.D., Behan-Pelletier, V., Herrick, J.E., Jones, H., Ritz, K., Six, J., Strong, D.R., Van der Putten, W.H. (Eds.), Soil Ecology and Ecosystem Services. Oxford University Press, Oxford, pp. 45–58.
Burton, G.A., De Zwart, D., Diamond, J., Dyer, S., Kapo, K.E., Liess, M., Posthuma, P., 2012. Making ecosystem reality checks the status quo. Environ. Toxicol. Chem. 31, 459–468.
Calder III, W.A., 1984. Size, Function and Life History. Harvard University Press, Cambridge, MA.
Carbone, C., Pettorelli, N., Stephens, P., 2011. The bigger they come, the harder they fall: body size and prey abundance influence predator–prey ratios. Biol. Lett. 7, 312–315.
Cardinale, B.J., 2011. Biodiversity improves water quality through niche partitioning. Nature 472, 86–89.
Cardinale, M., Svedang, H., 2004. Modelling recruitment and abundance of Atlantic cod, *Gadus morhua*, in the eastern Skagerrak-Kattegat (North Sea): evidence of severe depletion due to a prolonged period of high fishing pressure. Fish. Res. 69, 263–282.
Cardinale, B.J., Srivastava, D.S., Duffy, J., Wright, J.P., Downing, A.L., Sankaran, M., Jouseau, C., 2006. Effects of biodiversity on the functioning of trophic groups and ecosystems. Nature 443, 989–992.
Cardinale, B.J., Bennett, D.M., Nelson, C.E., Gross, K., 2009. Does productivity drive diversity or vice versa? A test of the multivariate productivity–diversity hypothesis in streams. Ecology 90, 1227–1241.
Carpenter, S.R., Kraft, C.E., Wright, R., He, X., Soranno, P.A., Hodgson, J.R., 1992. Resilience and resistance of a lake phosphorus cycle before and after food web manipulation. Am. Nat. 140, 781–798.
Carpenter, S.R., Cole, J.J., Hodgson, J.R., Kitchell, J.F., Pace, M.L., Bade, D., Cottingham, K.L., Essington, T.E., Houser, J.N., Schindler, D.E., 2001. Trophic cascades, nutrients, and lake productivity: whole-lake experiments. Ecol. Monogr. 71, 163–186.
Carpenter, S.R., Armbrust, E.V., Arzberger, P.W., Chapin III, F.S., Elser, J.J., Hackett, E.J., Ives, A.R., Kareiva, P.M., Leibold, M.A., Lundberg, P., Mangel, M., Merchant, N., et al., 2009. Accelerate synthesis in ecology and environmental sciences. Bioscience 59, 699–701.
Carpenter, S.R., Cole, J.J., Pace, M.L., Batt, R., Brock, W.A., Cline, T., Coloso, J., Hodgson, J.R., Kitchell, J.F., Seekel, D.A., Smith, L., Weidell, B., 2011. Early warnings of regime shifts: a whole-ecosystem experiment. Science 332, 1079–1082.
Castle, M.D., Blanchard, J.L., Jennings, S., 2011. Predicted effects of behavioural movement and passive transport on individual growth and community size structure in marine ecosystems. Adv. Ecol. Res. 45, 41–66.
Caswell, H., Cohen, J.E., 1991. Disturbance, interspecific interaction and diversity in metapopulations. Biol. J. Linn. Soc. 42, 193–218.

Cavalier-Smith, T., 1993. Kingdom protozoa and its 18 phyla. Microbiol. Rev. 57, 953–994.

Chapin III, F.S., Zavaleta, E.S., Eviner, V.T., Naylor, R.L., Vitousek, P.M., Reynolds, H.L., Hooper, D.U., Lavorel, S., Sala, O.E., Hobbie, S.E., Mack, M.C., Díaz, S., 2000. Consequences of changing biodiversity. Nature 405, 234–242.

Chase, J.M., 2000. Are there real differences among aquatic and terrestrial food webs? Trends Ecol. Evol. 15, 408–412.

Chown, S.L., Marais, E., Terblanche, J.S., Klok, C.J., Lighton, J.R.B., Blackburn, T.M., 2007. Scaling of insect metabolic rate is inconsistent with the nutrient supply network model. Funct. Ecol. 21, 282–290.

Clarke, A., Johnston, N.M., 1999. Scaling of metabolic rate with body mass and temperature in teleost fish. J. Anim. Ecol. 68, 893–905.

Cleveland, C.C., Liptzin, D., 2007. C:N:P stoichiometry in soil: is there a "Redfield ratio" for the microbial biomass? Biogeochemistry 85, 235–252.

Cohen, J.E., 1994. Marine and continental food webs: three paradoxes? Philos. Trans. R. Soc. Lond. B 343, 57–69.

Cohen, J.E., Carpenter, S.R., 2005. Species' average body mass and numerical abundance in a community food web: statistical questions in estimating the relationship. In: De Ruiter, P.C., Wolters, V., Moore, J.C. (Eds.), Dynamic Food Webs: Multispecies Assemblages, Ecosystem Development, and Environmental Change. Academic Press, San Diego, CA, pp. 137–156.

Cohen, J.E., Jonsson, T., Carpenter, S.R., 2003. Ecological community description using the food web, species abundance, and body size. Proc. Natl. Acad. Sci. U.S.A. 100, 1781–1786.

Cotner, J.B., Hall, E.K., Scott, J.T., Heldal, M., 2010. Freshwater bacteria are stoichiometrically flexible with a nutrient composition similar to seston. Front. Microbiol. 1, a132.

Cottingham, K.L., Brown, B.L., Lennon, J.T., 2001. Biodiversity may regulate the temporal variability of ecological systems. Ecol. Lett. 4, 72–85.

Courtright, E.M., Wall, D.H., Virginia, R.A., 2001. Determining habitat suitability for soil invertebrates in an extreme environment: the McMurdo Dry Valleys, Antarctica. Antarct. Sci. 13, 9–17.

Cousins, S.H., 1980. A trophic continuum derived from plant structure, animal size and a detritus cascade. J. Theor. Biol. 82, 607–618.

Cousins, S.H., 1996. Food webs: from the Lindeman paradigm to a taxonomic general theory of ecology. In: Polis, G.A., Winemiller, K.O. (Eds.), Food Webs: Integration of Patterns and Dynamics. Chapman & Hall, New York, pp. 243–251.

Crotty, F.V., Blackshaw, R.P., Murray, P.J., 2011. Tracking the flow of bacterially derived ^{13}C and ^{15}N through soil faunal feeding channels. Rapid Commun. Mass Spectrom. 25, 1503–1513.

Crutzen, P.J., 2002. Geology of mankind. Nature 415, 23.

Damuth, J., 1981. Population density and body size in mammals. Nature 290, 699–700.

Damuth, J., 1987. Interspecific allometry of population density in mammals and other animals: the independence of body mass and population energy-use. Biol. J. Linn. Soc. 31, 193–246.

Damuth, J., 1991. Of size and abundance. Nature 351, 268–269.

Dawson, S.C., Hagen, K.D., 2009. Mapping the protistan 'rare biosphere'. J. Biol. 8, a105.

De Visser, S., Freymann, B., Olff, H., 2011. The Serengeti food web: empirical quantification and analysis of topological changes under increasing human impact. J. Anim. Ecol. 80, 465–475.

De Vries, F.T., Hoffland, E., Van Eekeren, N., Brussaard, L., Bloem, J., 2006. Fungal/bacterial ratios in grasslands with contrasting nitrogen management. Soil Biol. Biochem. 38, 2092–2103.

Díaz, S., Lavorel, S., De Bello, F., Quétier, F., Grigulis, K., Robson, T.M., 2007. Incorporating plant functional diversity effects in ecosystem service assessments. Proc. Natl. Acad. Sci. U.S.A. 104, 20684–20689.

Dossena, M., Grey, J., Montoya, J.M., Perkins, D.M., Trimmer, M., Woodward, G., Yvon-Durocher, G., 2012. Warming alters community size structure and ecosystem functioning. Proc. R. Soc. Lond. B 279, 3011–3019.

Du Rietz, G.E., 1931. Life forms of terrestrial flowering plants. Acta Phytogeogr. Suec. 3, 1–95.

Duarte, C.M., Middelburg, J., Caraco, N., 2005. Major role of marine vegetation on the oceanic carbon cycle. Biogeosciences 2, 1–8.

Dukes, J.S., Chiariello, N.R., Cleland, E.E., Moore, L.A., Shaw, M.R., Thayer, S., Tobeck, T., Mooney, H.A., Field, C.B., 2005. Responses of grassland production to single and multiple global environmental changes. PLoS Biol. 3, e319.

Dunne, J.A., 2009. Food webs. In: Meyers, R.A. (Ed.), Complex Networks and Graph Theory—Encyclopedia of Complexity and Systems Science, 3661–3682. http://www.springer.com/978-0-387-75888-6.

Dunne, J.A., Williams, R.J., 2009. Cascading extinctions and community collapse in model food webs. Philos. Trans. R. Soc. Lond. B 364, 1711–1723.

Dunne, J.A., Williams, R.J., Martinez, N.D., 2002. Network structure and biodiversity loss in food webs: robustness increases with connectance. Ecol. Lett. 5, 558–567.

Dunne, J.A., Williams, R.J., Martinez, N.D., 2004. Network structure and robustness of marine food webs. Mar. Ecol. Prog. Ser. 273, 291–302.

Dunne, J.A., Williams, R.J., Martinez, N.D., Wood, R.A., Erwin, D.H., 2008. Compilation and network analyses of Cambrian food webs. PLoS Biol. 6, e102.

Duplisea, D.E., Drgas, A., 1999. Sensitivity of a benthic, metazoan, biomass size spectrum to differences in sediment granulometry. Mar. Ecol. Prog. Ser. 177, 73–81.

Dyer, S.D., White-Hull, C.E., Wang, X., Johnson, T.D., Carr, G.J., 1998. Determining the influence of habitat and chemical factors on instream biotic integrity for a Southern Ohio watershed. J. Aquat. Ecosyst. Stress Recov. 6, 91–110.

Ebenman, B., Jonsson, T., 2005. Using community viability analysis to identify fragile systems and keystone species. Trends Ecol. Evol. 20, 568–575.

Elser, J., 2006. Biological stoichiometry: a chemical bridge between ecosystem ecology and evolutionary biology. Am. Nat. 168, S25–S35.

Elser, J.J., Urabe, J., 1999. The stoichiometry of consumer-driven nutrient recycling: theory, observations, and consequences. Ecology 80, 735–751.

Elser, J.J., Luecke, C., Brett, M.T., Goldman, C.R., 1995. Effects of food web compensation after manipulation of rainbow trout in an oligotrophic lake. Ecology 76, 52–69.

Elser, J.J., Chrzanowski, T.H., Sterner, R.W., Mills, K.H., 1998. Stoichiometric constraints on food-web dynamics: a whole-lake experiment on the Canadian Shield. Ecosystems 1, 120–136.

Elser, J.J., Sterner, R.W., Galford, A.E., Chrzanowski, T.H., Findlay, D.L., Mills, K.H., Paterson, M.J., Stainton, M.P., Schindler, D.W., 2000. Pelagic C:N:P stoichiometry in a eutrophied lake: responses to a whole-lake food-web manipulation. Ecosystems 3, 293–307.

Elser, J.J., Fagan, W.F., Kerkhoff, A.J., Swenson, N.G., Enquist, B.J., 2010. Biological stoichiometry of plant production: metabolism, scaling and ecological response to global change. New Phytol. 186, 593–608.

Enquist, B.J., West, G.B., Charnov, E.L., Brown, J.H., 1999. Allometric scaling of production and life-history variation in vascular plants. Nature 401, 907–911.

Enquist, B.J., Allen, A.P., Brown, J.H., Gillooly, J.F., Kerkhoff, A.J., Niklas, K.J., Price, C.A., West, G.B., 2007. Does the exception prove the rule? Nature 445, E9.

Ernest, S.K.M., Enquist, B.J., Brown, J.H., Charnov, E.L., Gillooly, J.F., Savage, V.M., White, E.P., Smith, F.A., Hadly, E.A., Haskell, J.P., Lyons, S.K., Maurer, B.A., et al., 2003. Thermodynamic and metabolic effects on the scaling of production and population energy use. Ecol. Lett. 6, 990–995.

Estes, J.A., Terborgh, J., Brashares, J.S., Power, M.E., Berger, J., Bond, W.J., Carpenter, S.R., Essington, T.E., Holt, R.D., Jackson, J.B.C., Marquis, R.J., Oksanen, L., et al., 2011. Trophic downgrading of planet Earth. Science 333, 301–306.

Fahey, T.J., Woodbury, P.B., Battles, J.J., Goodale, C.L., Hamburg, S.P., Ollinger, S.V., Woodall, C.W., 2010. Forest carbon storage: ecology, management, and policy. Front. Ecol. Environ. 8, 245–252.

FAO, 2000. The State of the World's Fisheries and Aquaculture 2000. Food and Agriculture Organization of the United Nations, Rome.

Feld, C.K., Birk, S., Bradley, D.C., Hering, D., Kail, J., Marzin, A., Melcher, A., Nemitz, D., Pedersen, M.L., Pletterbauer, F., Pont, D., Verdonschot, P.F.M., et al., 2011. From natural to degraded rivers and back again: a test of restoration ecology theory and practice. Adv. Ecol. Res. 44, 119–210.

Fenchel, T., Finlay, B.J., 1983. Respiration rates in heterotrophic, free-living protozoa. Microb. Ecol. 9, 99–122.

Finlay, B.J., 1998. The global diversity of protozoa and other small species. Intern. J. Parasitol. 28, 29–48.

Finlay, B.J., 2002. Global dispersal of free-living microbial eukaryote species. Science 296, 1061–1063.

Flombaum, P., Sala, O.E., 2012. Effects of plant species traits on ecosystem processes: experiments in the Patagonian steppe. Ecology 93, 227–234.

Flotemersch, J.E., 2001. Fish assessment methods. In: Flotemersch, J.E., Autrey, B.C., Cormier, S.M. (Eds.), Comparisons of Boating and Wading Methods Used to Assess the Status of Flowing Waters. EPA/600/R-00/108. U.S. Environmental Protection Agency, Cincinnati, OH, pp. 68–82.

Foissner, W., 2006. Biogeography and dispersal of micro-organisms: a review emphasizing protists. Acta Protozool. 45, 111–136.

Foissner, W., 2008. Protist diversity and distribution: some basic considerations. Biodivers. Conserv. 17, 235–242.

Forster, J., Hirst, A.G., Woodward, G., 2011. Growth and development rates have different thermal responses. Am. Nat. 178, 668–678.

Fortunel, C., Garnier, E., Joffre, R., Kazakou, E., Quested, H., Grigulis, K., Lavorel, S., Ansquer, P., Castro, H., Cruz, P., Doležal, J., Eriksson, O., et al., 2009. Plant functional traits capture the effects of land use change and climate on litter decomposability of herbaceous communities in Europe and Israel. Ecology 90, 598–611.

Friberg, N., Bonada, N., Bradley, D.C., Dunbar, M.J., Edwards, F.K., Grey, J., Hayes, R.B., Hildrew, A.G., Lamouroux, N., Trimmer, M., Woodward, G., 2011. Biomonitoring of human impacts in freshwater ecosystems: the good, the bad, and the ugly. Adv. Ecol. Res. 44, 2–68.

Fridley, J.D., 2001. The influence of species diversity on ecosystem productivity: how, where, and why? Oikos 93, 514–526.

Froese, R., Pauly, D. (Eds.), 2005. FishBase, http://www.fishbase.org.

Fry, B., Quiñones, R.B., 1994. Biomass spectra and stable isotope indicators of trophic level in zooplankton of the northwest Atlantic. Mar. Ecol. Prog. Ser. 112, 201–204.

Gams, W., 1992. The analysis of communities of saprophytic microfungi with special reference to soil fungi. In: Winterhoff, W. (Ed.), Fungi in Vegetation Science. Kluwer Academic Publishers, Dordrecht, pp. 183–223.

Gardner, R.H., Kemp, W.M., Kennedy, V.S., Petersen, J.E. (Eds.), 2001. Scaling Relations in Experimental Ecology. Columbia University Press, New York.

Garnier, E., Cortez, J., Billès, G., Navas, M.-L., Roumet, C., Debussche, M., Laurent, G., Blanchard, A., Aubry, D., Bellmann, A., Neill, C., Toussaint, J.-P., 2004. Plant functional markers capture ecosystem properties during secondary succession. Ecology 85, 2630–2637.

Gerritsen, J., Barbour, M.T., King, K., 2000. Apples, oranges, and ecoregions: on determining pattern in aquatic assemblages. J. N. Am. Benthol. Soc. 19, 487–496.
Ghilarov, A.M., 2000. Ecosystem functioning and intrinsic value of biodiversity. Oikos 90, 408–412.
Gilljam, D., Thierry, A., Edwards, F.K., Figueroa, D., Ibbotson, A.T., Jones, J.I., Lauridsen, R.B., Petchey, O.L., Woodward, G., Ebenman, B., 2011. Seeing double: size-based and taxonomic views of food web structure. Adv. Ecol. Res. 45, 67–133.
Glazier, D.S., 2005. Beyond the '3/4-power law': variation in the intra- and interspecific scaling of metabolic rate in animals. Biol. Rev. 80, 611–662.
Glazier, D.S., 2010. A unifying explanation for diverse metabolic scaling in animals and plants. Biol. Rev. 85, 111–138.
Glibert, P.M., Harrison, J., Heil, C., Seitzinger, S., 2006. Escalating worldwide use of urea: a global change contributing to coastal eutrophication. Biogeochemistry 77, 441–463.
Goldbogen, J.A., Calambokidis, J., Croll, D.A., McKenna, M.F., Oleson, E., Potvin, J., Pyenson, N.D., Schorr, G., Shadwick, R.E., Tershy, B.R., 2012. Scaling of lunge-feeding performance in rorqual whales: mass-specific energy expenditure increases with body size and progressively limits diving capacity. Funct. Ecol. 26, 216–226.
Goldstein, R.M., Meador, M.R., 2004. Comparisons of fish species traits from small streams to large rivers. Trans. Am. Fish. Soc. 133, 971–983.
Goldstein, R.M., Simon, T.P., 1999. Toward a united definition of guild structure for feeding ecology of North American freshwater fishes. In: Simon, T.P. (Ed.), Assessing the Sustainability and Biological Integrity of Water Resources Using Fish Communities. CRC Press, Boca Raton, FL, pp. 123–202.
Gonzalez, A., Lawton, J.H., Gilbert, F.S., Blackburn, T.M., Evans-Freke, I.I., 1998. Metapopulation dynamics, abundance, and distribution in a microecosystem. Science 281, 2045–2047.
González, A.L., Fariña, J.M., Kay, A.D., Pinto, R., Marquet, P.A., 2011. Exploring patterns and mechanisms of interspecific and intraspecific variation in body elemental composition of desert consumers. Oikos 120, 1247–1255.
Grime, J.P., 1973. Competitive exclusion in herbaceous vegetation. Nature 242, 344–347.
Grime, J.P., 1977. Evidence for existence of 3 primary strategies in plants and its relevance to ecological and evolutionary theory. Am. Nat. 111, 1169–1194.
Grime, J.P., 1979. Plant Strategies and Vegetation Processes. Wiley, Chichester, UK.
Grime, J.P., 1997. Biodiversity and ecosystem function: the debate deepens. Science 277, 1260–1261.
Gross, K., Cardinale, B.J., 2007. Does species richness drive community production or vice versa? Reconciling historical and contemporary paradigms in competitive communities. Am. Nat. 170, 207–220.
Guénard, G., Von Der Ohe, P.C., De Zwart, D., Legendre, P., Lek, S., 2011. Using phylogenetic information to predict species tolerances to toxic chemicals. Ecol. Appl. 21, 3178–3190.
Hagen, M., Kissling, W.D., Rasmussen, C., De Aguiar, M.A.M., Brown, L., Carstensen, D.W., Alves-Dos-Santos, I., Dupont, Y.L., Edwards, F.K., Genini, J., Guimarães Jr., P.K., Jenkins, G.B., et al., 2012. Biodiversity, species interactions and ecological networks in a fragmented world. Adv. Ecol. Res. 46, 89–210.
Hall, S.R., 2009. Stoichiometrically explicit food webs: feedbacks between resource supply, elemental constraints, and species diversity. Annu. Rev. Ecol. Evol. Syst. 40, 503–528.
Hartnett, D.C., Wilson, G.W.T., 1999. Mycorrhizae influence plant community structure and diversity in tallgrass prairie. Ecology 80, 1187–1195.
Havens, K.E., 1992. Scale and structure in natural food webs. Science 257, 1107–1109.
Hawkins, C.P., 2006. Quantifying biological integrity by taxonomic completeness: its utility in regional and global assessments. Ecol. Appl. 16, 1277–1294.

Hector, A., Schmid, B., Beierkuhnlein, C., Caldeira, M.C., Diemer, M., Dimitrakopoulos, P.G., Finn, J.A., Freitas, H., Giller, P.S., Good, J., Harris, R., Högberg, P., et al., 1999. Plant diversity and productivity experiments in European grasslands. Science 286, 1123–1127.
Hector, A., Loreau, M., Schmid, B., Beierkuhnlein, C., Caldeira, M.C., Diemer, M., Dimitrakopoulos, P.G., Finn, J.A., Freitas, H., Giller, P.S., Good, J., Harris, R., 2002. Biodiversity manipulation experiments: studies replicated at multiple sites. In: Loreau, M., Naeem, S., Inchausti, P. (Eds.), Biodiversity and Ecosystem Functioning: Synthesis and Perspectives. Oxford University Press, Oxford, pp. 36–46.
Hedin, L.O., 2006. Plants on a different scale. Nature 436, 399–400.
Hein, M., Pedersen, M.F., Sandjensen, K., 1995. Size-dependent nitrogen uptake in micro- and macroalgae. Mar. Ecol. Prog. Ser. 118, 247–253.
Hendriks, A.J., Mulder, C., 2008. Scaling of offspring number and mass to plant and animal size: model and meta-analysis. Oecologia 155, 705–716.
Hendriks, A.J., Mulder, C., 2012. Delayed logistic and Rosenzweig–MacArthur models with allometric parameter setting estimate population cycles at lower trophic levels well. Ecol. Complex. 9, 43–54.
Hendrixson, H.A., Sterner, R.W., Kay, A.D., 2007. Elemental stoichiometry of freshwater fish in relation to phylogeny, allometry and ecology. J. Fish Biol. 70, 121–140.
Hildrew, A.G., 2009. Sustained research on stream communities: a model system and the comparative approach. Adv. Ecol. Res. 41, 175–312.
Hill, M.O., 1973. Diversity and evenness: a unifying notation and its consequences. Ecology 54, 427–432.
Hirose, T., Werger, M.J.A., 1987. Maximizing daily canopy photosynthesis with respect to the leaf nitrogen allocation pattern in the canopy. Oecologia 72, 520–526.
Hladyz, S., Gessner, M.O., Giller, P.S., Pozo, J., Woodward, G., 2009. Resource quality and stoichiometric constraints on stream ecosystem functioning. Freshw. Biol. 54, 957–970.
Hladyz, S., Åbjörnsson, K., Cariss, H., Chauvet, E., Dobson, M., Elosegi, A., Ferreira, V., Fleituch, T., Gessner, M.O., Giller, P.S., Graça, M.A.S., Gulis, V., et al., 2011. Stream ecosystem functioning in an agricultural landscape: the importance of terrestrial-aquatic linkages. Adv. Ecol. Res. 44, 211–276.
Hodgson, J.G., Wilson, P.J., Hunt, R., Grime, J.P., Thompson, K., 1999. Allocating C-S-R plant functional types: a soft approach to a hard problem. Oikos 85, 282–294.
Hooper, D.U., Chapin III, F.S., Ewel, J.J., Hector, A., Inchausti, P., Lavorel, S., Lawton, J.H., Lodge, D.M., Loreau, M., Naeem, S., Schmid, B., Setälä, H., et al., 2005. Effects of biodiversity on ecosystem functioning: a consensus of current knowledge and needs for future research. Ecol. Monogr. 75, 3–35.
Höppener-Ogawa, S., Leveau, J.H., Hundscheid, M.P., van Veen, J.A., de Boer, W., 2009. Impact of *Collimonas* bacteria on community composition of soil fungi. Environ. Microbiol. 11, 1444–1452.
Hubbell, S.P., 2001. The Unified Neutral Theory of Biodiversity and Biogeography. Princeton University Press, Princeton, NJ.
Hunt, H.W., Wall, D.H., 2002. Modeling the effects of loss of soil biodiversity on ecosystem function. Glob. Change Biol. 8, 32–49.
Hunting, E.R., Whatley, M.H., Van der Geest, H.G., Mulder, C., Kraak, M.H.S., Breure, A.M., Admiraal, W., 2012. Invertebrate footprints on detritus processing, bacterial community structure and spatiotemporal redox profiles. Freshw. Sci. 31, 724–732.
Hurlbert, S.H., 1971. The nonconcept of species diversity: a critique and alternative parameters. Ecology 52, 577–586.
Hurlbert, A.H., Jetz, W., 2010. More than "More Individuals": the nonequivalence of area and energy in the scaling of species richness. Am. Nat. 176, E50–E65.

Huston, M.A., 1994. Biological Diversity: The Coexistence of Species on Changing Landscapes. Cambridge University Press, Cambridge, UK.
Huston, M.A., 1997. Hidden treatments in ecological experiments: re-evaluating the ecosystem function of biodiversity. Oecologia 110, 449–460.
Huston, M.A., Wolverton, S., 2009. The global distribution of net primary production: resolving the paradox. Ecol. Monogr. 79, 343–377.
Huston, M.A., Wolverton, S., 2011. Regulation of animal size by eNPP, Bergmann's rule, and related phenomena. Ecol. Monogr. 81, 349–405.
Huston, M.A., Aarssen, L.W., Austin, M.P., Cade, B.S., Fridley, J.D., Garnier, E., Grime, J.P., Hodgson, J., Lauenroth, W.K., Thompson, K., Vandermeer, J.H., Wardle, D.A., 2000. No consistent effect of plant diversity on productivity. Science 289, 1255a.
Hutchinson, G., 1978. An Introduction to Population Ecology. Yale University Press, New Haven, CT.
Isaac, N.J.B., Carbone, C., 2010. Why are metabolic scaling exponents so controversial? Quantifying variance and testing hypotheses. Ecol. Lett. 13, 728–735.
Ives, A.R., Carpenter, S.R., Dennis, B., 1999. Community interaction webs and zooplankton responses to planktivory manipulations. Ecology 80, 1405–1421.
Jackson, J.B.C., Kirby, M.X., Berger, W.H., Bjorndal, K.A., Botsford, L.W., Bourque, B.J., Bradbury, R.H., Cooke, R., Erlandson, J., Estes, J.A., Hughes, T.P., Kidwell, S., et al., 2001. Historical overfishing and the recent collapse of coastal ecosystems. Science 293, 629–638.
Jacob, U., Thierry, A., Brose, U., Arntz, W.E., Berg, S., Brey, T., Fetzer, I., Jonsson, T., Mintenbeck, K., Möllmann, C., Petchey, O., Riede, J.O., et al., 2011. The role of body size in complex food webs: a cold case. Adv. Ecol. Res. 45, 181–223.
Jenkins, D.G., Brescacin, C.R., Duxbury, C.V., Elliott, J.A., Evans, J.A., Grablow, K.R., Hillegass, M., Lyon, B.N., Metzger, O.G.A., Olandese, M.L., Pepe, D., Silvers, G.A., et al., 2007. Does size matter for dispersal distance? Glob. Ecol. Biogeogr. 16, 415–425.
Jennings, S., 2005. Size-based analyses of aquatic food webs. In: Belgrano, A., Scharler, U., Dunne, J., Ulanowicz, R.E. (Eds.), Aquatic Food Webs: An Ecosystem Approach. Oxford University Press, Oxford, pp. 86–97.
Jennings, S., Blanchard, J., 2004. Fish abundance with no fishing: predictions based on macroecological theory. J. Anim. Ecol. 73, 632–642.
Jennings, S., Mackinson, S., 2003. Abundance–body mass relationships in size-structured food webs. Ecol. Lett. 6, 971–974.
Jennings, S., Greenstreet, S.P.R., Reynolds, J.D., 1999. Structural change in an exploited fish community: a consequence of differential fishing effects on species with contrasting life histories. J. Anim. Ecol. 68, 617–627.
Jeschke, J.M., Kopp, M., Tollrian, R., 2004. Consumer-food systems: why type I functional responses are exclusive to filter feeders. Biol. Rev. 79, 337–349.
Johnson, N.C., 2010. Resource stoichiometry elucidates the structure and function of arbuscular mycorrhizas across scales. New Phytol. 185, 631–647.
Jonsson, T., Cohen, J.E., Carpenter, S.R., 2005. Food webs, body size and species abundance in ecological community description. Adv. Ecol. Res. 36, 1–84.
Jørgensen, C.B., 1966. Biology of Suspension Feeding. Pergamon Press, Oxford.
Kaspari, M., 2001. Taxonomic level, trophic biology and the regulation of local abundance. Glob. Ecol. Biogeogr. 10, 229–244.
Kaspari, M., 2004. Using the metabolic theory of ecology to predict global patterns of abundance. Ecology 85, 1800–1802.
Kaspari, M., 2005. Global energy gradients and size in colonial organisms: worker mass and worker number in ant colonies. Proc. Natl. Acad. Sci. U.S.A. 102, 5079–5083.

Kaspari, M., Weiser, M.D., 2012. Energy, taxonomic aggregation, and the geography of ant abundance. Ecography 35, 65–72.
Kaspari, M., Pickering, J., Longino, J.T., Windsor, D., 2001. The phenology of a Neotropical ant assemblage: evidence for continuous and overlapping reproduction. Behav. Ecol. Sociobiol. 50, 382–390.
Kaspari, M., Yuan, M., Alonso, L., 2003. Spatial grain and the causes of regional diversity gradients in ants. Am. Nat. 161, 459–477.
Kattge, J., Díaz, S., Lavorel, S., Prentice, I.C., Leadley, P., Bönisch, G., Garnier, E., Westoby, M., Reich, P.B., Wright, I.J., Cornelissen, J.H.C., Violle, C., et al., 2011. TRY—a global database of plant traits. Glob. Change Biol. 17, 2905–2935.
Kerkhoff, A.J., Enquist, B.J., 2007. The implications of scaling approaches for understanding resilience and reorganization in ecosystems. Bioscience 57, 489–499.
Kerkhoff, A.J., Enquist, B.J., Elser, J.J., Fagan, W.F., 2005. Plant allometry, stoichiometry and the temperature-dependence of primary productivity. Glob. Ecol. Biogeogr. 14, 585–598.
Killen, S.S., Atkinson, D., Glazier, D.S., 2010. The intraspecific scaling of metabolic rate with body mass in fishes depends on lifestyle and temperature. Ecol. Lett. 13, 184–193.
Klironomos, J.N., McCune, J., Hart, M., Neville, J., 2000. The influence of arbuscular mycorrhizae on the relationship between plant diversity and productivity. Ecol. Lett. 3, 137–141.
Krivtsov, V., Griffiths, B., Liddell, K., Garside, A., Salmond, R., Bezginova, T., Thompson, J., 2011. Soil nitrogen availability is reflected in the bacterial pathway. Pedosphere 21, 26–30.
Lack, D., 1954. The Natural Regulation of Animal Numbers. Clarendon, Oxford.
Ladygina, N., Hedlund, K., 2010. Plant species influence microbial diversity and carbon allocation in the rhizosphere. Soil Biol. Biochem. 42, 162–168.
Larrañaga, A., Layer, K., Basaguren, A., Pozo, J., Woodward, G., 2010. Consumer body composition and community structure in a stream is altered by pH. Freshw. Biol. 55, 670–680.
Lauridsen, R.B., Edwards, F.K., Bowes, M.J., Woodward, G., Hildrew, A.G., Ibbotson, A.T., Jones, J.I., 2012. Consumer-resource elemental imbalances in a nutrient rich stream. Freshw. Sci. 31, 408–422.
Lavorel, S., Garnier, E., 2002. Predicting the effects of environmental changes on plant community composition and ecosystem functioning: revisiting the Holy Grail. Funct. Ecol. 16, 545–556.
Lavorel, S., McIntyreb, S., Landsberg, J., Forbes, T.D.A., 1997. Plant functional classifications: from general groups to specific groups based on response to disturbance. Trends Ecol. Evol. 12, 474–478.
Lavorel, S., Harrington, R., Storkey, J., Díaz, S., De Bello, F., Bardgett, R.D., Dolédec, S., Feld, C., Roux, X.L., Berg, M.P., Cornelissen, J.H., Hance, T., et al., 2009. RUBICODE—how trait linkages within and across trophic levels underlie the vulnerability of ecosystem services. FP6, Thematic Area: Global Change and Ecosystems. European Commission, DG Research, Bruxelles.
Layer, K., Riede, J.O., Hildrew, A.G., Woodward, G., 2010. Food web structure and stability in 20 streams across a wide pH gradient. Adv. Ecol. Res. 42, 265–299.
Layer, K., Hildrew, A.G., Jenkins, G.B., Riede, J., Rossiter, S.J., Townsend, C.R., Woodward, G., 2011. Long-term dynamics of a well-characterised food web: four decades of acidification and recovery in the Broadstone Stream model system. Adv. Ecol. Res. 44, 69–117.
Leaper, R., Raffaelli, D., 1999. Defining the abundance body-size constraint space: data from a real food web. Ecol. Lett. 2, 191–199.
Ledger, M.E., Harris, R.M.L., Armitage, P.D., Milner, A.M., 2012. Climate change impacts on community resilience: evidence from a drought disturbance experiment. Adv. Ecol. Res. 46, 211–258.

Lepš, J., 2001. Species-pool hypothesis: limits to its testing. Folia Geobot. 36, 45–52.
Lindeman, R.L., 1942. The trophic-dynamic aspect of ecology. Ecology 23, 399–418.
Lipton, D.S., Blanchar, R.W., Blevins, D.G., 1987. Citrate, malate, and succinate concentration in exudates from P-sufficient and P-stressed *Medicago sativa* L. seedlings. Plant Physiol. 85, 315–317.
Loreau, M., 2010. Linking biodiversity and ecosystems: towards a unifying ecological theory. Philos. Trans. R. Soc. Lond. B 365, 49–60.
Loreau, M., Hector, A., 2001. Partitioning selection and complementarity in biodiversity experiments. Nature 412, 72–76.
Loreau, M., Naeem, S., Inchausti, P., Bengtsson, J., Grime, J.P., Hector, A., Hooper, D.U., Huston, M.A., Raffaelli, D., Schmid, B., Tilman, D., Wardle, D.A., 2001. Biodiversity and ecosystem functioning: current knowledge and future challenges. Science 294, 804–808.
Lovegrove, B.G., 2000. The zoogeography of mammalian basal metabolic rate. Am. Nat. 156, 201–219.
Lynch, J.P., Ho, M.D., 2005. Rhizoeconomics: carbon costs of phosphorus acquisition. Plant Soil 269, 45–56.
Lynch, J.M., Whipps, J.M., 1990. Substrate flow in the rhizosphere. Plant Soil 129, 1–10.
Lyons, J., 1989. Correspondence between the distribution of fish assemblages in Wisconsin streams and Omernik ecoregions. Am. Midl. Nat. 122, 163–182.
Lyons, K.G., Schwartz, M.W., 2001. Rare species loss alters ecosystem function–invasion resistance. Ecol. Lett. 4, 358–365.
MacArthur, R., 1955. Fluctuations of animal populations and a measure of community stability. Ecology 36, 533–536.
Maestre, F.T., Quero, J.L., Gotelli, N.J., Escudero, A., Ochoa, V., Delgado-Baquerizo, M., García-Gómez, M., Bowker, M.A., Soliveres, S., Escolar, C., García-Palacios, P., Berdugo, M., et al., 2012. Plant species richness and ecosystem multifunctionality in global drylands. Science 335, 214–218.
Makarieva, A.M., Gorshkov, V.G., Li, B.L., Chown, S.L., Reich, P.B., Gavrilov, V.M., 2008. Mean mass-specific metabolic rates are strikingly similar across life's major domains: evidence for life's metabolic optimum. Proc. Natl. Acad. Sci. U.S.A. 105, 16994–16999.
Makino, W., Cotner, J.B., Sterner, R.W., Elser, J.J., 2003. Are bacteria more like plants or animals? Growth rate and resource dependence of bacterial C:N:P stoichiometry. Funct. Ecol. 17, 121–130.
Mann, C.C., 1991. Extinction: are ecologists crying wolf? Science 253, 736–738.
Marquet, P.A., Labra, F.A., Maurer, B.A., 2004. Metabolic ecology: linking individuals to ecosystems. Ecology 85, 1794–1796.
Marquet, P.A., Quiñones, R.A., Abades, S., Labra, F., Tognelli, M., Arim, M., Rivadeneira, M., 2005. Scaling and power-laws in ecological systems. J. Exp. Biol. 208, 1749–1769.
Martinez, N.D., 1991. Artifacts or attributes? Effects of resolution on the Little Rock Lake food web. Ecol. Monogr. 61, 367–392.
Martinez, N.D., 1992. Constant connectance in community food webs. Am. Nat. 139, 1208–1218.
Martinez, N.D., 1993. Effects of scale on food web structure. Science 260, 242–243.
Martinson, H.M., Schneider, K., Gilbert, J., Hines, J.E., Hambäck, P.A., Fagan, W.F., 2008. Detritivory: stoichiometry of a neglected trophic level. Ecol. Res. 23, 487–491.
May, R.M., 1972. Will a large complex system be stable? Nature 238, 413–414.
McGill, B.J., Enquist, B.J., Weiher, E., Westoby, M., 2006. Rebuilding community ecology from functional traits. Trends Ecol. Evol. 21, 178–185.
Meehan, T.D., 2006. Energy use and animal abundance in litter and soil communities. Ecology 87, 1650–1658.

Menge, D.N.L., Field, C.B., 2007. Simulated global changes alter phosphorus demand in annual grassland. Glob. Change Biol. 13, 2582–2591.

Millennium Ecosystem Assessment, 2005. Ecosystems and Human Well Being. Island Press. Encyclopedia of Earth. http://www.eoearth.org/article/Millennium_Ecosystem_Assessment_Synthesis_Reports (Retrieved 10 December 2010).

Mohr, C.D., 1940. Comparative populations of game, fur and other mammals. Am. Midl. Nat. 24, 581–584.

Montoya, J.M., Pimm, S.L., Solé, R.V., 2006. Ecological networks and their fragility. Nature 442, 259–264.

Moore, J.C., Berlow, E.L., Coleman, D.C., De Ruiter, P.C., Dong, Q., Hastings, A., Johnson, N.C., McCann, K.S., Melville, K., Morin, P.J., Nadelhoffer, K., Rosemond, A.D., et al., 2004. Detritus, trophic dynamics and biodiversity. Ecol. Lett. 7, 584–600.

Moorhead, D.L., Wall, D.H., Virginia, R.A., Parsons, A.N., 2002. Distribution and life-cycle of *Scottnema lindsayae* (Nematoda) in Antarctic soils: a modeling analysis of temperature responses. Polar Biol. 25, 118–125.

Mori, S., Yamaji, K., Ishida, A., Prokushkin, S.G., Masyagina, O.V., Hagihara, A., Hoque, A.T.M.R., Suwa, R., Osawa, A., Nishizono, T., Ueda, T., Kinjo, M., et al., 2010. Mixed-power scaling of whole-plant respiration from seedlings to giant trees. Proc. Natl. Acad. Sci. U.S.A. 107, 1447–1451.

Mulder, C., 2010. Soil fertility controls the size-specific distribution of eukaryotes. Ann. N.Y. Acad. Sci. 1195, E74–E81.

Mulder, C., Elser, J.J., 2009. Soil acidity, ecological stoichiometry and allometric scaling in grassland food webs. Glob. Change Biol. 15, 2730–2738.

Mulder, C., Vonk, J.A., 2011. Nematode traits and environmental constraints in 200 soil systems: scaling within the 60–6,000 μm body size range. Ecology 92, 2004 [Ecological Archives E092-171].

Mulder, C., Van Wijnen, H.J., Van Wezel, A.P., 2005a. Numerical abundance and biodiversity of below-ground taxocenes along a pH gradient across the Netherlands. J. Biogeogr. 32, 1775–1790.

Mulder, C., Cohen, J.E., Setälä, H., Bloem, J., Breure, A.M., 2005b. Bacterial traits, organism mass, and numerical abundance in the detrital soil food web of Dutch agricultural grasslands. Ecol. Lett. 8, 80–90.

Mulder, C., Den Hollander, H.A., Hendriks, A.J., 2008. Aboveground herbivory shapes the biomass distribution and flux of soil invertebrates. PLoS One 3, e3573.

Mulder, C., Den Hollander, H.A., Vonk, J.A., Rossberg, A.G., Jagers op Akkerhuis, G.A.J.M., Yeates, G.W., 2009. Soil resource supply influences faunal size-specific distributions in natural food webs. Naturwissenschaften 96, 813–826.

Mulder, C., Helder, J., Vervoort, M.T.W., Vonk, J.A., 2011a. Gender matters more than intraspecific trait variation. Ecol. Evol. 1, 386–391.

Mulder, C., Boit, A., Bonkowski, M., De Ruiter, P.C., Mancinelli, G., Van der Heijden, M.G.A., Van Wijnen, H.J., Vonk, J.A., Rutgers, M., 2011b. A belowground perspective on Dutch agroecosystems: how soil organisms interact to support ecosystem services. Adv. Ecol. Res. 44, 277–357.

Mulder, C., Vonk, J.A., Den Hollander, H.A., Hendriks, A.J., Breure, A.M., 2011c. How allometric scaling relates to soil abiotics. Oikos 120, 529–536.

Naeem, S., 2008. Advancing realism in biodiversity research. Trends Ecol. Evol. 23, 414–416.

Naeem, S., Chapin III, F.S., Costanza, R., Ehrlich, P., Golley, F., Hooper, D., Lawton, J.H., O'Neill, R., Mooney, H., Sala, O., Symstad, A., Tilman, D., 1999. Biodiversity and ecosystem functioning: maintaining natural life support processes. Issues Ecol. 4, 4–12.

Nakazawa, T., Ushio, M., Kondoh, M., 2011. Scale dependence of predator-prey mass ratio: determinants and applications. Adv. Ecol. Res. 45, 269–302.

O'Gorman, E.J., Emmerson, M.C., 2010. Manipulating interaction strengths and the consequences for trivariate patterns in a marine food web. Adv. Ecol. Res. 42, 301–419.

O'Gorman, E.J., Enright, R.A., Emmerson, M.C., 2008. Predator diversity enhances secondary production and decreases the likelihood of trophic cascades. Oecologia 158, 557–567.

OHIO-EPA, 2008. Biological and Water Quality Survey of the Scioto Brush Creek Basin. State of Ohio Environmental Protection Agency, Division of Surface Water, Columbus, OH. OEPA Report DSW/EAS 2008-4-6.

Olesen, J.M., Dupont, Y.L., O'Gorman, E., Ings, T.C., Layer, K., Melian, C.J., Troejelsgaard, K., Pichler, D.E., Rasmussen, C., Woodward, G., 2010. From Broadstone to Zackenberg: space, time and hierarchies in ecological networks. Adv. Ecol. Res. 42, 1–71.

Orth, R.J., Carruthers, T.J.B., Dennison, W.C., Duarte, C.M., Fourqurean, J.W., Heck, K.L., Hughes, A.R., Kendrick, G.A., Kenworthy, W.J., Olyarnik, S., Short, F.T., Waycott, M., et al., 2006. A global crisis for seagrass ecosystems. Bioscience 56, 987–996.

Perkins, D.M., McKie, B.G., Malmqvist, B., Gilmour, S.G., Reiss, J., Woodward, G., 2010. Environmental warming and biodiversity-ecosystem functioning in freshwater microcosms: partitioning the effects of species identity, richness and metabolism. Adv. Ecol. Res. 43, 177–209.

Perner, J., Voigt, W., 2007. Measuring the complexity of interaction webs using vertical links between functional groups. Agric. Ecosyst. Environ. 120, 192–200.

Perrings, C., Folke, C., Mäler, K.-G., 1992. The ecology and economics of biodiversity loss: the research agenda. Ambio 21, 201–211.

Perrings, C., Naeem, S., Ahrestani, F.S., Bunker, D.E., Burkill, P., Canziani, G., Elmqvist, T., Fuhrman, J.A., Jaksic, F.M., Kawabata, Z., Kinzig, A., Mace, G.M., et al., 2011. Ecosystem services, targets, and indicators for the conservation and sustainable use of biodiversity. Front. Ecol. Environ. 9, 512–520.

Peters, R.H., 1983. The Ecological Implications of Body Size. Cambridge University Press, Cambridge, UK.

Pfisterer, A.B., Schmid, B., 2002. Diversity-dependent production can decrease the stability of ecosystem functioning. Nature 416, 84–86.

Pimm, S.L., 1980. Food web design and the effect of species deletion. Oikos 35, 139–149.

Pimm, S.L., Lawton, J.H., Cohen, J.E., 1991. Food web patterns and their consequences. Nature 350, 669–674.

Polis, G.A., 1998. Stability is woven by complex webs. Nature 395, 744–745.

Polis, G.A., 1999. Why are parts of the world green? Multiple factors control productivity and the distribution of biomass. Oikos 86, 3–15.

Pope, J.G., Shepherd, J.G., Webb, J., 1994. Successful surf-riding on size spectra: the secret of survival in the sea. Philos. Trans. R. Soc. Lond. B 343, 41–49.

Pope, J.G., Rice, J.C., Daan, N., Jennings, S., Gislason, H., 2006. Modelling an exploited marine fish community with 15 parameters: results from a simple size-based model. ICES J. Mar. Sci. 63, 1029–1044.

Posthuma, L., De Zwart, D., 2006. Predicted effects of toxicant mixtures are confirmed by changes in fish species assemblages in Ohio, USA, rivers. Environ. Toxicol. Chem. 25, 1094–1105.

Prothero, J., 1986. Scaling of energy metabolism in unicellular organisms: a re-analysis. Comp. Biochem. Physiol. 83A, 243–248.

Ptacnik, R., Moorthi, S.D., Hillebrand, H., 2010. Hutchinson reversed, or why there need to be so many species. Adv. Ecol. Res. 43, 1–43.

Purdy, K.J., Hurd, P.J., Moya-Larano, J., Trimmer, M., Woodward, G., 2010. Systems biology for ecology: from molecules to ecosystems. Adv. Ecol. Res. 43, 87–149.
Quince, C., Higgs, P.G., McKane, A.J., 2005. Deleting species from model food webs. Oikos 110, 283–296.
Raunkiaer, C., 1934. The Life Forms of Plants and Statistical Plant Geography. Oxford University Press, Oxford.
Redfield, A.C., 1958. The biological control of chemical factors in the environment. Am. Sci. 46, 205–221.
Reich, P.B., 2000. Do tall trees scale physiological heights? Trends Ecol. Evol. 15, 41–42.
Reich, P.B., Tjoelker, M.G., Machado, J.-L., Oleksyn, J., 2006. Universal scaling of respiratory metabolism, size and nitrogen in plants. Nature 439, 457–461.
Reich, P.B., Oleksyn, J., Wright, I.J., Niklas, K.J., Hedin, L., Elser, J.J., 2010. Evidence of a general 2/3-power law of scaling leaf nitrogen to phosphorus among major plant groups and biomes. Proc. R. Soc. Lond. B 277, 877–883.
Reiss, J., Bridle, J., Montoya, J.M., Woodward, G., 2009. Emerging horizons in biodiversity and ecosystem functioning research. Trends Ecol. Evol. 24, 505–514.
Reiss, J., Bailey, R.A., Cássio, F., Woodward, G., Pascoal, C., 2010. Assessing the contribution of micro-organisms and macrofauna to biodiversity-ecosystem functioning relationships in freshwater microcosms. Adv. Ecol. Res. 43, 150–176.
Reiss, J., Bailey, R.A., Perkins, D.M., Pluchinotta, A., Woodward, G., 2011. Testing effects of consumer richness, evenness and body size on ecosystem functioning. J. Anim. Ecol. 80, 1145–1154.
Reuman, D.C., Mulder, C., Raffaelli, D., Cohen, J.E., 2008. Three allometric relations of population density to body mass: theoretical integration and empirical tests in 149 food webs. Ecol. Lett. 11, 1216–1228.
Reuman, D.C., Mulder, C., Banašek-Richter, C., Cattin Blandenier, M.-F., Breure, A.M., Den Hollander, H., Kneitel, J.M., Raffaelli, D., Woodward, G., Cohen, J.E., 2009. Allometry of body size and abundance in 166 food webs. Adv. Ecol. Res. 41, 1–44.
Richardson, A.E., Barea, J.M., McNeill, A.M., Prigent-Combaret, C., 2009. Acquisition of phosphorus and nitrogen in the rhizosphere and plant growth promotion by microorganisms. Plant Soil 321, 305–339.
Romanuk, T.N., Zhou, Y., Brose, U., Berlow, E.L., Williams, R.J., Martinez, N.D., 2009. Predicting invasion success in complex ecological networks. Philos. Trans. R. Soc. Lond. B 364, 1743–1754.
Roscher, C., Thein, S., Schmid, B., Scherer-Lorenzen, M., 2008. Complementary nitrogen use among potentially dominant species in a biodiversity experiment varies between two years. J. Ecol. 96, 477–488.
Rossberg, A.G., Ishii, R., Amemiya, T., Itoh, K., 2008. The top-down mechanism for body-mass-abundance scaling. Ecology 89, 567–580.
Rossberg, A.G., Farnsworth, K.D., Satoh, K., Pinnegar, J.K., 2011. Universal power-law diet partitioning by marine fish and squid with surprising stability—diversity implications. Proc. R. Soc. Lond. B 278, 1617–1625.
Sala, O.E., Chapin III, F.S., Armesto, J.J., Berlow, E., Bloomfield, J., Dirzo, R., Huber-Sanwald, E., Huenneke, L.F., Jackson, R.B., Kinzig, A., Leemans, R., Lodge, D.M., et al., 2000. Global biodiversity scenarios for the year 2100. Science 287, 1770–1774.
Samanta, A., Ganguly, S., Hashimoto, H., Devadiga, S., Vermote, E., Knyazikhin, Y., Nemami, R.R., Myneni, R.B., 2010. Amazon forests did not green-up during the 2005 drought. Geophys. Res. Lett. 37, L05401.
Santos, P.F., Phillips, J., Whitford, W.G., 1981. The role of mites and nematodes in early stages of buried litter decomposition in a desert. Ecology 62, 664–669.

Savage, V.M., Gillooly, J.F., Woodruff, W.H., West, G.B., Allen, A.P., Enquist, B.J., Brown, J.H., 2004. The predominance of quarter-power scaling in biology. Funct. Ecol. 18, 257–282.

Schmid, B., Balvanera, P., Cardinale, B.J., Godbold, J., Pfisterer, A.B., Raffaelli, D., Solan, M., Srivastava, D.S., 2009. Consequences of species loss for ecosystem functioning: meta-analyses of data from biodiversity experiments. In: Naeem, S., Bunker, D.E., Hector, A., Loreau, M., Perrings, C. (Eds.), Biodiversity, Ecosystem Functioning, and Human Wellbeing: An Ecological and Economic Perspective. Oxford University Press, Oxford, pp. 14–29.

Schneider, K., Kay, A.D., Fagan, W.F., 2010. Adaptation to a limiting environment: the phosphorus content of terrestrial cave arthropods. Ecol. Res. 25, 565–577.

Schneider, F.D., Scheu, S., Brose, U., 2012. Body mass constraints on feeding rates determine the consequences of predator loss. Ecol. Lett. 15, 436–443.

Shin, Y.-J., Rochet, M.-J., Jennings, S., Field, J.G., Gislason, H., 2005. Using size-based indicators to evaluate the ecosystem effects of fishing. ICES J. Mar. Sci. 62, 384–396.

Short, F.T., Duarte, C.M., 2001. Methods for the measurement of seagrass growth and production. In: Short, F.T., Coles, R.G. (Eds.), Global Seagrass Research Methods. Elsevier, Amsterdam, pp. 155–182.

Simon, T.P. (Ed.), 1999. Assessing the Sustainability and Biological Integrity of Water Resources Using Fish Communities. CRC Press, Boca Raton, FL.

Solé, R.V., Montoya, J.M., 2006. Ecological network meltdown from habitat loss and fragmentation. In: Pascual, M., Dunne, J.A. (Eds.), Ecological Networks: Linking Structure to Dynamics in Food Webs. Oxford University Press, New York, pp. 305–323.

Solé, R.V., Manrubia, S.C., Benton, M., Kauffman, S., Bak, P., 1999. Criticality and scaling in evolutionary ecology. Trends Ecol. Evol. 14, 156–160.

Srinivasan, U.T., Dunne, J.A., Harte, J., Martinez, N.D., 2007. Response of complex food webs to realistic extinction sequences. Ecology 88, 671–682.

Srivastava, D.S., Lawton, J.H., 1998. Why more productive sites have more species: an experimental test of theory using tree-hole communities. Am. Nat. 152, 510–529.

Stachowicz, J.J., Best, R.J., Bracken, M.E., Graham, M.H., 2008. Complementarity in marine biodiversity manipulations: reconciling divergent evidence from field and mesocosm experiments. Proc. Natl. Acad. Sci. U.S.A. 105, 18842–18847.

Standen, V., 1980. Factors affecting the distribution of Enchytraeidae (Oligochaeta) in associations at peat and mineral sites in northern England. Bull. Ecol. 11, 599–608.

Standen, V., 1982. Associations of Enchytraeidae (Oligochaeta) in experimentally fertilized grasslands. J. Anim. Ecol. 51, 501–522.

Sterner, R.W., 2004. A one-resource "stoichiometry"? Ecology 85, 1813–1816.

Sterner, R.W., Elser, J.J., 2002. Ecological Stoichiometry: The Biology of Elements from Molecules to the Biosphere. Princeton University Press, Princeton, NJ.

Sterner, R.W., George, N.B., 2000. Carbon, nitrogen and phosphorus stoichiometry of cyprinid fishes. Ecology 81, 127–140.

Storch, D., Marquet, P.A., Brown, J.H. (Eds.), 2007. Scaling Biodiversity. Cambridge University Press, Cambridge, UK.

Suding, K.N., Goldberg, D.E., Hartman, K.M., 2003. Relationship among species traits: separating levels of response and indentifying linkages to abundance. Ecology 84, 1–16.

Sultan, S.E., Spencer, H.G., 2002. Metapopulation structure favors plasticity over local adaptation. Am. Nat. 160, 271–283.

Thingstad, T.F., 2000. Elements of a theory for the mechanisms controlling abundance, diversity, and biogeochemical role of lytic bacterial viruses in aquatic systems. Limnol. Oceanogr. 45, 1320–1328.

Thomas, J.A., Telfer, M.G., Roy, D.B., Preston, C.D., Greenwood, J.J.D., Asher, J., Fox, R., Clarke, R.T., Lawton, J.H., 2004a. Comparative losses of British butterflies, birds, and plants and the Global Extinction Crisis. Science 303, 1879–1881.

Thomas, C.D., Cameron, A., Green, R.E., Bakkenes, M., Beaumont, L.J., Collingham, Y.C., Erasmus, B.F.N., De Siqiuera, M.F., Grainger, A., Hannah, L., Hughes, L., Huntley, B., et al., 2004b. Extinction risk from climate change. Nature 427, 145–148.

Thornley, J.H.M., 2011. Plant growth and respiration re-visited: maintenance respiration defined—it is an emergent property of, not a separate process within, the system—and why the respiration: photosynthesis ratio is conservative. Ann. Bot. 108, 1365–1380.

Tilman, D., 1988. Plant Strategies and the Dynamics and Structure of Plant Communities. Princeton University Press, Princeton, NJ.

Tilman, D., HilleRisLambers, J., Harpole, S., Dybzinski, R., Fargione, J., Clark, C., Lehman, C., 2004. Does metabolic theory apply to community ecology? It's a matter of scale. Ecology 85, 1797–1799.

Toepfer, C.S., Williams, L.R., Martinez, A.D., Fisher, W.L., 1998. Fish and habitat heterogeneity in four streams in the central Oklahoma/Texas Plains Ecoregion. Proc. Okla. Acad. Sci. 78, 41–48.

Tomas, F., Abbott, J.M., Steinberg, C., Balk, M., Williams, S.L., Stachowicz, J.J., 2011. Plant genotype and nitrogen loading influence seagrass productivity, biochemistry, and plant-herbivore interactions. Ecology 92, 1807–1817.

Treonis, A.M., Wall, D.H., Virginia, R.A., 1999. Invertebrate biodiversity in Antarctic Dry Valley soils and sediments. Ecosystems 2, 482–492.

Treonis, A.M., Wall, D.H., Virginia, R.A., 2000. The use of anhydrobiosis by soil nematodes in the Antarctic Dry Valleys. Funct. Ecol. 14, 460–467.

Ulrich, W., Buszko, J., Czarnecki, A., 2005. The local interspecific abundance—body weight relationship of ground beetles: a counterexample to the common pattern. Pol. J. Ecol. 53, 585–589.

UNEP/CBD, 2004. Convention on Biological Diversity: Decisions Adopted by the Conference of the Parties to the Convention of Biological Diversity at Its Seventh Meeting. Kuala Lumpur, Malaysia. http://www.cbd.int/decisions/.

Valcu, M., Dale, J., Kempenaers, B., 2012. rangeMapper: a platform for the study of macroecology of life-history traits. Glob. Ecol. Biogeogr. 21, 945–951.

Voigt, W., Perner, J., Jones, T.H., 2007. Using functional groups to investigate community response to environmental changes: two grassland case studies. Glob. Change Biol. 13, 1710–1721.

Vonk, J.A., Stapel, J., 2008. Regeneration of nitrogen (^{15}N) from seagrass litter in tropical Indo-Pacific meadows. Mar. Ecol. Prog. Ser. 368, 165–175.

Vonk, J.A., Pijnappels, M.H.J., Stapel, J., 2008. *In situ* quantification of *Tripneustes gratilla* grazing and its effects on three co-occurring tropical seagrass species. Mar. Ecol. Prog. Ser. 360, 107–114.

Wagner, T., Bremigan, M.T., Cheruvelil, K.S., Soranno, P.A., Nate, N.A., Breck, J.E., 2007. A multilevel modeling approach to assessing regional and local landscape features for lake classification and assessment of fish growth rates. Environ. Monit. Assess. 130, 437–454.

Waide, R.B., Willig, M.R., Steiner, C.F., Mittelbach, G., Gough, L., Dodson, S.I., Juday, G.P., Parmenter, R., 1999. The relationship between productivity and species richness. Annu. Rev. Ecol. Syst. 30, 257–300.

Walker, G., Dorrell, R.G., Schlacht, A., Dacks, J.B., 2011. Eukaryotic systematics: a user's guide for cell biologists and parasitologists. Parasitology 138, 1638–1663.

Wall, D.H., 2007. Global change tipping points: above- and below-ground biotic interactions in a low diversity ecosystem. Philos. Trans. R. Soc. Lond. B 362, 2291–2306.

Wall, D.H., 2008. Biodiversity: extracting lessons from extreme soils. Soil Biol. 13, 71–84.

Wall, D.H., Bradford, M.A., St. John, M.G., Trofymow, J.A., Behan-Pelletier, V., Bignell, D.E., Dangerfield, J.M., Parton, W.J., Rusek, J., Voigt, W., Wolters, V., Gardel, H.Z., et al., 2008. Global decomposition experiment shows soil animal impacts on decomposition are climate-dependent. Glob. Change Biol. 14, 2661–2677.

Walter, H., 1964. Die Vegetation der Erde in Öko-physiologischer Betrachtung. Gustav Fischer Verlag, Jena.

Walters, C.J., Maguire, J.J., 1996. Lessons for stock assessment from the northern cod collapse. Rev. Fish Biol. Fish. 6, 125–137.

Wardle, D.A., 1999. How soil food webs make plants grow. Trends Ecol. Evol. 14, 418–420.

Wardle, D.A., 2002. Communities and Ecosystems: Linking the Aboveground and Belowground Components. Princeton University Press, Princeton, NJ.

Wardle, D.A., 2006. The influence of biotic interactions on soil biodiversity. Ecol. Lett. 9, 870–886.

Wardle, D.A., Bardgett, R.D., Klironomos, J.N., Setälä, H., Van der Putten, W.H., Wall, D.H., 2004. Ecological linkages between aboveground and belowground biota. Science 304, 1629–1633.

Wardle, D.A., Hyodo, F., Bardgett, R.D., Yeates, G.W., Nilsson, M.-C., 2011. Long-term aboveground and belowground consequences of red wood ant exclusion in boreal forest. Ecology 92, 645–656.

Watkinson, A.R., Freckleton, R.P., 1997. Quantifying the impact of arbuscular mycorrhiza on plant competition. J. Ecol. 85, 541–546.

Waycott, M., Duarte, C.M., Carruthers, T.J.B., Orth, R.J., Dennison, W.C., Olyarnik, S., Calladine, A., Fourqurean, J.W., Heck Jr., K.L., Hughes, A.R., Kendrick, G.A., Kenworthy, W.J., et al., 2009. Accelerating loss of seagrasses across the globe threatens coastal ecosystems. Proc. Natl. Acad. Sci. U.S.A. 106, 12377–12381.

Weiser, M.D., Sanders, N.J., Agosti, D., Andersen, A.N., Ellison, A.M., Fisher, B.L., Gibb, H., Gotelli, N.J., Gove, A.D., Gross, K., Guénard, B., Janda, M., et al., 2011. Canopy and litter ant assemblages share similar climate–species density relationships. Biol. Lett. 6, 769–772.

Werner, E.E., Gilliam, J.F., 1984. The ontogenetic niche and species interactions in size-structured populations. Annu. Rev. Ecol. Syst. 15, 393–425.

West, G.B., Brown, J.H., 2004. Life's universal scaling laws. Phys. Today 57, 36–43.

White, C.R., 2010. There is no single *p*. Nature 464, 691–692.

White, E.P., Ernest, S.K.M., Kerkhoff, A.J., Enquist, B.J., 2007. Relationships between body size and abundance in ecology. Trends Ecol. Evol. 22, 323–330.

White, C., Kendall, B.E., Gaines, S., Siegel, D.A., Costello, C., 2008. Marine reserve effects on fishery profit. Ecol. Lett. 11, 370–379.

Williams, R.J., Martinez, N.D., 2004. Limits to trophic levels and omnivory in complex food webs: theory and data. Am. Nat. 163, 458–468.

Winemiller, K.O., 1989. Must connectance decrease with species richness? Am. Nat. 134, 960–968.

Woodward, G., Hildrew, A.G., 2002. The impact of a sit-and-wait predator: separating consumption and prey emigration. Oikos 99, 409–418.

Woodward, G., Ebenman, B., Emmerson, M., Montoya, J.M., Olesen, J.M., Valido, A., Warren, P.H., 2005a. Body size in ecological networks. Trends Ecol. Evol. 20, 402–409.

Woodward, G., Speirs, D.C., Hildrew, A.G., 2005b. Quantification and resolution of a complex, size-structured food web. Adv. Ecol. Res. 36, 85–135.

Woodward, G., Blanchard, J., Lauridsen, R.B., Edwards, F.K., Jones, J.I., Figueroa, D., Warren, P.H., Petchey, O.L., 2010a. Individual-based food webs: species identity, body size and sampling effects. Adv. Ecol. Res. 43, 211–266.

Woodward, G., Perkins, D.M., Brown, L.E., 2010b. Climate change and freshwater ecosystems: impacts across multiple levels of organization. Philos. Trans. R. Soc. Lond. B 365, 2093–2106.

Woodward, G., Gessner, M.O., Giller, P.S., Gulis, V., Hladyz, S., Lecerf, A., Malmqvist, B., McKie, B.G., Tiegs, S.D., Cariss, H., Dobson, M., Elosegi, A., et al., 2012. Continental-scale effects of nutrient pollution on stream ecosystem functioning. Science 336, 1438–1440.

Wu, T., Ayres, E., Bardgett, R.D., Wall, D.H., Garey, J.R., 2011. Molecular study of worldwide distribution and diversity of soil animals. Proc. Natl. Acad. Sci. U.S.A. 108, 17720–17725.

Xu, L., Samanta, A., Costa, M.H., Ganguly, S., Nemani, R.R., Myneni, R.B., 2011. Widespread decline in greenness of Amazonian vegetation due to the 2010 drought. Geophys. Res. Lett. 38, L07402.

Yee, D.A., Juliano, S.A., 2007. Abundance matters: a field experiment testing the More Individuals Hypothesis for richness–productivity relationships. Oecologia 153, 153–162.

Yodzis, P., Innes, S., 1992. Body size and consumer-resource dynamics. Am. Nat. 139, 1151–1175.

Yoneyama, K., Xie, X., Kusumoto, D., Sekimoto, H., Sugimoto, Y., Takeuchi, Y., 2007. Nitrogen deficiency as well as phosphorus deficiency in sorghum promotes the production and exudation of 5-deoxystrigol, the host recognition signal for arbuscular mycorrhizal fungi and root parasites. Planta 227, 125–132.

Yoon, I., Williams, R.J., Levine, E., Yoon, S., Dunne, J.A., Martinez, N.D., 2004. Webs on the Web (WOW): 3D visualization of ecological networks on the WWW for collaborative research and education. SPIE 5295, 124–132.

Yvon-Durocher, G., Allen, A.P., Montoya, J.M., Trimmer, M., Woodward, G., 2010. The temperature dependence of the carbon cycle in aquatic systems. Adv. Ecol. Res. 43, 267–313.

Yvon-Durocher, G., Montoya, J.M., Trimmer, M., Woodward, G., 2011a. Warming alters the size spectrum and shifts the distribution of biomass in freshwater ecosystems. Glob. Change Biol. 17, 1681–1694.

Yvon-Durocher, G., Reiss, J., Blanchard, J., Ebenman, B., Perkins, D.M., Reuman, D.C., Thierry, A., Woodward, G., Petchey, O.L., 2011b. Across ecosystem comparisons of size structure: methods, approaches, and prospects. Oikos 120, 550–563.

Yvon-Durocher, G., Caffrey, J.M., Cescatti, A., Dossena, M., Del Giorgio, P., Gasol, J.M., Montoya, J.M., Pumpanen, J., Staehr, P.A., Trimmer, M., Woodward, G., Allen, A.P., 2012. Reconciling the temperature dependence of respiration across timescales and ecosystem types. Nature 487, 472–476.

Zobel, M., 1997. The relative role of species pools in determining plant species richness: an alternative explanation of species coexistence? Trends Ecol. Evol. 12, 266–269.

Biodiversity, Species Interactions and Ecological Networks in a Fragmented World

Melanie Hagen*, W. Daniel Kissling*, Claus Rasmussen*, Marcus A.M. De Aguiar[†], Lee E. Brown[‡], Daniel W. Carstensen*, Isabel Alves-Dos-Santos[§], Yoko L. Dupont*, Francois K. Edwards[¶], Julieta Genini[||], Paulo R. Guimarães Jr.[§], Gareth B. Jenkins[#], Pedro Jordano, Christopher N. Kaiser-Bunbury*, Mark E. Ledger[††], Kate P. Maia[§], Flavia M. Darcie Marquitti[§], Órla Mclaughlin[‡‡,§§], L. Patricia C. Morellato[||], Eoin J. O'Gorman[#], Kristian Trøjelsgaard*, Jason M. Tylianakis[¶¶], Mariana Morais Vidal[§], Guy Woodward[#], Jens M. Olesen*[,1]**

*Department of Bioscience, Aarhus University, Aarhus, Denmark
[†]Instituto de Física Gleb Wataghin, Universidade Estadual de Campinas, Campinas, São Paulo, Brazil
[‡]School of Geography, University of Leeds, Leeds, United Kingdom
[§]Departamento de Ecologia, Instituto de Biociências, Universidade de São Paulo, São Paulo, São Paulo, Brazil
[¶]Centre for Ecology and Hydrology, Wallingford, United Kingdom
[||]Departamento de Botânica, Laboratório de Fenologia, UNESP Univ Estadual Paulista, Rio Claro, São Paulo, Brazil
[#]School of Biological and Chemical Sciences, Queen Mary University of London, London, United Kingdom
**Integrative Ecology Group, Estación Biológica de Doñana, CSIC, Sevilla, Spain
[††]School of Geography, Earth and Environmental Sciences, University of Birmingham, Edgbaston, Birmingham, United Kingdom
[‡‡]Environmental Research Institute, University College Cork, Cork, Ireland
[§§]School of Biological, Earth, and Environmental Sciences, University College Cork, Cork, Ireland
[¶¶]School of Biological Sciences, University of Canterbury, Christchurch, New Zealand
[1]Corresponding author: e-mail address: jens.olesen@biology.au.dk

Contents

Advances in Ecological Research, Volume 46
ISSN 0065-2504
http://dx.doi.org/10.1016/B978-0-12-396992-7.00002-2

Abstract

Biodiversity is organised into complex ecological networks of interacting species in local ecosystems, but our knowledge about the effects of habitat fragmentation on such systems remains limited. We consider the effects of this key driver of both local and global change on both mutualistic and antagonistic systems at different levels of biological organisation and spatiotemporal scales.

There is a complex interplay of patterns and processes related to the variation and influence of spatial, temporal and biotic drivers in ecological networks. Species traits (e.g. body size, dispersal ability) play an important role in determining how networks respond to fragment size and isolation, edge shape and permeability, and the quality of the surrounding landscape matrix. Furthermore, the perception of spatial scale (e.g. environmental grain) and temporal effects (time lags, extinction debts) can differ markedly among species, network modules and trophic levels, highlighting the need to develop a more integrated perspective that considers not just nodes, but the structural role and strength of species interactions (e.g. as hubs, spatial couplers and determinants of connectance, nestedness and modularity) in response to habitat fragmentation.

Many challenges remain for improving our understanding: the likely importance of specialisation, functional redundancy and trait matching has been largely overlooked. The potentially critical effects of apex consumers, abundant species and super-generalists on network changes and evolutionary dynamics also need to be addressed in future research. Ultimately, spatial and ecological networks need to be combined to explore the effects of dispersal, colonisation, extinction and habitat fragmentation on network structure and coevolutionary dynamics. Finally, we need to embed network approaches more explicitly within applied ecology in general, because they offer great potential for improving on the current species-based or habitat-centric approaches to our management and conservation of biodiversity in the face of environmental change.

1. INTRODUCTION

The planet's ecosystems are losing biodiversity at an accelerating rate (Dyer et al., 2010; Fahrig, 2003; Gonzalez et al., 2011; Millennium Ecosystem Assessment, 2005) due to land-use change, deforestation, agricultural intensification, pollution, urbanisation, climate change and habitat fragmentation (Albrecht et al., 2007; Hanski, 2005; Ledger et al., 2012; Meerhoff et al., 2012; Mintenbeck et al., 2012; Tilman et al., 2001). The latter in particular could severely disrupt ecological networks and the goods and services they provide (e.g. pollination in mutualistic webs or biological control in food webs) as it is a rapidly growing phenomenon throughout the world, yet its impacts on the higher multispecies levels of organisation are still poorly understood.

A major challenge for predicting the consequences of changes on biodiversity is to understand the complexity of natural systems and the steps needed to conserve them in a rapidly changing world. Biodiversity is organised at local scales into complex networks of interacting species, which provide the ecosystem processes that ultimately underpin the goods and services of value to human societies (Rossberg, 2012). These *links* (italicised terms, see Glossary) among interacting species are often ignored in the context of global change even though they will disappear from local communities as a precursor to local (and ultimately global) extinctions (Albrecht et al., 2007; Fortuna and Bascompte, 2006; Sabatino et al., 2010; Tylianakis et al., 2007; Woodward et al., 2010a). Understanding the causes and consequences of the loss of species interactions therefore promises to provide critical new insights into ecological responses to perturbations (Mulder et al., 2012; Tylianakis et al., 2010).

The interplay between the abiotic environment and biotic *complexity* over space and time makes natural ecosystems seemingly difficult to understand. One simplifying approach is to study interactions among multiple species in the framework of ecological networks (e.g. Fortuna and Bascompte, 2008). These include both mutualistic (e.g. pollination, seed dispersal networks) and antagonistic (e.g. food webs, *host–parasitoid networks*) interactions, which could respond differently to disturbances, such as fragmentation, which in turn determines their stability in terms of resilience, resistance and robustness (Ings et al., 2009; Layer et al., 2010, 2011; Woodward et al., 2010a).

Landscape changes may be caused by physical processes, biotic drivers such as ecological engineers, and/or anthropogenic influences. Species will reshuffle their population sizes and some links between species might be rewired or break apart entirely (Tscharntke et al., 2005). Any seemingly restricted spatiotemporal disturbance may ripple throughout the network of interacting species, causing further (i.e. secondary) species and link perturbations. New data analytical tools, such as network analysis, now form an essential ingredient in the study of complex systems, with clear implications for biodiversity research (Heleno et al., 2009; Kremen and Hall, 2005; Tylianakis et al., 2008).

Habitat fragmentation is almost ubiquitous in both natural and human-modified landscapes (Fig. 1), with consequences for biodiversity and species interactions (Fahrig, 2003; Laurance et al., 2011; Tylianakis et al., 2007), which in turn has implications for the entire ecological network. It reduces habitat area and species connectivity, and the sizes and isolation of remaining *fragments* are particularly critical to the long-term conservation of biodiversity. Connectivity among fragments, the characteristics of the matrix, the availability of *corridors* for movement between fragments, and the permeability and structure of habitat edges are all important in this context and affect the structure, persistence and strength of species interactions (Fortuna and Bascompte, 2006). Certain species traits (e.g. body size, dispersal ability, degree of specialisation or trophic rank) are likely to be particularly crucial for assessing the higher-level consequences

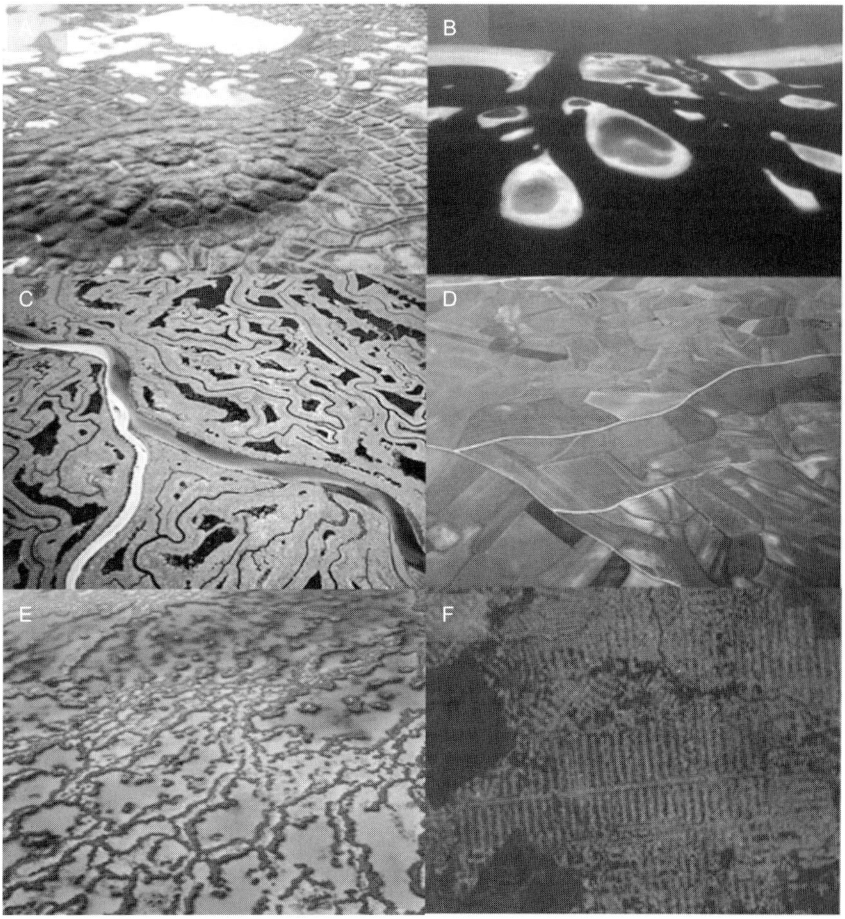

Figure 1 The ubiquity of fragmentation. Selected examples of common naturally and artificially fragmented habitats from terrestrial and aquatic ecosystems, with hard (aquatic–terrestrial) versus soft (aquatic–aquatic, terrestrial–terrestrial) boundaries. From top left to bottom right are (A) pingos in the arctic; (B) tropical atoll islands; (C) temperate river network and associated off-river habitats; (D) agricultural landscape in Spain; (E) a portion of the Great Barrier Reef and (F) forest clearance in Amazonia. (For colour version of this figure, the reader is referred to the online version of this chapter.)

of habitat fragmentation (Ewers and Didham, 2006), so functional attributes may be just as important as taxonomic diversity in this context. The invasion of functionally similar species, for example, may homogenise ecological processes (McKinney and Lockwood, 1999; Olden et al., 2004). Species at higher trophic levels, or with particular traits, that connect different fragments or network modules, may act as important spatial couplers or network stabilisers, essentially operating as network-level keystones.

Both the physical and biological worlds can be seen as networks (Gonzalez et al., 2011): a (spatial) landscape network of habitat fragments that provides the underlying matrix and habitat connectivity, and an ecological species interaction network, driven by ecological and evolutionary processes. Interactions between such different kinds of networks occur, but to date such multiple interdependent networks have mainly been studied outside ecology (Buldyrev et al., 2010) and the consequences of habitat fragmentation on these (often interdependent) biological–physical systems remain largely unexplored.

Here, we synthesise current knowledge about the consequences of habitat fragmentation on different types of biodiversity within ecological networks. We begin by introducing the major characteristics and types of ecological and *spatial networks*. We then review the spatial and temporal settings of habitat fragmentation, including fragment characteristics, habitat edges, matrix quality and permeability, spatial and temporal turnover of species and individuals, and different scales of fragmentation. We illustrate how habitat fragmentation effects depend on species traits, paying particular attention to both mutualistic (plant–pollinator, plant–frugivore, plant–ant) and antagonistic (host–parasitoid, food web) interactions, and we synthesise current knowledge on likely consequences for ecological networks and make suggestions about future research directions. Finally, we summarise possible applications for conservation, agriculture and applied ecology in general. Throughout the paper, we consider different kinds of interactions and networks across a range of spatiotemporal scales.

2. NETWORKS

2.1. Ecological networks

Networks contain nodes and their links: in ecology, *nodes* may be individuals, species populations, species, guilds, *functional groups* (e.g. body-size groups), entire communities, or even entire networks, and interactions can take many forms (e.g. plant–pollinator, plant–frugivore and predator–prey associations (Fig. 2)).

Links in an ecological network are defined in an interaction matrix. The coarsest measure of link strength is simply the occurrence (presence/absence

Figure 2 Examples of biotic interactions. (A) Carpenter bee (*Xylocopa flavorufa*) pollinating cowpea (*Vicia unguiculata*) in Western Kenya (photo: M. Hagen). (B) Sunbird (*Cinnyris jugularis*) pollinating palm inflorescences in Flores, Indonesia (photo: J. M. Olesen). (C) Day Gecko (*Phelsuma ornata*) pollinating *Gastonia mauritiana* in Mauritius (photo: C. Kaiser-Bunbury). (D) Long-tailed Macaque (*Macaca fascicularis*) consuming figs on Lombok, Indonesia (photo: J. M. Olesen). (E) Green Imperial Pigeon (*Ducula aenea*) consuming fruits of a palm (*Corypha taliera*) in Komodo, Indonesia (photo: J. M. Olesen). (F) Seed dispersal of *Casearia coriacea* by ants in Le Pétrin, Mauritius (photo: C. Kaiser-Bunbury). (G) Great Lizard Cuckoo (*Coccyzus merlini*) predating a snake in Cuba (photo: J. M. Olesen). (H) African lion (*Panthera leo*) 'resting' after a biotic interaction in Masai Mara, Kenya (photo: W. D. Kissling). (I) Crab spider predating a bumblebee (*Bombus* cf. *pascuorum*) in Liguria, Northern Italy (photo: C. Kaiser-Bunbury). (For colour version of this figure, the reader is referred to the online version of this chapter.)

data), within qualitative networks, although it can be measured in many ways (Berlow et al., 2004). For instance, for a plant–pollinator network, the links may represent the number of visitors to a plant, number of visits, number of pollen grains transferred to the stigma or number of pollen grains siring seeds, seedlings or reproductive individuals. For food webs, numerous measures and definitions have been described (see review by Berlow et al., 2004), whereas in *mutualistic networks* the interaction frequency is the norm (Vázquez et al., 2005). Both qualitative and quantitative interaction parameters allow not only the description of local community-level interactions, but also the modelling of multispecies interactions across larger scales (Kissling et al., 2012a).

Mutualistic and *antagonistic networks* represent the two main groups encountered in the ecological literature, and each has its own historical tradition (Olesen et al., 2012). Thus, antagonistic networks include 'traditional food webs' (typically larger consumers kill and eat many individual prey; e.g. Jacob et al., 2011; Layer et al., 2010, 2011; McLaughlin et al., 2010; O'Gorman et al., 2010), host–parasitoid networks (e.g. Henri and van Veen, 2011; Tylianakis et al., 2007), as well as less-familiar host–parasite or pathogen networks (e.g. Lafferty et al., 2008). Mutualistic networks include plant–flower visitor/pollinator (e.g. Memmott, 1999) and plant–frugivore/seed disperser networks (e.g. Donatti et al., 2011; Schleuning et al., 2011a), with less familiar forms including plant–ant networks (Guimarães et al., 2007) and host–symbiont interactions (e.g. gut microbiomes; Purdy et al., 2010). These categories are not exhaustive, but they represent main foci of current ecological network research (Ings et al., 2009). No doubt new forms of networks will appear as this rapidly growing research field expands its horizons further: for instance, interspecific competition within trophic levels has been largely ignored to date, except in the context of trophic niche partitioning within food webs, but such networks may become important, especially in the context of habitat fragmentation, where space rather than food may be limiting.

Food webs are traditionally divided into aquatic (freshwater and marine) and terrestrial (aboveground and belowground) systems, although some of the oldest food web studies included several habitats (e.g. Pimm and Lawton, 1980). These early ideas are now being revisited increasingly, with a focus upon 'spatial couplers', such as allochthonous inputs at the base of the food web, migratory top predators that link different local webs or species that have both an aquatic and terrestrial life history (Jonsson et al., 2005;

Layer et al., 2010; McCann et al., 2005a,b; O'Gorman and Emmerson, 2010; Woodward et al., 2005).

Mutualistic and antagonistic webs are inherently difficult to compare directly (e.g. in their responses to fragmentation) because they differ in their structure, dynamics and link type. The former are bipartite or *bimodal*, that is, consisting of two interacting sets of taxa, whereas the latter are multi-modal, that is, containing multiple trophic levels (e.g. producer–herbivore–predator). One way to approach this might be to slice food webs up according to pairs of interacting trophic levels into a series of bimodal networks, that is, plant–herbivore, herbivore–predator and so on. Alternatively, mutualistic networks, such as plant–pollinator networks, could be merged with other *bimodal networks*, for example, those of plant–herbivore or plant–fungi networks, to create networks of several interacting groups (see Fontaine et al., 2011; L. Kromann-Gallop, personal communication). Until such an analysis is made, it remains difficult to compare the properties of different kinds of networks directly (but see Olesen et al., 2006), although such comparisons are theoretically possible (Thebault and Fontaine, 2010), and we therefore address both types as separate cases throughout the paper.

2.1.1 Properties of mutualistic and antagonistic networks

Common measures of network structure include species and link numbers, connectance, and *linkage level distribution*, many of which are important because they make implicit connections between network complexity, stability and resource partitioning in ecology (Berlow et al., 2009; Elton, 1927; MacArthur, 1955; May, 1972, 1973; McCann et al., 1998; Warren, 1996; Williams and Martinez, 2000). These measures and their significance in networks have been discussed extensively elsewhere (Berlow et al., 2004; Ings et al., 2009; Olesen et al., 2010b), so we will not cover them in detail here. Instead, we provide a brief overview of the main concepts, with a specific focus on habitat fragmentation.

Networks also display recognisable substructural patterns, often in a fractal-like manner, such that they may contain repeating motifs, modules or *compartments* within the wider web (e.g. Olesen et al., 2007; Stouffer and Bascompte, 2010). For example, food webs can be decomposed into food chains, tritrophic chains and ultimately their pairwise individual feeding links, each of which may display its own response to habitat fragmentation (Woodward et al., 2012). These have received less attention than the whole-network measures of complexity (e.g. connectance), but in recent years considerable advances have been made, especially in the study of mutualistic webs. Substructures could be especially important in the context of habitat fragmentation, as they may represent some form of 'network fragmentation' related to spatial *compartmentalisation*. For instance, connector species that link modules might be species with large space requirements or long dispersal distances, that join otherwise spatially distinct subwebs. The same principles may apply through time: for instance, top predators move not only over wide distances but also tend to be relatively long-lived, linking seasonally or spatiotemporally fragmented subwebs together (Woodward and Hildrew, 2002a).

The two most common forms of network (sub)structure, nestedness and modularity, have been studied intensively (Bascompte et al., 2003; Lewinsohn et al., 2006; Olesen et al., 2007; Pimm, 1984). In a nested network, the links of *specialist* species are well-defined subsets of the links of *generalists* (Bascompte et al., 2003). Modularity describes subsets of species (modules) that are internally highly connected, but poorly connected to other such subsets of species (Olesen et al., 2007). Nestedness and modularity have often been regarded as mutually exclusive (Lewinsohn et al., 2005), but this is not necessarily true (Fortuna et al., 2010; Olesen et al., 2007). Link patterns in bimodal networks vary with presence of links and the frequency or intimacy of interactions between partners (Olesen et al., 2008). If link presence and intimacy are short and weak, the network may become nested and modular, such as in pollination and frugivory/seed dispersal networks, but if prolonged and tight, nestedness may be lost although modularity might be retained, such as in host–parasitoid and plant–ant *domatia* networks. Generalists and common species may be lost or 'forced' over evolutionary time towards being more specialised and rare. Interaction 'intruders' may also break into the latter networks, making them more nested. Such species are generalists and can also act as spatial couplers in otherwise fragmented networks, as seen in plant–ant domatia networks (Olesen et al., 2002).

2.1.2 Body size as a driver of ecological network structure

Body size is an important driver of structure and dynamics in many food webs (Arim et al., 2011; Melián et al., 2011; Nakazawa et al., 2011), especially in aquatic ecosystems (Jacob et al., 2011; Woodward et al., 2005), and can give rise to substructures, such as feeding hierarchies arising from gape-limited predation (Petchey et al., 2008; Woodward et al., 2010b). Recent explorations of so-called trivariate webs, in which feeding links are overlaid on mass-abundance plots, in marine (O'Gorman et al., 2010), freshwater (Jonsson et al., 2005; Layer et al., 2010; Woodward et al., 2012) and terrestrial (McLaughlin et al., 2010; Mulder et al., 2011) systems have revealed strong size structure. Typically, energy flows from many abundant, small resources to fewer, rarer and larger consumer species, with many webs containing one or a few apex predators but orders of magnitude more than basal species. These properties play an important stabilising role in the face of species loss and other perturbations (McLaughlin et al., 2010; O'Gorman et al., 2010), and could be especially important in fragmented habitats (Ledger et al., 2012; Woodward et al., 2012), where dispersal ability is also linked to body size. While seemingly ubiquitous in food webs, these patterns have yet to be described for mutualistic or host–parasitoid networks. More recently, body size, abundance, biomass and link data have been used to assess a range of substructural properties in aquatic food webs (Cohen et al., 2009), including tritrophic interactions (i.e. the smallest modular substructure beyond species pairs) and other recurring motifs (Woodward et al., 2012). Given that network substructure is likely to be related to both body size and spatiotemporal context, future work needs to focus on the potential impact of habitat fragmentation on the robustness of the underlying structural mechanisms in food webs and mutualistic networks, although species traits (e.g. abundance) other than size might be more important in the latter (but see Stang et al., 2006, 2009).

2.1.3 Species abundance as a driver of ecological network structure

Studies of ecological networks mostly focus on interactions among *species* (e.g. network references in Bascompte et al., 2003; Olesen et al., 2007). Individuals are the entities that are actually interacting, however, and as such their encounter rates, sensitive to habitat fragmentation, drive network structure (e.g. Petchey et al., 2010; Vázquez et al., 2009). For instance, flower abundance can account for much of the variation in linkage level of plants in pollination networks (Stang et al., 2006; but see Olesen et al., 2008). The importance of abundance for the functional roles of species in antagonistic networks is well known, but remains largely unexplored in mutualistic networks. Often a few common species engage in many interactions, and most rare species engage in few interactions (e.g. Memmott, 1999). This skewed structure affects several network metrics including nestedness, connectance and *asymmetry* (e.g. Blüthgen et al., 2008), although sampling artefacts need to be ruled out (Fischer and

Lindenmayer, 2002; Lewinsohn et al., 2006; Vázquez, 2005; Vázquez et al., 2007; Woodward et al., 2010b). The effects of spatiotemporal changes in abundances on network structure remain relatively underexplored, but they are potentially key issues in the context of habitat fragmentation.

2.1.4 Functional groups in ecological networks

Species within functional groups (Hobbs et al., 1995; Körner, 1993) may be redundant, which is critical to network persistence under species extinction scenarios (Kaiser-Bunbury et al., 2010; Memmott et al., 2004) and other perturbations (Aizen et al., 2008; Kaiser-Bunbury et al., 2011; Tylianakis et al., 2007). The species traits that determine functional groups in ecological networks can differ within and between types of networks. In pollination networks, functional diversity defined by morphological traits might be vital for the persistence of diverse plant communities (Fægri and van der Pijl, 1979; Fontaine et al., 2006) and can constrain interaction patterns (Stang et al., 2006). In addition, functional groups can also be defined by behavioural traits (e.g. generalist vs. specialist), lifespan and temporal activity (e.g. seasonality of occurrence), phylogeny (similar roles of closely related species) and place of origin (e.g. native vs. exotic), which can influence pollination rates and species interactions (Fishbein and Venable, 1996; Kandori, 2002; Raine and Chittka, 2005), or whole pollination networks (Lopezaraiza-Mikel et al., 2007). Still remarkably little is known about how relative abundance affects within-functional group competition for the same resources at the network level.

2.2. Spatial networks

The analysis of multispecies ecological networks in a spatially explicit setting is still in its infancy (Dale and Fortin, 2010; Kissling et al., 2012a), although other types of networks have been investigated in spatial and landscape ecology (Dale and Fortin, 2010). Here, nodes are considered as locations (such as lakes or habitat fragments) and links define the connections among them (Dale and Fortin, 2010). The nodes (e.g. habitat fragments) have spatial coordinates and additional attributes related to size, shape, habitat quality and so on. The links among them can be defined by their distance or weight (e.g. measures of similarity in species composition among locations). Links are usually bidirectional (i.e. symmetric), but they can also be unidirectional, for instance, when the connection between lakes is represented by water flow. Spatial networks can thus form a conceptual basis for adding functional interrelations to habitat connectedness and physical structure to ecosystems (Dale and Fortin, 2010; Urban et al., 2009).

In a habitat fragmentation framework, spatial networks can quantify the effects of losing nodes or links, for example, by mimicking the loss of habitat patches or dispersal corridors for a single species within a *meta-population* (Urban and Keitt, 2001; Urban et al., 2009). More complex measures of species-specific landscape features, such as least-cost paths that describe the movement of a species through a heterogeneous matrix habitat, can also be integrated (Fall et al., 2007). The analysis of spatial networks in a static landscape (Urban and Keitt, 2001) can be extended to capture dynamic landscape processes that influence the persistence of patchy populations (Fortuna et al., 2006). Island biogeography perspectives (MacArthur and Wilson, 1967) can also be applied where separate fragments are seen as ecological islands embedded in a matrix of varying hostility.

Fragments can be connected via species that are present in both, creating a bimodal (rather than a one-mode) network of fragments and species. Roles can then be assigned to species and fragments according to their *topological role* and position in the network (Carstensen and Olesen, 2009; Guimerà and Amaral, 2005). Carstensen et al. (2012) used such an approach on a large scale and identified island roles and modules on the basis of shared avifaunas (i.e. biogeographic regions) and island characteristics.

2.3. Combining spatial and ecological networks

Regardless of whether it is possible to estimate landscape connectivity for all interacting species or for only a few key species, an integrative approach between spatial and ecological networks is needed to evaluate population persistence in fragmented landscapes (Gonzalez et al., 2011). This depends not only on the amount of habitat and its distribution in the landscape, but also on the position of each species within the ecological network (Solé and Montoya, 2006). For instance, top predators are particularly vulnerable to extinction in fragmented landscapes (Holyoak, 2000). Both spatial and ecological networks have similar concepts and are analysed with similar tools (Gonzalez et al., 2011), and integrating these into a single framework offers a promising way to advance the field (Dale and Fortin, 2010; Fortuna and Bascompte, 2008; Gonzalez et al., 2011; Olesen et al., 2010b).

Following Dale and Fortin (2010), a 'graph of graphs' can represent ecological network properties (e.g. nestedness of a plant–animal network) as nodes of a spatial network. In this way, one possibility is to view each local population as a node in a network with two kinds of links: (i) dispersal of individuals between fragments (local populations) and (ii) interactions between individuals of different species (e.g. pollination). The first kind of link provides an evaluation of landscape connectivity or habitat availability (Pascual-Hortal and Saura, 2006) for each species and the second kind gives the role each species plays in the ecological network of species interactions, such as its degree, *centrality* or contribution to nestedness. In this way, a value of habitat availability at the landscape scale may be assigned to each species plus a measure of its role in the ecological network(s), information that can be combined to evaluate its persistence probability. Moreover, different spatial configurations of habitats in the landscape and different arrangements of ecological networks can be modelled to estimate the impacts of fragmentation on persistence probabilities.

Recent theoretical studies illustrate the potential of unexpected consequences of the interplay between spatial and ecological networks by exploring three-species food chains. As a simple example, we may consider a tritrophic chain (Hastings and Powell, 1991) where a top predator Z feeds on an intermediate predator Y and on a prey X, whereas Y feeds only on X, with interactions ordered by body size ($Z > Y > X$). The local extinction of Y in small patches jeopardises the survival of the large predator Z and may lead to a overpopulation of X. Examples of outbreaks in spatially distributed populations have indeed been described theoretically (Araújo and de Aguiar, 2007; Maionchi et al., 2006), showing that probable reduction in abundance of intermediate species may have important indirect ramifications for other species via their interactions in the ecological network. Recent experimental work shows that although intermediate species may be lost, it is often the larger species at the terminus of tritrophic chains that are especially prone to local extinctions due to habitat fragmentation, leading a reduction in the trophic level of the web as a whole (Woodward et al., 2012).

Theoretical studies further indicate that dynamical instabilities caused by large dispersal abilities of predators, relative to their prey, in spatial networks create abundance heterogeneities among otherwise equivalent fragments (Mimura and Murray, 1978; Nakao and Mikhailov, 2010; Rietkerk et al., 2004). These so-called Turing patterns (Murray, 1993; Rietkerk et al., 2004; Turing, 1952) represent the combined effect of species dispersal, interactions and spatial configuration. They may also have indirect consequences on other species by altering the composition of potential prey, predators, competitors and mutualistic partners in ecological networks among fragments. Such explorations of the interplay between spatial and ecological networks highlight the need to focus on understanding how fragmentation affects population dynamics within multispecies systems.

3. HABITAT FRAGMENTATION

3.1. General introduction

Habitat fragmentation is often defined as a process during which a large expanse of habitat is transformed into a number of patches of a smaller total area, isolated from each other by a matrix of habitats unlike the original (Wilcove et al., 1986). It increases discontinuity in the spatial patterning of resource availability, affecting the conditions for species occupancy, and ultimately individual fitness. Fragmentation can arise via both natural and anthropogenic processes in terrestrial and aquatic systems (Figs. 1 and 3). In the latter, fragmentation affects freshwaters (e.g. rivers and lakes) as well as marine systems (e.g. oceans, coral reefs, seagrass meadows, kelp forests, salt marshes and sea ice) (Box 1). In terrestrial systems, habitat fragmentation can be induced by many drivers, including lava flows and the conversion of forest to farmland (either grasslands or arable fields). Our focus is primarily on anthropogenic fragmentation of pristine habitats, which is occurring at an accelerating rate on a global scale. An illustrative example of the effect of habitat fragmentation in the Atlantic Rainforest of Brazil is provided in Box 2.

The effects of fragmentation on biodiversity depend on specific species traits and characteristics of the fragments and the surrounding matrix (Ewers and Didham, 2006; Fahrig, 2003; Henle et al., 2004). At least four effects form the basis of most quantitative measures of habitat fragmentation (Fahrig, 2003): (a) reduction in habitat amount, (b) increase in the number of fragments, (c) decrease in fragment size and (d) increase in fragment isolation. While habitat loss *per se* will reduce population sizes and, ultimately, the loss of species and their links (Bierregaard et al., 1992; Fahrig, 2003; Franklin and Forman, 1987; Saunders et al., 1991), fragmentation includes a much wider array of patterns and processes and far more complex consequences for biodiversity. We will review the importance of fragment characteristics (size and isolation, including connectivity and corridors), habitat edges (including *edge permeability* and geometry) and matrix quality, before discussing spatial and temporal turnover and the importance of scale.

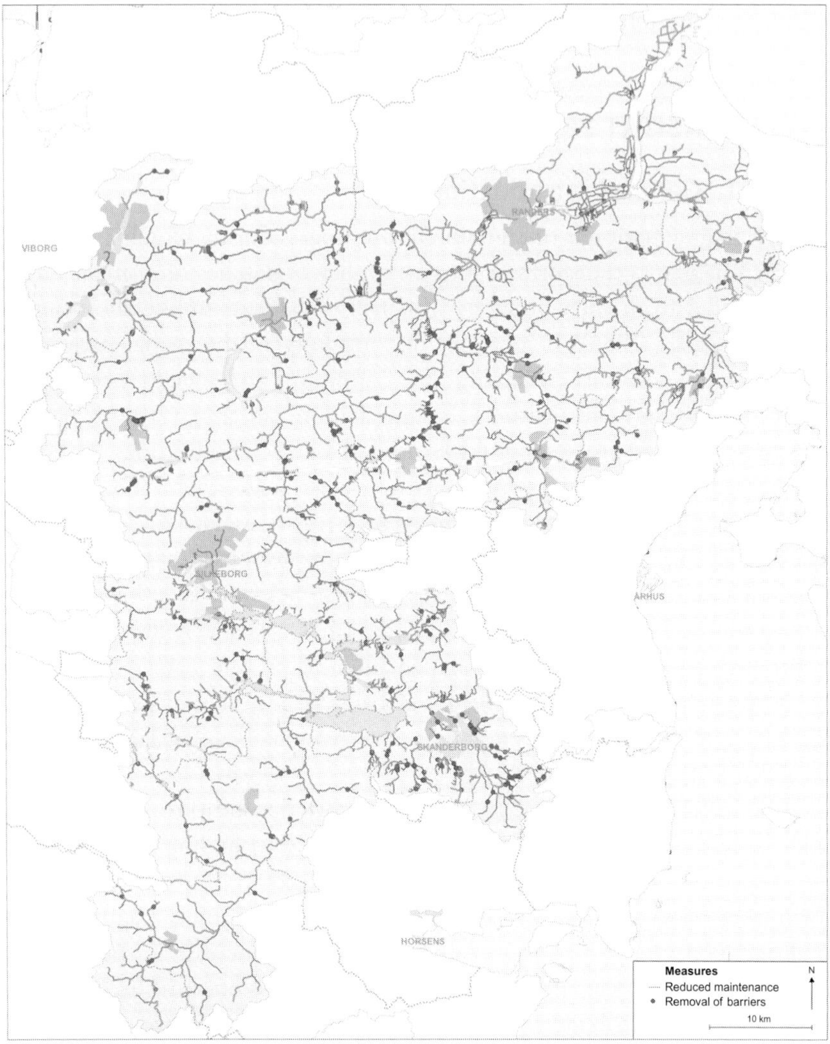

Figure 3 Anthropogenic fragmentation of a European river network (Denmark, E. Jutland, the city of Aarhus at the bay-center-right of map; map size: E-w 80km). Dots indicate physical barriers (weirs, dams, impoundments) to fish migration, a major source of human-mediated impacts (Feld et al., 2011). The map of the Gudenå catchment, is derived from the River Basin Management Plan, reproduced courtesy of The Danish Ministry of Environment. (For interpretation of the references to colour in this figure legend, the reader is referred to the online version of this chapter.)

BOX 1 Habitat Fragmentation in Aquatic Ecosystems

Fragmentation plays a key role in both terrestrial and aquatic ecosystems, including freshwater, estuarine and marine systems (e.g. oceans, coral reefs and seagrass meadows).

Freshwaters are commonly viewed as being bounded by hard edges as they are 'fragmented islands in a terrestrial sea' (Woodward and Hildrew, 2002a), but they also have soft boundaries within their borders (Figs. 5 and 16) imposed by chemical gradients such as pH or salinity, especially where they mix with coastal waters in estuaries. Human activity has accelerated the rate and extent of fragmentation in freshwaters, particularly by overabstraction of water by growing populations (Vörösmarty et al., 2010). Climate change is also to exacerbate hydrological droughts (Poff and Zimmerman, 2010) via reduced rainfall in many areas (Kundzewicz et al., 2008), potentially causing widespread habitat loss and fragmentation (Boulton, 2003; Lake, 2003; Ledger et al., 2011). During droughts, river flows decline, reducing the volume of wetted habitat (water width and depth) and altering habitat structure, increasing water temperature, reducing dissolved oxygen (Everard, 1996) and altering nutrient supply (Dahm et al., 2003). In some regions, droughts occur predictably as part of the natural hydrologic cycle and species are able to tolerate such conditions (Bonada et al., 2007), but elsewhere unpredictable drought fragmentation can have devastating effects on aquatic food webs (Ledger et al., 2011).

Marine systems such as oceans, coral reefs and seagrass meadows are also exposed to fragmentation. For instance, the open ocean might appear to be relatively homogenous, but there are distinct vertical and horizontal regions separated by physicochemical barriers, such as pycnoclines and frontal systems, which are more permeable to larger organisms (e.g. anadromous and catadromous fishes) than to the smaller organisms. Coral reefs experience increased rates of habitat loss and fragmentation due to dynamite fishing (Fox, 2004; Raymundo et al., 2007; Riegl and Luke, 1998; Wells, 2009), and coral bleaching is occurring with increasing frequency due to rising sea temperature (Oliver and Palumbi, 2009). The loss of structural complexity in these fragmented coral landscapes results in declining abundances and diversities of reef fish and mobile invertebrates (Bonin et al., 2011; Coker et al., 2009; Graham et al., 2007; Pratchett et al., 2008; Syms and Jones, 2000). Local extinctions are proportionally greater for resource specialists than generalists (Munday, 2004). Other marine systems include seagrass meadows, which form unique, productive and diverse ecosystems (Bostrom et al., 2006; Duarte and Chiscano, 1999). They are affected by fragmentation through dredging and boating effects, eutrophication, extreme weather events, urchin grazing and wasting disease (Bostrom et al., 2006; Orth et al., 2006; Rasmussen, 1977; Walker and

Continued

BOX 1 Habitat Fragmentation in Aquatic Ecosystems—cont'd

McComb, 1992; Walker et al., 2006). While many studies suggest that fragmentation of seagrass meadows has limited (Frost et al., 1999; Hirst and Attrill, 2008; MacReadie et al., 2009), inconsistent (Bell et al., 2001) or even positive (Eggleston et al., 1998; Hovel and Lipcius, 2001) impacts on epifaunal diversity and abundance, fragmentation beyond a threshold level can lead to rapid declines in species diversity and abundance (Reed and Hovel, 2006).

Other major marine habitats influenced by fragmentation include kelp forests, salt marshes and sea ice. Habitat loss in kelp forests reduces biomass and abundance of fish (Deza and Anderson, 2010). The die-off of salt marshes results in changes in the behaviour of key grazers (snails) as they seek shelter from predation by blue crabs (Griffin et al., 2011; Silliman et al., 2005). Finally, increased fragmentation of sea ice habitats results in declines in mating success and searching efficiency of top predators such as polar bears (Molnár et al., 2011) and in changes in phototrophic community structure and relative abundance of dominant marine taxa (Mueller et al., 2006).

BOX 2 Habitat Fragmentation and its Effect on Brazilian Atlantic Rainforest Trees

A good example of a biodiversity hotspot affected by fragmentation is the Brazilian Atlantic rainforest landscape, which is dominated by a mosaic of small forest fragments usually embedded in a heterogeneous matrix of urban and agricultural land (Ribeiro et al., 2009). The abundance and diversity of many taxa (including frogs, lizards, small mammals and birds) are generally positively affected by the surrounding matrix (Pardini et al., 2009, and see also Faria et al., 2006, 2007), whereas the richness and abundance of shade-tolerant trees are negatively affected and decline from large to small fragments (Pardini et al., 2009). This indicates that increasing landscape heterogeneity might allow the maintenance of higher diversity of animals, but that specialist tree species depend on the maintenance of native forest patches (Pardini et al., 2009; Ribeiro et al., 2009). In the more extreme scenario of a hyper-fragmented Northeast Brazilian Atlantic forest (i.e. a landscape composed of pastures, monoculture plantations and a few small native forest fragments), tree species and reproductive trait diversity are lost (Lopes et al., 2009; Oliveira et al., 2008), whereas early successional trees can proliferate in small forest remnants (Tabarelli et al., 2008). An expansion of pioneer species in the edge dominated habitats can be associated with changes in functional reproductive traits, diurnal pollination systems, and loss of long-distance flying pollinators, self-

BOX 2 Habitat Fragmentation and its Effect on Brazilian Atlantic Rainforest Trees—cont'd

incompatible breeding systems and large-seeded plant species. Furthermore, phenological trait mismatches can occur, due to shifts in the proportions annual versus supra-annual flowering (Lopes et al., 2009; Santos et al., 2008; Tabarelli et al., 2010). Modelling efforts predict a pervasive long-term trend towards vegetation dominated by early successional trees and impoverished tree species composition (Pütz et al., 2011), with important implications for plant–animal mutualistic networks. Specialised and long-distance moving connector species in mutualistic networks such as large pollinators (bees or hummingbirds) and seed dispersers (large birds) are likely to be particularly vulnerable due to reduced floral diversity and quality arising from the dominance of generalist pollination systems, and the large proportion of species that are wind dispersed or which have small fleshy fruits (e.g. Lopes et al., 2009; P. Morellato, unpublished data).

3.2. Fragment characteristics

Fragment characteristics are important for understanding fragmentation effects on biodiversity (Table 1). Apart from original habitat loss *per se* (Tilman et al., 1994), size (i.e. area) and degree of isolation of fragments are important properties (Fahrig, 2003). For some taxa such as butterflies, habitat heterogeneity seems to be a more important determinant of diversity than fragment size and isolation (Kivinen et al., 2006; Rundlöf and Smith, 2006; Weibull et al., 2000), and this may be true for other herbivorous insects as well.

The area needed to maintain populations is determined by fragment size, with smaller patches generally containing fewer individuals and species than larger patches (Debinski and Holt, 2000). The area effect on biodiversity can be predicted from species–area curves (Sabatino et al., 2010), and the set of species in smaller patches is often a fairly predictable subset of those in larger patches (nested structure; e.g. Ganzhorn and Eisenbeiss, 2001; Hill et al., 2011). Species richness in forest fragments in relation to fragment area (Brooks et al., 1997; Ewers and Didham, 2006) can mirror the classic species–area relationships known from island biogeography (MacArthur and Wilson, 1967). To some extent, temporal effects are also dependent on fragment size because what happens quickly in small fragments happens slowly in larger fragments (Terborgh et al., 1997).

Table 1 Fragment characteristics and animal and plant traits, which are relevant for assessing fragmentation effects on biodiversity

Trait	Importance of trait in relation to fragmentation	References
Trait at fragment level		
Size (area)	The size of fragments determines the area available for population and species persistence and influences extinction and immigration rates	Bender et al. (1998), Fahrig (2003), MacArthur and Wilson (1967)
Isolation	The degree of isolation of fragments represents the lack of habitat in the surrounding landscape and has an influence on the movement and dispersal of species among fragments	Ewers and Didham (2006), Fahrig (2003)
Shape	Convoluted fragment shapes can lead to increased turnover and variability in population size when compared to fragments that are compact in shape	Ewers and Didham (2006)
Edge effects	Edges of fragments affect microclimate and animal abundances	Laurance et al. (2011)
Matrix effects	The surrounding matrix mediates edge effects and influences animal (e.g. pollinator and seed disperser) movements	Laurance et al. (2011)
Animals		
Dispersal ability	Species with high mobility are more likely to survive in fragmented landscapes than species with low mobility. Low mobility or poor dispersal ability of species is thus expected to increase species-level fragmentation effects. For some butterflies, it has been shown that species with intermediate mobility are more likely to decline in abundance following habitat fragmentation than species with either high or low mobility	Ewers and Didham (2006), Thomas (2000)

Table 1 Fragment characteristics and animal and plant traits, which are relevant for assessing fragmentation effects on biodiversity—cont'd

Trait	Importance of trait in relation to fragmentation	References
Habitat specialisation	Habitat specialists are expected to be more affected by fragmentation than habitat generalists. The matrix tolerance of a species might play an important role here (e.g. forest generalist vs. habitat generalist)	Ewers and Didham (2006)
Trophic level	Higher trophic levels are predicted to be more strongly affected by habitat fragmentation than lower trophic levels	Ewers and Didham (2006), Milton and May (1976)
Dietary specialisation	Species with broad dietary niches might be less impacted by fragmentation than dietary specialists	Bommarco et al. (2010)
Gap-crossing ability	Species persistence in isolated fragments is strongly linked to gap-crossing ability	Lees and Peres (2009)
Body size	Body size constrains animal space use and home range size. Home range size is expected to increase with habitat fragmentation, and home ranges of larger species are more sensitive to habitat fragmentation than those of smaller species	Greenleaf et al. (2007), Haskell et al. (2002), Jetz et al. (2004), Laurance et al. (2011)
Sociality	Sociality can buffer against negative effects of fragmentation (e.g. social bees vs. solitary bees) or increase susceptibility to fragmentation (e.g. obligate mixed-flock feeders in Amazonian forest birds)	Aizen and Feinsinger (1994a,b), Bommarco et al. (2010), Laurance et al. (2011)

Continued

Table 1 Fragment characteristics and animal and plant traits, which are relevant for assessing fragmentation effects on biodiversity—cont'd

Trait	Importance of trait in relation to fragmentation	References
Plants		
Dispersal mode	Dispersal mode (e.g. abiotic vs. biotic) can be a key factor influencing species responses to habitat fragmentation	Montoya et al. (2008), Tabarelli et al. (1999), Tabarelli and Peres (2002)
Fruit/seed size	Large big-seeded fleshy fruits tend to have few dispersal agents and are likely to be more strongly affected by fragmentation than plant species with small fleshy fruits	Corlett (1998)
Pollination mode	Plants depending on animals for pollination are probably negatively affected by habitat fragmentation (specifically isolation) than wind-pollinated species	Aizen and Feinsinger (1994a,b), Fægri and van der Pijl (1979), Kolb and Diekmann (2005)
Breeding system	Characteristics of breeding systems, for example, the degree of protandry, self-incompatibility or sex ratios, might be affected by fragmentation	Jennersten (1988), Yu and Lu (2011)
Growth form	Specific growth forms (e.g. clonal plants) might be more strongly affected than others (e.g. annuals)	Dupré and Ehrlén (2002), Kolb and Diekmann (2005)
Seed bank	Long-lived seed banks may prevent species from going extinct in small habitat fragments	Dupré and Ehrlén (2002)

The list highlights some key traits but is not intended to be exhaustive.

Isolation restricts the movement and dispersal of species among fragments and depends on physical distance and matrix quality (Bender et al., 2003). Two aspects of fragment isolation are particularly important: connectivity and the availability of corridors. Connectivity is the degree to which the landscape permits or impedes movement among fragments (Taylor et al., 1993) and is a species and system-specific parameter (Taylor et al., 2006; Tischendorf and Fahrig, 2000). Its effect on biotic interactions (e.g. pollination services) is therefore a complex function of the individual responses of the different interacting species (for pollinators see e.g. Fenster and Dudash, 2001; Herrera, 1988; Horvitz and Schemske, 1990; Moeller, 2005; Ricketts et al., 2006). Some species may primarily be influenced by the distance to a fragment of a particular habitat, while others may be more influenced by the quality or availability of the resource (e.g. nesting sites) in adjacent habitats. One structural landscape characteristic of high importance for connectivity is the presence of corridors, which can be either natural or man-made. They are landscape elements that facilitate the movement of organisms among fragments, promoting biotic connectivity and synchrony (Hilty et al., 2006). Recent experiments have demonstrated that corridors play a key role in maintaining plant and animal populations and their interactions in fragmented landscapes, and that connected fragments retain more species from native biota than isolated ones (Damschem et al., 2006; Tewksbury et al., 2002). Their importance for biodiversity conservation is still a moot point (Gilbert-Norton et al., 2010; Noss, 1987; Simberloff and Cox, 1987; Simberloff et al., 1992), as in some systems (e.g. tropical rainforests) corridors and fragments dominated by secondary vegetation may be of limited value (Oliveira et al., 2008).

3.3. Habitat edges

Increased edge habitats, which may be natural (e.g. light gaps, rivers and landslides in natural forests) or anthropogenic, are prominent features of a fragmented landscape. Habitat edge and fragment shape are important determinants of biodiversity (Ewers and Didham, 2006; Laurance et al., 2011; Murcia, 1995), and strong effects on a variety of plant and animal species are well documented (e.g. Bach and Kelly, 2004; Davies et al., 2000; Gehlhausen et al., 2000; Laurance et al., 1998).

Three main physical and biological effects of edges are important in fragmented habitats (Murcia, 1995): (i) abiotic environmental changes across edges; (ii) biological effects related to changes in species in the edge and across the edge as a result of (i), and (iii) indirect biological effects, which relate to how changes in (ii) cascade up and affect species via their antagonistic and mutualistic interactions.

Changes in abundance across a habitat edge depend on the taxonomical/functional groups involved. Generalist species are often favoured in habitat edges, because they offer access to new habitats and resources (e.g. pollinators: Burgess et al., 2006, herbivores: Wirth et al., 2008, predators and nest predation: Chalfoun et al., 2002; Lidicker, 1999), whereas specialists typically decline (plants: Laurance et al., 1997, 2006a; Tabarelli et al., 2008, insectivorous birds: Restrepo and Gómez, 1998, vertebrates: Hansson, 1994, but see Pardini et al., 2009 for a multi-taxa approach).

Species that require different habitat types for different resources or life history stages (e.g. nesting, feeding and foraging) are expected to benefit from a structurally diverse habitat mosaic (including edges). For example, solitary bees that nest above-ground forage in agricultural landscapes, but nest in neighbouring natural habitats (Gathmann and Tscharntke, 2002). Aquatic insects often rely on trees as 'swarm-markers' for breeding once they have emerged from the water and crossed the aquatic–terrestrial boundary. Similarly, riparian vegetation provides the main source of energy to many stream food webs in the form of terrestrial leaf-litter, so the proximity to this edge can determine the trophic basis for production for the entire system (Hladyz et al., 2011b). Even predators can benefit from inputs from terrestrial edges, with such subsidies supporting some stream fishes at densities far beyond what in-stream production alone can support (Allen, 1951). Edges also influence seed banks and the quality, abundance and diversity of seed rain (Devlaeminck et al., 2005, Melo et al., 2006).

In forests, especially tropical ones, the increasing air temperature, light incidence and decreasing relative humidity towards the edge (Didham and Lawton, 1999; Kapos et al., 1997; Murcia, 1995) can affect plant reproduction by shifting phenology and boosting flower and fruit production (Burgess et al., 2006; Camargo et al., 2011; D'Eça Neves and Morellato, in press; Kato and Hiura, 1999; Murcia, 1995) (Fig. 4). In turn, important animal–plant interactions can be affected (Aizen and Feinsinger, 1994a,b; Cunningham, 2000; Fleury and Galetti, 2006; Galetti et al., 2006; Jordano and Schupp, 2000; Wright and Duber, 2001). Pollination rates at edges may decrease (Aizen and Feinsinger, 1994a,b; Burgess et al., 2006; Harris and Johnson, 2004; Hobbs and Yates, 2003), increase (Burgess et al., 2006), or may not change at all (Burgess et al., 2006), with implications for plant reproductive success

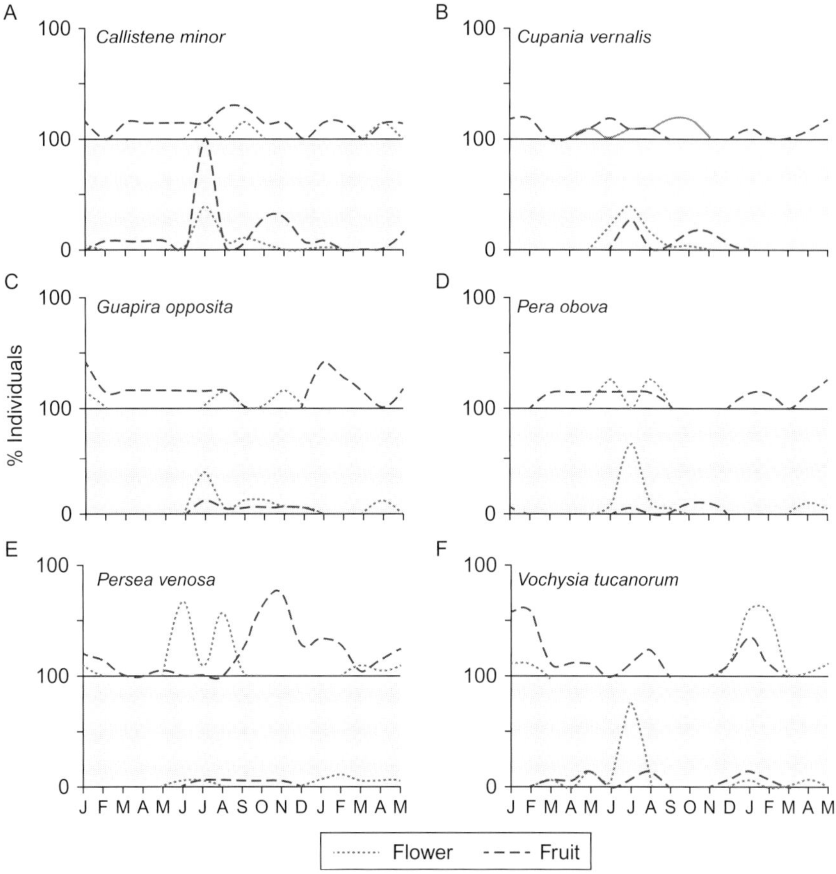

Figure 4 Phenological response of trees occurring in the edge (shaded graphics) and interior of a seasonal forest at Serra do Japi, Southeastern Brazil (after D'Eça Neves and Morellato, in press). Positive responses (higher proportion of trees) for flowering were detected in four of six species (Figure edge A to D). On the other hand, negative edge effects on fruiting were detected for four species (Figure edge C to F). Although the fruit production of the woody *Cupania vernalis* (Sapindaceae) was positively affected by fragment edge (Figure edge B), Guimarães and Cogni (2002) observed a higher seed predation of *C. vernalis* in the edges at the same study site. Therefore, differential phenological responses at the edges may change the visitation rates of pollinators, dispersers and seed consumers, making it hard to predict the reproductive outcome to the plant. (For colour version of this figure, the reader is referred to the online version of this chapter.)

(Burgess et al., 2006; Cunningham, 2000) and seed dispersal. The influence on the latter may be either positive due to differences in animal densities, foraging patterns, fruit display, plant size and vigour (Jordano and Schupp, 2000), or negative via limited animal movement at edges

(Restrepo et al., 1999). Furthermore, recruitment and predation of seeds in the forest interior might decrease relative to edges (Baldissera and Ganade, 2005; Fleury and Galetti, 2006; Jules and Rathcke, 1999; Restrepo and Vargas, 1999, but see Cunningham, 2000; Guimarães and Cogni, 2002).

Besides the capability of a species to perceive suitable habitat fragments and the connectivity of the landscape, its persistence in a fragmented landscape depends on its ability to cross the edge between fragment and matrix (Morris, 1997; Stamps et al., 1987a; Stevens et al., 2006). Habitat edges can be characterised as 'hard' or 'soft' according to their permeability. Hard edges are boundaries which dispersing individuals rarely (if ever) cross, although their permeability can vary with life history, for example, adults or juveniles (Fig. 5). Soft edges are more permeable: for example, bumblebees (*Bombus hortorum*) cross several habitat edges between meadows, fields and gardens and move widely within a mosaic landscape (Hagen et al., 2011). Changes in edge permeability (e.g. due to degradation of the landscape matrix around a fragment) can alter migration rates, as well as several other ecological and demographic processes. For instance, population densities within the fragment may be elevated, maturity delayed, and reproductive and growth rates reduced (Abramsky and Tracy, 1979, 1980; Gliwicz, 1980; Lidicker, 1985; Myers and Krebs, 1971; Stamps et al., 1987b).

Emigration rates (i.e. the proportion of dispersing individuals that leave the fragment) from habitat fragments are also determined by the edge-to-size ratio and the shape of the habitat edge (Nams, 2011). For instance, Hardt and Forman (1989) found forest herbivores to concentrate in the grassy areas where the edge intrudes into the forest. Some pollinating bee species (e.g. *Bombus lapidarius*; Rasmussen and Brødsgaard, 1992) avoid edges while foraging for pollen within fragments, while responses of birds to edges vary markedly among species and edge types (Sisk and Battin, 2002).

3.4. Matrix

The matrix surrounding fragments also influences their structure and dynamics (Brotons et al., 2003; Cook et al., 2002; Prevedello and Vieira, 2011; Prugh et al., 2008). Among forest fragments, matrix quality can range from a completely deforested agricultural landscape to mature secondary growth, varying immensely in hostility and permeability to each species. Matrix quality thus determines connectivity, dispersal and associated mortality rates, and its influence may even override those of

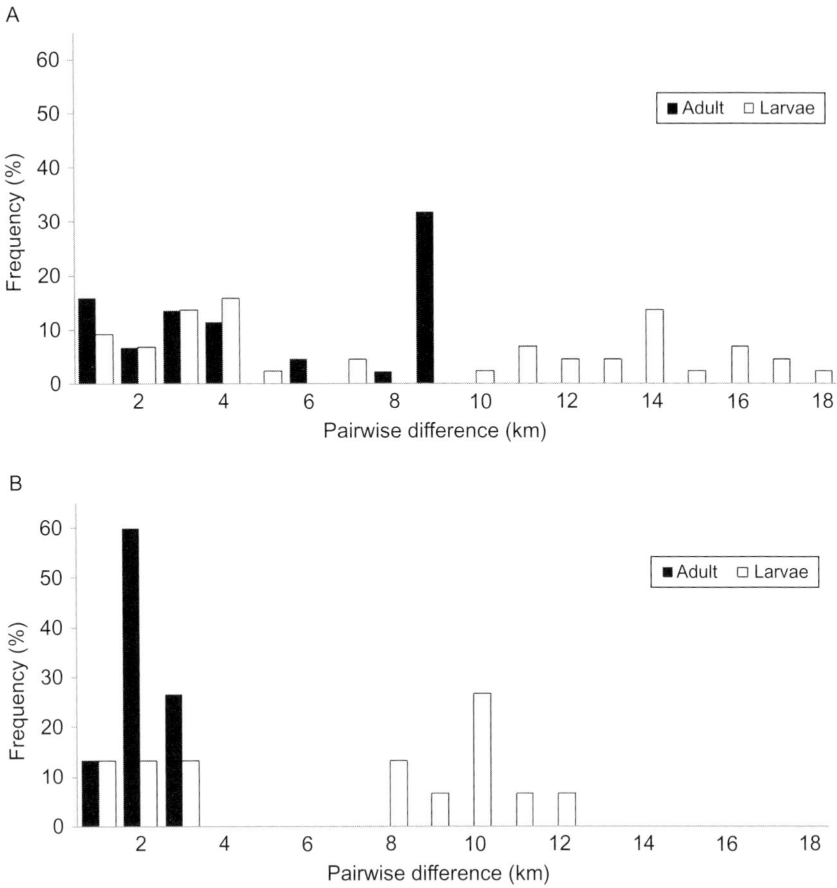

Figure 5 Frequency distribution of pairwise distances for all individual streams in the Ashdown Forest network, Sussex, UK (Fig. 16), for both adult winged insects (solid black bars) and immature aquatic insects and other solely aquatic organisms, including fishes, molluscs, Crustacea and other groups (white bars). (A) The River Medway network and (B) the River Ouse network. *Note*: inter-catchment exchange is not included here, since although it is feasible in the aerial phase, none of the solely aquatic taxa in these webs are able to cross the marine–freshwater boundary, which acts as a 'hard' boundary for all the insect species that dominate these food webs. Aquatic invertebrates are incapable of crossing from one network to the other, due to the lack of suitable corridors. Fewer than 1% of all fish species can make the transition between fresh and salt water (brown trout and common eels are the only notable exceptions within the river networks shown here), so for many taxa these two catchments are in reality separated by 100s of kilometres of an insurmountable physicochemical barrier even though the local webs may be just a few kilometres apart in the upper headwaters. There is also likely to be an evolutionary spatiotemporal component to fragmentation here, as these catchments have likely been flowing in different directions and hence effectively isolated for many taxa since the retreat of the ice sheets at the end of the last glaciation.

fragment area and isolation (Cook et al., 2002; Ewers and Didham, 2006). A high-quality matrix (e.g. forest regrowth) can minimise edge effects by supporting a proportion of the communities in the fragments (Laube et al., 2008; Pardini et al., 2009 and references therein).

A diverse and structurally complex, anthropogenic matrix may even harbour a significant fraction of the original biota, potentially reducing biodiversity loss (Lindenmayer and Luck, 2005; Pardini et al., 2009). For instance, in Western Kenyan rainforest, some bird species (11% out of 194 forest-dependent species; Bennun and Njoroge, 1999) also used the heterogeneous farmland close to the forest as feeding habitat, gaining access to additional food resources outside their core habitat (Laube et al., 2008). Thus, agroecosystems with a diverse habitat structure can have at least some capacity to compensate for forest loss. Indeed, several frugivorous bird species use native and exotic fruiting trees in the farmland around the same forest, increasing seedling establishment (Berens et al., 2008; Eshiamwata et al., 2006), suggesting the matrix can aid fragment regeneration and restoration (Fisher et al., 2010). Further, bee diversity is higher than in the nearby forest, so the farmland may even act as a 'pollinator rescue', supporting pollination services inside the forest (Hagen and Kraemer, 2010). Other studies have reported positive influences of natural forest on pollination interactions in farmland (e.g. Florida, USA: Artz and Waddington, 2006; North Queensland, Australia: Blanche et al., 2006).

Matrix quality can also be important for food webs. A recent study has shown how the invasion of the terrestrial edge habitat can cause a collapse in food web structure and ecosystem processes of an adjacent stream, by altering the porosity of energy flux across the ecotone (Hladyz et al., 2011a). Here, the native terrestrial matrix through which the stream would normally flow is either in the form of the mixed deciduous woodland climax community, or rough pasture maintained by low intensity farmland. The invasive tree *Rhododendron ponticum* forms dense, dark monocultures that outcompete native riparian plant species and cast a deep shade over the stream food webs. Invasions can occur within either of these starting conditions, although they are accelerated by anthropogenic disturbance along the aquatic–terrestrial fragment–matrix edge. Because the tough, leathery leaves of the invader are also a poor-quality food source, being very high in C:N and lignin content (Hladyz et al., 2009), they effectively shut down the detrital pathway at the base of the stream food web, which is normally fuelled by leaf-litter when the matrix is dominated by oak woodland. The invader also suppressed the alternative energy source supplied by algal–herbivore pathways that would otherwise dominate when the terrestrial matrix is rough pasture, by shading the stream channel (Hladyz et al., 2011a). Consequently, *invasive species* can harden the fragment–matrix *boundary*, by reducing the permeability of energy transfer. Additional recent evidence from a pan-European study suggests that riparian alterations tend to suppress animal–resource interactions at the base of stream food webs, increasing reliance on microbial-driven rather than invertebrate-driven processes (Hladyz et al., 2011b).

3.5. Spatial and temporal turnover of species and individuals

Spatial and temporal turnover in species composition among habitats in a fragmented landscape can be pronounced. For instance, Hagen and Kraemer (2010) found high turnover rates in bee species composition between open farmland, forest–farmland edge and forest interior: almost 50% of all bee species in this landscape mosaic occurred in all three habitat types, indicating a high edge permeability or a so-called soft edge.

In contrast, in a European meadow, pollinator species did not cross the edge into the adjacent forest, whereas herbivores and pathogens did (L. Kromann-Gallop, personal communication). Shifts in behaviour (e.g. flower visitation rates) may also occur among individuals of the same species of pollinator, leading to differences in fruit and seed set among habitats (Kaiser et al., 2008). Additionally, the roles of species in an ecological network (e.g. peripherals, connectors, module hubs and network hubs; Olesen et al., 2007) can change when crossing habitat borders (M. Hagen et al., unpublished data). Of 35 species (8 plant and 27 bee species) occurring in all three habitats in a forest–agriculture landscape, 23 (3 plants, 20 bees) had similar roles in all habitats, as did 11 (4 plants, 7 bees) species in two of the three habitats, and one plant had a different role in each habitat.

Due to physical changes at habitat edges, phenological shifts in interactions may arise, resulting in a complex interplay between spatial and temporal turnover. Edges and interiors may therefore differ in the timing of resource availability and network structure and dynamics. Unfortunately, detailed data remain scarce (Kato and Hiura, 1999; Ramos and Santos, 2005), but an increase in flower production at forest edges associated with high light incidence and temperatures have been reported for some species (Alberti and Morellato, 2010; Camargo et al., 2011; Fuchs et al., 2003; Kato and Hiura, 1999; Ramos and Santos, 2005). D'Eça Neves and

Morellato (in press) compared the phenology of tree species between forest edge and interior in Southeastern Brazil and found a higher proportion of reproductive trees along the forest edge (59% flowering and 73% fruiting) than inside the forest (47% flowering and 29% fruiting), and flowering and fruiting were more seasonal in the latter. As individual tree species can respond differently to edge effects (Fig. 4), the synchrony and degree of overlap between the interaction partners in an ecological network may be affected by this aspect of habitat fragmentation (e.g. Hegland et al., 2009; Memmott et al., 2007).

The predominance of generalism and seemingly high plasticity of interactions in many ecological networks may reduce the effects of spatial and temporal mismatches. The available literature, albeit scarce, indicates that pollination networks are fairly robust against such mismatches (see Hegland et al., 2009) and the same may be true for food webs, which are typically even more generalised (Ings et al., 2009). Plants and pollinators exposed to similar environmental changes may react in synchrony, decreasing the occurrence of mismatches (Hegland et al., 2009). In pollination networks, high turnover in species composition and interactions over time are well documented (Alarcón et al., 2008; Dupont and Olesen, 2009; Olesen et al., 2008; Petanidou et al., 2008), but the consequences of adding the spatial component of a fragmented landscape to temporal mismatches are virtually unknown.

3.6. Scales of habitat fragmentation

Fragmentation operates over many spatial and temporal scales (Levin, 1992), from tiny water bodies within individual plants (Phytotelmata; Box 3) to successional processes across entire landscapes, for instance, as stream networks develop following glacier retreat (Brown and Milner, 2012; Jacobsen et al., 2012; Woodward and Hildrew, 2002a). Individual organisms perceive the world at different spatial and temporal scales and thus will respond to fragment characteristics, habitat edges and *matrix permeability* in different ways. Within food webs, consumer–resource perceptual disparities may be pronounced, closely coupled to the relationship between body size and environmental grain: for example, single-celled algae and small invertebrates at the base of aquatic food webs are many orders of magnitude smaller than the large vertebrates at the top (e.g. Cohen et al., 2009; Layer et al., 2010). The immediate environment within which a diatom spends its (short)life attached to a substrate particle on a streambed is thus shaped largely by small-scale

BOX 3 Phytotelmata—Small Aquatic Worlds in a Highly Fragmented Landscape

Phytotelmata (from Ancient Greek, meaning 'plant ponds') are small water bodies within plants that exist as aquatic refugia within a much larger terrestrial ecosystem. Examples include tree holes, bamboo internodes, pitcher plants, tank bromeliads and water-retaining plant axils (Kitching, 2001). Phytotelmata have been intensively studied as they represent naturally replicated systems containing discrete communities and food webs within individual plants (Reuman et al., 2009). The macrofaunal assemblages they contain can range from 2 to 20 species (Kitching, 2001) and are often dominated by arthropods, although annelids, frog tadpoles and molluscs have also been recorded (Kitching, 2000). In addition, they contain a diverse range of microscopic life, including rotifers, protozoa and bacteria (Buckley et al., 2010; Kneitel and Miller, 2002).

Phytotelmata can be regarded as insular systems (Kitching, 2001), and they have been useful models for testing island biogeography theory (MacArthur and Wilson, 1967). An investigation of the macrofaunal diversity in epiphytic bromeliads shows that species richness increases with phytotelma size and physical habitat complexity (Armbruster et al., 2002; Buckley et al., 2010; Srivastava, 2006). Phytotelmata are extremely isolated as the surrounding matrix (e.g. terrestrial forest) is hostile. There is no connectance between phytotelmata via corridors, so the aquatic–terrestrial boundary presents a discrete hard edge between fragments. This can only be overcome in the adult phases of phytotelma-inhabiting species, for example, as winged phase of aquatic insects or after metamorphosis in tadpoles.

In addition to these hard edges, there can also be soft edges that act as secondary filters among separated phytotelmata. For instance, the physicochemical environment differs within each plant so that some hoverflies avoid bamboo internodes with low pH for oviposition (Kurihara, 1959) or mosquito larvae exhibit reduced survivorship with rising pH in tree holes (Carpenter, 1982). During extreme rainfall events, extensive flushing and recharging of the aquatic reservoir can occur and thus provide potential connectance among phytotelmata. The nutrient content (Carpenter, 1982) and pH (Clarke and Kitching, 1993) of phytotelmata can vary widely, and these varying levels of habitat restriction and fragmentation can create a 'hierarchy of fragmentation', with the imagines of phytotelm invertebrates being exposed to a less fragmented environment than the juvenile stages.

Phytotelmata fragmentation will have pronounced effects on the structure and function of ecological networks formed within such water bodies. Whilst there are examples of mutualistic interactions within pitcher plants (Clarke and Kitching, 1993), the vast majority of described phytotelma networks are antagonistic, and there is evidence for both bottom-up and top-down control within the food web (Hoekman et al., 2011; Kneitel and Miller, 2002). At least three discrete levels of fragmentation are apparent, from local to larger landscape scales (e.g. bromeliad leaf pools within a plant; phytotelmata within a single terrestrial matrix vs. multiple, fragmented terrestrial matrices).

forces related to fluid viscosity or nutrient diffusion, whereas the herbivores that eat it will be more influenced by factors such as availability of physical refugia from predators (who in turn operate at larger scales), channel discharge or water depth (Woodward and Hildrew, 2002a; Woodward et al., 2010a). Thus, the fragment size within which each species operates tends to increase up the food chain, and the species' perception of edges also changes. In terms of 'flow habitats' in stream ecosystems, individual diatoms will be strongly influenced by boundary layer effects within the nearest few millimetres, herbivorous macroinvertebrates will respond to near-bed velocity and microhabitats at the scale of centimetres to metres, and predatory fish will respond to the availability of suitable territories at the pool-riffle or macrohabitat scale. The largest, most mobile, migratory species may even respond at the scale of the entire river catchment (Woodward and Hildrew, 2002a).

Most fragmentation studies usually focus on a particular spatial scale: Doak et al. (1992) reviewed 61 primary research papers on the effects of habitat fragmentation on population structure of terrestrial arthropods, all of which were conducted at a single spatial scale. In general, studies that account for fragmentation on different spatial scales are rare (but see Garcia and Chacoff, 2007; Schleuning et al., 2011b; Stephens et al., 2003). Forest fragmentation (large-scale reduction of fragment size) can affect ecosystem processes indirectly by changes in biodiversity, whereas selective logging (local scale) influenced ecosystem processes (e.g. pollination and seed dispersal) by modifying local environmental conditions and resource distributions (Schleuning et al., 2011b).

Many long-term consequences only become apparent after many decades (Laurance et al., 2011), yet most studies of anthropogenic fragmentation have been conducted over much shorter periods (Ewers and Didham, 2006), which may not be sufficient to detect the full range of responses. Nevertheless, empirical studies suggest that time lags in species responses at such time scales are very common (Ewers and Didham, 2006; Laurance et al., 2006b). While population densities may increase in the short term as survivors are concentrated in remaining patches, in the long term, species abundance and richness decline (Debinski and Holt, 2000) because some can survive for up to several generations under unsuitable habitat conditions before eventually going extinct ('extinction debt'; Tilman et al., 1994). Extinction debts can be pronounced if many species are near the threshold capacity of the landscape that ensures meta-population persistence (Hanski and Ovaskainen, 2000). Time-lagged responses of species to fragmentation are not only observed for long-lived trees, but also for other organism groups such as vertebrates and insects (Ewers and Didham, 2006; Metzger et al., 2009). Considering such time lags thus becomes especially important for evaluating fragmentation effects on species interactions and ecological networks.

Time lags are most pronounced where generation times strongly differ between interacting or dependant species (Kissling et al., 2008, 2010). For instance, in climate change impact assessments, low dispersal rates and long generation times of woody plants can slow distributional responses, with important consequences for bird species that depend on such plants for habitat and food (Kissling et al., 2010). In a fragmentation context, the different generation times of invertebrates and vertebrates, parasites and hosts, and species from different trophic levels in plant–animal mutualistic systems might lead to contrasting responses of interacting species, thus disrupting existing networks. For instance, long-lived vascular plants in European grasslands showed time-delayed extinctions whereas short-lived butterflies did not, even after 40 years (Krauss et al., 2010). This suggests that interacting species (at different trophic levels) have different extinction debts, so co-extinctions associated with long-lived taxa might amplify future biodiversity loss even without any further fragmentation occurring.

Given the various levels of complexity and spatiotemporal scales involved, a hierarchical approach seems necessary for understanding the effects of habitat fragmentation on species interactions, ecological networks and community-level changes (Didham et al., 2012; Urban et al., 1987).

4. HABITAT FRAGMENTATION AND SPECIES TRAITS

In addition to landscape attributes, species traits also modulate the effects of fragmentation (Aguilar et al., 2006; Ewers and Didham, 2006; Fahrig, 2003; Henle et al., 2004). For instance, overall species richness of butterflies in Europe and America decreases with fragmentation, but those with low dispersal ability, a narrow larval feeding niche and low reproduction are most strongly affected (Öckinger et al., 2010). In addition, intraspecific variation in phenotypic traits may ultimately affect community patterns, such as the distribution of niche width (Bolnick et al., 2011). In general, seemingly contradictory responses might be better explained by considering the role of species traits (Ewers and Didham, 2006). In this section, we briefly review fragmentation-relevant traits for plants and animals and then highlight the potential importance of species trait combinations for understanding the consequences of fragmentation for biodiversity and ecological networks.

4.1. Plant traits

Important plant traits for persistence in fragmented landscapes include seed dispersal, pollination and breeding system, growth form and seed bank (Table 1). Two aspects of seed dispersal are particularly relevant: dispersal mode and fruit traits (e.g. fruit and seed size). The former (abiotic dispersal by wind or via animal vectors) can strongly influence how the relative abundance of tree species responds to habitat fragmentation (Fægri and van der Pijl, 1979; Montoya et al., 2008; Tabarelli and Peres, 2002; Tabarelli et al., 1999). Additionally, fruit traits that influence frugivore choice (fruit size, edibility of the peel, defensive chemistry, crop size and phenology: Buckley et al., 2006) will influence the responses of fleshy-fruited plants to habitat fragmentation. Large, big-seeded fruits, which are consumed by only a few vertebrate species, might be most vulnerable to fragmentation (Corlett, 1998), and fruit size and colour may be crucial for plant colonisation of habitat fragments (Shanahan et al., 2001), where certain trait combinations attract a specific set of animal dispersers (e.g. birds vs. bats).

Plants also differ in their dependency on pollinators (e.g. Aizen and Feinsinger, 2003; Bond, 1994), and this can determine their vulnerability to fragmentation. Certain plants traits are especially important to attract pollinators and to exclude floral reward robbers, for example, flowering phenology, amount and quality of pollen and nectar, and structural complexity of the flower. Habitat fragmentation may contract flowering periods because abundant plant species should have longer population-level phenophases than rarer species (but see Morellato, 2004), increasing the risk of losing pollinators, which could further reduce plant fitness (Aizen and Feinsinger, 1994a,b).

Within species, flower morphology can vary among habitats: certain plants in urban fragments have more, but smaller flower heads, which may decrease floral attractiveness and affect pollinator behaviour (Andrieu et al., 2009). Changes in pollinator behaviour could increase self-pollen deposition (Aizen and Feinsinger, 1994a,b), and drive a divergence in the evolution of floral traits in fragmented populations (Kingsolver et al., 2001; Pérez-Barrales et al., 2007). Demographic, environmental and genetic stochasticity are likely to be most pronounced in small fragments (Matthies et al., 2004; Willi et al., 2005), and the latter may trigger a loss of self-compatibility alleles due to genetic drift, increasing inbreeding and genetic erosion (Ellstrand and Elam, 1993; Lande, 1988; Menges, 1991; Young et al., 1996; but see Aizen and Feinsinger, 1994a,b).

Differences in breeding systems can affect plant species responses to fragmentation. For instance, the herb *Dianthus deltoides* (Caryophyllaceae) is protandrous (i.e. anthers open before stigmas ripen) but in small fragments it becomes homogamous (i.e. the male and female sexual parts ripen simultaneously), increasing the probability of self-fertilisation (Jennersten, 1988). Self-compatible plants are often facultatively dependent on pollinators, whereas self-incompatible species are obligate outcrossers, relying exclusively on pollinators (e.g. Aguilar et al., 2006). Sex ratios in dioecious species might also be sensitive to fragmentation. In China, populations of the dioecious tree *Pistacia chinensis* (Anacardiaceae) were surveyed on islands of different size in a recently flooded reservoir (Yu and Lu, 2011): small islands with poor soils had a male-biased sex ratio, whereas large and nutrient-rich islands had a stable 1:1 ratio. Such drops in effective population size on small islands could accelerate population extinction.

4.2. Animal traits

The key animal traits in relation to fragmentation are dispersal ability, niche width, body size and sociality (Table 1), with the first two being especially important (Bommarco et al., 2010; Ewers and Didham, 2006). Species with high dispersal ability are less likely to be affected by fragmentation (Hanski and Ovaskainen, 2000; Öckinger and Smith, 2007; Roland and Taylor, 1997). For example, solely aquatic invertebrates must swim long distances if they are to colonise new streams in a river system, encountering many potential barriers to dispersal (Fig. 5), whereas larvae with winged adult phases can reach these new habitat fragments relatively easily. Although the abundance of adult phases of aquatic invertebrates (such as stoneflies) decreases exponentially with distance from their 'home stream', with the rate of decline varying with matrix permeability (Fig. 6), only a few gravid females may be needed to (re)populate an entire food web due to high-density-dependent predation on early life stages (e.g. Hildrew et al., 2004). This can lead to increased genetic differentiation in adult populations at larger distances between streams, highlighting the potential for genetic-level impacts of soft versus hard barriers to dispersal (Fig. 7).

Species with a wider dietary or habitat niche will also be less susceptible to fragmentation. Generalists may survive in very small patches by using

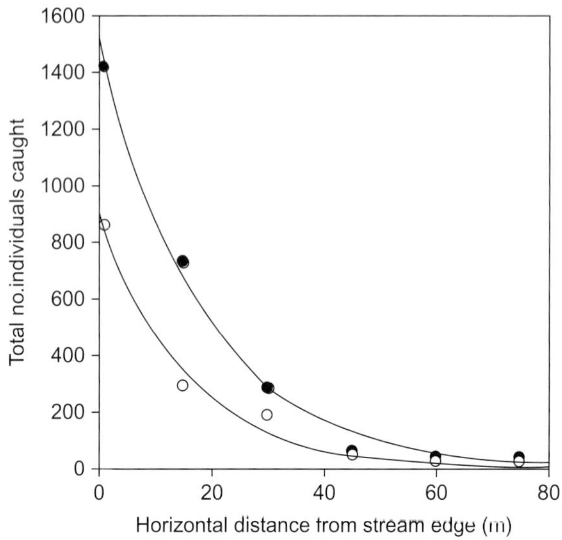

Figure 6 Lateral dispersal of winged adults of a common stonefly species (*Leuctra nigra*) from the stream edge through the terrestrial matrix (woodland—black circles; open land—white circles) within the Ashdown Forest, UK (see Fig. 16). Total number of males and females caught in passive Malaise traps are shown on the *y*-axis, with exponential declining models fitted for each habitat type [woodland: $y = 1.517\ \exp(-0.055*x$; $R^2 = 0.99$, $F = 665.2$, $p < 0.001$); open land: $y = 903\ \exp(-0.065*x$; $R^2 = 0.99$, $F = 324.6$, $p < 0.001$)]. *Redrawn after Petersen et al. (1999).*

resources in both the fragment and the surrounding matrix (Andren, 1994). Specialists might find their resources (e.g. specific food plants) retained in only a few fragments, and habitat specialisation can further restrict their distribution. Some specialists also have a narrow geographic range (Gaston, 1988; Roy et al., 1998) again increasing the vulnerability to fragmentation. Finally, the trophic rank of a species is important and those at higher trophic levels are expected to be more sensitive because of their lower carrying capacity (Didham et al., 1996; Hance et al., 2007; Holt, 2002; Kruess and Tscharntke, 1994; Steffan-Dewenter and Tscharntke, 1999; Steffan-Dewenter, 2003; Tscharntke et al., 2002; Tylianakis et al., 2007; van Nouhuys, 2005, Vanbergen et al., 2006) and there is evidence from experimental food webs that this is indeed the case, although it is just one of several determinants (Ledger et al., 2012; Woodward et al., 2012).

Body size is a key trait as it determines home range size and dispersal ability for many species (Castle et al., 2011; Greenleaf et al., 2007; Haskell et al., 2002; Jetz et al., 2004; Leck, 1979; Lindstedt et al., 1986; Milton and May, 1976; Schaffer, 1981; Willis, 1979), and large species are often

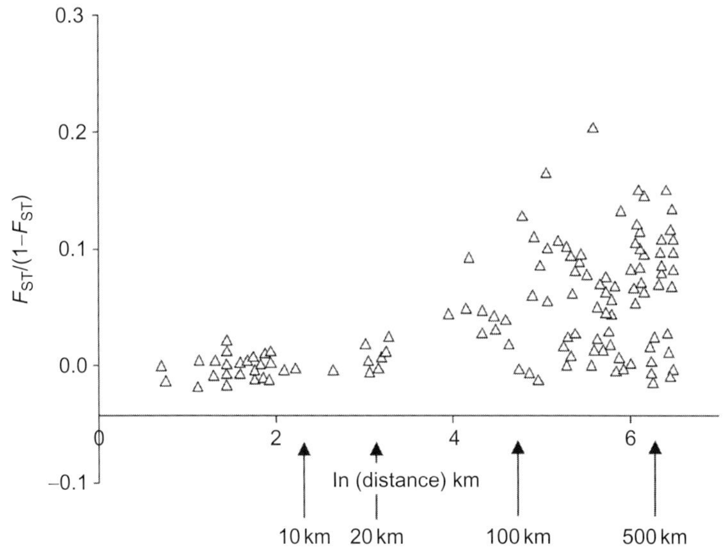

Figure 7 Genetic differentiation as an exponential function of geographical distance among 33 populations of a predatory caddisfly species (*Plectrocnemia conspersa*) across Britain, including 10 sites within the Ashdown Forest (Figs. 5, 6 and 16). Like all freshwater insects, this species has a larval aquatic phase and a winged terrestrial adult phase. The former are typically constrained to living in fragmented acid headwaters (where they are often top predators) within river networks, whereas the latter can disperse across land to connect otherwise isolated food webs. The genetic data above reveal panmictic populations at the regional catchment scale, with significant differentiation (measured as $F_{ST}/(1-F_{ST})$ based on allozyme frequency data) occurring only at larger scales of fragmentation. Even though dispersal across large distances is a rare event (e.g. Fig. 6), only a few gravid females may be needed to repopulate an entire food web due to high fecundity combined with strong density-dependent mortality early in the life cycle (Hildrew, 2009; Hildrew et al., 2004). *Redrawn after Wilcock et al. (2003).*

especially vulnerable—unless they are able to span the gaps between fragments (Crooks, 2002; Ewers and Didham, 2006; but see Laurance et al., 2011). In Amazonia, wide-ranging forest bird species (van Houtan et al., 2007) and primates (Boyle and Smith, 2010) are more vulnerable to fragmentation than those with smaller territoria, and species with limited spatial requirements such as small mammals, non-trap-lining hummingbirds and ants are generally less susceptible (Laurance et al., 2011). Besides body size, restricted mobility, resource specialisation, low annual survival rate, high population variability, and terrestrial foraging and nesting increase vulnerability among birds to fragmentation (Sieving and Karr, 1997).

Species that are large and/or rare are especially vulnerable to the effects of habitat fragmentation by drought in stream food webs (Ledger et al., 2012).

In bees (Box 4), relationships between habitat loss and species traits have been intensively studied (Krauss et al., 2009; Moretti et al., 2009; Steffan-Dewenter et al., 2006), with diet width and sociality being especially important (Aizen and Feinsinger, 1994a,b; Klein et al., 2003; Öckinger and Smith, 2007; Rundlöf et al., 2008; Steffan-Dewenter et al., 2002). Social bees are expected to outperform solitary taxa in harvesting resources because of their higher foraging and food-provision capacity (e.g. Bommarco et al., 2010) and communication systems (e.g. the waggle-dance in honeybees). Social bee species are always diet generalists, because their long-lasting colony needs food throughout the year, although there are differences between tropical and temperate areas. In temperate regions, wild social bees (*Bombus* spp.) appear to be less sensitive to habitat fragmentation than solitary bees (Steffan-Dewenter et al., 2002), whereas in the tropics solitary bees appear to be less sensitive to land-use change than social stingless bees (Aizen and Feinsinger, 1994a,b), probably due to their specialisation on forest as nesting habitat (Roubik, 1989, 2006). In bumblebees (*Bombus* spp.), long-tongued species have declined more than short-tongued ones due to changes in agricultural practices and habitat fragmentation (Bommarco et al., 2012; Dupont et al., 2011), and late-season species have declined more than early-season species (Fitzpatrick et al., 2007).

4.3. Species trait combinations

Any given species comprise a suite of traits, some of which are strongly correlated, whereas others may be orthogonal (Herrera, 2009). Data on individual traits of species, however, are insufficient for predicting fragmentation effects on biodiversity (Ewers and Didham, 2006): rather, their combination and the wider ecological context are both key here. For more detailed examples, see the textboxes on bees (Box 4) and avian frugivores (Box 5).

A combination of body size, diet, dispersal ability, habitat specialisation and sociality may be needed to predict species responses to fragmentation (Boyle and Smith, 2010; Milton and May, 1976). For instance, among European bees, large dietary generalists are less affected by fragment area than small generalists, whereas small specialists may be less affected than large specialists (Bommarco et al., 2010). In Amazonian forest

BOX 4 Bees as Network Nodes

Bee species vary widely in body size, foraging mode, social organisation, seasonal activity and specialisation on flower resources. These characteristics play an important role for the structure and dynamics of plant–pollinator networks. Large body size reflects the capacity to fly longer distances, and genetic markers and radio-tracking techniques are increasingly used to estimate actual flight distances (Darvill et al., 2004; Hagen et al., 2011). Small stingless bees (e.g. *Plebeia* and *Tetragonisca*) often fly distances of up to 1 km while flight distances for larger *Melipona* species can be >2 km (Araújo et al., 2004). The largest orchid bees (*Eulaema*; 18–31 mm) can fly up to 23 km, including the crossing of unsuitable matrix habitat such as open waters for several kilometres (Janzen, 1971; Wikelski et al., 2010). In this bee group, patrolling flights of males can reach long distances and thus disperse pollen among fragmented plant populations. Large bees (e.g. *Xylocopa*, *Bombus*, *Centris*, *Epicharis*, *Eulaema* and *Oxaea*) are hence particularly important for connecting habitat fragments. However, bee mobility also depends on the abundance of food resources and on the amount of floral awards.

Sociality and behavioural differences also affect bee species responses to habitat fragmentation. Not all are as highly eusocial as the honeybees (*Apis* spp.) and the stingless bees (Meliponini): most species are solitary (the female performs all tasks) or subsocial and semisocial (some cooperation among the females). These differences in social organisation can strongly influence network topology due to differences in the abundance of individuals in the nest (one, few, hundreds or thousands). Most eusocial species have perennial colonies, tend to be floral resource generalists and need resources throughout the year, at least in the Tropics, and these tend to be key species or hubs in ecological networks. Most bee species in the tropics are also multivoltine (multiple generations per year), and some (e.g. *Xylocopa*) are long-lived, which can affect the temporal dynamics of plant–pollinator interactions due to differences in abundance and phenophase length. In arctic or temperate regions, where climatic seasonality is pronounced and univoltine bee species are dominant, temporal dynamics in the structure of plant–pollinator networks have already been empirically demonstrated (Olesen et al., 2008).

Resource specialisation also influences network structure (e.g. the range of available nectar plants is broader than that of pollen plants) because nectar is mainly consumed by the adults whereas pollen is used in the brood cell to feed the larvae (Cane and Sipes, 2006). In general, oligolectic bees are recognised for their specialised floral niches whereas polylectic bees (e.g. social species such as *Apis*, *Bombus* and Meliponini) visit a wide range of plants (including flowers of different morphology, colour, size, etc.): within the interaction network, the latter species represent highly connected nodes.

BOX 5 Avian Frugivores and Seed Dispersal in a Fragmented World

Avian frugivores predominate in warm and wet climates of the world's tropical and subtropical regions (Fleming et al., 1987; Kissling et al., 2009). Of the >1200 frugivorous bird species worldwide, most (~50%) are found within the order Passeriformes (perching birds) (Kissling et al., 2009, 2012b), with a body mass of usually <200 g. Over 100 species of frugivores are also found in the orders Columbiformes (doves and pigeons), Psittaciformes (parrots) and Piciformes (woodpeckers, toucans, barbets, honeyguides) (Kissling et al., 2009, 2012b). The spatial distribution patterns vary among bird orders with frugivorous perching birds and parrots dominating in the Neotropics and frugivorous pigeons and hornbills prevailing in Southeast Asia (Kissling et al., 2009). Given the spatial heterogeneity of future land-use changes on bird distributions (Jetz et al., 2007) and the taxonomic and geographic differences in frugivores among regions (Kissling et al., 2009), the global consequences of habitat fragmentation for seed dispersal of fleshy-fruited plants are likely to be complex.

At the landscape scale, the effectiveness of seed dispersers is characterised by the quantity and quality of seed dispersal (Schupp et al., 2010), which in turn is dependent upon body size and associated life history behavioural traits. Due to their requirements for extensive home ranges, large frugivorous birds are especially extinction prone in small fragments (Renjifo, 1999; Uriarte et al., 2011). The ability to fly long distances allows large-bodied frugivores to connect habitat patches (Lees and Peres, 2009; Spiegel and Nathan, 2007). Habitat fragmentation can cause changes in the movement patterns of frugivores, with consequences for seed dispersal (Lenz et al., 2011), especially for plants with large, big-seeded fruits because their dispersal often only depends on one or a few large frugivores (Corlett, 1998; Guimarães et al., 2008). Seed dispersal effectiveness of plants with smaller fruit largely depends on the range of frugivore body sizes in the network, with smaller frugivores allowing for within-patch dispersal and larger frugivores for between-patch dispersal (Spiegel and Nathan, 2007). In addition to body size *per se*, gut retention times and movement velocities of frugivores also determine seed-dispersal distances (Schurr et al., 2009). The interplay of animal behaviour, plant and animal traits, and the specific characteristics of the landscape thus produce complex seed dispersal kernels (Morales and Carlo, 2006) and seed dispersal effectiveness landscapes (Schupp et al., 2010).

fragments, the most capable gap-crossers among birds are medium or large species of insectivores, frugivores and granivores, and these species dominate in small patches (Lees and Peres, 2009). Certain species trait combinations can amplify (or mitigate) vulnerability to fragmentation. For instance, on Barro Colorado Island (Panama), the largest bird was the Black-faced Antthrush (*Formicarius analis*), which also had low annual recruitment and survival rate, and this potent combination of traits, which are often combined in many other species, could explain why it went extinct particularly rapidly as its habitat fragmented (Sieving and Karr, 1997).

Trait matching between interacting plants and animals could affect higher-level responses to fragmentation. For instance, interactions in some plant–pollinator networks show size matching, that is, insect species with a long proboscis visit a wider range of flowers than do species with a short proboscis (e.g. Borrell, 2005; Corbet, 2000; Goldblatt and Manning, 2000; Harder, 1985; Stang et al., 2009).

Developing a combined trait-response framework could provide important future advances in assessing fragmentation effects in ecological networks. Additionally, interaction effects between fragment characteristics (see Section 3.2) and species traits could also be important. Network analysis offers a potentially powerful way to identify modules of species with similar responses to fragmentation, which then may be analysed with respect to their trait combinations (Verdú and Valiente-Banuet, 2011).

5. HABITAT FRAGMENTATION AND BIOTIC INTERACTIONS

In the previous sections, we have examined the importance of landscape structure (e.g. fragment characteristics, habitat edges, matrix) and species traits for assessing the consequences of habitat fragmentation on biodiversity: here, we turn our attention to impacts on species interactions, the strengths and outcomes of which (Fig. 2) vary spatially and over time. This spatial dependency arises because the probability of an encounter between predator and prey, pathogen and host, or mutualistic animals and their plants has a landscape context, and hence sensitivity to fragmentation.

5.1. Mutualistic plant–pollinator interactions

Pollination and, hence, plant reproduction can be strongly affected by habitat loss and fragmentation (Fægri and van der Pijl, 1979; Jennersten, 1988; Kearns et al., 1998; Olesen and Jain, 1994; Rathcke and Jules, 1993; Renner, 1998) (for examples see also Box 4).

Due to habitat fragmentation, pollinator communities could become more homogenous, and generalists (Ewers and Didham, 2006) and introduced species (e.g. Do Carmo et al., 2004) may replace natives and dominate interactions, potentially altering the reproductive output of the plant community. However, the effect on pollen dispersal and pollination effectiveness may strongly vary among species, without necessarily being related to a species habitat niche: habitat generalists and invasive pollinators can either be less (Didham et al., 1996; Do Carmo et al., 2004) or, in some cases, more effective pollinators than habitat specialists (Dick, 2001).

Fragmentation can isolate host plant patches, reducing genetic and ecological exchange among them. Although still little is known about precise flight distances and movement patterns of pollinators at the landscape scale (Hagen et al., 2011), body size influences the genetic connectance of, and pollen flow among, distant plant populations (Pasquet et al., 2008). The effective movement of pollinators may be tracked by paternity assignment of seeds and pollen (Lander et al., 2011), and the influence of landscape configuration on pollinator movement (e.g. for trap-lining species) can be incorporated into the analysis (Lander et al., 2011).

The reproductive output of plants can vary with pollinator composition, abundance and behaviour (Lamont and Barker, 1988; Lamont et al., 1993). Although visitation rates are expected to be influenced by habitat fragmentation, the results are inconclusive: some pollinators are more abundant in larger fragments (Sih and Baltus, 1987), some are equally abundant (Jennersten, 1988), while others are rarer in fragments (Sih and Baltus, 1987; Strickler, 1979). Temporal aspects such as phenological changes influence how fragmentation affects plant–pollinator interactions (Memmott et al., 2007). When fragmentation reduces plant species richness, food shortages could reduce pollinator diversity, especially among long-living insects, such as bumblebees (Memmott et al., 2007).

The local extinction of pollinators might not always have consequences for interacting plants, if redundant species can compensate. For instance, the Hawaiian tree *Freycinetia arborea* (Pandanaceae) was once pollinated by now extinct birds, but has recently been rescued from extinction by an introduced white-eye bird (*Zosterops* sp.) that replaces previous pollinator species (Cox, 1983). If redundancy is not evident, even the loss of single interactions can initiate waves of further extinctions (Nilsson et al., 1992; Olesen and Jain, 1994). For instance, the orchid *Cynorkis uniflora* is a mountain rock plant highly specialised upon a few pollinating sphingids in Madagascar (Nilsson et al., 1992). The host plants of the larvae of these sphingids are found in nearby forests, and the delicate orchid–pollinator adult/larva interactions are highly vulnerable to forest loss and fragmentation. The extinction of the pollinating hawkmoths can trigger the loss of orchids and initiate a 'cascade of linked extinctions' (Myers, 1986).

5.2. Mutualistic plant–frugivore interactions

Mutualistic interactions between fleshy-fruited plants and frugivores play a central role for assessing the consequences of habitat fragmentation on biodiversity, especially in the Tropics (Box 5). Frugivorous vertebrates are the focal seed dispersers because only very few invertebrates (e.g. ants, earthworms and grasshoppers) play this role (Duthie et al., 2006; Rico-Gray and Oliveira, 2007; Willems and Huijsmans, 1994).

Larger animal species are expected to be particularly sensitive to habitat fragmentation (Haskell et al., 2002), and there is supporting evidence of this for frugivorous birds (e.g. Sub-Andeans: Renjifo, 1999; Amazonia: Uriarte et al., 2011). The proportion of fruit in primate diets is positively correlated with home range size (Milton and May, 1976) and species persistence in forest fragments (Boyle and Smith, 2010). The disappearance of large frugivores thus decreases the probability of long-distance dispersal of fleshy-fruited plants from small patches and fragments (Fragoso, 1997; Fragoso et al., 2003; Spiegel and Nathan, 2007; Uriarte et al., 2011). The response of small-to–medium-sized frugivores to fragmentation is probably driven by species' habitat specialisation and matrix tolerance, and their ability of gap-crossing (Table 1). Compared to medium-sized frugivores, mesopredators (i.e. medium-sized carnivorous habitat generalists) move more freely between matrix and fragment (Terborgh et al., 1997).

The traits of fleshy-fruited plants determine frugivore choice and hence endozoochorous seed dispersal and the relationship between fruit size, consumer size and gape width is the key (Buckley et al., 2006; Burns and Lake, 2009; Jordano, 1995; Lord, 2004). Small fruits are typically consumed by a wide range of potential seed dispersers, including many species that thrive in small forest fragments and degraded landscapes (Corlett, 1998). However, large, big-seeded fruits tend to have fewer dispersers, and the very largest may depend on only one or a few species (Corlett, 1998). Consequently, these species are the specialists in the network and most vulnerable to fragmentation. More generally, the proportion of fleshy-fruited species is likely to decrease in smaller fragments (Tabarelli and Peres, 2002).

Beyond fruit size, the presence of an inedible pulp, defensive chemicals, crop size, fruit colour and fruiting phenology also influence frugivore choice (Buckley et al., 2006; Voigt et al., 2004; Willson and Whelan, 1990), but if and how they relate to fragmentation is currently unclear. Pre-and post-ingestion processing of fruit and movement of consumers determine seed-dispersal distances and plant establishment patterns (Buckley et al., 2006; Schurr et al., 2009; Spiegel and Nathan, 2007). The mean dispersal distance of endozoochorously dispersed seeds depends upon a combination of frugivore body size, mobility and gut retention time (Schurr et al., 2009). Large frugivores (e.g. the trumpeter hornbill *Bycanistes bucinator*) may change their movement patterns, with unimodal seed-dispersal distribution within forests but bimodal distribution in fragmented agricultural landscapes (Lenz et al., 2011). Individual fruiting trees, even exotic ones, in farmland may be important food sources for the frugivore community and thus represent foci for seed dispersal and forest regeneration, even in highly degraded landscapes (Berens et al., 2008; Fisher et al., 2010).

5.3. Mutualistic plant–ant interactions

Another type of mutualism that is important in a fragmentation context is the interaction between ants and plants in defensive mutualist systems (Box 6).

The intimacy of this interaction (i.e. the degree of biological association between individuals of interacting species) varies, and this could determine how plant–ant interactions respond to habitat fragmentation. Some plant–ant defensive mutualisms, such as extrafloral nectary-based mutualisms, are typical among free-living species (Guimarães et al., 2007), that is, each individual ant and plant can interact with dozens of partners from different species through its lifespan. These are therefore similar to most of the pollination and seed dispersal interactions with respect to degree of *interaction intimacy* (Guimarães et al., 2007). In contrast, many plant–ant mutualisms are symbiotic, that is, one individual plant hosts an ant colony and, as a consequence, individuals (the plant and the ant colony in this case) interact with one or a few partners through their lifetime (Fonseca and Benson, 2003; Fonseca and Ganade, 1996). Few studies have investigated how environmental change affects the network structure of plant–ant interactions (Diaz-Castelazo et al., 2010), but information about these mutualistic interactions is becoming increasingly available.

Key traits in extrafloral nectary interactions include ant body size (Chamberlain and Holland, 2009) and the distribution of ant and/or plant abundances (Chamberlain et al., 2010), which are likely to change with

BOX 6 Interactions Between Ants and Plants

Ants form one of the dominant groups in terrestrial ecosystems, and they interact in multiple ways with plants (Rico-Gray and Oliveira, 2007), as seed predators (Rico-Gray and Oliveira, 2007), leaf-harvesters (Oliveira et al., 1995; Pizo and Oliveira, 2000) and mutualistic partners (Christianini and Oliveira, 2009; Palmer et al., 2008; Rico-Gray and Oliveira, 2007). Although ant pollination is rare (Beattie, 1985; Gómez, 2000), ants are among the main seed dispersers of many plant species (Culver and Beattie, 1978; Pizo et al., 2005). In some tropical ecosystems, ants form gardens (Davidson, 1988), actively dispersing seeds of plants and nesting within the plant parts. Finally, ants are among the most conspicuous defensive mutualists of plants (Rico-Gray and Oliveira, 2007), which offer extrafloral nectar, other food resources and/or nesting sites such as domatia.

Recent studies of extrafloral nectary assemblages suggest ant body size and species abundance are important in shaping patterns of interactions: the number of interactions increases with ant body size (Chamberlain and Holland, 2009; Chamberlain et al., 2010). These results mirror those often reported in predator–prey interactions (Sinclair et al., 2003) and plant–frugivore mutualisms (Jordano, 2000). Several hypotheses suggest that the effects of ant body size are more indirect than direct, with larger ants interacting with more plant species than smaller ants because they (i) forage over a greater area, (ii) are more widely distributed or (iii) because of size-driven competition hierarchies (Chamberlain and Holland, 2009). In the latter scenario, larger ants, that often recruit fewer workers when foraging, are outcompeted by smaller recruitment-efficient ant species from the optimal resources, leading to an increase in the number of plants the larger ants interact with.

habitat fragmentation. The effects of fragmentation can differ among ant functional groups (Pacheco et al., 2009; Wirth et al., 2008), and it may even benefit some plant–ant networks, which often naturally occur in habitat edges (e.g. *Cecropia* spp and its ant partners). Predicting which ant or plant species will be affected, and how, requires an understanding of the underlying traits shaping these interactions. The challenge is that we still need to improve the taxonomy of a considerable fraction of ant species, and the natural history of many species still remains unknown. In this context, the phylogenetic relatedness of interacting species is a proxy for non-random trait distributions.

Understanding the ecological and evolutionary dynamics in these complex fragmented landscapes faces challenges similar to other kinds of interaction. For example, plant–ant interactions involve organisms that differ radically in how they perceive their environment. Ants are small, short-lived organisms, whereas plants are much larger and often longer-lived. Thus, they will perceive the effects of habitat fragmentation at distinct scales and will respond in different ways. Additionally, all plants and most ant colonies are essentially fixed in space, whereas most other plant–animal mutualisms involve a fixed individual (e.g. plant) and a mobile forager (e.g. pollinator). Thus, dispersal of both ants and plants is a between-generation process, which may lead to as yet unexplored *meta-community* dynamics that differ from other types of network. Moreover, plant–ant protective mutualisms are based on indirect benefits: plants benefit from a trophic cascade caused by ants attacking herbivores (Perfecto and Vandermeer, 2008; Vandermeer et al., 2010). Thus, if habitat fragmentation changes the intensity of herbivory, it also changes the fitness consequences of the mutualism (see Palmer et al., 2008).

5.4. Antagonistic interactions within food webs

While the previous sections have focused on mutualistic interactions, we now address antagonistic interactions, specifically food webs. Body size is a key determinant of predator–prey interactions in many food webs (Emmerson and Raffaelli, 2004; Woodward et al., 2005), with large predators typically consuming smaller resources (Layer et al., 2010; McLaughlin et al., 2010), especially in aquatic systems. As a result, trophic height tends to increase with body mass (Jonsson et al., 2005; O'Gorman and Emmerson, 2010), although predator–prey body mass ratios may decline (Brose et al., 2006; Jonsson and Ebenman, 1998; Mulder et al., 2011). Since large species are most susceptible to habitat fragmentation due to their perception and use of resources over larger distances (Holt, 1996) and their need for larger home ranges (Haskell et al., 2002), top predators should be especially prone to extinction. As they often exert strong effects within food webs, their loss could have severe implications for network structure and stability, although recent field experiments suggest that this might be primarily via direct effects of their loss from the system rather than more subtle indirect food web effects *per se* (Woodward et al., 2012).

Habitat fragmentation can reduce encounter rates and hence *interaction strengths* within food webs. This may ultimately decouple pairwise interactions, leading to a simpler and potentially more fragmented food web, since the starting point at which a food web assembles is the level of interactions among individuals. In many food webs, predators (and other non-predatory consumers) are often far from satiation as indicated by the high proportions of relatively empty predator guts compared with what they could consume if feeding rates were maximal (Woodward and Hildrew, 2002b). This suggests that encounter rate is a key determinant of the strength of predator–prey interactions and network structure (Petchey et al., 2010; Woodward et al., 2010b).

In a fragmented landscape, encounter rate can be influenced at different spatial and temporal scales, from short-term patch-scale aggregative responses of predators to their prey within particular fragments during disturbance events (e.g. Lancaster, 1996) to larger scale habitat-level effects that reflect longer-term depletion of prey by predators.

Handling time is also important for food web structure and dynamics, but it is difficult to envisage how it might be affected by fragmentation, as it seems likely to be relatively robust to this kind of disturbance (e.g. in contrast to the effect of temperature changes). Thus, encounter rate rather than handling time might change under increasing levels of fragmentation, and the relative importance of the two rates could be key for predicting the higher-level effects in food webs (e.g. Petchey et al., 2010).

As in mutualistic networks, the scale and environmental grain of fragmentation will also interact with species life histories to determine food web effects. For instance, in fresh waters undergoing fragmentation (e.g. temporary pools formed by the retreat or drying of waters from floodplains), food web interactions can be intensified in the short (i.e. intragenerational) term if predators and prey are concentrated in increasingly smaller patches. Conversely, fragmentation may weaken top-down effects in the longer (intergenerational) term if large predators are lost from small habitat patches. Here, meta-population and source–*sink* dynamics and the ability of predators and prey to recolonise isolated or small habitat patches may be key, and species traits such as body size, behaviour, life history and taxonomic identity will influence these dynamics (Ledger et al., 2012).

5.5. Antagonistic host–parasitoid interactions

Antagonistic host–parasitoid interactions can also be affected by habitat fragmentation, and the degree of specialisation of parasitoids on their host is likely to be critical aspect here. When the host is restricted to certain plant species or habitats, highly host–specific parasitoids will experience landscapes as islands within a sea of unusable matrix. Conversely, for a more generalist parasitoid, capable of using hosts from different habitats, the landscape represents a mosaic of variable-quality patches. Fragmentation should therefore have increasingly negative effects on more specialised parasitoids, and several empirical studies support this conclusion (moth parasitoids: Elzinga et al., 2007; aphid parasitoids: Rand and Tscharntke, 2007; leafminer parasitoids: Cagnolo et al., 2009; parasitoids of cavity-nesting bees and wasps: Holzschuh et al., 2010). These findings suggest that the effects of fragmentation on parasitoids will largely be mediated by altered host distributions, which are often coupled to plant densities (for herbivorous hosts) at the patch scale (Albrecht et al., 2007; Amarasekare, 2000; Cronin et al., 2004; Holzschuh et al., 2010; Kruess, 2003; Schnitzler et al., 2011; Vanbergen et al., 2007).

Although within-patch effects may be important in determining parasitoid densities, the location of refuge habitats, parasitoid attack rates and dispersal ability will determine parasitoid–host dynamics at a landscape scale (Mistro et al., 2009). The survival of a parasitoid meta-population will thus largely depend on individual dispersal abilities, and body size constraints might be important here (Roland and Taylor, 1997). Furthermore, dispersal limitation may moderate parasitoid–host interactions (Thies et al., 2005) because higher trophic levels are likely to be most negatively affected by fragmentation (Holt, 1997). The species-specific extent of dispersal limitation could ultimately determine the relative competitive success of different parasitoid species and how they experience the host landscape (van Nouhuys and Hanski, 2002). As a consequence of habitat fragmentation, attack and parasitism rates can change depending on the fragment isolation, matrix quality and the amount of suitable habitat in the landscape (Cronin, 2003; Kruess and Tscharntke, 2000; Roland and Taylor, 1997). The combination of within-patch effects (habitat quality, host abundance), landscape characteristics (fragment characteristics) and species traits (e.g. dispersal ability and body size) can thus ultimately produce a variety of outcomes for parasitoid–host interactions.

5.6. Summary of fragmentation effects on mutualistic and antagonistic interactions

The responses of biotic interactions to habitat fragmentation are complex, but several key themes arise repeatedly for both mutualistic and antagonistic interactions. The core question is how habitat fragmentation (e.g. fragment size and isolation) will change the links between species, and these are, in turn, a product of the functional traits of the interacting species (e.g. body size, dispersal ability, level of specialisation).

A key species trait is body size because it affects how species interact and their responses to habitat fragmentation. Its importance is evident in plant–pollinator interactions (e.g. proboscis length and size of floral structures), plant–frugivore interactions (e.g. gape width and fruit sizes), plant–ant interactions (e.g. size-driven competition hierarchies) and food webs (e.g. predator–prey mass ratios). Beyond general effects of body size and trophic rank on species interactions, the size of an animal (or plant) also correlates with a suite of other fragmentation-relevant traits. In particular, body size determines dispersal ability and movement distances of some taxa, a fundamental aspect for persistence in a fragmented landscape. Body size measures are often used as proxies for estimating movement distances indirectly, including body mass for birds and mammals (Haskell et al., 2002; Jetz et al., 2004), measures of wing shape in birds (Dawideit et al., 2009), and body length, intertegular span or wing span for insects (Cane, 1987; Greenleaf et al., 2007; Michener, 2007; Rogers et al., 1976). Similarly, fruit sizes can be used as a proxy for long-distance dispersal in fleshy-fruited plants, at least when body sizes of their extant vertebrate dispersers are correlated with seed dispersal effectiveness (*sensu* Schupp et al., 2010). Given the tremendous differences in body sizes among species involved in interactions (e.g. insects vs. vertebrates), responses of different-sized mutualists and antagonists should vary markedly even within the same level of fragmentation.

Specialisation also influences how fragmentation affects mutualistic and antagonistic interactions. The degree of habitat specialisation (e.g. forest dependence or matrix tolerance) is important because mutualistic and antagonistic interactions will change, as specialised species are lost as fragmentation proceeds. Dietary specialisation is particularly important in antagonistic interactions, but also in many mutualistic interactions. In this context, trophic redundancy may be key to buffering species losses. For instance, in mutualistic interactions, the functional loss of a species may be compensated by another species of similar size (cf. Zamora, 2000). As body-size distributions are typically skewed towards small species (Woodward et al., 2005), the potential for functional redundancy decreases with increasing body size (and trophic status). Consequently, large species may be functionally more important for conserving size-dependent ecosystem services, that is, seed dispersal and pollination in mutualistic networks, pest control by predators and biomass production for human consumption in fisheries (Rossberg, 2012).

A couple of other aspects, such as the role of animal behaviour, emerge as important drivers of how fragmentation will affect biotic interactions, but they might be specific to a particular interaction type. In plant–frugivore interactions, movement behaviour and gut retention times of frugivores will influence seed dispersal kernels at the landscape scale (Box 5). Flower and fruit handling behaviour are strongly species-specific and will alter pollination and seed dispersal effectiveness in mutualistic networks. Furthermore, differences in sociality (e.g. solitary vs. social bees) will influence spatiotemporal abundances of individuals and resource specialisation. To some extent, such behaviours are phylogenetically conserved, so taxonomic identity can provide important information in this regard. Unfortunately, in many instances, we still know little about the natural histories of interacting species and the importance of link strength, especially in tropical regions, which at present constrains our ability to generalise about fragmentation effects on mutualistic and antagonistic interactions.

6. EFFECTS OF HABITAT FRAGMENTATION ON DIFFERENT KINDS OF NETWORKS

6.1. General introduction

Habitat fragmentation influences biodiversity at different organisational levels, from individuals to species populations, communities and multispecies ecological networks (e.g. Didham et al., 1996; Hill et al., 2011; Krauss et al., 2010). To date, little is known about how ecological networks of interacting individuals and species change in response to habitat fragmentation. Here, we address potential consequences for the structure of mutualistic and antagonistic networks (rather than only interactions *per se*, see Section 5).

6.2. Mutualistic plant–pollinator networks

Pollination networks are the most species-rich of all mutualistic networks, globally involving 88% of all angiosperm species, at least 1 million insect species belonging to several orders, about 1000 species of birds, hundreds of lizards and perhaps more than 100 mammals (Carstensen and Olesen, 2009; Olesen and Valido, 2003; Ollerton et al., 2011; Box 7). This translates into a rich functional diversity with respect to body size, morphology, mobility, behaviour and breeding systems, which further leads to a wide variety of adaptive strategies for locating, accessing and exploiting resources. These strategies vary in space and phenotypic plasticity, further complicating our efforts to predict outcomes of ongoing habitat fragmentation processes.

First, habitat fragmentation reduces overall species abundance in pollination networks and then later species and link richness (e.g. Aizen et al., 2008; Morales and Aizen, 2006), for example, butterfly species richness and

BOX 7 Diversity and Mobility of Pollinators

Pollinators are known from a wide variety of invertebrate and vertebrate taxa, but not all groups are equally represented in all networks. There is considerable spatial variation, for example, bird pollination is rare on mainland in Europe whereas it is common on European islands (Dupont et al., 2004; Kraemer and Schmitt, 1999; Olesen, 1985; Olesen and Valido, 2003; Ollerton et al., 2003). Similarly, bat pollination is common in the Tropics but virtually unknown from temperate or arctic regions (Proctor et al., 1996; however, see Ecroyd, 1993). The taxonomic diversity of flower-visiting animals translates into a broad range of species traits. For example, body size may vary up to 2000-fold, from tiny insects (e.g. wasps with a body length of 0.2 mm) to large mammals (e.g. flying foxes, up to 400 mm in body length), while body size in plant–frugivore networks may typically vary over one or two orders of magnitude between small birds and mammals (Fleming et al., 1987). The high diversity of pollinators results in different strategies for accessing and exploiting floral resources and in a high variability of how species respond to environmental disturbances (Kearns, 2001). For instance, flies show very complex and varied life histories, with larval habitats ranging from predatory through saprophytic and parasitic. In contrast, bees rely on floral resources during all their life stages (Michener, 2007). Thus, in flies, larval food supply might be more important for responses to habitat fragmentation than flower availability to the adult forms (Bankowska, 1980).

Foraging distances of pollinators range from a few metres to several kilometres (excluding migration), and almost all taxonomic groups contain sedentary as well as highly mobile species. For insects, which comprise the largest and most diverse group of pollinators, large amplitudes of foraging ranges have been reported: small solitary bees may fly only a few hundred metres whereas larger species can fly 10–20 km (Box 4). Much less is known about space use and foraging ranges of other pollinator groups, although in syrphid flies, a species-rich group of important flower-visitors, a few species may migrate over hundreds of kilometres (Torp, 1994), while resident species tend to stay within a very limited area. Beetles, a relatively minor group among pollinators, tend to be sedentary and less mobile than other groups (Proctor et al., 1996). Butterflies can be classified into three mobility classes: sedentary, intermediately mobile

Continued

BOX 7 Diversity and Mobility of Pollinators—cont'd

and migrant species. While migrants may disperse hundreds to thousands of kilometres, sedentary species are very local, often limited to one patch of food plants (Pollard and Yates, 1993). A small group of flower-visitors are the lizards, which appear to be important for pollination on islands (Olesen and Valido, 2003). Little is known about their foraging ranges (Nyhagen et al., 2001), but for the common and widespread, generalist flower-visiting endemic gecko *Phelsuma ornata* in Mauritius, 89% of marked individuals were re-sighted on the next day less than 15 m from the place of release, while maximum dispersal range was <90 m (Nyhagen et al., 2001). The foraging range of nectarivorous birds depends both on body size and behaviour (Craig et al., 1981; Gill and Wolf, 1975). For hummingbirds, these interconnected attributes can translate into different community roles (Feinsinger, 1978). For instance, some species are trap-liners tracking spatially dispersed flower resources in a repeated route whereas other species are territorial and defend clumped resources, highlighting the potential for behavioural traits to determine the network consequences of fragmentation (Laurance, 2004).

composition per fragment decline with fragmentation (Öckinger et al., 2010). This process is called network contraction (Fig. 8; Valladares et al., 2012).

In pollination networks, abundance of species is positively correlated to their linkage level (Fig. 9; Olesen et al., 2008; Stang et al., 2009). During fragmentation, some pollination systems may disappear completely as abundance declines (Girao et al., 2007). In pollination networks, plants are generally longer-lived than their pollinators, resulting in an accumulation of time-delayed plant extinctions (Krauss et al., 2010). Thus, rare specialist pollinators (linkage level ≤ 2 links to other species), which constitute about half of all pollinator species in networks, are the first to go (Olesen, 2000). However, fewer pollinator species in a network does not necessarily compromise the fecundity of all plants, because the outcome depends also on the effectiveness of the pollinators (Perfectti et al., 2009). It can even be beneficial if the most abundant pollinators are the most effective, because other pollinators, which might be less efficient or less specialised pollinators or even nectar and pollen robbers, disappear (Genini et al., 2010). However, according to a supposed positive complexity–stability relationship, fewer species and links in pollination networks lower their disturbance resilience (e.g. Okuyama and Holland, 2008).

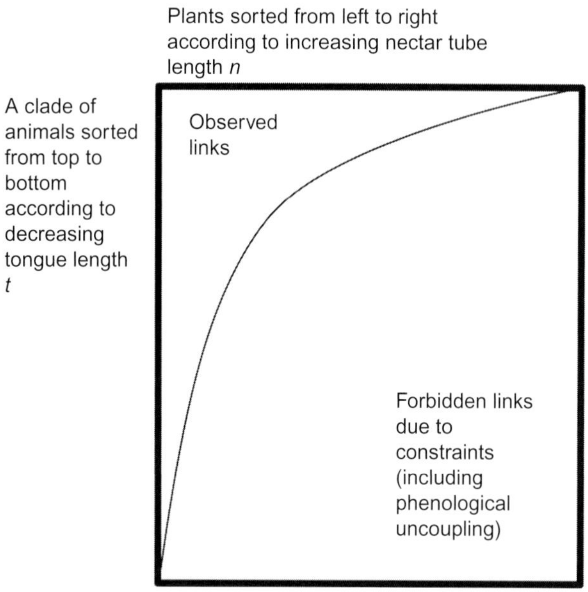

Figure 8 Pollination network contraction. During habitat fragmentation, pollinator specialists at the bottom left of the interaction matrix and plant specialists at the top right go extinct because of their low abundance. The first links to go extinct lie in concave bands running between lower left and upper right corners. Consequently, the matrix shrinks, that is, the links become more and more concentrated in the upper left corner of the matrix.

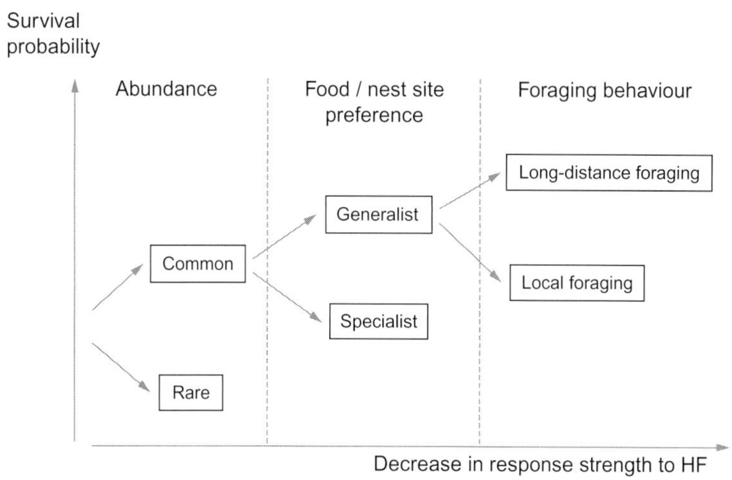

Figure 9 A simplified framework illustrating how the survival probability of pollinator species in response to fragmentation is hierarchically constrained by species traits. (For colour version of this figure, the reader is referred to the online version of this chapter.)

Local pollination networks trapped in single fragments tend to have higher connectance, because species number decreases and generalists are expected to survive better than specialists (Barbaro and van Halder, 2009; Girao et al., 2007; Koh, 2007; Steffan-Dewenter and Tscharntke, 2002; Williams, 2005; Williams et al., 2009; but see Ashworth et al., 2004). Furthermore, generalists may opportunistically switch or rewire their links depending on resource availability, making them less prone to secondary extinctions (Kaiser-Bunbury et al., 2010) by forcing new links closer to the upper left corner or the interaction matrix (Figs. 8 and 9). This will tend to make the pollinator community more homogenous (Ewers and Didham, 2006). Introduced species, which also tend to be generalists, tend to replace specialists, and this can influence the reproductive output of the plant community (e.g. Didham et al., 1996; Do Carmo et al., 2004) or more so (Dick, 2001).

6.2.1 Nestedness

The different ways networks are structured affect the dynamics of their communities and populations: identifying these patterns and their fundamental determinants makes it possible to predict the outcomes of habitat fragmentation. A distinctive property of mutualistic networks and food webs is their nested architecture (Fig. 8; Bascompte et al., 2003; Kondoh et al., 2010). Neutral models can be formulated to track interactions between two species with power law/lognormal (POLO) rank abundance distributions (Halloy and Barratt, 2007), that is, if individuals in two interacting species link randomly irrespectively of any species traits, except abundance ('the neutral theory of biodiversity'; Hubbell, 2001), then the link pattern becomes strongly nested, and even more so than in real networks. Abundance alone may explain 60–70% of nestedness in empirical networks (Krishna et al., 2008), although perturbations push communities away from a POLO distribution (Halloy and Barratt, 2007). The same neutral model with abundance variation also produces a nested pattern in plant–frugivore networks (Burns, 2006).

Abundance distributions show the importance of short-term disturbance regimes, whereas body-size distributions show more long-term community effects (Halloy and Barratt, 2007). Extending this to networks, certain nested link patterns to reflect systems at or close to equilibrium and deviations from such patterns may therefore be interpreted as a measure of disturbance: although this has yet to be tested formally, it could provide an important new biodiversity metric to gauge higher-level responses to environmental stressors.

6.2.2 Link switching

The strong effect of abundance is often evident, even in spite of the highly dynamical nature of linkage (Olesen et al., 2008, 2010b; Petanidou et al., 2008). An adaptive strategy to cope with spatiotemporal environmental dynamics is link switching or rewiring (Zhang et al., 2011). During network assembly or spatiotemporal changes in environmental conditions, linkage can become increasingly nested as species continuously switch or rewire partners to enhance their fitness gain from other species. Most often, these switches are to species with a higher abundance of more easily exploitable resources, that is, switches towards increasing abundance and trait matching ('The resource attraction principle'; Halloy, 1998). Thus, link switching can place a high selective premium on the ability of an individual to track resources by optimal diet choice and to exploit all resources above a given threshold quality (MacArthur and Pianka, 1966), that is, a more valuable resource becomes a more generalist node in the network, whereas a consumer with a lower choice threshold becomes a more generalist node (Kondoh et al., 2010).

When the landscape fragments, an increase in the intensity of fluctuations of species abundance is expected, and consequently, the ability to do link switching and resource tracking becomes increasingly critical. In mutualistic network models, including link switching into linkage assembly models increases the robustness of networks (Zhang et al., 2011). Consequently, species such as resource specialists that cannot track increasingly unpredictable resources are vulnerable to extinction. In networks, we have two kinds of specialists, ecological and evolutionary: the former because they are rare (or they feed on very few resources) and the latter because of their evolutionary history (low ability to switch resource). Thus, the loss of specialists disappear from networks during fragmentation may arise for different reasons.

6.2.3 Modularity

A commonly investigated linkage pattern in pollination networks is modularity (Olesen et al., 2007). The number of modules depends primarily on the size of the network. Modules may further have their own 'deeper' link pattern, for example, submodularity and subnestedness (Fig. 10). Modules are interconnected by species playing specific roles, viz., super-generalists or network hubs and connectors. Three per cent of species in plant–pollinator networks are super-generalists, linking to many species within and outside their own module; 11% are connectors with a few links, but a high proportion of these links to other modules (Olesen et al., 2007). In the early stages of fragmentation modules shrink in size, that is, the nestedness tails are 'cut off', and ultimately only the connectors and hubs are left leaving a topologically simplified network (Carvalheiro et al., 2011). This may initiate an irreversible transition phase or regime shift in network structure and dynamics (Kaiser-Bunbury et al., 2011), because at a certain size threshold modules begin to merge or even disappear. Through extinction and resource switching among generalists, the network slowly collapses by losing its modular structure.

6.2.4 Body size

Besides abundance *per se*, its close correlate body size also has a strong explanatory power of network properties. Body size is an important proxy for many ecological attributes in food webs (Woodward et al., 2005) and maybe also with respect to the response of pollinators to fragmentation. Body size has a huge span in pollinator communities, from tiny 1 mg parasitoids to the largest extant pollinator, the 3–4 kg Malagasy Black-White Ruffed Lemur (*Varecia variegata*), that is, a difference of six orders of magnitude. For comparison, in a lake food web, there may be difference of 10 orders of magnitude in body size (Woodward et al., 2005). However, the general relationship between linkage level (and thus network position) and body size in pollination network is not clear, although in Caribbean plant–pollinator networks, larger hummingbirds are more specialised than smaller hummingbirds (Dalsgaard et al., 2008).

An equivalent property of the flower is the extent to which floral rewards are accessible. Stang et al. (2006) reintroduced the term nectar-holder depth, that is, the depth from the opening of the flower and down to the surface of the nectar inside the flower. If a flower has a nectar-holder depth n, then legitimate pollinators have a tongue length $t \geq n$. If $t < n$, then the link between the species pair is 'forbidden', that is, morphologically constrained (Olesen et al., 2010a). Thus, the relationship between t and n becomes triangular, with generalist pollinators and plants having a high t and low n, respectively (Fig. 11; Corbet, 2000; Stang et al., 2006, 2007), as has been observed in several pollinator groups (Borrell, 2005). Tongue length and nectar-holder depth are both correlated with abundance, that is, abundant species have a high t or a low n. Since t and body size are positively correlated (Corbet, 2000; Stang et al., 2006) or, in fact, triangularly related (short-tongued pollinators vary considerably in body size, whereas long-tongued species are all large), large pollinators should, in theory, be more generalised. However, the evolution of a long tongue in insects

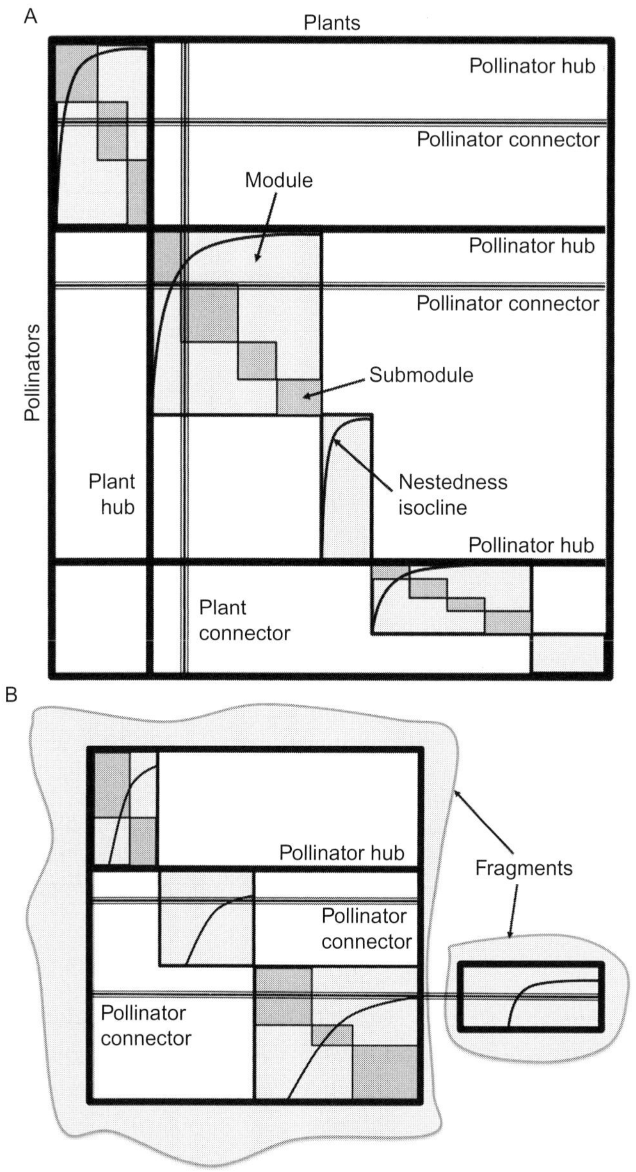

Figure 10 Expected scenario of the destruction of network modularity during habitat fragmentation. (A) An intact network in a non-fragmented landscape. The network has five modules, and three of these are submodular with several submodules (J. M. Olesen, unpublished data); two modules are so small that no modularity can be detected. However, the entire network and four of the modules have a level of nestedness that can be detected; this is indicated with the curved 'isoclines' *sensu* Atmar and Patterson (1993).

continued

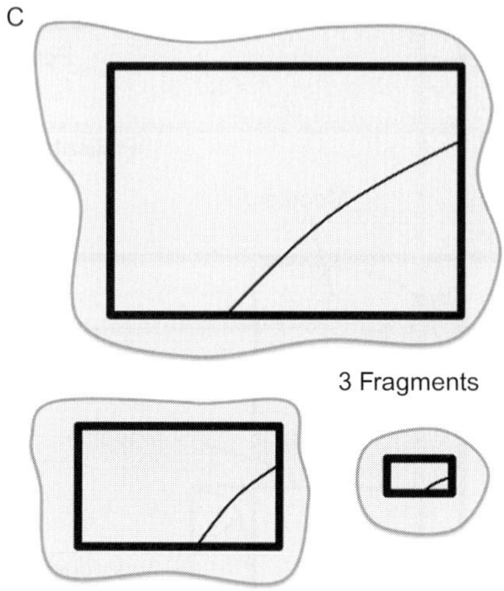

Figure 10—cont'd Most links are to the left of the isocline. Hubs and connectors are shown as bars. (B) The progressing habitat fragmentation has now caused the network to fragment as well. The network is present in two fragments: a large and a small one, and is only connected by one pollinator species. Many of the specialists of both pollinators and plants are gone and only three modules are left in the large fragment. The plant community has mainly lost its outcrossing herbs. The upper left two modules are the same as in (A), whereas the central one is the result of fusion of two modules in (A). This increases connectance as shown by the change in position and shape of the isoclines. A few submodules are still left. (C) The network has now got its modularity completely destroyed by habitat fragmentation. The entire network is now reduced to three single independent modules, each isolated in their own fragment. Most species remaining are generalists, and connectance is high. Many plants from (B) are still alive. They are selfing herbs and long-lived trees, and some of them constitute an extinction debt.

may be a generalist strategy as it allows pollinators to exploit a higher diversity of flowers. Borrell (2005) observed the same triangular relationship between tongue length of euglossine bees and nectar tube length (Fig. 11). In fact, the relationship is, upon closer inspection, more trapezoid-like, indicating that long-tongued bees may have problems with nectar extraction from shallow flowers and that super-generalists have an intermediate tongue length (Fig. 12). These details are, however, still poorly explored.

Stang et al. (2007) simulated extinction scenarios based on field data and found that if abundance is the only determinant then there is no difference in extinction risk between generalists and specialists, whereas an inclusion of

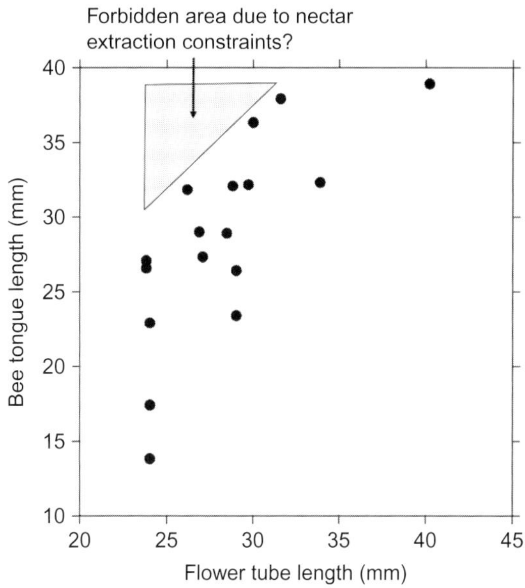

Figure 11 A triangular relationship between flower tube length (*n*, nectar-holder depth, *sensu* Stang et al. (2006)) and euglossine bee tongue length (*t*) (data from Borrell, 2005). Upon a closer look, the relationship may be trapezoid, because bees with the longest tongue have problems handling shallow flowers with easily accessible nectar.

nectar-holder depth and tongue length constraints gave an increased extinction risk with increasing *n* and decreasing *t*.

Body size is also related to mobility and reproduction (e.g. Greenleaf et al., 2007; Nieminen, 1996; Öckinger et al., 2010; van Nieuwstadt and Ruano Iraheta, 1996; Woodward et al., 2005). Expectations are that (i) highly mobile species are less affected by fragmentation than less mobile species; (ii) specialists require larger fragments to fulfil their demands and are also less likely to use the surrounding matrix than generalists and (iii) *r*-species are expected to suffer less from fragmentation than *K*-species, because of their higher reproductive output, which means relatively more emigrants to other fragments. All three hypotheses were confirmed in a study by Öckinger et al. (2010).

The mobility of pollinators affects their population dynamics, genetic structure and life history but also the other species with which they interact (Greenleaf et al., 2007), for example, large-bodied pollinators mediate a longer pollen flow, but also require more energy from their flowers. In many taxa, mobility increases non-linearly with body size (e.g. Steffan-Dewenter and Tscharntke, 1999). The specific movement pattern is of importance

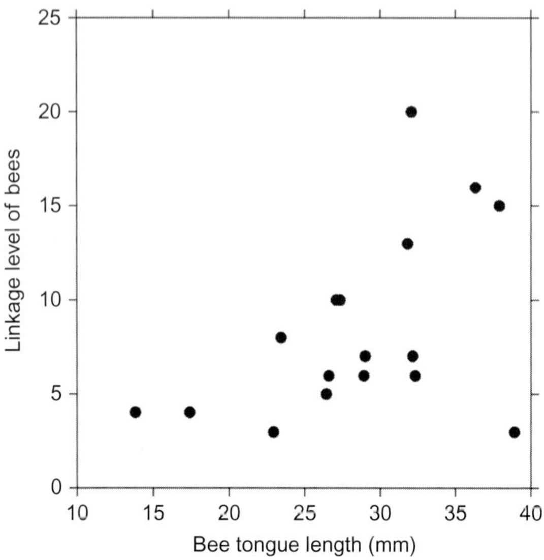

Figure 12 Linkage level, that is, number of visited plant species, of euglossine bees is an increasing function of their tongue length (data from Borrell, 2005). However, linkage level seems to peak between 30 and 35 mm, and the bees with longest tongue avoid some flowers (Fig. 11).

here and the influence of a certain landscape configuration on pollinator movement behaviour can also be incorporated into the analysis (Lander et al., 2011). Some species mediate a more linear pollen flow, for example, large bees and trap-lining hummingbirds, than others and such species may be key hubs or connectors.

In pollination networks, plants with limited modes of attracting pollinators over long distances suffer most from isolation. For example, visual cues tend to be more spatially restricted than scent, which can attract pollinators over considerable distances, for example, hawkmoths (Dudareva and Pichersky, 2006). Amongst generalist pollinators, those that can forage over longer distances due to morphological and behavioural traits can access distant, more isolated resources, and this increases their chance of persistence in fragmented pollination networks. It is important to highlight that this relationship occurs under increasing isolation scenarios, while habitat loss *per se* is likely to have the strongest adverse effects on large-bodied, long-distance flying animals with high resource requirements.

Plants differ in their dependence on pollinators and seed set by obligate selfers, for instance, should be unaffected by habitat fragmentation, whereas facultative selfers may be more affected in terms of seed quality than quantity. Wind-pollinated plants are also expected to suffer less from fragmentation than those pollinated by animals due to their long-distance pollen flow. An important determinant of linkage level in the network-participating plant community is flower morphology, especially level of flower openness (accessibility to the interior of the flower), which should increase with the number of pollinator species, although in reality the relationship is more complex (Olesen et al., 2007).

6.2.5 Four fragmentation scenarios

Assuming that the response of pollinators to habitat loss and fragment isolation is driven by body size, which could be true for some pollinators such as birds or specific bees, we can outline a simplified framework of how plant–pollinator networks will change in response to fragmentation (Fig. 13). In a system with large and poorly isolated fragments, a plant–pollinator network will consist of many links, including small-, medium- and large-bodied pollinators (Fig. 13A). If fragments become smaller in size, but are similarly isolated, resource availability and nesting sites will decline to critical levels, forcing species to move between fragments to maintain population sizes. Very small species with low resource requirements are more likely to survive, but species and link diversity of intermediate species with low mobility should decline due to a lack of resources within single fragments. Large species, however, should decline due to limited resource availability across fragments in the landscape (Fig. 13B). Maintaining large fragments but increasing the level of isolation will have a weak impact on small species as they can persist within fragments. Large species are likely to survive as they can move between distant fragments due to their large foraging ranges or dispersal abilities (Fig. 13C). The most affected species are expected to be those of intermediate size, with habitat requirements exceeding the fragment size but are unlikely to move the large distances between fragments. The worst-case scenario is that only small fragments remain that are separated by relatively large distances (Fig. 13D). Then, only some small and maybe intermediate generalist species will be able to persist and movements among fragments will be rare. As a consequence, the network is strongly depleted and highly skewed towards small species (Fig. 13D). Given this simplified framework, the number of links in a plant–pollinator network is expected to change in predictable ways as a consequence of habitat loss and isolation (see the two graphs to the right in Fig. 13).

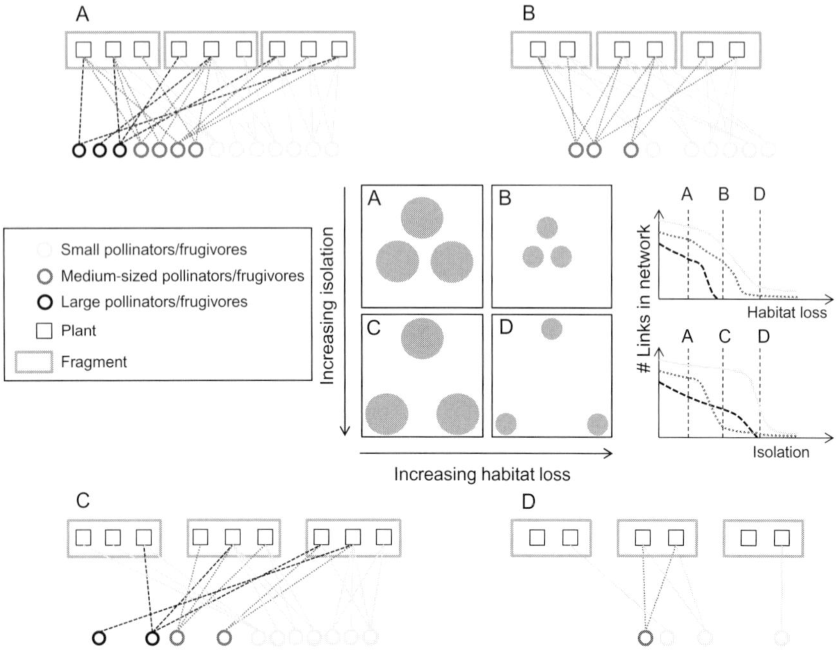

Figure 13 Simplified framework for the response of pollination and seed dispersal networks to habitat loss and isolation. The illustrated framework assumes that body size is the key trait for the response of pollinators and frugivores to fragmentation. In (A), a system with large and poorly isolated fragments contains a plant–pollinator/frugivore network with many links, including small-, medium-sized and large-bodied animals. In (B), fragments become smaller in size (but with a similar degree of isolation), resulting in a decline of small and intermediate species with low mobility and a loss of large-bodied species. In (C), large fragments have an increased level of isolation with weak impacts on small species and more pronounced effects on intermediate and large-bodied species. In (D), a landscape with small and isolated habitat fragments only sustains some small, and maybe intermediate, generalist pollinators or frugivores. The bipartite networks depict hypothetical pollination or seed dispersal networks covering the entire landscape. The number of plant species is kept constant. The two graphs on the right hand illustrate how the number of links in these plant–animal networks changes as a consequence of habitat loss and isolation under this simplified framework. (For colour version of this figure, the reader is referred to the online version of this chapter.)

6.3. Mutualistic plant–frugivore networks

Frugivores include a large diversity of taxa, from annelids to elephants, and fish and herps, spanning body masses from a few grams to several tonnes. Plants that produce fleshy fruits and rely on animals for seed dispersal are also diverse and differ in fruit size, seedling vigour, phenophase length and so on. In terms of the potential effects of habitat fragmentation, it matters which critical frugivore or plant life strategies are correlated within a network. For example, rare species might be more prone to local extinction following fragmentation (Davies et al., 2004) but they could be occupying peripheral positions in the network, or may be central species. Body mass influences population viability in fragmented landscapes (see e.g. Galetti et al., 2009 for mammals), but we are not aware of any studies to date that have mapped this onto plant–frugivore networks.

The overall response of such networks to fragmentation will depend on the array of species traits in the interacting assemblage. Differential responses and susceptibility among frugivore species will cause variation in incidence functions (Gilpin and Diamond, 1981) of each species across fragments in a complex landscape (Fig. 14), determining variation in survival probability in

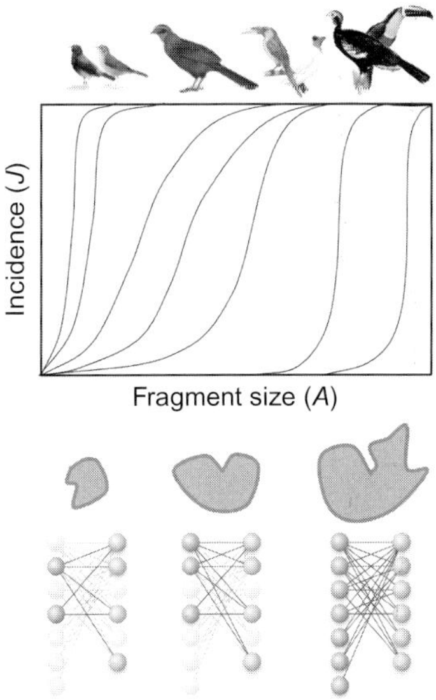

Figure 14 Incidence functions of frugivore species along a gradient of habitat loss (fragmentation). Incidence functions (top) represent the fraction of habitat patches of a given size where a frugivore species is present. Large-bodied frugivores will most likely disappear from small- and medium-sized fragments, while small-bodied frugivores would be the only species present in the small remnants. Variable incidence functions will thus result in differences in specific composition (species richness, relative abundance) of different fragments which, in turn, will cause large variations in network topology and structure (bottom). (For colour version of this figure, the reader is referred to the online version of this chapter.)

fragments of variable area. This will typically result in different richness and composition of the local plant–frugivore assemblages among fragments, with reciprocal influences between them (Kissling et al., 2007). Patterns of fragment occupation will be driven by colonisation/extinction dynamics, which will depend on how species respond to loss of habitat area and/or increasing distance and isolation among fragmented patches (Luck and Daily, 2003). While Fig. 14 illustrates the depauperation of frugivore assemblages, a similar scenario could be envisaged for fruiting plants, showing, for example, variable incidence functions associated with seed mass or fruit-size variation. The figure is inspired by trends in the composition of avian frugivore assemblages in the Atlantic forest of SE Brazil (Fadini et al., 2009; M. Galetti, personal communication; also see Estarada et al., 1993; Githiru et al., 2002; Graham, 2002). This highly fragmented landscape is impacted not only by habitat loss processes but also by different levels of hunting and poaching that, taken together, drive dramatic local changes in frugivore abundance across fragments (see e.g. Almeida-Neto et al., 2008; Galetti et al., 2009). Large tracts of Atlantic rainforest harbour reasonably complete frugivore assemblages and associated dispersal services to the plants (Fig. 14), yet the smaller fragments contain impoverished local communities that invariably lack the larger frugivores, such as toucans, large cracids and cotingids, whereas the dominant frugivores are thrushes and thraupids. The overall effect is highly transformed interaction networks in the fragments (Fig. 14, bottom) with reductions in degree, and potentially drastic increases in modularity due to loss of large super-generalist frugivores. This also reduces nestedness, largely due to the missing 'glueing' interactions that the generalists provide (Olesen et al., 2010a).

Plant–frugivore networks could exhibit similar responses to fragmentation to those described for pollination networks (Fig. 13), as the main relevant traits (e.g. body mass) are similar. The plant–frugivore networks in landscapes with large and well-connected fragments will harbour reasonably complete networks, with diverse interactions in nested assemblages (Fig. 13A). Most frugivorous birds, for instance, include generalised foragers with flocking behaviour and seasonally altitudinal migrants; many should have high mobility and dispersal abilities. Santos et al. (1999) reported that drastic alterations of local thrush assemblages in juniper fragments in central Spain mainly occur in the smallest fragments (also see Luck and Daily, 2003). If fragment area becomes reduced, but still maintaining good connectivity, some large species may still be lost because of reduced home range sizes and resource abundance. As for plant–pollinator networks, small species with reduced resource requirements are likely to survive (e.g. small avian frugivores with mixed diets, not relying extensively on fruit), while species and link diversity of small- and medium-bodied species with low mobility and large species should decline (Fig. 13B). With increased isolation, small species may persist in medium-sized and even small fragments, whereas the persistence of larger species will depend on their ability to disperse among fragments (Fig. 13C). As with the pollination networks, seed dispersal interaction networks in landscapes with both reduced fragment area and poor connectivity should be more prone to collapse (Fig. 13D; see e.g. Santos et al., 1999). Then, only some small, and maybe intermediate, generalist species will be able to persist and movement among fragments will be rare. The small fragments cannot support large species, and the network is again strongly biased towards a few small species (Fig. 13D).

Fragmentation and habitat loss will ultimately induce the loss of specific nodes (either plants or animals), reduced population densities of mutualistic partners, resulting in dramatic losses of important functional attributes. For example, in some Pacific islands, populations of flying foxes are periodically reduced by hurricanes to a point beyond which their capacity to disperse the seeds of big-seeded trees decreases dramatically (McConkey and Drake, 2006). Such functional losses will not take place at random, but will be concentrated in certain species, like larger frugivores and large-fruited plants. In summary, the main consequences of fragmentation for plant–frugivore networks will depend upon the extent that key traits determining susceptibility of species correlate (or match) with traits that define their functional roles in the network.

6.4. Mutualistic plant–ant networks

Symbiotic and free-living plant–ant mutualisms are organised in networks that differ markedly in their structure (Blüthgen et al., 2007; Fonseca and Ganade, 1996; Guimarães et al., 2007): for example, those that include extrafloral nectaries are often nested, whereas symbiotic, plant–ant networks are always strongly modular (Guimarães et al., 2007). These correlations between biological attributes and network structure can be used to infer likely responses to habitat fragmentation.

If habitat fragmentation affects ant species of distinct body sizes differently (see Section 5.3), the same will be true for the highly and poorly connected species. At present, the underlying mechanisms linking ant body size to the number of interactions and the degree of overlap among partners are unknown, making it difficult to predict the consequences for species networks, even if there is a clear body size-biased effect of habitat fragmentation on species composition (however, see Chamberlain and Holland, 2009). Thus, it is fundamental to develop a better understanding of how ant body size is related to network structure in plant–ant interactions to predict the fate of these networks facing habitat fragmentation.

Phylogeny is an important predictor of the structure of symbiotic networks (Fonseca and Ganade, 1996), which are composed of modules that often contain closely related ant and/or plant species. This strong association between phylogeny and network structure is predicted as a consequence of a 'complex coevolutionary handshaking' among interacting partners (Thompson, 2005). This relationship should enable responses of plant–ant networks to fragmentation to be predicted, if sensitive groups of taxa can be identified *a priori*: if the phylogenetic signal is very strong, such as in symbiotic plant–ant interactions, susceptibility traits and traits shaping the role of a species within a network are likely to be strongly correlated.

Key questions that need to be addressed include how nestedness will alter with changes in ant species richness and composition: the current evidence, although still limited, suggests the nested structure of extrafloral nectary networks to be robust to species turnover and invasions (Diaz-Castelazo et al., 2010). It is also important to understand how the strong modularity of symbiotic networks is affected by habitat fragmentation, which has the potential to cause the emergence, loss or even fusion of modules (e.g. via invasions of generalist ant species). In a fragmented landscape, one could imagine the creation of a mosaic of plant–ant networks varying in species composition and consequently in nestedness and modularity.

6.5. Antagonistic food webs

The effects of fragmentation on food webs have been surprisingly overlooked. In terrestrial systems, we can envisage fragmented networks in the classical biogeography sense when they are situated within islands within an aquatic matrix. An example of this comes from recent work carried out in Ireland (McLaughlin et al., 2010). The Gearagh woodland, located in the floodplain of the River Lee in County Cork, is composed of a complicated braided river system composed of approximately 13 channels, each 1–7 m wide. The main channels are stabilised by tree roots, which create a mosaic of small islands due to the accumulation of detrital material and fallen trees over time. A food web study, examining the trophic structure of the invertebrate community on series of 16 islands, ranging in size from 4.5 to

40.8 m^2 found that, on average, the larger islands contained more species and links than the smaller islands, and network structure consequently differed markedly among fragments (Fig. 15).

Fragmentation of food webs can also occur in other lateral (i.e. across landscape) and temporal dimensions, as well as via fractal branching pattern

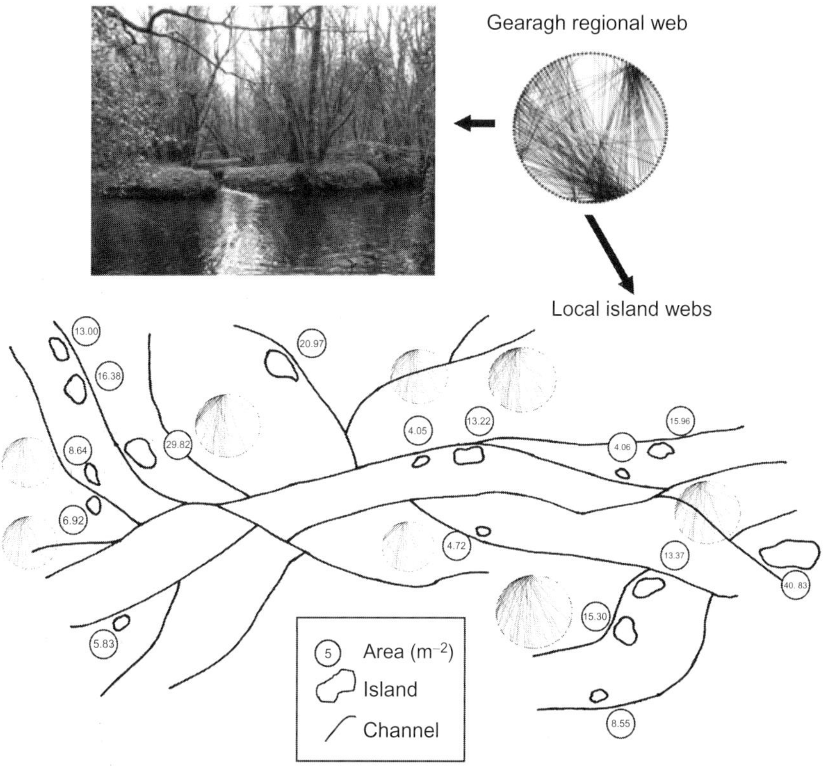

Figure 15 Schematic representation of the riverine network with the Gearagh forest, Ireland. Individual islands are inserted beside the river channel in which they were located (McLaughlin et al., 2010). The Gearagh is a complicated braided river system composed of approximately 13 channels, each 1–7 m wide. The study site was comprised of a small proportion of these channels. The stabilising effect of the tree roots within the main river channels, in conjunction with the accumulation of detrital material and tree falls, has resulted in the above mosaic of small islands. The diameter of the web from each island is scaled linearly with species richness: the larger webs are found in the larger fragments. *Note*: each web contains the same number and positioning of nodes as in the global web: solid black nodes represent macroinvertebrate taxa present within the depicted web and grey nodes indicate taxa present in the global web but absent from the depicted web. (For colour version of this figure, the reader is referred to the online version of this chapter.)

dimensions (e.g. in river networks) (Box 1; Fig. 16). Additionally, vertical fragmentation, which is even more rarely considered, can occur, such as in mountainous regions (Box 8; Fig. 17).

The loss of large consumers at higher trophic levels due to habitat fragmentation should result in a decreased overall trophic height of the food web, driven by shorter food chains (e.g. Byrnes et al., 2011; O'Gorman and Emmerson, 2009; Woodward et al., 2012). This could also lead to an increase in the proportion of top consumers relative to intermediate species, as the latter are effectively promoted to the termini of food

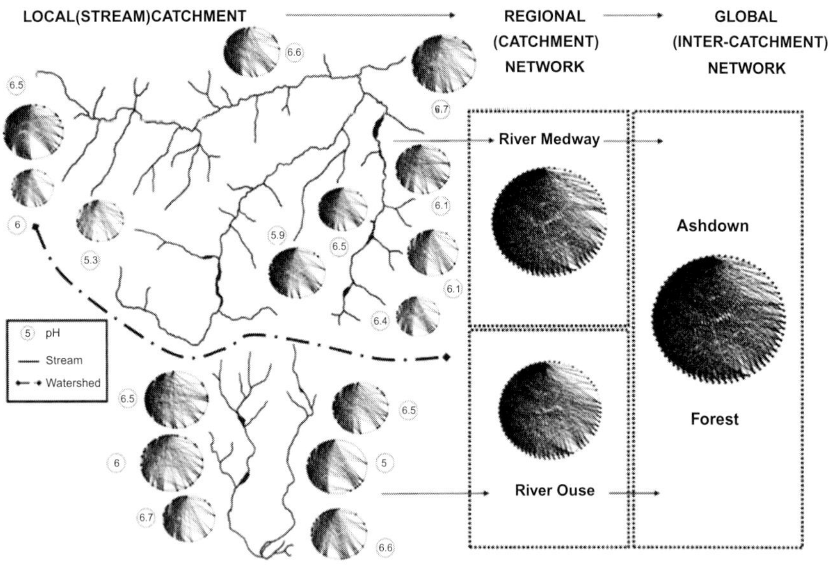

Figure 16 Ecological network structure of stream food webs from the Ashdown Forest, UK, shown from local to regional to global networks. *Note*: each web contains the same number and positioning of nodes as in the global web: solid black nodes represent macroinvertebrate taxa present within the depicted web and grey nodes indicate taxa present in the global web but absent from the depicted web (Fig. 15). Web diameter has been scaled to the number of nodes as a % of those in the global web: thus the smallest web also contains the fewest species. All streams are headwaters of either (a) River Medway or (b) River Ouse, which are separated into discrete watersheds (separated by the dashed east–west line) that flow predominantly either north or south into the sea. Individual networks are constrained by the 'hard' boundary of the water's edge and the 'soft' boundary of a physiochemical gradient (indicated by mean stream pH, within circles adjacent to each web). All individual streams can be viewed as a fragmented component of the catchment network, which in turn is a component of the global network. The increasing complexity of the network can be seen as the number of nodes and consequently the number of interactions increases once the fragmented nature of the landscape and habitat is discounted.

BOX 8 Fragmentation of Mountainous Aquatic Food Webs

Habitat fragmentation is typically considered in lateral (i.e. across landscape) and temporal dimensions, but vertical fragmentation is also possible, for example, in mountain ecosystems. At high altitudes, glacier retreat and changes to the magnitude of snowpack accumulation and their duration are likely to cause major changes to aquatic ecological networks within an already fragmented landscape (e.g. Brown et al., 2007, 2012; Finn et al., 2010; Jacobsen et al., 2012; Milner et al., 2009). There are strong upstream to downstream gradients in aquatic biological assemblages in these systems, driven predominantly by changes in stream water temperature and the geomorphological stability of the river channel (Milner et al., 2001). Consequently, alpine river food webs are highly fragmented along even short distances (kilometres), with high turnover of species, food web links and species' contributions to secondary production (e.g. Fig. 17). In non-glacial mountain rivers, altitudinal pressure effects on the saturation of dissolved oxygen can impart major effects on community composition (Jacobsen, 2008). Montane aquatic ecosystems that rely on meltwater are particularly susceptible to fragmentation, particularly in situations where decreases in meltwater production lead to drying of some river sections (e.g. Malard et al., 2006). Natural occurrences of river ecological network fragmentation are also evident where lakes introduce discontinuities into the system (Milner et al., 2011; Monaghan et al., 2005). Alpine lakes lead to notable changes in community composition and the relative abundance of morphological and biological traits relative to the nearby flowing waters, but may be insufficient to prevent insect dispersal and thus genetic differentiation within river valleys (Monaghan et al., 2002). Fish may be restricted to lower altitudes due to thermal or geomorphological barriers (e.g. falls, canyons; Evans and Johnston, 1980), thus preventing their upstream migration to avoid warming. Therefore, the more productive and species-rich aquatic food webs at lower altitude sites (e.g. Fig. 17) may fragment as some mobile organisms such as invertebrates are able to migrate to higher altitudes. The immigration of 'lowland' species to higher altitudes may also upset the balance of these food webs, causing fragmentation but also succession. Additionally, at higher latitudes, there may be fragmentation as the range of some amphibians (e.g. Pyrenean Brook Newt, *Calotriton asper*) expands from currently clear water habitat (Parc National des Pyrénées, 2005) into glacier-fed rivers that are receiving less meltwater (and proportionally more groundwater) with glacier retreat.

chains as the largest higher-level predators are lost (see O'Gorman and Emmerson, 2010; Woodward et al., 2012). Loss of large species at high trophic levels is also likely to result in reduced linkage density (Montoya et al., 2005; O'Gorman et al., 2010) and connectance (O'Gorman and Emmerson, 2010) within local networks, as well as reduced

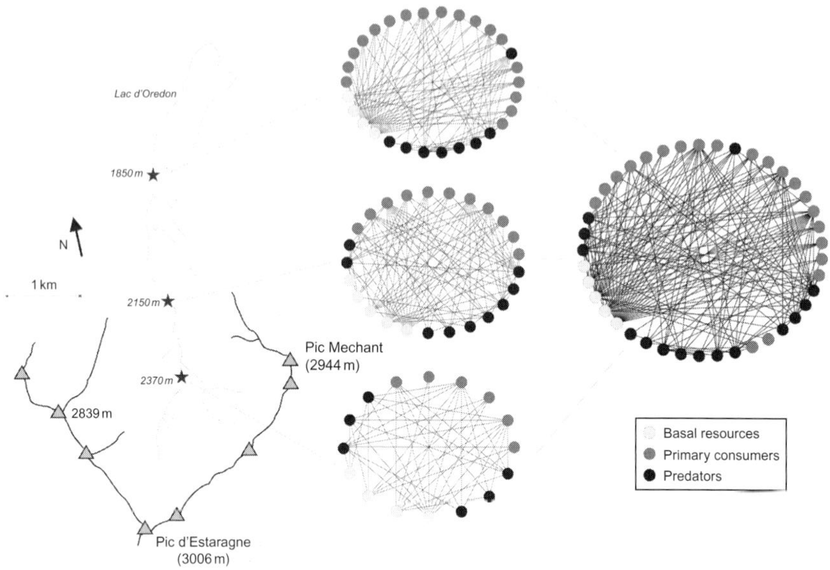

Figure 17 Stream benthic food webs along an altitudinal gradient in the Estaragne catchment, French Pyrénées. Light grey circles denote basal resources; dark grey denotes primary consumers; black denotes predators. Three food webs are displayed for (i) 2370 m altitude, maximum water temperature (T_{max}) = 4.5 °C, no. species (S) = 16, no. links (L) = 46, secondary production ($2P$) = 4.9 g m^{-2} year^{-1}; (ii) 2150 m altitude, T_{max} = 8.5 °C, S = 25, L = 93, $2P$ = 6.55 g m^{-2} year^{-1} and (iii) 1850 m altitude, T_{max} = 138 °C, S = 30, L = 87, $2P$ = 7.6 g m^{-2} year^{-1}. The individual food webs are fragmented as the individual study sites are separated by soft boundaries. Together, these food webs combine to a composite web of 41 species with 164 links. *Figures redrawn from Lavandier and Décamps (1983) and Lavandier and Céréghino (1995).*

compartmentalisation, which could make the web less robust to secondary extinctions (Dunne et al., 2002), although this is not necessarily the case if there is high redundancy in the system (Woodward et al., 2012). Large species may have weak per unit biomass interactions with their prey and high functional uniqueness (O'Gorman et al., 2011), so their extinction could increase the overall interaction strength within the system. This may reduce stability (see McCann et al., 1998; Neutel et al., 2002), while loss of functional trait diversity will alter ecosystem process rates and functioning (Petchey and Gaston, 2006).

Body-mass-driven extinctions due to habitat fragmentation may cause an overall increase in the predator–prey body mass ratio, assuming that larger predators eat prey closer to their own body mass (Brose et al., 2006). Smaller predator–prey body mass ratios have been linked to longer food chains due to their stabilising properties (Jennings and Warr, 2003; Jonsson and Ebenman, 1998; however see Mulder et al., 2009; Reuman et al., 2009), so increases could raise the probability of catastrophic phase shifts or total collapse. Conversely, in systems where large predators are considerably larger than their prey (e.g. fish eating invertebrates vs. invertebrates feeding on other invertebrates), the loss of these consumers could increase stability of the food web, as appears to be the case in headwater streams where fish are lost due to habitat loss and fragmentation arising from chemical and/or physical barriers (Layer et al., 2010, 2011).

The response of freshwater food webs to fragmentation by droughts (Box 1) has been characterised recently by manipulating flows in a series of artificial stream mesocosms (Ledger et al., 2008, 2011, 2012; Woodward et al., 2012; Fig. 18). These model systems reflected the abiotic conditions, biodiversity and food web properties of natural streams (Brown et al., 2011; Harris et al., 2007; Ledger et al., 2009). The results of this fragmentation experiment revealed some dramatic impacts on the food webs: consistent with the *higher trophic rank hypothesis* (e.g. Holt, 1996), top predators' production declined by >90%. Among the primary consumers, production of shredder detritivores was also suppressed (by 69%), whereas the base of the food web was relatively unaffected (Ledger et al., 2011, 2012). Contrasting responses were evident among functional groups, ranging from extirpation to irruptions in the case of small midge larvae, although production of most species was suppressed. The ratio of production to biomass increased, reflecting a shift in production from large, long-lived, taxa to smaller taxa with faster life cycles (Ledger et al., 2011). Fragmentation by drought caused high mortality and the partial collapse of the food web from the top-down (Ledger et al., 2012) as well as reversing successional dynamics of benthic algal assemblages (i.e. basal resources), with effective colonists replacing competitive dominants (Ledger et al., 2008, 2012). The general shift in biomass flux from large to small species could not fully compensate for the overall biomass flux. Many other network characteristics (e.g. connectance) were, however, conserved, suggesting some higher-level properties might be conserved even when exposed to extreme perturbations (Woodward et al., 2012).

Fragmentation can also affect marine food webs (Box 1). Coral bleaching creates fragments of surviving coral surrounded by reef pavement and coral rubble, with consequences for top-down control as average food chains shorten, generalist species proliferate and phase shifts may occur (Hughes, 1994). Simulations of fragmentation processes in Caribbean coral reefs indicate that species losses due to body size or diet constraints will lead to decreases in

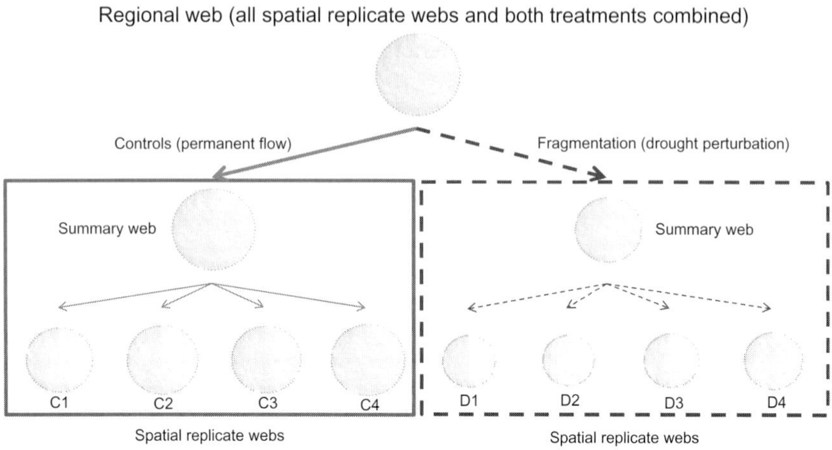

Figure 18 Impacts of habitat fragmentation caused by drought in experimental stream food webs: results from a long-term field experiment in artificial streams (Brown et al., 2011; Ledger et al., 2008, 2009, 2011). Drought can have patchy effects in river networks and individual stream channels can be viewed as fragmented patches in the wider riverscape. Note the two experimental treatments (monthly drought disturbance vs. permanent flow) were randomised spatially among the eight stream channels, but are grouped into two blocks here for illustrative purposes. The diameter of the circular webs is scaled according to species richness relative to the global web for the combined network. Solid nodes represent species present in a given web; open nodes represent those found in the global but not in local web (Figs. 16 and 17). Droughts simplified the networks with marked impacts on large rare species high in the web. (For colour version of this figure, the reader is referred to the online version of this chapter.)

number of links and changes in connectance and food chain length (Fig. 19). Human-induced fragmentation in seagrass food webs could further lead to fewer trophic groups and top predators, lower maximum trophic levels, shorter food chains and prey-dominated communities (Coll et al., 2011). In kelp forests, habitat loss and fragmentation due to storms simplify marine food webs, mainly by decreasing diversity and complexity at higher trophic levels, resulting in shorter food chains (Byrnes et al., 2011). The effects of habitat fragmentation on food webs, although little studied, can be pronounced.

6.6. Antagonistic host–parasitoid networks

Besides food webs, several examples from other multitrophic systems give an indication of how antagonistic host–parasitoid networks may be affected by fragmentation (Cronin, 2004; Kruess, 2003; Thies et al., 2005). However,

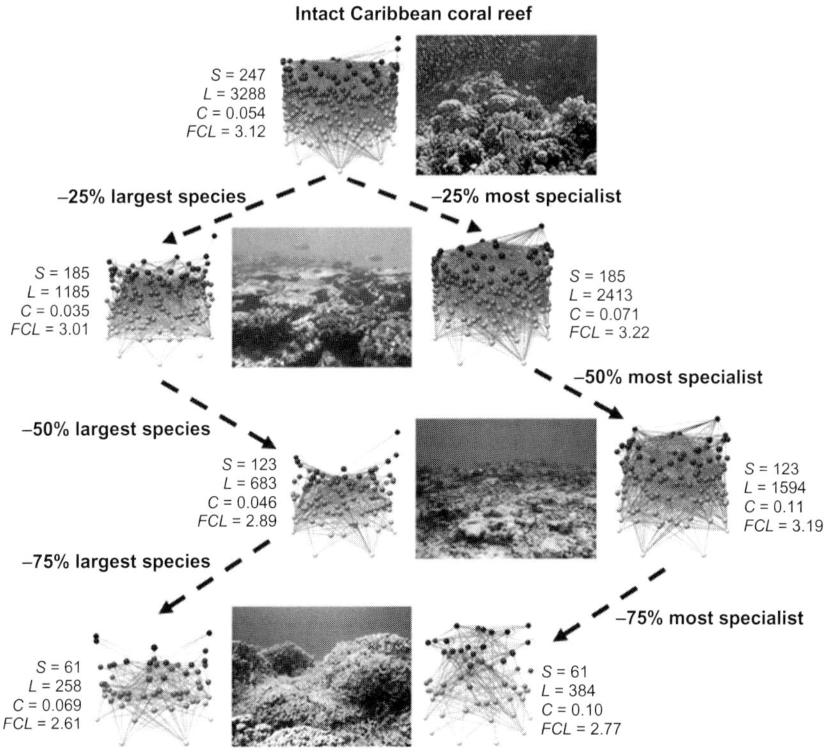

Figure 19 Simulated consequences of fragmentation-driven extinction scenarios on the network properties of a Caribbean coral reef. As species (*S*) are lost according to body size, the number of links (*L*) in the web decreases exponentially, leading to unpredictable fluctuations in connectance (*C*) and a linear decrease in mean food chain length (FCL). As species are lost according to diet specialisation, *L* decreases linearly, leading to an overall increase in *C* and FCL until a critical threshold is reached and the system undergoes a phase shift to a new state (e.g. macroalgae dominated). Coral reef photos are used by kind permission of José Eduardo Silva, Stephen Leahy, Nick Graham and James Acker (respective photo credits, from top to bottom). (For colour version of this figure, the reader is referred to the online version of this chapter.)

because species respond differently to fragmentation effects, it is currently not possible to predict whether some will compensate for others, and therefore how overall parasitoid–host network structure will be affected, although progress is being made in this area. For instance, in restored and adjacent intensively managed meadows, the abundance and parasitism rates of bee hosts decreased with increasing distance from restored meadows and the diversity of interactions declined more steeply than the diversity of species (Albrecht et al., 2007). This suggests a strong impact of habitat fragmentation on trophic networks and that interaction diversity might decline more rapidly than species diversity in fragmented systems. Another study examined host–parasitoid networks of specialist leafminers and their parasitoids on individual oak (*Quercus robur*, Fagaceae) trees in different landscape contexts (Kaartinen and Roslin, 2011). Isolated patches had fewer species and different composition than well-connected patches, but the quantitative metrics of network structure (interaction evenness, linkage density, connectance, generality or vulnerability) were unaffected, indicating some degree of functional compensation across species. More case studies are now needed to test the generality of fragmentation effects in host–parasitoid networks.

6.7. General effects of habitat fragmentation on network properties

The examples above illustrate that the properties of mutualistic and antagonistic networks can be strongly affected by habitat fragmentation, although this field is still very much in its infancy (Burkle and Alarcón, 2011; Fortuna and Bascompte, 2006; Gonzalez et al., 2011). Simulation studies indicate that mutualistic networks can be buffered to some extent against habitat fragmentation (Fortuna and Bascompte, 2006). Real communities might persist for longer but start to decay sooner than randomly generated *in silico* communities, with resilience against fragmentation being provided by degree or link heterogeneity (Jordano et al., 2003), nestedness (Bascompte et al., 2003), compensatory responses and/or redundancy (Ledger et al., 2012).

Species and link richness vary with habitat area, with the latter seemingly being more sensitive to fragmentation than the former (Sabatino et al., 2010), that is, as a local habitat shrinks, interactions are lost faster than species. This might be related to a reduced abundance of species (without initially going extinct), which reduces interaction probability (encounter rate). It might also be a consequence of several species having more than just one interaction, although ecological networks are highly skewed (Jordano, 1987). Habitat fragmentation influences the strength and timing of species interactions, which can cause cascading secondary extinctions in networks (Solé and Montoya, 2006; Terborgh et al., 2001; Tylianakis et al., 2008).

Nestedness and other network structure parameters are often determined by relative species abundances (Krishna et al., 2008). Given that habitat fragmentation reduces abundance (Hadley and Betts, 2012), nestedness should change with increasing fragmentation. Fragment area and trophic level or dietary guild identity are likely to influence the degree of nestedness in fragmented landscapes (Hill et al., 2011). Furthermore, effects of vegetational aggregation (clustering of plants in a landscape) and mobility of species can affect network properties, especially in antagonistic and plant–frugivore networks, while these influences on plant–pollinator network structure may be less pronounced (Morales and Vázquez, 2008).

Habitat fragmentation can also influence network substructure (modularity or compartmentalisation) and the extinction of top consumers may disconnect spatially segregated ecological networks and thus increase modularity. The opposite effect may be triggered by the invasion of hyper-generalist species, which connect distinct modules and reduce modularity in fragmented landscapes (Aizen et al., 2008). At some point, the local network must reach a critical level, below which modularity no longer exists. Thus, the modular structure disintegrates before the local network disappears completely. Using a spatial network approach, modularity analysis may lump similar fragments together based on their constituent species (for a biogeographical example, see Carstensen et al., 2012). Fragments within the same landscape might therefore have more similar dynamics and trajectories of change in species composition than those in other landscapes ('landscape-divergence hypothesis'; Laurance et al., 2007), which could be tested with modularity analyses if data from several fragmented landscapes are available.

Fragment size and isolation affect the composition of ecological networks: while large areas can support most interactions needed for normal functioning, small fragments will contain only a core group of species and fewer important interactions (see Section 7). The degree of specialisation of a species will determine whether it can persist, with generalist mutualists being least likely to suffer extinction (Fortuna and Bascompte, 2006). Matrix quality also determines the impact of fragmentation on networks as it defines landscape permeability. Network susceptibility will thus depend on species composition, interaction types and landscape properties (Bender and Fahrig, 2005): one could argue that large fragments have a higher conservation value due to the increased likelihood of modularity, which reduces the risk of the spread of disturbances.

7. HABITAT FRAGMENTATION IN A META-NETWORK CONTEXT

Meta-population ecologists envision a natural landscape as consisting of suitable habitat patches (fragments) containing local species populations, connected through dispersal (Hanski, 1998). Local extinction and colonisation create a dynamic state (Hanski and Simberloff, 1997), determined by the isolation of the patches (including matrix permeability) and the reproductive potential of each population. Likewise, the extinction probability in a given patch is related to its isolation (how likely the patch is to receive immigrants), area (small patches often have smaller populations, which are more vulnerable to stochasticity) and quality (MacArthur and Wilson, 1963; Hanski, 1998, 1999). Thus, patches are often divided into sources and sinks, depending on whether the populations are producing an excess of individuals or are relying on a net input to persist (Hanski and Simberloff, 1997).

Single-species meta-population models have been extended to models of two or more interacting species, which, through antagonistic or mutualistic interactions, modify the dynamics of each other, alongside traditional meta-population dynamics (extinction and colonisation) (Hanski, 1999; Nee et al., 1997; Prakash and de Roos, 2004). Intriguingly, Nee and May (1992) demonstrated that species interactions (superior competitor and inferior coloniser vs. inferior competitor and superior coloniser) may change species composition in remnant patches in a fragmenting landscape. The complexity of the mathematical models describing the dynamics of meta-populations increases rapidly as more species are added (Klausmeier, 2001), but in reality, habitat fragmentation affects whole communities of multiple species interacting simultaneously.

With an implicit reference to meta-populations and meta-communities (Hanski, 1999; Hanski and Gilpin, 1997), meta-networks can be defined as a set of spatially distributed local networks connected by species dispersal and influenced by colonisation and extinction dynamics (Fig. 20). These meta-networks can be considered as a combination of spatial and ecological networks (see Section 2) in a meta-population context. To date, little work has been done in this field, although such approaches offer a promising means for assessing (1) dispersal and movement between local networks, (2) the colonisation and extinction of species in local networks and (3) implications of habitat fragmentation on the topology of local networks.

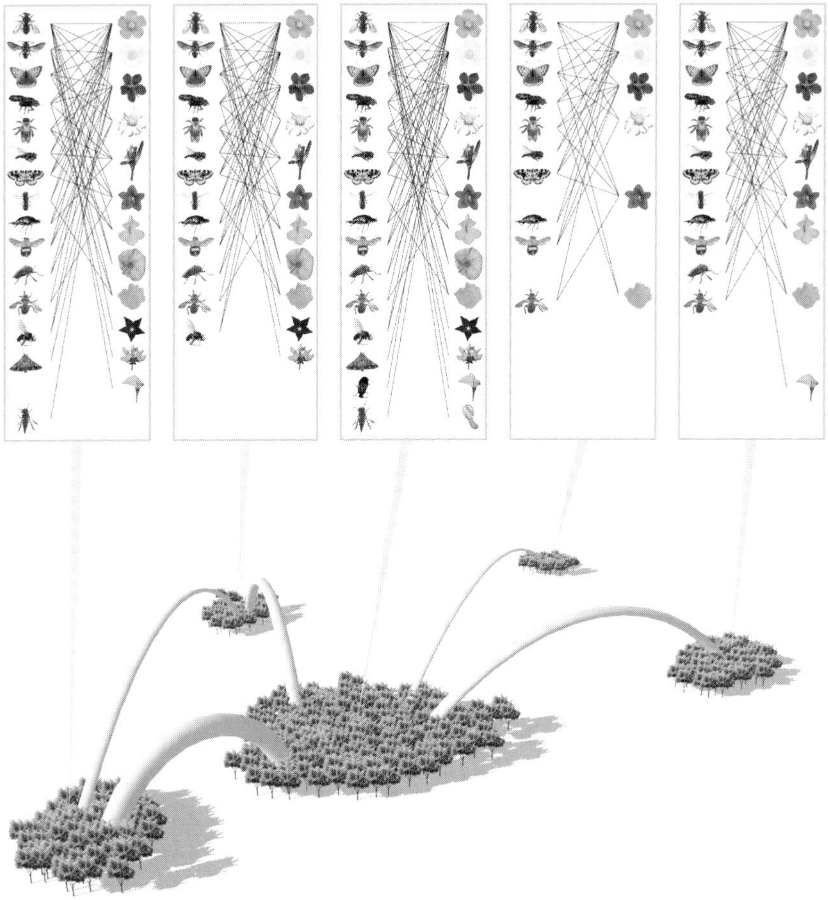

Figure 20 Ecological networks in a meta-network context. A fragmented landscape consists of local habitat fragments separated by a more or less permeable matrix. Within each habitat fragment, networks of interacting species can be found which differ in their structure and degree of complexity. A fragmented landscape usually contains fragments of different sizes at different degrees of isolation. Here, it is illustrated how a big fragment, containing many interacting species, may support minor fragments via species dispersal and thereby contribute to the maintenance of species composition and local network structure. However, small fragments are not able to support all species and isolated fragments are less likely to receive immigrants, and thus, some species and interactions (hence, links) will be lost. The thickness of the 'bridges' between fragments represents the relative degree of species movement between them. In some cases, dispersal might be bidirectional while in others (especially between large and small fragments) movement might be unidirectional, that is, from a source to a sink. Note that the most specialised species are likely to be the most vulnerable. A different effect on network structure will emerge if criteria other than specialisation are used. (For colour version of this figure, the reader is referred to the online version of this chapter.)

7.1. Meta-networks and dispersal

Dispersal and movement of species among patches may be density-dependent or density-independent (Hansson, 1991; Kuussaari et al., 1996; Sæther et al., 1999 and references therein). Low-density dispersal may, for example, be due to a failure in locating mates or specialised mutualists. When locating specialised mutualists, it is the density of the interacting partner that is critical for moving and dispersing. High-density dispersal, on the other hand, may be a result of resource competition among conspecifics or other species. Here again, the network approach offers promise, as it not only specify who is interacting with whom, but also who is interacting with the same partner and thereby, potentially, competing for the same resources. If a landscape becomes more fragmented over evolutionary relevant time scales, increased (mean and long-distance) dispersal rates will be selected for. For example, some sphingid male hawkmoths have evolved a strong olfactory sense enhancing their dispersal success and experienced meta-network-level selection for increased dispersal rates (Hanski, 1999).

Within a meta-population, dispersal may be unidirectional, that is, from a source to a sink (Pulliam, 1988), and analogies may be drawn with meta-networks (Fig. 20). In meta-population theory, a population is regarded as a source, if the intrinsic rate of increase (r) of the population is $r > 0$, and a sink if $r < 0$ (Leibold et al., 2004; Pulliam, 1988). However, a local network could be a source for some species but a sink for others (Pulliam, 1988). Thus, when assigning the label source or sink to a local network, a better approach might be to look at the overall intrinsic rate of increase for all the species. As such, a local network could be regarded as a source, if it has a net increase in species ($R > 0$, where R equals the number of species with $r > 0$ minus the number of species with $r < 0$), and a sink, if it has a net loss of species ($R < 0$), while neglecting immigration. The immigration of species is necessary to maintain both the species composition and interaction structure. Thus, for the network to persist, the rescue (Brown and Kodric-Brown, 1977) of individual species is essential. If some species go extinct, effects may cascade out to other parts of the local network, reducing the r of other species (either directly or indirectly), and triggering further cascading extinctions (e.g. Palmer et al., 2008).

7.2. Meta-networks and extinction

From a meta-network perspective, extinction and colonisation can be envisaged on several organisational levels, for example, the interaction-, species-, local network-, meta-network-, local patch- and regional level.

In an extreme case, an entire local patch might disappear, and with it the complete local network with its species and interactions.

As a local patch shrinks, some species and links will go extinct (Pauw, 2007; Rodríguez-Cabal et al., 2007), the consequences of which will depend on the network and ecosystem type. For instance, in antagonistic networks, mesopredator release (Crooks and Soulé, 1999) may trigger secondary extinctions. In contrast, in a mutualistic system, the loss of interactions could have negative effects on the immediate interaction partners, if there is limited functional redundancy among species (cf. Zamora, 2000). Since many species are taking part in both mutualistic and antagonistic interactions simultaneously (Fontaine et al., 2011), foreseeing the outcome of species loss on local networks is a challenging task.

Although reduction of habitat area does not always result in complete extinctions, it often reduces species abundances (Fahrig, 2003), with detrimental consequences for mutualistic partners (or consumers in food webs). A reduced abundance would, all else being equal, result in a reduced interaction frequency. Within pollination networks, this can lower plant fecundity (Pauw, 2007); in food webs, it can reduce predation pressure. Additionally, interactions might disappear if interaction partners are not lost but reduced to encounter probabilities approaching zero. Depending on whether the involved species have alternative partners, interaction extinction may lead to local species loss. If all local patches decrease sufficiently in area, the meta-network eventually fragments.

7.3. Meta-networks and colonisation

Both the abundance of the individual populations and the local species richness influence colonisation success. The more abundant and diverse the species are in the local habitat, the more difficult it is to colonise the local network, due to community closure, for instance (Hanski, 1999; MacArthur and Wilson, 1963). However, generality and competitiveness of the existing species and the area, isolation and quality of the local patch are also important factors (MacArthur and Wilson, 1963). Thus, the traits of both residents and colonists and fragment characteristics determine colonisation. For example, generalisation among the resident species may make it more difficult for colonists to find a vacant resource that is not already exploited. On the other hand, there may be many potential interaction partners, as predicted by the theory of preferential attachment (Barabási et al., 1999; Jordano et al., 2003; Olesen et al.,

2008). Thus, the effects of generalisation depend to a large extent on which community (mode) is exhibiting this trait (e.g. plant or pollinator trophic level in a bimodal network) and which is colonising.

It becomes more difficult to invade local networks that are characterised by a large number of generalist species, which might partly explain slow recovery of freshwater food webs from acidification (Layer et al., 2010, 2011). Networks consisting of many pairwise mutualistic interactions, for example, having tightly coevolved traits, might be more resistant to colonisation because species might be better able to compete for their resources.

The seminal work on island biogeography theory (MacArthur and Wilson, 1963) and later elaborated by other authors (e.g. Brown and Kodric-Brown, 1977; Whittaker et al., 2008) is especially relevant in this context: patches that are close to a source of dispersing species will, all else being equal, receive more colonisers and be less prone to extinction as they are more likely to be rescued (Brown and Kodric-Brown, 1977). As such, patches close to a source should therefore be better able to retain network structure than distant patches (of equal size).

As an additional consideration, bipartite ecological networks consisting of plant–pollinators or plant–seed dispersers contain both mutualistic and competitive interactions. On the one hand, plants and animals are involved in mutualistic interactions that might range from facultative to obligate, while pollinators interact competitively for resources (Goulson, 2003), as do some plants for pollinators (Morales and Traveset, 2009; Vamosi et al., 2006). Other plants do not compete (Hegland and Totland, 2008; Ollerton et al., 2003) or may even facilitate the pollinators of other species (Sargent et al., 2011). In cases of competition, the immigration, colonisation and extinction processes are governed by both antagonistic and mutualistic events depending on whether the interaction is related to similar nodes in the network. As a consequence, the simultaneous integration of both antagonistic and mutualistic network models (Klausmeier, 2001; Nee et al., 1997) might be needed. This will dramatically complicate any modelling process, especially when dealing with ecological networks of natural sizes (in a database of 54 community-wide pollination networks, species richness ranges from 16 to 952 species with a median of 105; Trøjelsgaard and Olesen, in press and similar-sized food webs are listed in Ings et al., 2009). Like extinction probability, colonisation ability will depend on species traits, including body size, mobility and generality. Generalists are often considered relatively good colonisers (Hanski, 1999), as are larger animals (Hoekstra and Fagan, 1998; Lomolino, 1985; Sutherland et al., 2000).

8. EFFECTS OF HABITAT FRAGMENTATION ON THE COEVOLUTIONARY DYNAMICS OF NETWORKS

So far we have focused upon the ecological consequences of habitat fragmentation, while only briefly touching on evolutionary processes. For example, we implicitly assume that a species is more likely to die out due to the loss of its mutualistic partners or prey instead of natural selection leading to rapid evolution of new interactions (e.g. Rezende et al., 2007). However, there is increasing evidence that human-driven evolutionary change can occur on very short ('ecological') time scales (Darimont et al., 2009), which has implications for ecological networks. A first step in this direction might be to use the geographic mosaic theory of coevolution (GMTC) (Thompson, 2005) to describe how selection will vary across fragmented landscapes and how that might influence species interactions and ecological networks.

8.1. The geographic mosaic theory of coevolution

The GMTC assumes that the evolutionary dynamics of species interactions are affected by the spatial configuration of potentially interacting populations (Thompson, 2005). GMTC models assume that (i) species interact in discrete habitat patches, (ii) selective pressures associated with interactions vary across space (hereafter geographic selective mosaics) and (iii) gene flow mixes traits among populations (Gomulkiewicz et al., 2000; Nuismer and Doebeli, 2004; Nuismer and Thompson, 2006; Nuismer et al., 1999, 2000).

Geographical mismatches among potentially interacting species, geographically selective mosaics and gene flow will lead to unique evolutionary dynamics that cannot be predicted by single-site models. Space is a key component of this theory, affecting evolutionary dynamics in three ways. First, geographical variation in genotype distributions among populations will alter fitness. Second, space generates geographic selective mosaics where there is spatial variation in the function that connects the fitness of genotype in one species with that of its interacting partner. The geographic selective mosaics occur if the fitness and, consequently, selective pressures are determined by an interaction of two genotypes (G) and by the environment (E) (i.e. $G \times G \times E$). Third, the spatial configuration of sites will affect gene flow across populations (Nuismer et al., 2000).

8.2. Habitat fragmentation and its effects on basic components of GMTC

Habitat fragmentation could affect the GMTC for two-species interactions through its basic components: the patches, species interactions, gene flow and by changing the environment in which the interactions occur. The resulting poorly connected patches will be smaller than natural patches. The within-patch variation will increase due to contrasting selection and stochastic genetic variation in the many fragmented subpopulations of a given species. In this sense, the unique (biotic) history of each fragment might lead to an equally unique combination of abiotic factors that might affect the selective pressures on the interaction.

If the landscape is perceived by a given species as a composition of isolated fragments, a break-up of interactions in some patches is expected. For instance, the local extinction of some top predators in rainforest fragments can lead to the loss of key predator–prey interactions that can affect the whole ecosystem via trophic cascades (Terborgh et al., 2001). The same is true for some large frugivores, whose extinction may lead to the loss of key interactions with large-seeded plants (Guimarães et al., 2008). On the other hand, new interactions could also be created by invasive species that might be able to persist in the fragments but not in the original connected environment, as open-habitat species may eventually use secondary forest fragments or species that were present before fragmentation 'rewire' their interactions due to some interacting partner loss. At present, the consequences of losing (or gaining) such key species on the selective pressures associated to interactions remain virtually unknown from a fragmentation perspective.

Habitat fragmentation could also alter the relevance of certain interactions, via changes in abundances of interacting species. Species abundance shapes ecological networks and common species are often also highly connected (e.g. Krishna et al., 2008). Changes in abundance due to fragmentation may, in turn, affect the selective pressures associated with particular interactions. An additional related factor is the reduction of gene flow across patches, which might ultimately have major consequences on species evolution and coevolution (Nuismer et al., 1999). Mathematical models of GMTC suggest that gene flow can have unexpected evolutionary consequences for local adaptation in pairwise interactions (Nuismer et al., 1999).

8.3. Habitat fragmentation and selection mosaics in ecological networks

Examining the GMTC in a fragmentation setting is especially challenging in species-rich networks, because the dynamical consequences of network structure are not simply the sum of the dynamics of pairwise interactions. For instance, if a network has N species in a continuous habitat, there are 2^N possible combinations of species for any given habitat fragment. Again, simplification is the route to address this challenge and we need to learn first more about which are the most relevant components of ecological networks to understand how they affect the speed and direction of evolutionary change.

Recent studies are starting to explore the role of species ecological networks in shaping evolutionary dynamics. For instance, ecological networks of interacting species might favour the maintenance of high levels of trait diversity (Fontaine et al., 2011). Explorations of the evolutionary dynamics in species ecological networks by integrating field data, evolutionary models and tools derived from statistical mechanics are still in their infancy. In mutualistic networks, evolutionary dynamics appear to be shaped mainly by a few super-generalist species that interact with multiple modules (Olesen et al., 2007). Such species shape the evolution and coevolution in these networks in multiple ways (Guimarães et al., 2011). First, they increase the frequency of evolutionary cascades through a small-world effect, by reducing path length between species within the network. Second, they create asymmetric dependencies among species, reducing the potential of reciprocal selection. Third, they impose similar selective pressures over multiple components of the network, promoting convergence in species traits (Guimarães et al., 2011). The hypothesised effects of super-generalists provide the first steps in predicting the potential evolutionary consequences of habitat fragmentation in ecological networks.

Changes in species composition will be particularly relevant if super-generalists are affected. For instance, the probability of local extinction increases with body size (Gaston and Blackburn, 1995), which is itself often positively associated with generalisation in both antagonistic predator–prey and mutualistic seed dispersal interactions. Thus, size-based extinctions are more likely to lead to the extinction of super-generalists and this could conceivably lead to an increase in the role of reciprocal selection. Furthermore, it could reduce the frequency of evolutionary cascades, ultimately favouring trait dissimilarity (i.e. mismatches) within interacting assemblages. In contrast, the introduction of generalist exotic species, such as honeybees, may favour convergence among plants (Guimarães et al., 2011). Therefore, habitat fragmentation may change the evolutionary dynamics within species networks, especially if super-generalists die out or invade newly fragmented habitats.

If the degree of habitat loss and fragmentation leads to a set of very small and disconnected fragments, each should have tiny and semi-autonomous networks with little dispersal among them (Fig. 20). These networks would be unlikely to contain super-generalist species that rely upon a diversity of partners to survive. Species that specialise on a few partners, such as large-seeded plants that use large vertebrates for dispersal, will also be absent (Da Silva and Tabarelli, 2000). Consequently, these tiny networks should contain species with relatively homogeneous interaction patterns, with no one species dominating evolutionary or coevolutionary processes in the network. Moreover, divergence in population traits due to local adaptation may occur if these small networks are also isolated. Finally, the role of species across networks is not fixed, although we still know little about this (but see Marquitti, 2011). Changing the abiotic and biotic features in a given patch, habitat fragmentation could alter both the ecology and evolution of interacting species. For example, forest fragmentation might suppress the population of a super-generalist species, transforming it to a peripheral species in the network and consequently reducing its ecological relevance and as well affecting evolutionary trajectories within the entire community.

Predicting evolutionary consequences of fragmentation on networks is still limited by a relative lack of both data and a mature theoretical framework. Theoretical studies using two-species models suggest that the coevolutionary dynamics may be qualitatively changed because of gene flow (Nuismer et al., 1999), and the potential for new evolutionary dynamics is even higher in a species-rich and fragmented network. The challenge ahead is to develop approaches to model these complex dynamics in ways that allow hypotheses to be tested in the field.

9. APPLICATIONS IN CONSERVATION AND AGRICULTURE

The effects of habitat loss and fragmentation on biodiversity are evident on a global scale, and researchers and managers must develop ways to understand and mitigate them (Bazelet and Samways, 2011). For instance, many European bird species have declined as agricultural intensification has resulted in the increasing fragmentation and isolation of natural habitats (Donald et al., 2001; Tscharntke et al., 2005), and yet the consequences of losing these often key species from mutualistic or antagonistic networks are still largely unknown. What is clear, however, is that the effects of habitat fragmentation are not evenly distributed within or among networks (e.g. Cagnolo et al., 2009).

The growing appreciation that the importance of network structure for ecosystem stability and functioning recognises that it is linked intrinsically to applied goals, such as biodiversity conservation (Kaiser-Bunbury et al., 2010; Tylianakis et al., 2010) or agricultural production (e.g. MacFadyen et al., 2011). Yet for network approaches to become fully integrated into ecosystem management, two objectives must be met. First, a conceptual challenge will be to demonstrate that complex network approaches add value to current practices. Underpinning this is the need to identify which specific attributes of networks require the greatest attention and which offer the best yield-to-effort reward. Second, a variety of practical hurdles need to be overcome, both in the quantification of network attributes using empirical data that can be feasibly obtained and in the application of concepts to practice (Tylianakis et al., 2010).

Gathering conceptual support for the adoption of network tools is the easier of these two objectives. The importance of network structure for properties such as system stability, and recognition that this can be altered even when species richness is not (e.g. Tylianakis et al., 2007), suggests that landscape degradation may be altering ecosystems in ways that cannot be detected by simple species-centric measurements. Furthermore, species cannot survive without their interacting partners, so there is an inherent need to consider the resources and mutualists of any species we wish to conserve. Additionally, the extinction sequence of species and interactions from a network during the fragmentation process (e.g. Sabatino et al., 2010) could provide guidance on the order in which species should be (re)introduced during restoration (Feld et al., 2011). A network perspective can also help predict the indirect effects of species additions or deletions (Carvalheiro et al., 2008). A major challenge now is to identify the most relevant aspects of network architecture for agriculture and conservation within fragmented landscapes, whilst taking into account the huge complexity of these networks.

One promising avenue in this context is to focus on some key components (e.g. species, links, functional roles, modules), as identified via network analysis, that are needed for the system to function 'normally'. For example, evidence is growing that super-generalists are the backbone of many networks, potentially governing their ecological and evolutionary dynamics (Guimarães et al., 2011; Olesen et al., 2007), which could provide clues as to how best to conserve or restore fragmented landscapes. There is also plenty of evidence that top predators can have cascading effects in marine, terrestrial and freshwater ecosystems worldwide (Estes et al., 2011), and many of these are also highly generalised. The reintroduction of locally extinct generalists may assist the restoration of previous ecosystem states whereas the removal of non-native super-generalists may be the first step needed to restore fragments and landscapes to their prior condition.

In addition to the presence or absence of apex consumers and super-generalists, several other network metrics can be important from a conservation perspective. Tylianakis et al. (2010) argued that conservation could focus on network attributes that confer stability or maximise rates of ecosystem functioning. Nestedness, compartmentalisation, *degree distributions*, interaction diversity and the presence of weak links are all potentially useful metrics, but some of these are sensitive to sampling effort. Thus, the best approach to conservation of complex networks could involve the monitoring and/or restoration of a suite of network metrics, at least if preserving stability and functioning are the primary objectives (Tylianakis et al., 2010). These would likely include measures of connectedness (such as connectance or link density), which would relate to functional redundancy and the probability of secondary extinctions following species loss. Furthermore, compartmentalisation or modularity (particularly to avoid the spread of pollutants or perturbations) and nestedness (to maintain robustness of functioning following local extinctions) are likely to be key network properties for restoration and conservation.

Despite being important in theory, measuring network metrics accurately and manipulating them empirically remains a hurdle to the implementation of a more 'link-focused' management. Simulations of sampling can help reveal which metrics may be least sensitive to sampling effort (Nielsen and Bascompte, 2007; Tylianakis et al., 2010), and these may be the optimal candidates for biomonitoring. A number of questions still need to be addressed before network conservation can be put into practice. At the most basic level, we need to know how the survival or conservation of a species in a fragmented landscape is affected by its biotic context, that is, the number and kinds of links connecting that species to others within the network. Second, we need to identify the traits of species that determine their role within the network, so that we can begin to predict and restore network structure. For example, species traits such as body size and morphology (e.g. Stang et al., 2007, 2009; Woodward et al., 2005) are known to influence network structure, and techniques have recently been developed to calculate the contribution of a species to network nestedness and persistence (Saavedra et al., 2011). As ecologists further unravel these traits, we can start to move towards developing a predictive framework for network architecture given community-wide traits of species (Gilljam et al., 2011; Petchey et al., 2008; Woodward et al., 2010b). Third, we need to better understand the relationship between physical structure and network architecture. Evidence that complex habitat structures can impede the realisation of potential interactions (Laliberté and Tylianakis, 2010) requires consideration in the restoration of complex (e.g. forest) habitats and provides a potential avenue for reducing the impact of undesirable or strong destabilising interactions.

10. CONCLUSIONS

Our synthesis provides ample evidence that the consequences of habitat fragmentation for biotic communities and ecological networks are highly complex, but that does not mean they are unpredictable. At least five components of this complexity become immediately apparent. First, there is spatial complexity in the fragmentation process due to variation in landscape structure in terms of fragment size and isolation, connectivity, matrix quality, edge permeability and geometry. Second, fragmentation can affect the temporal dynamics of interacting taxa (e.g. flowering and fruiting phenologies), and long-term consequences on interacting species may become apparent only after several decades. For instance, time lags will increase the probability of co-extinctions, especially when generation times strongly differ between interacting taxa. Third, responses by fragmentation- and network-relevant traits differ among species. The perception of fragmentation (e.g. environmental grain) by individual species, key traits and complexes (e.g. body size in food webs), and trait matching between interacting species might be particularly relevant for assessing the consequences of fragmentation. Fourth, there is complexity in the biological and analytical details of networks, which differ in type (e.g. mutualistic vs. antagonistic; bimodal vs. multi-modal). Effects of dispersal, colonisation and extinction need to be integrated (e.g. in meta-networks). Fifth, there is an evolutionary component to network responses to habitat fragmentation. The geographic settings of habitat configuration and selective mosaics might lead to rapid evolutionary changes, even at short 'ecological' time scales. Finally, these five complexity components may interact, creating potential synergies.

How can we usefully address and simplify this extreme complexity that originates from different spatial and temporal scales and organisational levels? First, we need to understand how individual links among interacting species are affected by habitat fragmentation, both in a spatial and temporal setting. These include phenologies and encounter rates and how they vary across space, time and levels of fragmentation. Second, there is overwhelming evidence that species are not equally important for ecosystem functioning and that a few exert disproportionate effects. These include large species at high trophic levels (e.g. top predators), abundant species and super-generalists. Such species can provide the structural backbones of ecological networks, shape evolutionary dynamics or initiate cascades of network changes. Thus, one way to circumvent the apparent complexity is to focus initially on understanding how fragmentation affects these key species and their links. Third, we need to gauge the extent of functional redundancy in ecological networks and to what extent habitat fragmentation disproportionally affects functionally unique species. This includes a better understanding of the role of specialisation, functional grouping and trait matching in ecological networks. Finally, we need to understand in more detail how network properties (e.g. connectance, linkage level, nestedness, modularity) and the roles of species in networks (e.g. hubs, connectors, spatial couplers) are affected by habitat fragmentation. This will become particularly interesting as we begin to link different types of networks, for example, when combining spatial with ecological networks or when moving from simple networks to meta- and *super-networks*.

There is a clear need to consider ecological and evolutionary processes of multispecies interactions in a network context to understand how habitat fragmentation affects biodiversity. Such an approach will become increasingly feasible as the availability of large databases, appropriate software and comparative studies continue to increase apace. We envisage a hierarchical approach to understand how individuals, populations, pairwise interactions, ecological networks and ultimately networks of networks are affected by fragmentation. For network approaches to become integrated into conservation, agriculture and ecosystem management, we need to find ways to simplify the inherent complexity and to measure and monitor management-relevant network properties. A link-based management approach has great potential to aid biodiversity conservation and restoration by highlighting the immense importance of biotic interactions and ecological network stability for ecosystem functioning.

ACKNOWLEDGEMENTS

This paper was developed and written during and after two workshops sponsored by the Danish Agency for Science, Technology and Innovation (FNU) under the international network call (Application title: 'Ecological Network Analysis in an Agricultural Landscape'). In addition, we acknowledge support from the Danish Council for Independent Research | Natural Sciences (M. H.; J. M. O.; and via a starting independent researcher grant (11-106163) to W. D. K.), the Carlsberg Foundation (C. R.), the Villum Kann Rasmussen Foundation (W. D. K., VKR09b-141 to J.- C. Svenning), the Brazilian Council for Science CNPq (M. A. M. D. A., L. P. C. M.) and the Swiss National Science Foundation (C. N. K. -B.). M. A. M. D. A., J. G., P. R. G., F. M. D. M., K. P. M., L. P. C. M. and M. M. V. were partially supported by Fundação de Amparo a Pesquisa do Estado de São Paulo (FAPESP), J. M. T was funded by a Rutherford Discovery Fellowship, and M. M. V. was partially supported by CAPES. G. B. J. and E. J. O. were supported by UK Natural Environment Research Council grants awarded to G. W. (Ref: NE/I528069/1; NE/I009280/1). We are most grateful to C. Mulder for his many corrections and comments.

APPENDIX

Methods for Ashdown Forest case study of food webs in fragmented river networks

A.1 Site description and food web construction

Ashdown Forest in Sussex, UK (National Grid Reference TQ 520300) contains the spring-fed headwaters of two rivers, the Ouse—which flows south into the English Channel—and the Medway, which flows north and joins the Thames estuary. The catchments of both streams lie in the centre of the Weald in SE England, on hills of soft, fine sandstone (Ashdown Sands). Further description of the site can be found in Townsend et al. (1983). Sixteen streams were sampled in this study, and pH was recorded in 1976 and 1994, and an average value was calculated for each stream. Five randomly dispersed Surber samples (sample-unit area 0.0625 m^2; mesh aperture 330 μm) were collected from each of the 16 streams in October 1976, 1984 and 1994 (Gjerløv et al., 2003; Townsend et al., 1987) (total *n* sample-units = 240). The benthos was disturbed to a depth of approximately 5 cm and all macroinvertebrates collected were preserved in the field and subsequently sorted. Taxonomic identification was standardised to the highest common level of resolution (usually to species) across all webs (Woodward et al., 2002a). Several of the more difficult to identify taxa were aggregated: for example, all members of the Tanypodinae subfamily were presented as a single node. Feeding links were taken from direct observed interactions (gut contents analysis) in Broadstone Stream (Woodward et al., 2010b) and elsewhere within these two river networks (e.g. Layer et al., 2010, 2011), and this dataset was augmented with feeding links inferred from known interactions described in the literature from different systems (Brose et al., 2006; Gilljam et al., 2011; Lancaster et al., 2005; Warren, 1996; Woodward et al., 2008, 2010b). Additional feeding link data were supplied by F. Edwards (unpublished data).

GLOSSARY

Note that some of the terms in this glossary have alternate meanings, and some also have general and specific definitions (e.g. complexity) in different disciplines (e.g. in food webs vs. mutualistic networks; in landscape ecology vs. ecological network ecology), which can lead to potential misunderstandings when undertaking interdisciplinary research. We have highlighted these with '⋆', below.

Antagonistic network (p. 96) A network with associations between organisms in which one benefits at the expense of the other, for example, food webs, host–parasitoid networks and competitive networks.

⋆Asymmetry (p. 199) In a network context, a property of nested assemblages (e.g. mutualistic networks). Specialist plants interact just with generalist animals, while generalist animals use a broad range of host plants, including both specialists and generalists. It also refers to inequality of strong and weak interactions between species or nodes, competition or energy flow within a network.

Bimodal networks (p. 97) Pollination and seed dispersal networks are by definition bimodal (bipartite or two-mode), linking two sets of taxa (e.g. flower-visitors and plants, or frugivores and plants). They are often best represented by two-level bipartite graphs. Host–parasitoid networks or food webs that consider just two trophic levels also fall under this definition.

Boundary (p. 117) A border (or edge) between contrasting habitat patches that delimits the spatial heterogeneity of a landscape.

Centrality (p. 102) A measure of the importance of a node as a focal point within a network. There are various types of centrality measures for any node within a network, such as degree (the number of nodes that a focal node is connected to), closeness (the inverse sum of shortest distances to all other nodes from a focal node) and betweenness (the degree to which a node lies on the shortest path between two other nodes).

Coevolutionary dynamics (p. 91) Coevolution is the process of reciprocal evolutionary change between interacting species, driven by natural selection. This may lead to coevolutionary dynamics, whereby changes in gene frequency in one species trigger reciprocal changes in the other interacting species.

Compartment (p. 97) An assemblage of species within a network. Specific definitions vary depending on the point of view of the constituent organisms. Density view: an assemblage of species that are highly connected to each other. Predator view: an assemblage of species that share a large number of prey. Prey view: an assemblage of species that share a large number of predators. See also *module* below.

Compartmentalisation (p. 98) The development of groups of species or 'topological compartments' that have a higher probability of interacting with one another than with other species in the network. See also *modularity* below.

***Complexity (p. 92)** Property or set of properties that characterise systems composed of many interacting parts or elements. In organised complexity, the non-random or correlated interaction among the parts generates emergent properties, that is, properties not carried or dictated by individual parts. In ecology, complexity can be used as a general term (to describe a large number of interacting nodes) or with a more specific definition, for example, the average number of trophic links per species within a whole food web.

Connectance (p. 91) The proportion of all possible interactions within a system that are realised. This is typically measured as directed connectance, the proportion of documented directed links out of the maximum number of possible directed links in the food web, that is, the number of links (L) divided by the number of species (S) squared, L/S^2.

Connectors (p. 143) Species that link different modules within a network together. For example, large-bodied species, which disperse widely and thus link subwebs together (e.g. avian predators in a fragmented landscape).

Corridor (p. 92) Long, thin strips of habitat that connect otherwise isolated habitat patches. They reduce local extinction risk by connecting isolated populations and by promoting gene flow.

Degree distribution/linkage level distribution (p. 97, 174) The probability distribution of the number of links per node, typically measured over an entire network.

Domatium(-a) (p. 98) Specialised chamber(s) in different plant parts, providing refuge for predatory arthropods.

Ecological network (p. 91) A representation of biotic interactions in a multispecies community, in which pairs of species or other forms of taxonomic or functional aggregates (nodes) are connected when they are interacting (links), both directly and indirectly (e.g. sharing the same resource but not directly linked). There are three broad categories—food webs, host–parasitoid and mutualistic networks.

***Edge (p. 90)** In a landscape context, the (artificial) boundaries of habitat fragments. Also used as a synonym for link in network analysis, highlighting the need for clarity when using this term in interdisciplinary studies.

Edge permeability (p. 103) The extent to which a species can move through a physical border, for example, from a fragment to the surrounding matrix. A 'hard' edge contains an impenetrable boundary which dispersing individuals virtually never cross, for example, a physical barrier such as an ocean surrounding an island. A 'soft' edge is more permeable to emigrating individuals than a hard edge, for example, the boundary between a meadow and a garden.

Environmental grain (p. 91) The scale of environmental variation (temporal or spatial), relative to the temporal/spatial scales of activity of the organisms, that is, a description of the organism's 'perception' of its own environment.

Fragments (p. 92) Habitat that was once continuous but has become divided into discrete patches. Fragments are separated by and embedded within areas (matrix) with abiotic and biotic properties different from the previously continuous habitat (see *habitat fragmentation* below).

Functional group (p. 94) A group of species or taxa with a similar response to a given factor. This may also include trophic species, groups of taxa that share the same set of predators and prey.

Functional redundancy (p. 91) The idea that some species perform similar roles in communities and ecosystems and may therefore be substitutable with little impact on system properties.

Generalist (p. 98) A species that is able to thrive in a wide variety of environmental conditions and/or can make use of a variety of different resources.

Habitat fragmentation (p. 90) A process during which a large expanse of habitat is transformed into a number of smaller patches of smaller total area, isolated from each other by a matrix of habitats unlike the original. The effects of this process may include some, but not all of the following: (1) reduction in habitat amount, (2) increase in number of habitat patches, (3) decrease in size of habitat patches and (4) increase in isolation of patches.

Higher trophic rank hypothesis (p. 159) Species found at higher trophic levels tend to have a stronger relationship with area than species found at lower trophic levels as they have larger space and resource requirements. As such, species found at high trophic levels should have a higher susceptibility towards habitat fragmentation.

Host–parasitoid networks (p. 92) A specific form of antagonistic ecological network in which parasitoids benefit and subsist off their hosts. They may also contain information about hyperparasitoids (parasitoids that attack other parasitoids). These networks often involve a high degree of specialisation.

Hub (p. 91) Highly linked species within their own module of a network.

Interaction intimacy (p. 132) Degree of biological association between individuals of interacting species, for example, host–parasite during all of their life or only part of their lifespan.

***Interaction strength (p. 134)** The magnitude of the effect of one species on another mediated by their pairwise interaction. This can be measured in a variety of ways, including experimental and theoretical approaches, or using allometric body-size scaling relationships.

Invasive species (p. 117) Species that arrive, become established and subsequently disperse in a community where they did not previously exist in historical time.

Link (p. 91) The pairwise interaction between two nodes in a network.

Linkage level (p. 97) Number of links per species.

***Matrix (p. 91)** A landscape that has undergone fragmentation, often leading to a heterogenous habitat. Also quantifies the pairwise interactions between multiple species in a network, for example, qualitative (presence/absence of an interaction) or quantitative (coefficients reflecting interaction strengths, such as the Community or Jacobian matrix). The different meanings of this term in different fields of ecology highlight the importance of clarity in interdisciplinary studies.

Matrix permeability (p. 118) The property of a habitat matrix that describes the extent to which species can move through it, that is, between fragmented habitat patches.

Meta-populations/meta-communities (p. 101, 164) Potentially unstable local populations inhabiting discrete habitat patches, which persist at a larger scale via dispersal.

Module/modularity (p. 91) Ecological networks consist of link-dense and link-sparse areas. Link-dense regions are termed compartments or modules. Species within a module are linked more tightly together than they are to species in other modules. The extent to which species interactions are organised into modules is termed the modularity of the network. Modularity may reflect habitat heterogeneity, divergent selection regimes and phylogenetic clustering of closely related species.

Mutualistic networks (p. 96) Networks where both groups benefit from each other. Examples include plant–animal interactions (typically pollinators, frugivores, ants), plant–mycorrhizal systems, coral–zooxanthellae associations and many other networks involving microbial endosymbionts. These networks do not exist on multiple trophic levels, unlike antagonistic networks.

★Nestedness (p. 91) A distinctive pattern of mutualistic community assembly showing two characteristics, namely, asymmetrical specialisation (specialists interacting with generalists) and a generalist core (generalists interacting with generalists). Nestedness occurs when specialist species interact with a proper subset of species with which more generalised species interact. Nestedness can also describe niche overlap in antagonistic networks: for instance, where predator diets are arranged hierarchically on the basis of body size in food webs.

Nodes (p. 94) In an ecological network, nodes mostly refer to species or trophic groups of species. In a broader context, however, nodes can also refer to individuals, populations, functional groups (e.g. body-size or feeding groups), guilds, communities or even entire networks.

Sink (p. 135) A habitat in which mortality exceeds production and is reliant on immigration to maintain population levels.

Spatial network (p. 94) A network or weighted spatial graph, where the nodes have a location and the links have lengths and also a magnitude or weight.

Specialist (p. 98) A species that can only thrive in a narrow range of environmental conditions and/or has a limited diet.

Super-generalist (p. 91) Species with a very high level of generalisation compared to co-existing species. In a network context, they will have a much higher linkage level and centrality than the other species. They are often super-abundant, density-compensating island species.

Super-network (p. 176) Expanding the network study from looking at single bipartite networks to multiple bipartite networks (e.g. plant–pollinator, plant–herbivore and plant–pathogen networks).

Topological role (peripherals, connectors, module and network hubs) (p. 101) Functional role of a node in the network in relation to the modular structure.

REFERENCES

Abramsky, Z., Tracy, C.R., 1979. Habitat selection in two species of short-horned grasshoppers. Oecologia 38, 359–374.

Abramsky, Z., Tracy, C.R., 1980. Relation between home range-size and regulation of population size in *Microtus ochrogaster*. Oikos 34, 347–355.

Aguilar, R., Ashworth, L., Galetto, L., Aizen, M.A., 2006. Plant reproductive susceptibility to habitat fragmentation: review and synthesis through a meta-analysis. Ecol. Lett. 9, 968–980.

Aizen, M.A., Feinsinger, P., 1994a. Forest fragmentation, pollination and plant reproduction in a Chaco dry forest, Argentina. Ecology 75, 330–351.

Aizen, M.A., Feinsinger, P., 1994b. Habitat fragmentation, native insect pollinators and feral honey-bees in Argentine Chaco Serrano. Ecol. Appl. 4, 378–392.

Aizen, M.A., Feinsinger, P., 2003. Bees not to be? Responses of insect pollinator faunas and lower pollination to habitat fragmentation. In: Bradshaw, G.A., Marquet, P.A. (Eds.), How Landscapes Change: Human Disturbance and Ecosystem Fragmentation in the Americas. Springer-Verlag, Berlin, pp. 111–129.

Aizen, M.A., Morales, C.L., Morales, J.M., 2008. Invasive mutualists erode native pollination webs. PLoS Biol. 6, 396–403.

Alarcón, R., Waser, N.M., Ollerton, J., 2008. Year-to-year variation in the topology of a plant-pollinator interaction network. Oikos 117, 1796–1807.

Alberti, L.F., Morellato, L.P.C., 2010. Variation on fruit production of *Nectandra megapotamica* (Lauraceae) trees on the edge and interior of a semideciduous forest: a case study. Naturalia (São José do Rio Preto) 33, 57–68.

Albrecht, M., Duelli, P., Schmid, B., Müller, C.B., 2007. Interaction diversity within quantified insect food webs in restored and adjacent intensively managed meadows. J. Anim. Ecol. 76, 1015–1025.

Allen, K.R., 1951. The Horokiwi Stream: a study of a trout population. N. Z. Mar. Dept. Fish. Bull. 10-10a, 1–231.

Almeida-Neto, M., Campassi, F., Galetti, M., Jordano, P., Oliveira-Filho, A., 2008. Vertebrate dispersal syndromes along the Atlantic forest: broad-scale patterns and macroecological correlates. Glob. Ecol. Biogeogr. 17, 503–513.

Amarasekare, P., 2000. Coexistence of competing parasitoids on a patchily distributed host, local *vs.* spatial mechanisms. Ecology 81, 1286–1296.

Andren, H., 1994. Effects of habitat fragmentation on birds and mammals in landscapes with different proportions of suitable habitat—a review. Oikos 71, 355–366.

Andrieu, E., Dornier, A., Rouifed, S., Schatz, B., Cheptou, P.-O., 2009. The town Crepis and the country Crepis: how does fragmentation affect a plant-pollinator interaction? Acta Oecol. 35, 1–7.

Araújo, S.B.L., de Aguiar, M.A.M., 2007. Pattern formation, outbreaks and synchronization in food chains with two and three species. Phys. Rev. E 75, 061908.

Araújo, E.D., Costa, M., Chaud-Netto, J., Fowler, H.G., 2004. Body size and flight distance in stingless bees (Hymenoptera: Meliponini): inference of flight range and possible ecological implications. Braz. J. Biol. 64 (3B), 563–568.

Arim, M., Berazategui, M., Barreneche, J.M., Ziegler, L., Zarucki, M., Abades, S.R., 2011. Determinants of density-body size scaling within food webs and tools for their detection. Adv. Ecol. Res. 45, 1–39.

Armbruster, P., Hutchinson, R.A., Cotgreave, P., 2002. Factors influencing community structure in a South American tank bromeliad fauna. Oikos 96, 225–234.

Artz, D.R., Waddington, K.D., 2006. The effects of neighbouring tree islands on pollinator density and diversity and on pollination of a wet prairie species, *Asclepias lanceolata* (Apocynaceae). J. Ecol. 94, 597–608.

Ashworth, L., Aguilar, R., Galetto, L., Aizen, M.A., 2004. Why do pollination generalist and specialist plant species show similar reproductive susceptibility to habitat fragmentation? J. Ecol. 92, 717–719.

Atmar, W., Patterson, B.D., 1993. The measure of order and disorder in the distribution of species in fragmented habitat. Oecologia 96, 373–382.

Bach, C.E., Kelly, D., 2004. Effects of forest edges on herbivory in a New Zealand mistletoe, Alepis flavida. N. Z. J. Ecol. 28, 195–205.

Baldissera, R., Ganade, G., 2005. Predação de sementes ao longo de uma borda de Floresta Ombrófila Mista e pasta-gem. Acta Bot. Bras. 19, 161–165.

Bankowska, R., 1980. Fly communities of the family Syrphidae in natural and anthropogenic habitats of Poland. Memorabilia Zoologica. 33, 3–93.

Barabási, A.L., Albert, R., Jeong, H., 1999. Mean-field theory for scale-free random networks. Physica A 272, 173–187.

Barbaro, L., van Halder, I., 2009. Linking bird, carabid beetle and butterfly life-history traits to habitat fragmentation in mosaic landscapes. Ecography 32, 321–333.

Bascompte, J., Jordano, P., Melián, C.J., Olesen, J.M., 2003. The nested assembly of plant-animal mutualistic networks. Proc. Natl. Acad. Sci. USA 100, 9383–9387.

Bazelet, C.S., Samways, M.J., 2011. Relative importance of management *vs.* design for implementation of large-scale ecological networks. Landsc. Ecol. 26, 341–353.

Beattie, A.J., 1985. Ant pollination. In: Beck, E., Birks, H.J.E., Connor, E.F. (Eds.), The Evolutionary Ecology of Ant-Plant Mutualisms. Cambridge University Press, Cambridge, pp. 96–109.

Bell, S.S., Brooks, R.A., Robbins, B.D., Fonseca, M.S., Hall, M.O., 2001. Faunal response to fragmentation in seagrass habitats: implications for seagrass conservation. Biol. Conserv. 100, 115–123.

Bender, D.J., Fahrig, L., 2005. Matrix structure obscures the relationship between interpatch movement and patch size and isolation. Ecology 86, 1023–1033.

Bender, D.J., Contreras, T.A., Fahrig, L., 1998. Habitat loss and population decline: a meta-analysis of the patch size effect. Ecology 79, 517–533.

Bender, D.J., Tischendorf, L., Fahrig, L., 2003. Using patch isolation metrics to predict animal movement in binary landscapes. Landsc. Ecol. 18, 17–39.

Bennun, L., Njoroge, P., 1999. Important Bird Areas in Kenya. Nature Kenya, Nairobi.

Berens, D.G., Farwig, N., Schaab, G., Boehning-Gaese, K., 2008. Exotic guavas are foci of forest regeneration in Kenyan farmland. Biotropica 40, 104–112.

Berlow, E.L., Neutel, A.-M., Cohen, J.E., de Ruiter, P.C., Ebenman, B., Emmerson, M., Fox, J.W., Jansen, V.A.A., Jones, J.I., Kokkoris, G.D., Logofet, D.O., McKane, A.J., et al., 2004. Interaction strengths in food webs: issues and opportunities. J. Anim. Ecol. 73, 585–598.

Berlow, E.L., Dunne, J.A., Martinez, N.D., Stark, P.B., Williams, R.J., Brose, U., 2009. Simple prediction of interaction strengths in complex food webs. Proc. Natl. Acad. Sci. USA 106, 187–191.

Bierregaard, R.O., Lovejoy, T.E., Kapos, V., Dos Santos, A.A., Hutchings, R.W., 1992. The biological dynamics of tropical rainforest fragments. Bioscience 42, 859–866.

Blanche, K.R., Ludwig, J.A., Cunningham, S.A., 2006. Proximity to rainforest enhances pollination and fruit set in orchards. J. Appl. Ecol. 43, 1182–1187.

Blüthgen, N., Menzel, F., Hovestadt, T., Fiala, B., Blüthgen, N., 2007. Specialization, constraints and conflicting interests in mutualistic networks. Curr. Biol. 17, 341–346.

Blüthgen, N., Fründ, J., Vázquez, D.P., Menzel, F., 2008. What do interaction network metrics tell us about specialization and biological traits? Ecology 89, 3387–3399.

Bolnick, D.I., Amarasekare, P., Araújo, M.S., Bürger, R., Levine, J.M., Novak, M., Rudolf, V.H.W., Schreiber, S.J., Urban, M.C., Vasseur, D.A., 2011. Why intraspecific trait variation matters in community ecology. Trends Ecol. Evol. 26, 183–192.

Bommarco, R., Biesmeijer, J.C., Meyer, B., Potts, S.G., Poyry, J., Roberts, S.P.M., Steffan-Dewenter, I., Ockinger, E., 2010. Dispersal capacity and diet breadth modify the response of wild bees to habitat loss. Proc. R. Soc. B 277, 2075–2082.

Bommarco, R., Lundin, O., Smith, H.G., Rundlöf, M., 2012. Drastic historic shifts in bumble-bee community composition in Sweden. Proc. R. Soc. B 279, 309–315.

Bonada, N., Dolédec, S., Statzner, B., 2007. Taxonomic and biological trait differences of stream macroinvertebrate communities between Mediterranean and temperate regions: implications for future climatic scenarios. Glob. Change Biol. 13, 1658–1671.

Bond, W.J., 1994. Do mutualisms matter—assessing the impact of pollinator and disperser disruption on plant extinction. Philos. Trans. R. Soc. B 344, 83–90.

Bonin, M.C., Almany, G.R., Jones, G.P., 2011. Contrasting effects of habitat loss and fragmentation on coral-associated reef fishes. Ecology 92, 1503–1512.

Borrell, B.J., 2005. Long tongues and loose niches: evolution of euglossine bees and their nectar flowers. Biotropica 37, 664–669.

Bostrom, C., Jackson, E.L., Simenstad, C.A., 2006. Seagrass landscapes and their effects on associated fauna: a review. Estuar. Coast. Shelf Sci. 68, 383–403.

Boulton, A.J., 2003. Parallels and contrasts in the effects of drought on stream macroinvertebrate assemblages. Freshw. Biol. 48, 1173–1185.

Boyle, S.A., Smith, A.T., 2010. Can landscape and species characteristics predict primate presence in forest fragments in the Brazilian Amazon? Biol. Conserv. 143, 1134–1143.
Brooks, T.M., Pimm, S.L., Collar, N.J., 1997. Deforestation predicts the number of threatened birds in insular southeast Asia. Conserv. Biol. 11, 382–394.
Brose, U., Jonsson, T., Berlow, E.L., Warren, P., Banašek-Richter, C., Bersier, L.F., Blanchard, J.L., Brey, T., Carpenter, S.R., Blandenier, M.F.C., Cushing, L., Dawah, H.A., et al., 2006. Consumer-resource body-size relationships in natural food webs. Ecology 87, 2411–2417.
Brotons, L., Monkkonen, M., Martin, J.L., 2003. Are fragments islands? Landscape context and density-area relationships in boreal forest birds. Am. Nat. 162, 343–357.
Brown, J.H., Kodric-Brown, A., 1977. Turnover rates in insular biogeography: effect of immigration on extinction. Ecology 58, 445–449.
Brown, L.E., Milner, A.M., 2012. Rapid retreat of glacial ice reveals invertebrate community assembly dynamics. Glob. Change Biol. http://dx.doi.org/10.1111/j.1365-2486.2012.02675.x.
Brown, L.E., Hannah, D.M., Milner, A.M., 2007. Vulnerability of alpine stream biodiversity to shrinking glaciers and snowpacks. Glob. Change Biol. 13, 958–966.
Brown, L.E., Edwards, F.K., Milner, A.M., Woodward, G., Ledger, M.E., 2011. Food web complexity and allometric scaling relationships in stream mesocosms, implications for experimentation. J. Anim. Ecol. 80, 884–895.
Buckley, Y.M., Anderson, S., Catterall, C.P., Corlett, R.T., Engel, T., Gosper, C.R., Nathan, R., Richardson, D.M., Setter, M., Spiegel, O., Vivian-Smith, G., Voigt, F.A., et al., 2006. Management of plant invasions mediated by frugivore interactions. J. Appl. Ecol. 43, 848–857.
Buckley, H.L., Miller, T.E., Ellison, A.M., Gotelli, N.J., 2010. Local- to continental-scale variation in the richness and composition of an aquatic food web. Glob. Ecol. Biogeogr. 19, 711–723.
Buldyrev, S.V., Parshani, R., Paul, G., Stanley, H.E., Havlin, S., 2010. Catastrophic cascade of failures in interdependent networks. Nature 464, 1025–1028.
Burgess, V.J., Kelly, D., Robertson, A.W., Ladley, J.J., 2006. Positive effects of forest edges on plant reproduction: literature review and a case study of bee visitation to flowers of *Peraxilla tetrapetala* (Loranthaceae). N. Z. J. Ecol. 30, 179–190.
Burkle, L.A., Alarcón, R., 2011. The future of plant-pollinator diversity: understanding interaction networks across time, space and global change. Am. J. Bot. 98, 528–538.
Burns, K.C., 2006. A simple null model predicts fruit-frugivore interactions in a temperate rainforest. Oikos 115, 427–432.
Burns, K.C., Lake, B., 2009. Fruit-frugivore interactions in two southern hemisphere forests: allometry, phylogeny and body size. Oikos 118, 1901–1907.
Byrnes, J.E., Reed, D.C., Cardinale, B.J., Cavanaugh, K.C., Holbrook, S.J., Schmitt, R.J., 2011. Climate-driven increases in storm frequency simplify kelp forest food webs. Glob. Change Biol. 17, 2513–2524.
Cagnolo, L., Valladares, G., Salvo, A., Cabido, M., Zak, M., 2009. Habitat fragmentation and species loss across three interacting trophic levels: effects of life-history and food-web traits. Conserv. Biol. 23, 1167–1175.
Camargo, M.G.G., Souza, R.M., Reys, P., Morellato, L.P.C., 2011. Effects of environmental conditions associated to the cardinal orientation on the reproductive phenology of the cerrado savanna tree *Xylopia aromatic* (Annonaceae). An. Acad. Bras. Cienc. 83, 1007–1020.
Cane, J.H., 1987. Estimation of bee size using intertegular span (Apoidea). J. Kansas Entomol. Soc. 60, 145–147.
Cane, J.H., Sipes, S., 2006. Floral specialization by bees: analytical methods and a revised lexicon for oligolecty. In: Waser, N.M., Ollerton, J. (Eds.), Plant-Pollinator Interactions:

From Specialization to Generalization. University of Chicago Press, Chicago, IL, pp. 99–122.
Carpenter, S.R., 1982. Stemflow chemistry, effects on population dynamics of detritivorous mosquitoes in tree-hole ecosystems. Oecologia 53, 1–6.
Carstensen, D.W., Olesen, J.M., 2009. Wallacea and its nectarivorous birds: nestedness and modules. J. Biogeogr. 36, 1540–1550.
Carstensen, D.W., Dalsgaard, B., Svenning, J.C., Rahbek, C., Fjeldså, J., Sutherland, W.J., Olesen, J.M., 2012. Biogeographical modules and island roles: a comparison of Wallacea and the West Indies. J. Biogeogr. 39, 739–749.
Carvalheiro, L.G., Barbosa, E.R.M., Memmott, J., 2008. Pollinator networks, alien species and the conservation of rare plants: *Trinia glauca* as a case study. J. Appl. Ecol. 45, 1419–1427.
Carvalheiro, L.G., Veldtman, R., Shenkute, A.G., Tesfay, G.B., Pirk, C.W.W., Donaldson, J.S., Nicolson, S.W., 2011. Natural and within-farmland biodiversity enhances crop productivity. Ecol. Lett. 14, 251–259.
Castle, M.D., Blanchard, J.L., Jennings, S., 2011. Predicted effects of behavioural movement and passive transport on individual growth and community size structure in marine ecosystems. Adv. Ecol. Res. 45, 41–66.
Chalfoun, A.D., Thompson, F.R., Ratnaswamy, M.J., 2002. Nest predators and fragmentation: a review and meta-analysis. Conserv. Biol. 16, 306–318.
Chamberlain, S.A., Holland, J.N., 2009. Body size predicts degree in ant-plant mutualistic networks. Funct. Ecol. 23, 196–202.
Chamberlain, S.A., Kilpatrick, J.R., Holland, J.N., 2010. Do extrafloral nectar resources, species abundances and body sizes contribute to the structure of ant-plant mutualistic networks? Oecologia 164, 741–750.
Christianini, A.V., Oliveira, P.S., 2009. The relevance of ants as seed rescuers of a primarily bird-dispersed tree in the neotropical cerrado savanna. Oecologia 160, 735–745.
Clarke, C.M., Kitching, R.L., 1993. The metazoan food webs of six Bornean *Nepenthes* species. Ecol. Entomol. 18, 7–16.
Cohen, J.E., Schittler, D.N., Raffaelli, D.G., Reuman, D.C., 2009. Food webs are more than the sum of their tritrophic parts. Proc. Natl. Acad. Sci. USA 106, 22335–22340.
Coker, D.J., Pratchett, M.S., Munday, P.L., 2009. Coral bleaching and habitat degradation increase susceptibility to predation for coral-dwelling fishes. Behav. Ecol. 20, 1204–1210.
Coll, M., Schmidt, A., Romanuk, T., Lotze, H.K., 2011. Food-web structure of seagrass communities across different spatial scales and human impacts. PLoS One 6, e22591.
Cook, W.M., Lane, K.T., Foster, B.L., Holt, R.D., 2002. Island theory, matrix effects and species richness patterns in habitat fragments. Ecol. Lett. 5, 619–623.
Corbet, S.A., 2000. Butterfly nectaring flowers, butterfly morphology and flower form. Entomol. Exp. Appl. 96, 289–298.
Corlett, R.T., 1998. Frugivory and seed dispersal by vertebrates in the Oriental (Indomalayan) region. Biol. Rev. 73, 413–448.
Cox, P.A., 1983. Extinction of the Hawaiian avifauna resulted in a change of pollinators for the ieie, *Freycinetia arborea*. Oikos 41, 195–199.
Craig, J.L., Stewart, A.M., Douglas, M.E., 1981. The foraging of New Zealand honeyeaters. N. Z. J. Zool. 8, 87–91.
Cronin, J.T., 2003. Matrix heterogeneity and host-parasitoid interactions in space. Ecology 84, 1506–1516.
Cronin, J.T., 2004. Host-parasitoid extinction and colonization in a fragmented prairie landscape. Oecologia 139, 503–514.
Cronin, J.T., Haynes, K.J., Dillemuth, F., 2004. Spider effects on planthopper mortality: dispersal, and spatial population dynamics. Ecology 85, 2134–2143.

Crooks, K.R., 2002. Relative sensitivities of mammalian carnivores to habitat fragmentation. Conserv. Biol. 16, 488–502.

Crooks, K.R., Soulé, M.E., 1999. Mesopredator release and avifaunal extinctions in a fragmented system. Nature 400, 563–566.

Culver, D.C., Beattie, A.J., 1978. Myrmecochory in *Viola*: dynamics of seed-ant interactions in some West Virginia species. J. Ecol. 66, 53–72.

Cunningham, S.A., 2000. Effects of habitat fragmentation on the reproductive ecology of four plant species in Mallee Woodland. Conserv. Biol. 14, 758–768.

D'Eça Neves, F.F., Morellato, L.P.C., in press. Efeitos de borda na fenologia de árvores em floresta semidecídua de altitude na Serra do Japi, SP. In: Vasconcellos Neto, J., Polli, P.R. (Eds.), *Novos Olhares da Serra do Japi*. Editora da Unicamp, Campinas.

Da Silva, J.M.C., Tabarelli, M., 2000. Tree species impoverishment and the future flora of the Atlantic Forest of northeast Brazil. Nature 404, 72–74.

Dahm, C.N., Baker, M.A., Moore, D.I., Thibault, J.R., 2003. Coupled biogeochemical and hydrological responses of streams and rivers to drought. Freshw. Biol. 48, 1219–1231.

Dale, M.R.T., Fortin, M.-J., 2010. From graphs to spatial graphs. Annu. Rev. Ecol. Syst. 41, 21–38.

Dalsgaard, B., Martín, A.M.G., Olesen, J.M., Timmermann, A., Andersen, L.H., Ollerton, J., 2008. Pollination networks and functional specialization: a test using Lesser Antillean plant-hummingbird assemblages. Oikos 117, 789–793.

Damschem, E.I., Haddad, N.M., Orrock, J.L., Tewksbury, J.J., Levey, D.J., 2006. Corridors increase plant species richness at larger scales. Science 313, 1284–1286.

Darimont, C.T., Carlson, S.M., Kinnison, M.T., Paquet, P.C., Reimchen, T.E., Wilmers, C.C., 2009. Human predators outpace other agents of trait change in the wild. Proc. Natl. Acad. Sci. USA 106, 952–954.

Darvill, B., Knight, M.E., Goulson, D., 2004. Use of genetic markers to quantify bumblebee foraging range and nest density. Oikos 107, 471–478.

Davidson, D.W., 1988. Ecological studies of neotropical ant gardens. Ecology 69, 1138–1152.

Davies, K.F., Margules, C.R., Lawrence, J.F., 2000. Which traits of species predict population declines in experimental forest fragments? Ecology 81, 1450–1461.

Davies, K.F., Margules, C.R., Lawrence, J.F., 2004. A synergistic effect puts rare, specialized species at greater risk of extinction. Ecology 85, 265–271.

Dawideit, B.A., Phillimore, A.B., Laube, I., Leisler, B., Böhning-Gaese, K., 2009. Ecomorphological predictors of natal dispersal distances in birds. J. Anim. Ecol. 78, 388–395.

Debinski, D.M., Holt, R.D., 2000. A survey and overview of habitat fragmentation experiments. Conserv. Biol. 14, 342–355.

Devlaeminck, R., Bossuyt, B., Hermy, M., 2005. Inflow of seeds through the forest edge: evidence from seed bank and vegetation patterns. Plant Ecol. 176, 1–17.

Deza, A.A., Anderson, T.W., 2010. Habitat fragmentation, patch size, and the recruitment and abundance of kelp forest fishes. Mar. Ecol. Prog. Ser. 416, 229–240.

Diaz-Castelazo, C., Guimarães, P.R., Jordano, P., Thompson, J.N., Marquis, R.J., Rico-Gray, V., 2010. Changes of a mutualistic network over time: reanalysis over a 10-year period. Ecology 91, 793–801.

Dick, C.W., 2001. Genetic rescue of remnant tropical trees by an alien pollinator. Proc. R. Soc. B 268, 2391–2396.

Didham, R.K., Lawton, J.H., 1999. Edge structure determines the magnitude of changes in microclimate and vegetation structure in tropical forest fragments. Biotropica 31, 17–30.

Didham, R.K., Ghazoul, J., Stork, N., Davis, A.J., 1996. Insects in fragmented forests: a functional approach. Trends Ecol. Evol. 11, 255–260.

Didham, R.K., Kapos, V., Ewers, R.M., 2012. Rethinking the conceptual foundations of habitat fragmentation research. Oikos 121, 161–170.

Do Carmo, R.M., Franceschinelli, E.V., da Silveira, F.A., 2004. Introduced honeybees (*Apis mellifera*) reduce pollination success without affecting the floral resource taken by native pollinators. Biotropica 36, 371–376.

Doak, D.F., Marino, P.C., Kareiva, P.M., 1992. Spatial scale mediates the influence of habitat fragmentation on dispersal success—implications for conservation. Theor. Popul. Biol. 41, 315–336.

Donald, P.F., Green, R.E., Heath, M.F., 2001. Agricultural intensification and the collapse of Europe's farmland bird populations. Proc. R. Soc. B 268, 25–29.

Donatti, C.I., Guimarães, P.R., Galetti, M., Pizo, M.A., Marquitti, F.M.D., Dirzo, R., 2011. Analysis of a hyper-diverse seed dispersal network: modularity and underlying mechanisms. Ecol. Lett. 14, 773–781.

Duarte, C.M., Chiscano, C.L., 1999. Seagrass biomass and production: a reassessment. Aquat. Bot. 65, 159–174.

Dudareva, N., Pichersky, E. (Eds.), 2006. Biology of Floral Scent. CRC Press, Boca Raton.

Dunne, J.A., Williams, R.J., Martinez, N.D., 2002. Network structure and biodiversity loss in food webs: robustness increases with connectance. Ecol. Lett. 5, 558–567.

Dupont, Y.L., Olesen, J.M., 2009. Ecological modules and roles of species in heathland plant-insect flower visitor networks. J. Anim. Ecol. 78, 346–353.

Dupont, Y.L., Hansen, D.M., Rasmussen, J.T., Olesen, J.M., 2004. Evolutionary changes in nectar sugar composition associated with switches between bird and insect pollination: the Canarian bird-flower element revisited. Funct. Ecol. 18, 670–676.

Dupont, Y.L., Trøjelsgaard, K., Olesen, J.M., 2011. Scaling down from species to individuals: a flower-visitation network between individual honeybees and thistle plants. Oikos 120, 170–177.

Dupré, C., Ehrlén, J., 2002. Habitat configuration, species traits and plant distributions. J. Ecol. 90, 796–805.

Duthie, C., Gibbs, G., Burns, K.C., 2006. Seed dispersal by weta. Science 311, 1575.

Dyer, L.A., Walla, T.R., Greeney, H.F., Stireman III, J.O., Hazen, R.F., 2010. Diversity of interactions: a metric for studies of biodiversity. Biotropica 42, 281–289.

Ecroyd, C., 1993. In search of the wood rose. For. Bird 1993, 24–28.

Eggleston, D.B., Etherington, L.L., Elis, W.E., 1998. Organism response to habitat patchiness: species and habitat-dependent recruitment of decapod crustaceans. J. Exp. Mar. Biol. Ecol. 223, 111–132.

Ellstrand, N.C., Elam, D.R., 1993. Population genetic consequences of small population size: implications for plant conservation. Annu. Rev. Ecol. Syst. 24, 217–242.

Elton, C.S., 1927. Animal Ecology. Macmillian, New York.

Elzinga, J.A., van Nouhuys, S., van Leeuwen, D.J., Biere, A., 2007. Distribution and colonization ability of three parasitoids and their herbivorous host in a fragmented landscape. Basic Appl. Ecol. 8, 75–88.

Emmerson, M.C., Raffaelli, D., 2004. Predator-prey body size, interaction strength and the stability of a real food web. J. Anim. Ecol. 73, 399–409.

Eshiamwata, G.W., Berens, D.G., Bleher, B., Dean, W.R.J., Böhning-Gaese, K., 2006. Bird assemblages in isolated *Ficus* trees in Kenyan farmland. J. Trop. Ecol. 22, 723–726.

Estes, J.A., Terborgh, J., Brashares, J.S., Power, M.E., Berger, J., Bond, W.J., Carpenter, S.R., Essington, T.E., Holt, R.D., Jackson, J.B.C., Marquis, R.J., Oksanen, L., et al., 2011. Trophic downgrading of planet Earth. Science 333, 301–306.

Evans, W.A., Johnston, B., 1980. Fish Migration and Fish Passage: A Practical Guide to Solving Fish Passage Problems. U.S. Forest Service, Washington, DC, 163pp.

Everard, M., 1996. The importance of periodic droughts for maintaining diversity in the freshwater environment. Freshw. Forum 7, 33–50.

Ewers, R.M., Didham, R.K., 2006. Confounding factors in the detection of species responses to habitat fragmentation. Biol. Rev. 81, 117–142.

Fadini, R.F., Fleury, M., Donatti, C.I., Galetti, M., 2009. Effects of frugivore impoverishment and seed predators on the recruitment of a keystone palm. Acta Oecol. 35, 188–196.

Faegri, K., Pijl, L.V.D., 1979. The principles of Pollination Ecology. Pergamon Press, Oxford.

Fahrig, L., 2003. Effects of habitat fragmentation on biodiversity. Annu. Rev. Ecol. Syst. 34, 487–515.

Fall, A., Fortin, M.-J., Manseau, M., O'Brien, D., 2007. Spatial graphs: principles and applications for habitat connectivity. Ecosystems 10, 448–461.

Faria, D., Laps, R.R., Baumgarten, J., Cetra, M., 2006. Bat and bird assemblages from forests and shade cacao plantations in two contrasting landscapes in the Atlantic rainforest of southern Bahia, Brazil. Biodivers. Conserv. 15, 587–612.

Faria, D., Paciencia, M.L.B., Dixo, M., Laps, R.R., Baumgarten, J., 2007. Ferns, frogs, lizards, birds and bats in forest fragments and shade cacao plantations in two contrasting landscapes in the Atlantic forest, Brazil. Biodivers. Conserv. 16, 2335–2357.

Feinsinger, P., 1978. Ecological interactions between plants and hummingbirds in a successional tropical community. Ecol. Monogr. 48, 269–287.

Feld, C.K., Birk, S., Bradley, D.C., Hering, D., Kail, J., Marzin, A., Melcher, A., Nemitz, D., Pederson, M.L., Pletter Bauer, F., Pont, D., Verdonschot, P.F.M., et al., 2011. From natural to degraded rivers and back again: a test of restoration ecology theory and practice. Adv. Ecol. Res. 44, 119–210.

Fenster, C.B., Dudash, M.R., 2001. Spatiotemporal variation in the role of hummingbirds as pollinators of *Silene virginica*. Ecology 82, 844–851.

Finn, D.S., Rasanen, K., Robinson, C.T., 2010. Physical and biological changes to a lengthening stream gradient following a decade of rapid glacial recession. Glob. Change Biol. 16, 3314–3326.

Fischer, J., Lindenmayer, D.B., 2002. Treating the nestedness temperature calculator as a "black box" can lead to false conclusions. Oikos 99, 193–199.

Fishbein, M., Venable, D.L., 1996. Diversity and temporal change in the effective pollinators of *Asclepias tuberosa*. Ecology 77, 1061–1073.

Fisher, J., Stott, J., Law, B.S., 2010. The disproportionate value of scattered trees. Biol. Conserv. 143, 1564–1567.

Fitzpatrick, Ú., Murray, T.E., Paxton, R.J., Breen, J., Cotton, D., Santorum, V., Brown, M.F., 2007. Rarity and decline in bumblebees—a test of causes and correlates in the Irish fauna. Biol. Conserv. 136, 185–194.

Fleming, T.H., Breitwisch, R., Whitesides, G.H., 1987. Patterns of tropical vertebrate frugivore diversity. Annu. Rev. Ecol. Syst. 18, 91–109.

Fleury, M., Galetti, M., 2006. Forest fragment size and microhabitat effects on palm seed predation. Biol. Conserv. 131, 1–13.

Fonseca, G., Benson, P.J., 2003. Biodiversity conservation demands open access. PLoS Biol. 1, e46.

Fonseca, G.R., Ganade, G., 1996. Asymmetries, compartments and null interactions in an Amazonian ant-plant community. J. Anim. Ecol. 65, 339–347.

Fontaine, C., Dajoz, I., Meriguet, J., Loreau, M., 2006. Functional diversity of plant-pollinator interaction webs enhances the persistence of plant communities. PLoS Biol. 4, 129–135.

Fontaine, C., Guimarães Jr., P.R., Kéfi, S., Loeuille, N., Memmott, J., van der Putten, W.H., van Veen, F.J.F., Thébault, E., 2011. The ecological and evolutionary implications of merging different types of networks. Ecol. Lett. 14, 1170–1181.

Fortuna, M.A., Bascompte, J., 2006. Habitat loss and the structure of plant-animal mutualistic networks. Ecol. Lett. 9, 281–286.

Fortuna, M.A., Bascompte, J., 2008. The network approach in ecology. In: Valladares, F., Camacho, A., Elosegi, A., Gracia, C., Estrada, M., Senar, J.C., Gili, J.P. (Eds.), Unity in Diversity: Reflections on Ecology After the Legacy of Ramon Margalef. Fundación BBVA, Bilbao, pp. 371–393.
Fortuna, M.A., Gómez-Rodríguez, C., Bascompte, J., 2006. Spatial network structure and amphibian persistence in stochastic environments. Proc. R. Soc. B 273, 1429–1434.
Fortuna, M.A., Stouffer, D.B., Olesen, J.M., Jordano, P., Mouillot, D., Krasnov, B.R., Poulin, R., Bascompte, J., 2010. Nestedness versus modularity in ecological networks: two sides of the same coin? J. Anim. Ecol. 79, 811–817.
Fox, H.E., 2004. Coral recruitment in blasted and unblasted sites in Indonesia: assessing rehabilitation potential. Marine Ecol. Prog. Ser. 269, 131–139.
Fragoso, J., 1997. Tapir-generated seed shadows: scale-dependent patchiness in the Amazon rain forest. J. Ecol. 85, 519–529.
Fragoso, J., Silvius, K., Correa, L., 2003. Long-distance seed dispersal by tapirs increases seed survival and aggregates tropical trees. Ecology 84, 1998–2006.
Franklin, J.E., Forman, R.T.T., 1987. Creating landscape patterns by forest cutting: ecological consequences and principles. Landsc. Ecol. 1, 5–18.
Frost, M.T., Rowden, A.A., Attrill, M.J., 1999. Effect of habitat fragmentation on the macroinvertebrate infaunal communities associated with the seagrass *Zostera marina* L. Aquat. Conserv. 9, 255–263.
Fuchs, E.J., Lobo, J.A., Quesada, M., 2003. Effects of forest fragmentation and flowering phenology on the reproductive success and mating patterns of the tropical dry forest tree *Pachira quinata*. Conserv. Biol. 17, 149–157.
Galetti, M., Donatti, C., Pires, A., Guimarães, P., Jordano, P., 2006. Seed survival and dispersal of an endemic Atlantic forest palm: the combined effects of defaunation and forest fragmentation. Bot. J. Linn. Soc. 151, 141–149.
Galetti, M., Giacomini, H.C., Bueno, R.S., Bernardo, C.S., Marques, R.M., Bovendorp, R.S., et al., 2009. Priority areas for the conservation of Atlantic forest large mammals. Biol. Conserv. 142, 1229–1241.
Ganzhorn, J.U., Eisenbeiss, B., 2001. The concept of nested species assemblages and its utility for understanding effects of habitat fragmentation. Basic Appl. Ecol. 2, 87–95.
Garcia, D., Chacoff, N.P., 2007. Scale-dependent effects of habitat fragmentation on hawthorn pollination, frugivory, and seed predation. Conserv. Biol. 21, 400–411.
Gaston, K.J., 1988. Patterns in local and regional dynamics of moth populations. Oikos 53, 49–57.
Gaston, K.J., Blackburn, T.M., 1995. Birds, body size and the threat of extinction. Philos. Trans. R. Soc. B 347, 205–212.
Gathmann, A., Tscharntke, T., 2002. Foraging ranges of solitary bees. J. Anim. Ecol. 71, 757–764.
Gehlhausen, S.M., Schwartz, M.W., Augspurger, C.K., 2000. Vegetation and microclimatic edge effects in two mixed-mesophytic forest fragments. Plant Ecol. 147, 21–35.
Genini, J., Morrellato, L.P.C., Guimarães Jr., P.R., Olesen, J.M., 2010. Cheaters in mutualism networks. Biol. Lett. 6, 494–497.
Gilbert-Norton, L., Wilson, R., Stevens, J.R., Beard, K.H., 2010. A meta-analytic review of corridor effectiveness. Conserv. Biol. 24, 660–668.
Gill, F.B., Wolf, L.L., 1975. Economics of feeding territoriality in Golden-winged sunbird. Ecology 56, 333–345.
Gilljam, D., Thierry, A., Edwards, F.K., Figueroa, D., Ibbotson, A.T., Jones, J.I., Lauridsen, R.B., Petchey, O.L., Woodward, G., Ebenman, B., 2011. Seeing double: size-based versus taxonomic views of food web structure. Adv. Ecol. Res. 45, 67–134.
Gilpin, M.E., Diamond, J.M., 1981. Immigration and extinction probabilities for individual species: relation to incidence functions and species colonization curves. Proc. Natl. Acad. Sci. USA 78, 392–396.

Girao, L.C., Lopes, A.V., Tabarelli, M., Bruna, E.M., 2007. Changes in tree reproductive traits reduce functional diversity in a fragmented Atlantic forest landscape. PLoS One 2, e908.

Githiru, M., Lens, L., Bennur, L.A., Ogol, C., 2002. Effects of site and fruit size on the composition of avian frugivore assemblages in a fragmented Afrotropical forest. Oikos 96, 320–330.

Gjerløv, C., Hildrew, A.G., Jones, J.I., 2003. Mobility of stream invertebrates in relation to disturbance and refugia: a test of habitat templet theory. J. N. Am. Benthol. Soc. 22, 207–223.

Gliwicz, J., 1980. Island populations of rodents: their organization and functioning. Biol. Rev. 55, 109–138.

Goldblatt, P., Manning, J.C., 2000. The long-proboscid fly pollination system in southern Africa. Ann. Mo. Bot. Gard. 87, 146–170.

Gómez, J.M., 2000. Effectiveness of ants as pollinators of *Lobularia maritima*: effects on main sequential fitness of the host plant. Oecologia 122, 90–97.

Gomulkiewicz, R., Thompson, J.N., Holt, R.D., Nuismer, S.L., Hochberg, M.E., 2000. Hot spots, cold spots, and the geographic mosaic theory of coevolution. Am. Nat. 156, 156–174.

Gonzalez, A., Rayfield, B., Lindo, Z., 2011. The disentangled bank: how loss of habitat fragments and disassembles ecological networks. Am. J. Bot. 98, 503–516.

Goulson, D., 2003. Effects of introduced bees on native ecosystems. Annu. Rev. Ecol. Syst. 34, 1–26.

Graham, C., 2002. Use of fruiting trees by birds in continuous forest and riparian forest remnants in Los Tuxtlas, Veracruz, Mexico. Biotropica 34, 589–597.

Graham, N.A.J., Wilson, S.K., Jennings, S., Polunin, N.V.C., Robinson, J., Bijoux, J.P., Daw, T.M., 2007. Lag effects in the impacts of mass coral bleaching on coral reef fish, fisheries, and ecosystems. Conserv. Biol. 21, 1291–1300.

Greenleaf, S.S., Williams, N.M., Winfree, R., Kremen, C., 2007. Bee foraging ranges and their relationship to body size. Oecologia 153, 589–596.

Griffin, J.N., Butler, J., Soomdat, N.N., Brun, K.E., Chejanorski, Z.A., Silliman, B.R., 2011. Top predators suppress rather than facilitate plants in a trait-mediated tri-trophic cascade. Biol. Lett. 7, 510–513.

Guimarães, P.R., Cogni, R., 2002. Seed cleaning of *Cupania vernalis* (Sapindaceae) by ants: edge effect in a highland forest in South-East Brazil. J. Trop. Ecol. 18, 303–307.

Guimarães, P.R., Rico-Gray, V., Oliveira, P.S., Izzo, T.J., dos Reis, S.F., Thompson, J.N., 2007. Interaction intimacy affects structure and coevolutionary dynamics in mutualistic networks. Curr. Biol. 17, 1797–1803.

Guimarães, P.R., Galetti, M., Jordano, P., 2008. Seed dispersal anachronisms: rethinking the fruits extinct megafauna ate. PLoS One 3, e1745.

Guimarães, P.R., Jordano, P., Thompson, J.N., 2011. Evolution and coevolution in mutualistic networks. Ecol. Lett. 14, 877–885.

Guimerà, R., Amaral, L.A.N., 2005. Cartography of complex networks: modules and universal roles. J. Stat. Mech. 2005 (2), P02001. http://dx.doi.org/10.1088/1742-5468/2005/02/P02001.

Hadley, A.S., Betts, M.G., 2012. The effects of landscape fragmentation on pollination dynamics: absence of evidence not evidence of absence. Biol. Rev. 87, 526–544.

Hagen, M., Kraemer, M., 2010. Agricultural surroundings support flower-visitor networks in an Afrotropical rain forest. Biol. Conserv. 143, 1654–1663.

Hagen, M., Wikelski, M., Kissling, W.D., 2011. Space use of bumblebees (*Bombus* spp.) revealed by radio-tracking. PLoS One 6, e19997.

Halloy, S.R.P., 1998. A theoretical framework for abundance distributions in complex systems. Complex. Int. 6, 1–12.

Halloy, S.R.P., Barratt, B.I.P., 2007. Patterns of abundance and morphology as indicators of ecosystem status: a meta-analysis. Ecol. Complex. 4, 128–147.
Hance, T., Baaren, J., van Vernon, P., Boivin, G., 2007. Impact of extreme temperatures on parasitoids in a climate change perspective. Annu. Rev. Entomol. 52, 107–126.
Hanski, I., 1998. Metapopulation dynamics. Nature 396, 41–49.
Hanski, I., 1999. Metapopulation Ecology. Oxford University Press Inc., New York.
Hanski, I., 2005. Landscape fragmentation, biodiversity loss and the societal response—the longterm consequences of our use of natural resources may be surprising and unpleasant. EMBO Rep. 6, 388–392.
Hanski, I., Gilpin, M.E., 1997. Metapopulation Biology, Ecology, Genetics and Evolution. Academic Press, San Diego, CA.
Hanski, I., Ovaskainen, O., 2000. The metapopulation capacity of a fragmented landscape. Nature 404, 755–758.
Hanski, I., Simberloff, D., 1997. The metapopulation approach, its history, conceptual domain and application to conservation. In: Hanski, I.A., Gilpin, M.E. (Eds.), Metapopulation Biology, Ecology, Genetics and Evolution. Academic Press, San Diego, CA, pp. 5–26.
Hansson, L., 1991. Dispersal and connectivity in metapopulations. Biol. J. Linn. Soc. 42, 89–103.
Hansson, L., 1994. Vertebrate distributions relative to clear-cut edges in a boreal forest landscape. Landsc. Ecol. 9, 105–115.
Harder, L.D., 1985. Morphology as a predictor of flower choice by bumble bees. Ecology 66, 198–210.
Hardt, R.A., Forman, R.T.T., 1989. Boundary form effects on woody colonization of reclaimed surface mines. Ecology 70, 1252–1260.
Harris, F.L., Johnson, S.D., 2004. The consequences of habitat fragmentation for plant-pollinator mutualisms. Int. J. Trop. Insect Sci. 24, 29–43.
Harris, R.M.L., Milner, A.M., Armitage, P.D., Ledger, M.E., 2007. Replicability of physicochemistry and macroinvertebrate assemblages in stream mesocosms, implications for experimental research. Freshw. Biol. 52, 2434–2443.
Haskell, J.P., Ritchie, M.E., Olff, H., 2002. Fractal geometry predicts varying body size scaling relationships for mammal and bird home ranges. Nature 418, 527–530.
Hastings, A., Powell, T., 1991. Chaos in a three-species food chain. Ecology 72, 896–903.
Hegland, S.J., Totland, Ø., 2008. Is the magnitude of pollen limitation in a plant community affected by pollinator visitation and plant species specialisation levels? Oikos 117, 883–891.
Hegland, S.J., Nielsen, A., Lázaro, A., Bjerknes, A.-L., Totland, Ø., 2009. How does climate warming affect plant-pollinator interactions? Ecol. Lett. 12, 184–195.
Heleno, R.H., Ceia, R.S., Ramos, J.A., Memmott, J., 2009. Effects of alien plants on insect abundance and biomass: a food-web approach. Conserv. Biol. 23, 410–419.
Henle, K., Davies, K.F., Kleyer, M., Margules, C., Settele, J., 2004. Predictors of species sensitivity to fragmentation. Biodivers. Conserv. 13, 207–251.
Henri, D.C., van Veen, F.J.F., 2011. Body size, life history and the structure of host-parasitoid networks. Adv. Ecol. Res. 45, 135–180.
Herrera, C.M., 1988. Variation in mutualisms, the spatiotemporal mosaic of a pollinator assemblage. Biol. J. Linn. Soc. 35, 95–125.
Herrera, C.M., 2009. Multiplicity in Unity. Plant Subindividual Variation and Interactions with Animals. University of Chicago Press, Chicago.
Hildrew, A.G., 2009. Sustained research on stream communities: a model system and the comparative approach. Adv. Ecol. Res. 41, 175–312.
Hildrew, A.G., Woodward, G., Winterbottom, J.H., Orton, S., 2004. Strong density-dependence in a predatory insect: larger scale experiments in a stream. J. Anim. Ecol. 73, 448–458.

Hill, J.K., Gray, M.A., Khen, C.V., Benedick, S., Tawatao, N., Hamer, K.C., 2011. Ecological impacts of tropical forest fragmentation: how consistent are patterns in species richness and nestedness? Philos. Trans. R. Soc. B 366, 3265–3276.
Hilty, J.A., Lidicker Jr., W.Z., Merenlender, A.M., 2006. Corridor Ecology: The Science and Practice of Linking Landscapes for Biodiversity Conservation. Island Press, Washington, DC.
Hirst, J.A., Attrill, M.J., 2008. Small is beautiful: an inverted view of habitat fragmentation in seagrass beds. Estuar. Coast. Shelf Sci. 78, 811–818.
Hladyz, S., Gessner, M.O., Giller, P.S., Pozo, J., Woodward, G., 2009. Resource quality and stochiometric constraints in a stream food web. Freshw. Biol. 54, 957–970.
Hladyz, S., Åbjörnsson, K., Cariss, H., Chauvet, E., Dobson, M., Elosegi, A., Ferreira, V., Fleituch, T., Gessner, M.O., Giller, P.S., Graça, M.A.S., Gulis, V., et al., 2011a. Stream ecosystem functioning in an agricultural landscape: the importance of terrestrial-aquatic linkages. Adv. Ecol. Res. 44, 211–276.
Hladyz, S., Åbjörnsson, K., Giller, P.S., Woodward, G., 2011b. Impacts of an aggressive riparian invader on community structure and ecosystem functioning in stream food webs. J. Appl. Ecol. 48, 443–452.
Hobbs, R.J., Yates, C.J., 2003. Impacts of ecosystem fragmentation on plant populations: generalising the idiosyncratic. Aust. J. Bot. 51, 471–488.
Hobbs, R.J., Richardson, D.M., Davis, G.W., 1995. Mediterranean-type ecosystems, opportunities and constraints for studying the function of biodiversity. In: Davis, G.W., Richardson, D.M. (Eds.), Mediterranean-Type Ecosystem—The Function of Biodiversity. Springer-Verlag, Berlin, pp. 1–32.
Hoekman, D., Dreyer, J., Jackson, R.D., Townsend, P.A., Gratton, C., 2011. Lake to land subsidies: experimental addition of aquatic insects increases terrestrial arthropod densities. Ecology 92, 2063–2072.
Hoekstra, H.E., Fagan, W.F., 1998. Body size, dispersal ability and compositional disharmony, the carnivore-dominated fauna of the Kuril Islands. Divers. Distrib. 4, 135–149.
Holt, R.D., 1996. Food webs in space: an island biogeographic perspective. In: Polis, G.A., Winemiller, K.O. (Eds.), Food Webs, Integration of Patterns and Dynamics. Chapman and Hall, London, pp. 313–323.
Holt, R.D., 1997. From metapopulation dynamics to community structure some consequences of spatial heterogeneity. In: Hanski, I.A., Gilpin, M.E. (Eds.), Metapopulation Biology. Academic Press, San Diego, CA, pp. 149–164.
Holt, R.D., 2002. Food webs in space: on the interplay of dynamic instability and spatial processes. Ecol. Res. 17, 261–273.
Holyoak, M., 2000. Habitat subdivision causes changes in food web structure. Ecol. Lett. 3, 509–515.
Holzschuh, A., Steffan-Dewenter, I., Tscharntke, T., 2010. How do landscape composition and configuration, organic farming and fallow strips affect the diversity of bees, wasps and their parasitoids? J. Anim. Ecol. 79, 491–500.
Horvitz, C.C., Schemske, D.W., 1990. Spatiotemporal variation in insect mutualists of a neotropical herb. Ecology 71, 1085–1097.
Hovel, K.A., Lipcius, R.N., 2001. Habitat fragmentation in a seagrass landscape: patch size and complexity control blue crab survival. Ecology 82, 1814–1829.
Hubbell, S.P., 2001. The Unified Neutral Theory of Biodiversity and Biogeography. Princeton University Press, Princeton, MA.
Hughes, T.P., 1994. Catastrophes, phase-shifts, and large-scale degradation of a Caribbean coral-reef. Science 265, 1547–1551.
Ings, T.C., Montoya, J.M., Bascompte, J., Blüthgen, N., Brown, L., Dormann, C.F., Edwards, F., Figueroa, D., Jacob, U., Jones, J.I., Lauridsen, R.B., Ledger, M.E., et al., 2009. Ecological networks—beyond food webs. J. Anim. Ecol. 78, 253–269.

Jacob, U., Thierry, A., Brose, U., Arntz, W.E., Berg, S., Brey, T., Fetzer, I., Jonsson, T., Mintenbeck, K., Mollmann, C., Petchey, O., Riede, J.O., et al., 2011. The role of body size in complex food webs: a cold case. Adv. Ecol. Res. 45, 181–223.

Jacobsen, D., 2008. Low oxygen pressure as a driving factor for the altitudinal decline in taxon richness of stream macroinvertebrates. Oecologia 154, 795–807.

Jacobsen, D., Milner, A.M., Brown, L.E., Dangles, O., 2012. Biodiversity under threat in glacier-fed river systems. Nat. Clim. Change. 2, 361–364.

Janzen, D.H., 1971. Euglossine bees as long-distance pollinators of tropical plants. Science 171, 203–205.

Jennersten, O., 1988. Pollination in *Dianthus deltoides* (Caryophyllaceae): effects of habitat fragmentation on visitation and seed set. Conserv. Biol. 2, 359–366.

Jennings, S., Warr, K.J., 2003. Smaller predator-prey body size ratios in longer food chains. Proc. R. Soc. B 270, 1413–1417.

Jetz, W., Carbone, C., Fulford, J., Brown, J.H., 2004. The scaling of animal space use. Science 306, 266–268.

Jetz, W., Wilcove, D.S., Dobson, A.P., 2007. Projected impacts of climate and land-use change on the global diversity of birds. PLoS Biol. 5, 1211–1219.

Jonsson, T., Ebenman, B., 1998. Effects of predator-prey body size ratios on the stability of food chains. J. Theor. Biol. 193, 407–417.

Jonsson, T., Cohen, J.E., Carpenter, S.R., 2005. Food webs, body size, and species abundance in ecological community description. Adv. Ecol. Res. 36, 1–84.

Jordano, P., 1987. Patterns of mutualistic interactions in pollination and seed dispersal—connectance, dependence asymmetry and coevolution. Am. Nat. 129, 657–677.

Jordano, P., 1995. Angiosperm fleshy fruits and seed dispersers—a comparative analysis of adaptation and constraints in plant-animal interactions. Am. Nat. 145, 163–191.

Jordano, P., 2000. Fruits and frugivory. In: Fenner, M. (Ed.), Seeds: The Ecology of Regeneration in Natural Plant Communities. Commonwealth Agricultural Bureau International, Wallingford, UK, pp. 125–166.

Jordano, P., Schupp, E.W., 2000. Seed disperser effectiveness, the quantity component and patterns of seed rain for *Prunus mahaleb*. Ecol. Monogr. 70, 591–615.

Jordano, P., Bascompte, J., Olesen, J.M., 2003. Invariant properties in coevolutionary networks of plant-animal interactions. Ecol. Lett. 6, 69–81.

Jules, E.S., Rathcke, B.J., 1999. Mechanisms of reduced *Trillium* recruitment along edges of old-growth forest fragments. Conserv. Biol. 13, 784–793.

Kaartinen, R., Roslin, T., 2011. Shrinking by numbers: landscape context affects the species composition but not the quantitative structure of local food webs. J. Anim. Ecol. 80, 622–631.

Kaiser, C.N., Hansen, O.M., Müller, C.B., 2008. Habitat structure affects reproductive success of the rare endemic tree *syzygium mamillatum* (Myrtaceae) in restored and unrestored sites in Mauritius. Biotropica 40, 86–94.

Kaiser-Bunbury, C.N., Muff, S., Memmott, J., Müller, C.B., Caflisch, A., 2010. The robustness of pollination networks to the loss of species and interactions, a quantitative approach incorporating pollinator behaviour. Ecol. Lett. 13, 442–452.

Kaiser-Bunbury, C.N., Traveset, A., Hansen, D.M., 2010. Conservation and restoration of plant-animal mutualisms on oceanic islands. Pl. Ecol. Evol. Syst. 12, 131–143.

Kaiser-Bunbury, C.N., Valentin, T., Mougal, J., Matatiken, D., Ghazoul, J., 2011. The tolerance of island pollination networks to alien plants. J. Ecol. 99, 202–213.

Kandori, I., 2002. Diverse visitors with various pollinator importance and temporal change in the important pollinators of *Geranium thunbergii* (Geraniaceae). Ecol. Res. 17, 283–294.

Kapos, V., Wandelli, E., Camargo, J.L., Ganade, G., 1997. Edge-related changes in environmental and plant responses due to forest fragmentation in Central Amazonia. In:

Laurance, W.F., Bierregaard, R.O. (Eds.), Tropical Forest Remnants. Chicago University Press, Chicago, IL, pp. 33–44.

Kato, E., Hiura, T., 1999. Fruit set in *Styrax obassia* (Styracaceae): the effect of light availability, display size and local floral density. Am. J. Bot. 86, 495–501.

Kearns, C.A., 2001. North American, dipteran pollinators: assessing their value and conservation status. Conserv. Ecol. 5, 5.

Kearns, C.A., Inouye, D.W., Waser, N.M., 1998. Endangered mutualisms: the conservation of plant-pollinator interactions. Annu. Rev. Ecol. Syst. 29, 83–112.

Kingsolver, J.G., Hoekstra, H.E., Hoekstra, J.M., Berrigan, D., Vignieri, S.N., Hill, C.E., Hoang, A., Gibert, P., Beerli, P., 2001. The strength of phenotypic selection in natural populations. Am. Nat. 157, 245–261.

Kissling, W.D., Rahbek, C., Böhning-Gaese, K., 2007. Food plant diversity as broad-scale determinant of avian frugivore richness. Proc. R. Soc. B 274, 799–808.

Kissling, W.D., Field, R., Böhning-Gaese, K., 2008. Spatial patterns of woody plant and bird diversity: functional relationships or environmental effects? Glob. Ecol. Biogeogr. 17, 327–339.

Kissling, W.D., Böhning-Gaese, K., Jetz, W., 2009. The global distribution of frugivory in birds. Glob. Ecol. Biogeogr. 18, 150–162.

Kissling, W.D., Field, R., Korntheuer, H., Heyder, U., Böhning-Gaese, K., 2010. Woody plants and the prediction of climate-change impacts on bird diversity. Philos. Trans. R. Soc. B 365, 2035–2045.

Kissling, W.D., Dormann, C.F., Groeneveld, J., Hickler, T., Kühn, I., McInerny, G.J., Montoya, J.M., Römermann, C., Schiffers, K., Schurr, F.M., Singer, A., Svenning, J.-C., et al., 2012a. Towards novel approaches to modelling biotic interactions in multispecies assemblages at large spatial extents. J. Biogeogr. http://dx.doi.org/10.1111/j.1365-2699.2011.02663.x.

Kissling, W.D., Sekercioglu, C.H., Jetz, W., 2012b. Bird dietary guild richness across latitudes, environments and biogeographic regions. Glob. Ecol. Biogeogr. 21, 328–340.

Kitching, R.L., 2000. Food Webs and Container Habitats: The Natural History and Ecology of Phytotelmata. Cambridge University Press, Cambridge, UK p. 431.

Kitching, R.L., 2001. Food webs in phytotelmata: "bottom-up" and "top-down" explanations for community structure. Annu. Rev. Entomol. 46, 729–760.

Kivinen, S., Luoto, M., Kuussaari, M., Helenius, J., 2006. Multi-species richness of boreal agricultural landscapes: effects of climate, biotope, soil and geographical location. J. Biogeogr. 33, 862–875.

Klausmeier, C.A., 2001. Habitat destruction and extinction in competitive and mutualistic metacommunities. Ecol. Lett. 4, 57–63.

Klein, A.-M., Steffan-Dewenter, I., Tscharntke, T., 2003. Pollination of *Coffea canephora* in relation to local and regional agroforestry management. J. Appl. Ecol. 40, 837–845.

Kneitel, J.M., Miller, T.E., 2002. Resource and top-predator regulation in the pitcher plant (*Sarracenia purpurea*) inquiline community. Ecology 83, 680–688.

Koh, L.P., 2007. Impacts of land use change on South-east Asian forest butterflies: a review. J. Appl. Ecol. 44, 703–713.

Kolb, A., Diekmann, M., 2005. Effects of life-history traits on responses of plant species to forest fragmentation. Conserv. Biol. 19, 929–938.

Kondoh, M., Kato, S., Sakato, Y., 2010. Food webs are built up with nested subwebs. Ecology 91, 3123–3130.

Körner, C., 1993. Scaling from species to vegetation, the usefulness of functional groups. In: Schulze, E.D., Mooney, H.A. (Eds.), Biodiversity and Ecosystem Function. Springer-Verlag, Berlin, pp. 117–140.

Kraemer, M., Schmitt, U., 1999. Possible pollination by hummingbirds in *Anthurium sanguineum* Engl. (Araceae). Plant Syst. Evol. 217, 333–335.

Krauss, J., Alfert, T., Steffan-Dewenter, I., 2009. Habitat area but not habitat age determines wild bee richness in limestone quarries. J. Appl. Ecol. 46, 194–202.

Krauss, J., Bommarco, R., Guardiola, M., Heikkinen, R.K., Helm, A., Kuussaari, M., Lindborg, R., Öckinger, E., Pärtel, M., Pino, J., Pöyry, J., Raatikainen, K.M., et al., 2010. Habitat fragmentation causes immediate and time-delayed biodiversity loss at different trophic levels. Ecol. Lett. 13, 597–605.

Kremen, C., Hall, G., 2005. Managing ecosystem services: what do we need to know about their ecology? Ecol. Lett. 8, 468–479.

Krishna, A., Guimarães Jr., P.R., Jordano, P., Bascompte, J., 2008. A neutral-niche theory of nestedness in mutualistic networks. Oikos 117, 1609–1618.

Kruess, A., 2003. Effects of landscape structure and habitat type on a plant-herbivore-parasitoid community. Ecography 26, 283–290.

Kruess, A., Tscharntke, T., 1994. Habitat fragmentation, species loss and biological control. Science 264, 1581–1584.

Kruess, A., Tscharntke, T., 2000. Species richness and parasitism in a fragmented landscape: experiments and field studies with insects on *Vicia sepium*. Oecologia 122, 129–137.

Kundzewicz, Z.W., Mata, L.J., Arnell, N.W., Döll, P., Jimenez, B., Miller, K., Oki, T., Sem, Z., Shiklomanov, I., 2008. The implications of projected climate change for freshwater resources and their management. Hydrol. Sci. J. 53, 3–10.

Kurihara, Y., 1959. Synecological analyses of the biotic community in microcosm. IV. Studies on the relations of Diptera larvae to pH in bamboo containers. Sci. Rep. Tohuku Univ. Ser. IV Biol. 25, 165–171.

Kuussaari, M., Nieminen, M., Hanski, I., 1996. An experimental study of migration in the Glanville Fritillary Butterfly *Melitaea cinxia*. J. Anim. Ecol. 65, 791–801.

Lafferty, K.D., Allesina, S., Arim, M., Briggs, C.J., De Leo, G., Dobson, A.P., Dunne, J.A., Johnson, P.T.J., Kuris, A.M., Marcogliese, D.J., Martinez, N.D., Memmott, J., et al., 2008. Parasites in food webs: the ultimate missing links. Ecol. Lett. 11, 533–546.

Lake, P.S., 2003. Ecological effects of perturbation by drought in flowing water. Freshw. Biol. 48, 1161–1172.

Laliberté, E., Tylianakis, J.M., 2010. Deforestation homogenizes tropical parasitoid-host networks. Ecology 91, 1740–1747.

Lamont, B.B., Barker, M.J., 1988. Seed bank dynamics of a serotinous, non-sprouting *Banksia* species. Aust. J. Bot. 36, 193–203.

Lamont, B.B., Klinkhamer, P.G.L., Witkowski, E.T.F., 1993. Population fragmentation may reduce fertility to zero in *Banksia goodii*—a demonstration of the Allee effect. Oecologia 94, 446–450.

Lancaster, J., 1996. Scaling the effects of predation and disturbance in a patchy environment. Oecologia 107, 321–331.

Lancaster, J., Bradley, D.C., Hogan, A., Waldron, S., 2005. Intraguild omnivory in predatory stream insects. J. Anim. Ecol. 74, 619–629.

Lande, R., 1988. Genetics and demography in biological conservation. Science 241, 1455–1460.

Lander, T.A., Bebber, D.P., Choy, C.T., Harris, S.A., Boshier, D.H., 2011. The Circe principle explains how resource-rich land can waylay pollinators in fragmented landscapes. Curr. Biol. 9, 1302–1307.

Laube, I., Breitbach, N., Böhning-Gaese, K., 2008. Avian diversity in a Kenyan agroecosystem: effects of habitat structure and proximity to forest. J. Ornithol. 149, 181–191.

Laurance, S.G.W., 2004. Responses of understory rain forest birds to road edges in Central Amazonia. Ecol. Appl. 14, 1344–1357.

Laurance, W.F., Laurance, S.G.W., Ferreira, L.V., Rankin-de Merona, J.M., Gascon, C., Lovejoy, T.E., 1997. Biomass collapse in Amazonian forest fragments. Science 278, 1117–1118.

Laurance, W.F., Ferreira, L.V., Rankin-De Merona, J.M., Laurance, S.G., 1998. Rain forest fragmentation and the dynamics of Amazonian tree communities. Ecology 79, 2032–2040.

Laurance, W.F., Nascimento, H.E.M., Laurance, S.G., Andrade, A.C., Fearnside, P.M., Ribeiro, J.E.L., Capretz, R.L., 2006a. Rain forest fragmentation and the proliferation of successional trees. Ecology 87, 469–482.

Laurance, W.F., Nascimento, H.E.M., Laurance, S.G., Andrade, A., Ribeiro, J.E.L.S., Giraldo, J.P., Lovejoy, T.E., Condit, R., Chave, J., Harms, K.E., D'Angelo, S., 2006b. Rapid decay of tree-community composition in Amazonian forest fragments. Proc. Natl. Acad. Sci. USA 103, 19010–19014.

Laurance, W.F., Nascimento, H.E.M., Laurance, S.G., Andrade, A., Ewers, R.M., Harms, K.E., Luizão, R.C.C., Ribeiro, J.E., 2007. Habitat fragmentation, variable edge effects and the landscape-divergence hypothesis. PLoS One 2, e1017.

Laurance, W.F., Camargo, J.L.C., Luizão, R.C.C., Laurance, S.G., Pimm, S.L., Bruna, E.M., Stouffer, P.C., Williamson, G.B., Benítez-Malvido, J., Vasconcelos, H.L., van Houtan, K.S., Zartman, C.E., et al., 2011. The fate of Amazonian forest fragments: a 32-year investigation. Biol. Conserv. 144, 56–67.

Lavandier, P., Céréghino, R., 1995. Use and partition of space and resources by two coexisting *Rhyacophila* species (Trichoptera) in a high mountain stream. Hydrobiologia 300/301, 157–162.

Lavandier, P., Décamps, H., 1983. Un torrent d'altitude dans les Pyrénées, l'Estaragne. In: Bourliere, F., Lamotte, M. (Eds.), Ecosystemes Limniques. Masson, Paris, pp. 81–111.

Layer, K., Riede, J.O., Hildrew, A.G., Woodward, G., 2010. Food web structure and stability in 20 streams across a wide pH gradient. Adv. Ecol. Res. 42, 265–299.

Layer, K., Hildrew, A.G., Jenkins, G.B., Riede, J., Rossiter, S.J., Townsend, C.R., Woodward, G., 2011. Long-term dynamics of a well-characterised food web: four decades of acidification and recovery in the Broadstone Stream model system. Adv. Ecol. Res. 44, 69–117.

Leck, C.F., 1979. Avian extinctions in an isolated tropical wet-forest preserve, Ecuador. Auk 96, 343–352.

Ledger, M.E., Harris, R.M.L., Armitage, P.D., Milner, A.M., 2008. Disturbance frequency influences patch dynamics in stream benthic algal communities. Oecologia 155, 809–819.

Ledger, M.E., Harris, R.M.L., Armitage, P.D., Milner, A.M., 2009. Realism of model ecosystems: an evaluation of physicochemistry and macroinvertebrate assemblages in artificial streams. Hydrobiologia 617, 91–99.

Ledger, M.E., Edwards, F.K., Brown, L.E., Milner, A.M., Woodward, G., 2011. Impact of simulated drought on ecosystem biomass production: an experimental test in stream mesocosms. Glob. Change Biol. 17, 2288–2297.

Ledger, M.E., Harris, R.M.L., Armitage, P.D., Milner, A.M., 2012. Climate change impacts on community resilience: experimental evidence from a drought disturbance experiment. Adv. Ecol. Res. 46, 211–258.

Lees, A.C., Peres, C.A., 2009. Gap-crossing movements predict species occupancy in Amazonian forest fragments. Oikos 118, 280–290.

Leibold, M.A., Holyoak, M., Mouquet, N., Amarasekare, P., Chase, J.M., Hoopes, M.F., Holt, R.D., Shurin, J.B., Law, R., Tilman, D., Loreau, M., Gonzalez, A., 2004. The metacommunity concept: a framework for multi-scale community ecology. Ecol. Lett. 7, 601–613.

Lenz, J., Fiedler, W., Caprano, T., Friedrichs, W., Gaese, B.H., Wikelski, M., Böhning-Gaese, K., 2011. Seed-dispersal distributions by trumpeter hornbills in fragmented landscapes. Proc. R. Soc. Lond. B 278, 2257–2264.

Levin, S.A., 1992. The problem of pattern and scale in ecology. Ecology 73, 1943–1967.

Lewinsohn, T.M., Novotny, V., Basset, Y., 2005. Insects on plants: diversity of herbivore assemblages revisited. Annu. Rev. Ecol. Syst. 36, 597–620.

Lewinsohn, T.M., Prado, P.I., Jordano, P., Bascompte, J., Olesen, J.M., 2006. Structure in plant-animal interaction assemblages. Oikos 113, 174–184.

Lidicker, W., 1985. An overview of dispersal in non-volant small mammals. Contrib. Mar. Sci. Suppl. 27, 359–375.

Lidicker, W., 1999. Responses of mammals to habitat edges: an overview. Landsc. Ecol. 14, 333–343.

Lindenmayer, D.B., Luck, G., 2005. Synthesis: thresholds in conservation and management. Biol. Conserv. 124, 351–354.

Lindstedt, S.L., Miller, B.J., Buskirk, S.W., 1986. Home range, time, and body size in mammals. Ecology 67, 413–418.

Lomolino, M.V., 1985. Body size of mammals on islands: the island rule reexamined. Am. Nat. 125, 310–316.

Lopes, A.V., Girão, L.C., Santos, B.A., Peres, C.A., Tabarelli, M., 2009. Long-term erosion of tree reproductive trait diversity in edge-dominated Atlantic forest fragments. Biol. Conserv. 142, 1154–1165.

Lopezaraiza-Mikel, M.E., Hayes, R.B., Whalley, M.R., Memmott, J., 2007. The impact of an alien plant on a native plant pollinator network: an experimental approach. Ecol. Lett. 10, 539–550.

Lord, J.M., 2004. Frugivore gape size and the evolution of fruit size and shape in southern hemisphere floras. Austral Ecol. 29, 430–436.

Luck, G.W., Daily, G.C., 2003. Tropical countryside bird assemblages: richness, composition, and foraging differ by landscape context. Ecol. Appl. 13, 235–247.

MacArthur, R., 1955. Fluctuations of animal populations and a measure of community stability. Ecology 36, 533–536.

MacArthur, R.H., Pianka, E.R., 1966. On optimal use of a patchy environment. Am. Nat. 100, 603–609.

MacArthur, R.H., Wilson, E.O., 1963. An equilibrium theory of insular zoogeography. Evolution 17, 373–387.

MacArthur, R.H., Wilson, E.O., 1967. The Theory of Island Biogeography. Princeton University Press, Princeton, MA.

MacFadyen, S., Gibson, R.H., Symondson, W.O.C., Memmott, J., 2011. Landscape structure influences modularity patterns in farm food webs: consequences for pest control. Ecol. Appl. 21, 516–524.

Macreadie, P.I., Hindell, J.S., Jenkins, G.P., Connolly, R.M., Keough, M.J., 2009. Fish responses to experimental fragmentation of seagrass habitat. Conserv. Biol. 23, 644–652.

Maionchi, D.O., dos Reis, S.F., de Aguiar, M.A.M., 2006. Chaos and pattern formation in a spatial tritrophic food chain. Ecol. Modell. 191, 291–303.

Malard, F., Uehlinger, U., Zah, R., Tockner, K., 2006. Flood-pulse and riverscape dynamics in a braided glacial river. Ecology 87, 704–716.

Marquitti, F.M.D., 2011. Interaction networks between frugivorous bats and plants, geographical variation and niche conservatism. MSc Thesis.Universidade Estadual de Campinas, Brazil. http://www.bibliotecadigital.unicamp.br/document/?code=000796920.

Matthies, D., Bräuer, I., Maibom, W., Tscharntke, T., 2004. Population size and the risk of local extinction: empirical evidence from rare plants. Oikos 105, 481–488.

May, R.M., 1972. Will a large complex system be stable? Nature 238, 413–414.

May, R.M., 1973. Stability in randomly fluctuating versus deterministic environments. Am. Nat. 107, 621–650.

McCann, K.S., Hastings, A., Huxel, G.R., 1998. Weak trophic interactions and the balance of nature. Nature 395, 794–798.
McCann, K.S., Rasmussen, J.R., Umbanhowar, J., 2005a. The dynamics of spatially coupled food webs. Ecol. Lett. 8, 513–523.
McCann, K.S., Rasmussen, J.R., Umbanhowar, J., Humphries, M., 2005b. The role of space, time and variability in food web dynamics. In: de Ruiter, P.C., Wolters, V., Moore, J.C. (Eds.), Dynamic Food Webs, Multispecies Assemblages, Ecosystem Development and Environmental Change. Elsevier and Academic Press, Burlington, MA, pp. 56–70.
McConkey, K., Drake, D., 2006. Flying foxes cease to function as seed dispersers long before they become rare. Ecology 87, 271–276.
McKinney, M.L., Lockwood, J.L., 1999. Biotic homogenization: a few winners replacing many losers in the next mass extinction. Trends Ecol. Evol. 14, 450–453.
McLaughlin, O.B., Jonsson, T., Emmerson, M.C., 2010. Temporal variability in predator-prey relationships of a forest floor food web. Adv. Ecol. Res. 42, 171–264 3.
Meerhoff, M., Mello, F.T-d., Kruk, C., Alonso, C., González-Bergonzoni, I., Pablo Pacheco, J., Lacerot, G., Arim, M., Beklioğlu, M., Brucet, S., Goyenola, G., Iglesias, C., Mazzeo, N., Kosten, S., Jeppesen, E., et al., 2012. Environmental warming in shallow lakes: a review of potential changes in community structure as evidenced from spacefor-time substitution approaches. Adv. Ecol. Res. 46, 259–350.
Melián, C.J., Vilas, C., Baldó, F., González-Ortegón, E., Drake, P., Williams, R.J., 2011. Eco-evolutionary dynamics of individual-based food webs. Adv. Ecol. Res. 45, 225–268.
Melo, F.P.L., Dirzo, R., Tabarelli, M., 2006. Biased seed rain in forest edges: evidence from the Brazilian Atlantic forest. Biol. Conserv. 132, 50–60.
Memmott, J., 1999. The structure of a plant-pollinator food web. Ecol. Lett. 2, 276–280.
Memmott, J., Waser, N.M., Price, M.V., 2004. Tolerance of pollination networks to species extinctions. Proc. R. Soc. Lond. B 271, 2605–2611.
Memmott, J., Craze, P.G., Waser, N.M., Price, M.V., 2007. Global warming and the disruption of plant-pollinator interactions. Ecol. Lett. 10, 710–717.
Menges, E.S., 1991. The application of minimum viable population theory to plants. In: Falk, D.A., Holsinger, K.E. (Eds.), Genetics and Conservation of Rare Plants. Oxford University Press, New York, pp. 45–61.
Metzger, J.P., Martensen, A.C., Dixo, M., Bernacci, L.C., Ribeiro, M.C., Teixeira, A.M.G., Pardini, R., 2009. Time-lag in biological responses to landscape changes in a highly dynamic Atlantic forest region. Biol. Conserv. 142, 1166–1177.
Michener, C.D., 2007. The Bees of the World, second ed. John Hopkins University Press, Baltimore, MD.
Millennium Ecosystem Assessment, 2005. Ecosystems and Human Wellbeing: Synthesis. Island Press, Washington, DC.
Milner, A.M., Brittain, J.E., Castella, E., Petts, G.E., 2001. Trends in macroinvertebrate community structure in glacier-fed rivers in relation to environmental conditions, a synthesis. Freshw. Biol. 46, 1833–1847.
Milner, A.M., Brown, L.E., Hannah, D.M., 2009. Hydroecological response of river systems to shrinking glaciers. Hydrol. Process 23, 62–77.
Milner, A.M., Robertson, A.L., Brown, L.E., Sonderland, S., McDermott, M., Veal, A.J., 2011. Evolution of a stream ecosystem in recently deglaciated terrain. Ecology 92, 1924–1935.
Milton, K., May, M.L., 1976. Body-weight, diet and home range area in primates. Nature 259, 459–462.
Mimura, M., Murray, J.D., 1978. On a diffusive prey-predator model which exhibits patchiness. J. Theor. Biol. 75, 249–262.
Mintenbeck, K., Barrera-Oro, E.R., Brey, T., Jacob, U., Knust, R., Mark, F.C., Moreira, E., Strobel, A., Arntz, W.E., et al., 2012. Impact of Climate Change on Fishes in Complex Antarctic Ecosystems. Adv. Ecol. Res. 46, 351–426.

Mistro, D.C., Rodrigues, L.A., Varriale, M.C., 2009. The role of spatial refuges in coupled map lattice model for host-parasitoid systems. Bull. Math. Biol. 71, 1934–1953.

Moeller, D.A., 2005. Pollinator community structure and sources of spatial variation in plant-pollinator interactions in *Clarkia xantiana* ssp. *xantiana*. Oecologia 142, 28–37.

Molnár, P.K., Derocher, A.E., Klanjscek, T., Lewis, M.A., 2011. Predicting climate change impacts on polar bear litter size. Nat. Commun. 8, 186.

Monaghan, M.T., Spaak, P.T., Robinson, C.T., Ward, J.V., 2002. Population genetic structure of three alpine stream insects, influences of gene flow, demographics, and habitat fragmentation. J. N. Am. Benthol. Soc. 21, 114–131.

Monaghan, M.T., Robinson, C.T., Spaak, P.T., Ward, J.V., 2005. Macroinvertebrate diversity in fragmented alpine streams, implications for freshwater conservation. Aquat. Sci. 67, 454–464.

Montoya, J.M., Emmerson, M.C., Woodward, G., 2005. Perturbations and indirect effects in complex food webs. In: de Ruiter, P.C., Wolters, V., Moore, J.C. (Eds.), Dynamic Food Webs, Multispecies Assemblages, Ecosystem Development, and Environmental Change. Academic Press, Amsterdam, pp. 369–380.

Montoya, D., Zavala, M.A., Rodriguez, M.A., Purves, D.W., 2008. Animal versus wind dispersal and the robustness of tree species to deforestation. Science 320, 1502–1504.

Morales, C.L., Aizen, M.A., 2006. Invasive mutualisms and the structure of plant-pollinator interactions in the temperate forests of north-west Patagonia, Argentina. J. Ecol. 94, 171–180.

Morales, J.M., Carlo, T.S., 2006. The effects of plant distribution and frugivore density on the scale and shape of dispersal kernels. Ecology 87, 1489–1496.

Morales, C.L., Traveset, A., 2009. A meta-analysis of impacts of alien *vs.* native plants on pollinator visitation and reproductive success of co-flowering native plants. Ecol. Lett. 12, 716–728.

Morales, J.M., Vázquez, D.P., 2008. The effect of space in plant-animal mutualistic networks: insights from a simulation study. Oikos 117, 1362–1370.

Morellato, L.P.C., 2004. Phenology, sex ratio, and spatial distribution among dioecious species of *Trichilia* (Meliaceae). Plant Biol. 6, 491–497.

Moretti, M., de Bello, F., Roberts, S.P.M., Potts, S.G., 2009. Taxonomical *vs.* functional responses of bee communities to fire in two contrasting climatic regions. J. Anim. Ecol. 78, 98–108.

Morris, D.W., 1997. Optimally foraging deer mice in prairie mosaics: a test of habitat theory and absence of landscape effects. Oikos 80, 31–42.

Mueller, D.R., Vincent, W.F., Jefries, M.O., 2006. Environmental gradients, fragmented habitats and microbiota of a northern ice shelf cryoecosystem, Ellesmere Island, Canada. Arct. Antarct. Alp. Res. 38, 593–607.

Mulder, C., den Hollander, H.A., Vonk, J.A., Rossberg, A.G., op Akkerhuis, G.A., Yeates, G.W., 2009. Soil resource supply influences faunal size-specific distributions in natural food webs. Naturwissenschaften 96, 813–826.

Mulder, C., Boit, A., Bonkowski, M., de Ruiter, P.C., Mancinelli, G., van der Heijden, M.G.A., van Wijnen, H.J., Vonk, J.A., Rutgers, M., 2011. A belowground perspective on Dutch agroecosystems: how soil organisms interact to support ecosystem services. Adv. Ecol. Res. 44, 277–358.

Mulder, C., Boit, A., Mori, S., Vonk, J.A., Dyer, S.D., Faggiano, L., Geisen, S., González, A.L., Kaspari, M., Lavorel, S., Marquet, P.A., Rossberg, A.G., Sterner, R.W., Voigt, W., Wall, D.H., et al., 2012. Distributional (in)congruence of biodiversity ecosystem functioning. Adv. Ecol. Res. 46, 1–88.

Munday, P.L., 2004. Habitat loss, resource specialization, and extinction on coral reefs. Glob. Change Biol. 10, 1642–1647.

Murcia, C., 1995. Edge effects in fragmented forests, implications for conservation. Trends Ecol. Evol. 10, 58–62.
Murray, J.D., 1993. Mathematical Biology, second ed. Springer-Verlag, New York.
Myers, N., 1986. Tropical deforestation and a mega-extinction spasm. In: Soulé, M.E. (Ed.), Conservation Biology: The Science of Scarcity and Diversity. Sinauer Associates, Sunderland, MA, pp. 394–409.
Myers, J.H., Krebs, C.J., 1971. Genetic, behavioral, and reproductive attributes of dispersing field voles, *Microtus pennsylvanicus* and *Microtus ochrogaster*. Ecol. Monogr. 41, 53–78.
Nakao, H., Mikhailov, A.S., 2010. Turing patterns in network-organized activator-inhibitor systems. Nat. Phys. 6, 544–550.
Nakazawa, T., Ushio, M., Kondoh, M., 2011. Scale dependence of predator-prey mass ratio: determinants and applications. Adv. Ecol. Res. 45, 269–302.
Nams, V.O., 2011. Emergent properties of patch shapes affect edge permeability to animals. PLoS One 6, e21886.
Nee, S., May, R.M., 1992. Dynamics of metapopulations, habitat destruction and competitive coexistence. J. Anim. Ecol. 61, 37–40.
Nee, S., May, R.M., Hassell, M.P., 1997. Two-species metapopulation models. In: Hanski, I.A., Gilpin, M.E. (Eds.), Metapopulation Biology. Academic Press, San Diego, CA, pp. 123–147.
Neutel, A.M., Heesterbeek, J.A.P., de Ruiter, P.C., 2002. Stability in real food webs: weak links in long loops. Science 296, 1120–1123.
Nielsen, A., Bascompte, J., 2007. Ecological networks, nestedness and sampling effort. J. Ecol. 95, 1134–1141.
Nieminen, M., 1996. Risk of population extinction in moths: effect of host plant characteristics. Oikos 76, 475–484.
Nilsson, L.A., Rabakonandrianina, E., Razananaivo, R., Randriamanindry, J.-J., 1992. Long pollinia on eyes: hawk-moth pollination of *Cynorkis uniflora* Lindley (Orchidaceae) in Madagascar. Bot. J. Linn. Soc. 109, 145–160.
Noss, R.F., 1987. Corridors in real landscapes: a reply to Simberloff and Cox. Conserv. Biol. 1, 159–164.
Nuismer, S.L., Doebeli, M., 2004. Genetic correlations and the coevolutionary dynamics of three-species systems. Evolution 58, 1165–1177.
Nuismer, S.L., Thompson, J.N., 2006. Coevolutionary alternation in antagonistic interactions. Evolution 60, 2207–2217.
Nuismer, S.L., Thompson, J.N., Gomulkiewicz, R., 1999. Gene flow and geographically structured coevolution. Proc. R. Soc. Lond. B Biol. Sci. 266, 605–609.
Nuismer, S.L., Thompson, J.N., Gomulkiewicz, R., 2000. Coevolutionary clines across selection mosaics. Evolution 54, 1102–1115.
Nyhagen, D.F., Kragelund, C., Olesen, J.M., Jones, C.G., 2001. Insular interactions between lizards and flowers, flower visitation by an endemic Mauritian gecko. J. Trop. Ecol. 17, 755–761.
Öckinger, E., Smith, H.G., 2007. Semi-natural grasslands as population sources for pollinating insects in agricultural landscapes. J. Appl. Ecol. 44, 50–59.
Öckinger, E., Schweiger, O., Crist, T.O., Debinski, D.M., Krauss, J., Kuussaari, M., Petersen, J.D., Pöyry, J., Settele, J., Summerville, K.S., Bommarco, R., 2010. Life-history traits predict species responses to habitat area and isolation: a cross-continental synthesis. Ecol. Lett. 13, 969–979.
O'Gorman, E.J., Emmerson, M.C., 2009. Perturbations to trophic interactions and the stability of complex food webs. Proc. Natl. Acad. Sci. USA 106, 13393–13398.
O'Gorman, E.J., Emmerson, M.C., 2010. Manipulating interaction strengths and the consequences for trivariate patterns in a marine food web. Adv. Ecol. Res. 42, 301–419.

O'Gorman, E.J., Jacob, U., Jonsson, T., Emmerson, M.C., 2010. Interaction strength, food web topology and the relative importance of species in food webs. J. Anim. Ecol. 79, 682–692.

O'Gorman, E.J., Yearsley, J.M., Crowe, T.P., Emmerson, M.C., Jacob, U., Petchey, O.L., 2011. Loss of functionally unique species may gradually undermine ecosystems. Proc. R. Soc. B 278, 1886–1893.

Okuyama, T., Holland, J.N., 2008. Network structure properties mediate the stability of mutualistic communities. Ecol. Lett. 11, 208–216.

Olden, J.D., LeRoy Poff, N., Douglas, M.R., Douglas, M.E., Fausch, K.D., 2004. Ecological and evolutionary consequences of biotic homogenization. Trends Ecol. Evol. 19, 18–24.

Olesen, J.M., 1985. The Macaronesian bird-flower element and its relation to bird and bee opportunists. Bot. J. Linn. Soc. 91, 395–414.

Olesen, J.M., 2000. Exactly how generalised are pollination interactions? In: Totland, Ø., Armbruster, W.S., Fenster, C., Molau, U., Nilsson, L.A., Olesen, J.M., Ollerton, J., Philipp, M., Ågren, J. (Eds.), The Scandinavian Association for Pollination Ecology Honours Knut Fægri. Norwegian Academy of Science and Letters, Oslo, pp. 161–178.

Olesen, J.M., Jain, S., 1994. Fragmented plant populations and their lost interactions. In: Loeschcke, V., Tomiuk, J., Jain, S. (Eds.), Conservation Genetics. Birkhäuser, Basel, pp. 417–426.

Olesen, J.M., Valido, A., 2003. Lizards as pollinators and seed dispersers, an island phenomenon. Trends Ecol. Evol. 18, 177–181.

Olesen, J.M., Eskildsen, L.I., Venkatasamy, S., 2002. Invasion of pollination networks on oceanic islands: importance of invader complexes and endemic super generalists. Divers. Distrib. 8, 181–192.

Olesen, J.M., Bascompte, J., Dupont, Y.L., Jordano, P., 2006. The smallest of all worlds: pollination networks. J. Theor. Biol. 240, 270–276.

Olesen, J.M., Bascompte, J., Dupont, Y.L., Jordano, P., 2007. The modularity of pollination networks. Proc. Natl. Acad. Sci. USA 104, 19891–19896.

Olesen, J.M., Bascompte, J., Elberling, H., Jordano, P., 2008. Temporal dynamics in a pollination network. Ecology 89, 1573–1582.

Olesen, J.M., Bascompte, J., Dupont, Y.L., Elberling, H., Rasmussen, C., Jordano, P., 2010a. Missing and forbidden links in mutualistic networks. Proc. R. Soc. B 278, 725–732.

Olesen, J.M., Dupont, Y.L., O'Gorman, E., Ings, T.C., Layer, K., Melián, C.J., Trøjelsgaard, K., Pichler, D.E., Rasmussen, C., Woodward, G., 2010b. From Broadstone to Zackenberg: space, time and hierarchies in ecological networks. Adv. Ecol. Res. 42, 1–69.

Olesen, J.M., Dupont, Y.L., Hagen, M., Rasmussen, C., Trøjelsgaard, K., 2012. Structure and dynamics of pollination networks: the past, present and future. In: Patiny, S. (Ed.), Evolution of Plant-Pollinator Relationships. Cambridge University Press, Cambridge, pp. 374–391.

Oliveira, P.S., Galetti, M., Pedroni, F., Morellato, L.P.C., 1995. Seed cleaning by *Mycocepurus goeldii* ants (Attini) facilitates germination in *Hymenaea courbaril* (Caesalpiniaceae). Biotropica 27, 518–522.

Oliveira, M.A., Santos, A.M.M., Tabarelli, M., 2008. Profound impoverishment of the large-tree stand in a hyper-fragmented landscape of the Atlantic forest. For. Ecol. Manage. 256, 1910–1917.

Oliver, T.A., Palumbi, S.R., 2009. Distributions of stress-resistant coral symbionts match environmental patterns at local but not regional scales. Mar. Ecol. Prog. Ser. 378, 93–103.

Ollerton, J., Johnson, S.D., Cranmer, L., Kellie, S., 2003. The pollination ecology of an assemblage of grassland asclepiads in South Africa. Ann. Bot. 92, 807–834.

Ollerton, J., Winfree, R., Tarrant, S., 2011. How many flowering plants are pollinated by animals? Oikos 120, 321–326.

Orth, R.J., Carruthers, T.J.B., Dennison, W.C., Duarte, C.M., Fourqurean, J.W., Heck Jr., K.L., Hughes, A.R., Kendrick, G.A., Kenworthy, W.J., Olyarnik, S., Short, F.T., Waycott, M., et al., 2006. A global crisis for seagrass ecosystems. Bioscience 56, 987–996.

Pacheco, R., Silva, R.R., Morini, M.S., Brandão, C.R.F., 2009. A comparison of the leaf-litter ant fauna in a secondary Atlantic forest with an adjacent pine plantation in southeastern Brazil. Neotrop. Entomol. 38, 55–65.

Palmer, T.M., Stanton, M.L., Young, T.P., Goheen, J.R., Pringle, R.M., Karban, R., 2008. Breakdown of an ant-plant mutualism follows the loss of large herbivores from an African savanna. Science 319, 192–195.

Parc National des Pyrénées, 2005. Document d'Objectifs de la Zone Spéciale de Conservation: Estaubé, Gavarnie, Troumouse, Barroude. Parc National des Pyrénées, Tarbes, France.

Pardini, R., Faria, D., Accacio, G.M., Laps, R.R., Mariano-Neto, E., Paciencia, M.L.B., Dixo, M., Baumgarten, J., 2009. The challenge of maintaining Atlantic forest biodiversity: a multi-taxa conservation assessment of specialist and generalist species in an agro-forestry mosaic in southern Bahia. Biol. Conserv. 142, 1178–1190.

Pascual-Hortal, L., Saura, S., 2006. Comparison and development of new graph-based landscape connectivity indices: towards the priorization of habitat patches and corridors for conservation. Landsc. Ecol. 21, 959–967.

Pasquet, R.M.S., Peltier, A., Hufford, M.B., Oudin, E., Saulnier, J., Paul, L., Knudsen, J.T., Herren, H.R., Gepts, P., 2008. Long-distance pollen flow assessment through evaluation of pollinator foraging range suggests transgene escape distances. Proc. Natl. Acad. Sci. USA 105, 13456–13461.

Pauw, A., 2007. Collapse of a pollination web in small conservation areas. Ecology 88, 1759–1769.

Pérez-Barrales, R., Arroyo, J., Armbruster, W.S., 2007. Differences in pollinator faunas may generate geographic differences in floral morphology and integration in *Narcissus papyraceus* (Amaryllidaceae). Oikos 116, 1904–1918.

Perfecto, I., Vandermeer, J., 2008. Biodiversity conservation in tropical ecosystems: a new paradigm. Ann. N. Y. Acad. Sci. 1134, 173–200.

Perfectti, F., Gómez, J.M., Bosch, J., 2009. The functional consequences of diversity in plant-pollinator interactions. Oikos 118, 1430–1440.

Petanidou, T., Kallimanis, A.S., Tzanopoulos, J., Sgardelis, S.P., Pantis, J.D., 2008. Long-term observation of a pollination network, fluctuation in species and interactions, relative invariance of network structure and implications for estimates of specialization. Ecol. Lett. 11, 564–575.

Petchey, O.L., Gaston, K.J., 2006. Functional diversity: back to basics and looking forward. Ecol. Lett. 9, 741–758.

Petchey, O.L., Beckerman, A., Riede, J., Warren, P., 2008. Size, foraging, and food web structure. Proc. Natl. Acad. Sci. USA 105, 4191–4196.

Petchey, O.L., Brose, U., Rall, B.C., 2010. Predicting the effects of temperature on food web connectance. Philos. Trans. R. Soc. B 365, 2081–2091.

Petersen, I., Winterbottom, J.H., Orton, S., Friberg, N., Hildrew, A.G., 1999. Emergence and lateral dispersal of adult stoneflies and caddis flies from Broadstone Stream. Freshw. Biol. 42, 401–416.

Pimm, S.L., 1984. The complexity and stability of ecosystems. Nature 307, 321–326.

Pimm, S.L., Lawton, J.H., 1980. Are food webs divided into compartments? J. Anim. Ecol. 49, 879–898.

Pizo, M.A., Oliveira, P.S., 2000. The use of fruits and seeds by ants in the Atlantic forest of southeast Brazil. Biotropica 32, 851–861.

Pizo, M.A., Passos, L., Oliveira, P.S., 2005. Ants as seed dispersers of fleshy diaspores in Brazilian Atlantic forests. In: Forget, P.-M., Lambert, J.E., Hulme, P.E., Wall, S.B.V. (Eds.), Seed Fate. CAB int. Walling ford, UK, pp. 315–330.

Poff, N.L., Zimmerman, J.K.H., 2010. Ecological responses to altered flow regimes: a literature review to inform the science and management of environmental flows. Freshw. Biol. 55, 194–205.

Pollard, E., Yates, T.J., 1993. Monitoring Butterflies for Ecology and Conservation. Chapman and Hall, London.

Prakash, S., de Roos, A.M., 2004. Habitat destruction in mutualistic metacommunities. Theor. Popul. Biol. 65, 153–163.

Pratchett, M.S., Munday, P.L., Wilson, S.K., Graham, N.A.J., Cinner, J.E., Bellwood, D.R., Jones, G.P., Polunin, N.V.C., McClanahan, T.R., 2008. Effects of climate-induced coral bleaching on coral-reef fishes—ecological and economic consequences. Oceanogr. Mar. Biol. 46, 251–296.

Prevedello, J., Vieira, M., 2011. Does the type of matrix matter? A quantitative review of the evidence. Biodivers. Conserv. 19, 1205–1223.

Proctor, M., Yeo, P., Lack, A., 1996. The Natural History of Pollination. Harper Collins, London.

Prugh, L.R., Hodges, K.E., Sinclair, A.R.E., Brashares, J.S., 2008. Effect of habitat area and isolation on fragmented animal populations. Proc. Natl. Acad. Sci. USA 105, 20770–20775.

Pulliam, H.R., 1988. Sources, sinks and population regulation. Am. Nat. 132, 652–661.

Purdy, K.J., Hurd, P.J., Moya-Laraño, J., Trimmer, M., Woodward, G., 2010. Systems biology for ecology: from molecules to ecosystems. Adv. Ecol. Res. 43, 87–149.

Pütz, S., Groeneveld, J., Alves, L.F., Metzger, J.P., Huth, A., 2011. Fragmentation drives tropical forest fragments to early successional states: a modelling study for Brazilian Atlantic forests. Ecol. Modell. 222, 1986–1997.

Raine, N.E., Chittka, L., 2005. Comparison of flower constancy and foraging performance in three bumblebee species (Hymenoptera: Apidae: *Bombus*). Entomol. Gen. 28, 81–89.

Ramos, F.N., Santos, F.A.M., 2005. Phenology of *Psychotria tenuinervis* (Rubiaceae) in Atlantic forest fragments: fragment and habitat scales. Can. J. Bot. 83, 1305–1316.

Rand, T.A., Tscharntke, T., 2007. Contrasting effects of natural habitat loss on generalist and specialist aphid natural enemies. Oikos 116, 1353–1362.

Rasmussen, E., 1977. The wasting disease of eelgrass *Zostera marina* and its effects on environmental factors and fauna. In: McRoy, C.P., Helfferich, C. (Eds.), Seagrass Ecosystems: A Scientific Perspective. Dekker, New York, pp. 1–52.

Rasmussen, I.R., Brødsgaard, B., 1992. Gene flow inferred from seed dispersal and pollinator behavior compared to DNA analysis of restriction site variation in a patchy population of *Lotus corniculatus*. Oecologia 89, 277–283.

Rathcke, B.J., Jules, E.S., 1993. Habitat fragmentation and plant-pollinator interactions. Curr. Sci. 65, 273–277.

Raymundo, L.J., Maypa, A.P., Gomez, E.D., Cadiz, P., 2007. Can dynamite-blasted reefs recover? A novel, low-tech approach to stimulating natural recovery in fish and coral populations. Mar. Pollut. Bull. 54, 1009–1019.

Reed, B.J., Hovel, K.A., 2006. Seagrass habitat disturbance: how loss and fragmentation of eelgrass (*Zostera marina*) influences epifaunal community of San Diego Bay, California, USA. Mar. Ecol. Prog. Ser. 326, 115–131.

Renjifo, L.M., 1999. Composition changes in a subandean avifauna after long-term forest fragmentation. Conserv. Biol. 13, 1124–1139.
Renner, S.S., 1998. Effects of habitat fragmentation on plant-pollinator interactions in the tropics. In: Newbery, D.M., Prins, H.H.T., Brown, N.D. (Eds.), Dynamics of Tropical Communities. Blackwell, London, pp. 339–360.
Restrepo, C., Gómez, N., 1998. Responses of understory birds to anthropogenic edges in a neotropical montane forest. Ecol. Appl. 8, 170–183.
Restrepo, C., Vargas, A., 1999. Seeds and seedlings of two neotropical montane understory shrubs respond differently to anthropogenic edges and treefall gaps. Oecologia 119, 419–426.
Restrepo, C., Gómez, N., Heredia, S., 1999. Anthropogenic edges, treefall gaps and fruit-frugivore interactions in a neotropical montane forest. Ecology 80, 668–685.
Reuman, D.C., Mulder, C., Banašek-Richter, C., Blandenier, M.-F.C., Breure, A.M., Hollander, H.d., Kneitel, J.M., Raffaelli, D., Woodward, G., Cohen, J.E., 2009. Allometry of body size and abundance in 166 food webs. Adv. Ecol. Res. 41, 1–44.
Rezende, E., Lavabre, J.E., Guimarães, P.R., Jordano, P., Bascompte, J., 2007. Non-random coextinctions in phylogenetically structured mutualistic networks. Nature 448, 925–928.
Ribeiro, M.C., Metzger, J.P., Martensen, A.C., Ponzoni, F.J., Hirota, M.M., 2009. The Brazilian Atlantic Forest: how much is left and how is the remaining forest distributed? Implications for conservation. Biol. Conserv. 142, 1141–1153.
Ricketts, T., Williams, N.M., Mayfield, M.M., 2006. Connectivity and ecosystem services: crop pollination in agricultural landscapes. In: Crooks, K.R., Sanjayan, M. (Eds.), Connectivity Conservation. Cambridge University Press, Cambridge, pp. 255–289.
Rico-Gray, V., Oliveira, P.S., 2007. The Ecology and Evolution of Ant-Plant Interactions. University of Chicago Press, Chicago, IL.
Riegl, B., Luke, K.E., 1998. Ecological parameters of dynamited reefs in the northern Red Sea and their relevance to reef rehabilitation. Mar. Pollut. Bull. 37, 488–498.
Rietkerk, M., Dekker, S.C., de Ruiter, P.C., van de Koppel, J., 2004. Self-organized patchiness and catastrophic shifts in ecosystems. Science 305, 1926–1929.
Rodríguez-Cabal, M.A., Aizen, M.A., Novaro, A.J., 2007. Habitat fragmentation disrupts a plant-disperser mutualism in the temperate forest of South America. Biol. Conserv. 139, 195–202.
Rogers, L.E., Hinds, W.T., Buschbom, R.L., 1976. General weight *vs.* length relationship for insects. Ann. Entomol. Soc. Am. 69, 387–389.
Roland, J., Taylor, P.D., 1997. Insect parasitoid species respond to forest structure at different spatial scales. Nature 386, 710–713.
Rossberg, A.G., 2012. A complete analytic theory for structure and dynamics of populations and communities spanning wide ranges in body size. Adv. Ecol. Res. 46, 427–522.
Roubik, D.W., 1989. Ecology and Natural History of Tropical Bees. Cambridge University Press, Cambridge.
Roubik, D.W., 2006. Stingless bee nesting biology. Apidologie 37, 124–143.
Roy, D.B., Quinn, R.M., Gaston, K.J., 1998. Coincidence in the distribution of butterflies and their food plants. Ecography 21, 279–289.
Rundlöf, M., Smith, H.G., 2006. The effect of organic farming on butterfly diversity depends on landscape context. J. Appl. Ecol. 43, 1121–1127.
Rundlöf, M., Nilsson, H., Smith, H.G., 2008. Interacting effects of farming practice and landscape context on bumble-bees. Biol. Conserv. 141, 417–426.
Saavedra, S., Stouffer, D.B., Uzzi, B., Bascompte, J., 2011. Strong contributors to network persistence are the most vulnerable to extinction. Nature 478, 233–235.
Sabatino, M., Maceira, N., Aizen, M.A., 2010. Direct effects of habitat area on interaction diversity in pollination webs. Ecol. Appl. 20, 1491–1497.
Sæther, B.-E., Engen, S., Lande, R., 1999. Finite metapopulation models with density-dependent migration and stochastic local dynamics. Proc. R. Soc. B 266, 113–118.

Santos, T., Telleria, J., Virgos, E., 1999. Dispersal of Spanish juniper *Juniperus thurifera* by birds and mammals in a fragmented landscape. Ecography 22, 193–204.

Santos, B.A., Peres, C.A., Oliveira, M.A., Grillo, A., Alves-Costa, C.P., Tabarelli, M., 2008. Drastic erosion in functional attributes of tree assemblages in Atlantic forest fragments of northeastern Brazil. Biol. Conserv. 141, 249–260.

Sargent, R.D., Kembel, S.W., Emery, N.C., Forrestel, E.J., Ackerly, D.D., 2011. Effect of local community phylogenetic structure on pollen limitation in an obligately insect-pollinated plant. Am. J. Bot. 98, 283–289.

Saunders, D.A., Hobbs, R.J., Margules, C.R., 1991. Biological consequences of ecosystem fragmentation: a review. Conserv. Biol. 5, 18–32.

Schaffer, M.L., 1981. Minimum population sizes for species conservation. Bioscience 31, 131–134.

Schleuning, M., Blüthgen, N., Flörchinger, M., Braun, J., Schaefer, H.M., Böhning-Gaese, K., 2011a. Specialization and interaction strength in a tropical plant-frugivore network differ among forest strata. Ecology 92, 26–36.

Schleuning, M., Farwig, N., Peters, M.K., Bergsdorf, T., Bleher, B.R., Brandl, R., Dalitz, H., Fischer, G., Freund, W., Gikungu, M.W., Hagen, M., Garcia, F.H., et al., 2011b. Forest fragmentation and selective logging have inconsistent effects on multiple animal-mediated ecosystem processes in a tropical forest. PLoS One 6, e27785.

Schnitzler, F.-R., Hartley, S., Lester, P.J., 2011. Trophic-level responses differ at plant, plot and fragment levels in urban native forest fragments: a hierarchical analysis. Ecol. Entomol. 36, 241–250.

Schupp, E.W., Jordano, P., Gómez, J.M., 2010. Seed dispersal effectiveness revisited: a conceptual review. New Phytol. 188, 333–353.

Schurr, F.M., Spiegel, O., Steinitz, O., Trakhtenbrot, A., Tsoar, A., Nathan, R., 2009. Long-distance seed dispersal. Annu. Plant Rev. 38, 204–237.

Shanahan, M., Harrison, R.D., Yamuna, R., Boen, W., Thornton, I.W.B., 2001. Colonization of an island volcano, Long Island, Papua New Guinea, and an emergent island, Motmot, in its caldera lake. V. Colonization by figs (*Ficus* spp.), their dispersers and pollinators. J. Biogeogr. 28, 1365–1377.

Sieving, K.E., Karr, J.R., 1997. Avian extinction and persistence mechanisms in lowland Panama. In: Laurance, W.F., Bierregaard, R.O. (Eds.), Tropical Forest Remnants. University of Chicago Press, Chicago, IL, pp. 156–170.

Sih, A., Baltus, M.-S., 1987. Patch size, pollinator behavior, and pollinator limitation in catnip. Ecology 68, 1679–1690.

Silliman, B.R., van de Koppel, J., Bertness, M.D., Stanton, L., Mendelsohn, I., 2005. Drought, snails, and large-scale die-off of southern U.S. salt marshes. Science 310, 1803–1806.

Simberloff, D., Cox, J., 1987. Consequences and costs of conservation corridors. Conserv. Biol. 1, 63–71.

Simberloff, D., Farr, J.A., Cox, J., Mehlman, D.W., 1992. Movement corridors, conservation bargains or poor investments? Conserv. Biol. 6, 493–504.

Sinclair, A.R.E., Mduma, S., Brashares, J.S., 2003. Patterns of predation in a diverse predator-prey system. Nature 425, 288–290.

Sisk, T.D., Battin, J., 2002. Habitat edges and avian ecology: geographic patterns and insights for western landscapes. Stud. Avian Biol. 25, 30–48.

Solé, R.V., Montoya, J.M., 2006. Ecological network meltdown from habitat loss and fragmentation. In: Pascual, M., Dunne, J. (Eds.), Ecological Networks: Linking Structure to Dynamics in Food Webs. Oxford University Press, New York, pp. 305–323.

Spiegel, O., Nathan, R., 2007. Incorporating dispersal distance into the disperser effectiveness framework: frugivorous birds provide complementary dispersal to plants in a patchy environment. Ecol. Lett. 10, 718–728.

Srivastava, D.S., 2006. Habitat structure, trophic structure and ecosystem function: interactive effects in a bromeliad-insect community. Oecologia 149, 493–504.
Stamps, J.A., Buechner, M., Krishnan, V.V., 1987a. The effects of edge permeability and habitat geometry on emigration from patches of habitat. Am. Nat. 129, 533–552.
Stamps, J.A., Buechner, M., Krishnan, V.V., 1987b. The effects of habitat geometry on territorial defense costs: intruder pressure in bounded habitats. Am. Zool. 27, 307–325.
Stang, M., Klinkhamer, P.G.L., van der Meijden, E., 2006. Size constraints and flower abundance determine the number of interactions in a plant-flower visitor web. Oikos 112, 111–121.
Stang, M., Klinkhamer, P.G.L., van der Meijden, E., 2007. Asymmetric specialization and extinction risk in plant-flower visitor webs: a matter of morphology or abundance? Oecologia 151, 442–453.
Stang, M., Klinkhamer, P.G.L., Waser, N.M., Stang, I., van der Meijden, E., 2009. Size-specific interaction patterns and size matching in a plant-pollinator interaction web. Ann. Bot. 103, 1459–1469.
Steffan-Dewenter, I., 2003. Importance of habitat and landscape context for species richness of bees and wasps in fragmented orchard meadows. Conserv. Biol. 17, 1036–1044.
Steffan-Dewenter, I., Tscharntke, T., 1999. Effects of habitat isolation on pollinator communities and seed set. Oecologia 121, 432–440.
Steffan-Dewenter, I., Tscharntke, T., 2002. Insect communities and biotic interaction on fragmented calcareous grasslands: a mini review. Biol. Conserv. 104, 275–284.
Steffan-Dewenter, I., Munzenberg, U., Burger, C., Thies, C., Tscharntke, T., 2002. Scale-dependent effects of landscape context on three pollinator guilds. Ecology 83, 1421–1432.
Steffan-Dewenter, I., Klein, A.M., Gaebele, V., Alfert, T., Tscharntke, T., 2006. Bee diversity and plant-pollinator interactions in fragmented landscapes. In: Waser, N.M., Ollerton, J. (Eds.), Plant-Pollinator Interactions: From Specialization to Generalization. University of Chicago Press, Chicago, IL, pp. 387–407.
Stephens, S.E., Koons, D.N., Rotella, J.J., Willey, D.W., 2003. Effects of habitat fragmentation on avian nesting success: a review of the evidence at multiple spatial scales. Biol. Conserv. 115, 101–110.
Stevens, V.M., Leboulengé, È., Wesselingh, R.A., Baguette, M., 2006. Quantifying functional connectivity, experimental assessment of boundary permeability for the natterjack toad (*Bufo calamita*). Oecologia 150, 161–171.
Stouffer, D.B., Bascompte, J., 2010. Understanding food-web persistence from local to global scales. Ecol. Lett. 13, 154–161.
Strickler, K.L., 1979. Specialization and foraging efficiency of solitary bees. Ecology 60, 998–1009.
Sutherland, G.D., Harestad, A.S., Price, K., Lertzman, K.P., 2000. Scaling of natal dispersal distances in terrestrial birds and mammals. Conserv. Ecol. 4, 16.
Syms, C., Jones, G.P., 2000. Disturbance, habitat structure, and the dynamics of a coral-reef fish community. Ecology 81, 2714–2729.
Tabarelli, M., Peres, C.A., 2002. Abiotic and vertebrate seed dispersal in the Brazilian Atlantic forest: implications for forest regeneration. Biol. Conserv. 106, 165–176.
Tabarelli, M., Mantovani, W., Peres, C.A., 1999. Effects of habitat fragmentation on plant guild structure in the montane Atlantic forest of southeastern Brazil. Biol. Conserv. 91, 119–127.
Tabarelli, M., Lopes, A.V., Peres, C.A., 2008. Edge-effects drive tropical forest fragments towards an early-successional system. Biotropica 40, 657–661.
Tabarelli, M., Aguiar, A.V., Girão, L.C., Peres, C.A., Lopes, A.V., 2010. Effects of pioneer tree species hyperabundance on forest fragments in Northeastern Brazil. Conserv. Biol. 24, 1654–1663.

Taylor, P.D., Fahrig, L., Henein, K., Merriam, G., 1993. Connectivity is a vital element of landscape structure. Oikos 68, 571–573.
Taylor, P.D., Fahrig, L., With, K.A., 2006. Landscape connectivity: a return to the basics. In: Crooks, K.R., Sanjayan, M. (Eds.), Connectivity Conservation. Cambridge University Press, New York, pp. 29–43.
Terborgh, J., Lopez, L., Tello, J., Yu, D., Bruni, A.R., 1997. Transitory states in relaxing ecosystems of land bridge islands. In: Laurance, W.F., Bierregaard, R.O. (Eds.), Tropical Forest Remnants: Ecology, Management, and Conservation of Fragmented Communities. University of Chicago Press, Chicago, IL, pp. 256–274.
Terborgh, J., Lopez, L., Nuñez, P., Rao, M., Shahabuddin, G., Orihuela, G., Riveros, M., Ascanio, R., Adler, G.H., Lambert, T.D., Balbas, L., 2001. Ecological meltdown in predator-free forest fragments. Science 294, 1923–1926.
Tewksbury, J.J., Levey, D.J., Haddad, N.M., Sargent, S., Orrock, J.L., Weldon, A., Danielson, B.J., Brinkerhoff, J., Damschen, E.I., Townsend, P., 2002. Corridors affect plants, animals, and their interactions in fragmented landscapes. Proc. Natl. Acad. Sci. USA 99, 12923–12926.
Thebault, E., Fontaine, C., 2010. Stability of ecological communities and the architecture of mutualistic and trophic networks. Science 329, 853–856.
Thies, C., Roschewitz, I., Tscharntke, T., 2005. The landscape context of cereal aphid-parasitoid interactions. Proc. R. Soc. Lond. B 272, 203–210.
Thomas, C.D., 2000. Dispersal and extinction in fragmented landscapes. Proc. R. Soc. Lond. B 267, 139–145.
Thompson, J.N., 2005. The Geographic Mosaic of Coevolution. University of Chicago Press, Chicago, IL.
Tilman, D., May, R.M., Lehman, C.L., Nowak, M.A., 1994. Habitat destruction and the extinction debt. Nature 371, 65–66.
Tilman, D., Fargione, J., Wolff, B., D'Antonio, C., Dobson, A., Howarth, R., Schindler, D., Schlesinger, W.H., Simberloff, D., Swackhamer, D., 2001. Forecasting agriculturally driven global environmental change. Science 292, 281–284.
Tischendorf, L., Fahrig, L., 2000. On the usage and measurement of landscape connectivity. Oikos 90, 7–19.
Torp, E., 1994. Danmarks Svirrefluer, vol. 6. Apollo, Stenstrup, Denmark.
Townsend, C.R., Hildrew, A.G., Francis, J.E., 1983. Community structure in some southern English streams: the influence of physiochemical factors. Freshw. Biol. 19, 521–544.
Townsend, C.R., Hildrew, A.G., Schofield, K., 1987. Persistence of stream invertebrate communities in relation to environmental variability. J. Anim. Ecol. 56, 597–613.
Trøjelsgaard, K., Olesen, J.M., 2012. Macroecology of pollination networks. Global Ecol. Biogeogr http://dx.doi.org/10.1111/j.1466-8238.2012.00777.x.
Tscharntke, T., Steffan-Dewenter, I., Kruess, A., Thies, C., 2002. Characteristics of insect populations on habitat fragments: a mini review. Ecol. Res. 17, 229–239.
Tscharntke, T., Klein, A.M., Kruess, A., Steffan-Dewenter, I., Thies, C., 2005. Landscape perspectives on agricultural intensification and biodiversity-ecosystem service management. Ecol. Lett. 8, 857–874.
Turing, A.M., 1952. The chemical basis of morphogenesis. Proc. R. Soc. Lond. B 237, 37–72.
Tylianakis, J.M., Tscharntke, T., Lewis, O.T., 2007. Habitat modification alters the structure of tropical host-parasitoid food webs. Nature 445, 202–205.
Tylianakis, J.M., Didham, R.K., Bascompte, J., Wardle, D.A., 2008. Global change and species interactions in terrestrial ecosystems. Ecol. Lett. 11, 1351–1363.
Tylianakis, J.M., Laliberté, E., Nielsen, A., Bascompte, J., 2010. Conservation of species interaction networks. Biol. Conserv. 143, 2270–2279.

Urban, D.L., Keitt, T., 2001. Landscape connectivity: a graph-theoretic perspective. Ecology 82, 1205–1218.
Urban, D.L., O'Neill, R.V., Shugart, H.H., 1987. Landscape ecology: a hierarchical perspective can help scientists understand spatial patterns. Bioscience 37, 119–127.
Urban, D.L., Minor, E.S., Treml, E.A., Schick, R.S., 2009. Graph models of habitat mosaics. Ecol. Lett. 12, 260–273.
Uriarte, M., Anciaes, M., Da Silva, M.T.B., Rubim, P., Johnson, E., Bruna, E.M., 2011. Disentangling the drivers of reduced long-distance seed dispersal by birds in an experimentally fragmented landscape. Ecology 92, 924–937.
Valladares, G., Cagnolo, L., Salvo, A., 2012. Forest fragmentation leads to food web contraction. Oikos 121, 299–305.
Vamosi, J.C., Knight, T.M., Steets, J.A., Mazer, S.J., Burd, M., Ashman, T.-L., 2006. Pollination decays in biodiversity hotspots. Proc. Natl. Acad. Sci. USA 103, 956–961.
van Houtan, K.S., Pimm, S.L., Halley, J.M., Bierregaard, R.O., Lovejoy, T.E., 2007. Dispersal of Amazonian birds in continuous and fragmented forest. Ecol. Lett. 10, 219–229.
van Nieuwstadt, M.G.L., Ruano Iraheta, C.E., 1996. Relation between size and foraging range in stingless bees (Apidae, Meliponinae). Apidologie 27, 219–228.
van Nouhuys, S., 2005. Effects of habitat fragmentation at different trophic levels in insect communities. Ann. Zool. Fennici 42, 433–447
van Nouhuys, S., Hanski, I., 2002. Colonization rates and distances of a host butterfly and two specific parasitoids in a fragmented landscape. J. Anim. Ecol. 71, 639–650.
Vanbergen, A.J., Hails, R.S., Watt, A.D., Jones, T.H., 2006. Consequences for host-parasitoid interactions of grazing-dependent habitat heterogeneity. J. Anim. Ecol. 75, 789–801.
Vanbergen, A.J., Jone, T.H., Hail, R.S., Watt, A.D., Elston, D.A., 2007. Consequences for a host-parasitoid interaction of host-plant aggregation, isolation and phenology. Ecol. Entomol. 32, 419–427.
Vandermeer, J., Perfecto, I., Philpott, S.M., 2010. Ecological complexity and pest control in organic coffee production: uncovering an autonomous ecosystem service. Bioscience 60, 527–537.
Vázquez, D.P., 2005. Degree distribution in plant-animal mutualistic networks: forbidden links or random interactions? Oikos 108, 421–426.
Vázquez, D.P., Morris, W.F., Jordano, P., 2005. Interaction frequency as a surrogate for the total effect of animal mutualists on plants. Ecol. Lett. 8, 1088–1094.
Vázquez, D.P., Melián, C.J., Williams, N.M., Blüthgen, N., Krasnov, B.R., Poulin, R., 2007. Species abundance and asymmetric interaction strength in ecological networks. Oikos 116, 1120–1127.
Vázquez, D.P., Chacoff, N.P., Cagnolo, L., 2009. Evaluating multiple determinants of the structure of plant-animal mutualistic networks. Ecology 90, 2039–2046.
Verdú, M., Valiente-Banuet, A., 2011. The relative contribution of abundance and phylogeny to the structure of plant facilitation networks. Oikos 120, 1351–1356.
Voigt, F.A., Bleher, B., Fietz, J., Ganzhorn, J.U., Schwab, D., Böhning-Gaese, K., 2004. A comparison of morphological and chemical fruit traits between two sites with different frugivore assemblages. Oecologia 141, 94–104.
Vörösmarty, C.J., McIntyre, P.B., Gessner, M.O., Dudgeon, D., Prusevich, A., Green, P., Glidden, S., Bunn, S.E., Sullivan, C.A., Liermann, C.R., Davies, P.M., 2010. Global threats to human water security and river biodiversity. Nature 467, 555–561.
Walker, D.I., McComb, A.J., 1992. Seagrass degradation in Australian coastal waters. Mar. Pollut. Bull. 25, 191–195.
Walker, D.I., Kendrick, G.A., McComb, A.J., 2006. Decline and recovery of seagrass ecosystems: the dynamics of change. In: Larkum, A.W.D., Orth, R.J., Duarte, C.M. (Eds.), Seagrasses: Biology, Ecology and Conservation. Springer-Verlag, Dordrecht, pp. 551–565.

Warren, P.H., 1996. Dispersal and destruction in a multiple habitat system: an experimental approach using protist communities. Oikos 77, 317–325.
Weibull, A.C., Bengtsson, J., Nohlgren, E., 2000. Diversity of butterflies in the agricultural landscape: the role of farming system and landscape heterogeneity. Ecography 24, 743–750.
Wells, S., 2009. Dynamite fishing in northern Tanzania—pervasive, problematic and yet preventable. Mar. Pollut. Bull. 58, 20–23.
Whittaker, R.J., Triantis, K.A., Ladle, R.J., 2008. A general dynamic theory of oceanic island biogeography. J. Biogeogr. 35, 977–994.
Wikelski, M., Moxley, J., Eaton-Mordas, A., López-Uribe, M.M., Holland, R., Moskowitz, D., Roubik, D.W., Kays, R., 2010. Large-range movements of neotropical orchid bees observed via radio telemetry. PLoS One 5, e10738.
Wilcock, H.R., Nichols, R.A., Hildrew, A.G., 2003. Genetic population structure and neighbourhood population size estimates of the caddisfly *Plectrocnemia conspersa*. Freshw. Biol. 48, 1813–1824.
Wilcove, D.S., McLellan, C.H., Dobson, A.P., 1986. Habitat fragmentation in the temperate zone. In: Soulé, M.E. (Ed.), Conservation Biology: The Science of Scarcity and Diversity. Sinauer Associates, Sunderland, MA, pp. 237–256.
Willems, J.H., Huijsmans, K.G.A., 1994. Vertical seed dispersal by earthworms: a quantitative approach. Ecography 17, 124–130.
Willi, Y., van Buskirk, J., Fischer, M., 2005. A threefold genetic allee effect: population size affects cross-compatibility, inbreeding depression and drift load in the self-incompatible *Ranunculus reptans*. Genetics 169, 2255–2265.
Williams, P.H., 2005. Does specialization explain rarity and decline among British bumblebees? a response to Goulson *et al.* Biol. Conserv. 122, 33–43.
Williams, R.J., Martinez, N.D., 2000. Simple rules yield complex food webs. Nature 404, 180–183.
Williams, P.H., Colla, S., Xie, Z., 2009. Bumblebee vulnerability: common correlates of winners and losers across three continents. Conserv. Biol. 23, 931–940.
Willis, E.O., 1979. The composition of avian communities in remanescent wood lots in southeastern Brazil. Pap. Avul. Zool. 33, 1–25.
Willson, M.F., Whelan, C.J., 1990. The evolution of fruit color in fleshy-fruited plants. Am. Nat. 136, 790–809.
Wirth, R., Meyer, S.T., Leal, I.R., Tabarelli, M., 2008. Plant-herbivore interactions at the forest edge. Progr. Bot. 69, 423–448 1.
Woodward, G., Hildrew, A.G., 2002a. Food web structure in riverine landscapes. Freshw. Biol. 47, 777–798.
Woodward, G., Hildrew, A.G., 2002b. Differential vulnerability of prey to an invading top predator: integrating field surveys and laboratory experiments. Ecol. Entomol. 27, 732–744.
Woodward, G., Ebenman, B., Emmerson, M., Montoya, J.M., Olesen, J.M., Valido, A., Warren, P.H., 2005. Body size in ecological networks. Trends Ecol. Evol. 20, 402–409.
Woodward, G., Papantoniou, G., Edwards, F., Lavridsen, R.B., 2008. Trophic trickles and cascades in a complex food web: impacts of a keystone predator on stream community structure and ecosystem processes. Oikos 117, 683–692.
Woodward, G., Benstead, J.P., Beveridge, O.S., Blanchard, J., Brey, T., Brown, L.E., Cross, W.F., Friberg, N., Ings, T.C., Jacob, U., Jennings, S., Ledger, M.E., et al., 2010a. Ecological networks in a changing climate. Adv. Ecol. Res. 42, 71–138.
Woodward, G., Blanchard, J., Lauridsen, R.B., Edwards, F.K., Jones, J.I., Figueroa, D., Warren, P.H., Petchey, O.L., 2010b. Individual-based food webs: species identity, body size and sampling effects. Adv. Ecol. Res. 43, 211–266.

Woodward, G., Brown, L.E., Edwards, F.K., Hudson, L.N., Milner, A.M., Reuman, D.C., Ledger, M.E., 2012. Climate change impacts in multispecies systems: drought alters food web size-structure in a field experiment. Philos. Trans. R. Soc. Lond. B in press 87, 526–544.

Wright, S.J., Duber, H.C., 2001. Poachers and forest fragmentation alter seed dispersal, seed survival, and seedling recruitment in the palm *Attalea butyraceae*, with implications for tropical tree diversity. Biotropica 33, 583–595.

Young, A., Boyle, T., Brown, T., 1996. The population genetic consequences of habitat fragmentation for plants. Trends Ecol. Evol. 11, 413–418.

Yu, L., Lu, J., 2011. Does landscape fragmentation influence sex ratio of dioecious plants? A case study of *Pistacia chinensis* in the Thousand-Island Lake region of China. PLoS One 6, e22903.

Zamora, R., 2000. Functional equivalence in plant-animal interactions: ecological and evolutionary consequences. Oikos 88, 442–447.

Zhang, F., Hui, C., Terblanche, J.S., 2011. An interaction switch predicts the nested architecture of mutualistic networks. Ecol. Lett. 14, 797–803.

Climate Change Impacts on Community Resilience: Evidence from a Drought Disturbance Experiment

Mark E. Ledger*[,1], Rebecca M.L. Harris*, Patrick D. Armitage[†], Alexander M. Milner*[,‡]

*School of Geography, Earth and Environmental Sciences, University of Birmingham, Edgbaston, Birmingham, United Kingdom
[†]Freshwater Biological Association River Laboratory, East Stoke, Wareham, Dorset, United Kingdom
[‡]Institute of Arctic Biology, University of Alaska, Fairbanks, Alaska, USA
[1]Corresponding author: e-mail address: m.e.ledger@bham.ac.uk

Contents

Advances in Ecological Research, Volume 46
ISSN 0065-2504
http://dx.doi.org/10.1016/B978-0-12-396992-7.00003-4

Abstract

Climate change is expected to alter disturbance regimes with profound effects on the structure and dynamics of ecological communities. In many regions, climate models forecast shifts in precipitation patterns that will exacerbate droughts in rivers and streams, yet ecological impacts on freshwater ecosystems remain poorly understood. We report the results of a stream mesocosm experiment designed to test the effect of drought on the resilience of replicate macroinvertebrate communities, via direct manipulation of flows. Drying disturbances applied at high (monthly) and low (quarterly) frequency over 21-months had contrasting effects on the structure and temporal dynamics of the communities. Macroinvertebrates were resilient to low-frequency disturbance, sustaining abundant and diverse communities, which developed over experimental time. By comparison, high-frequency disturbance exceeded the capacity for recovery, skewing community structure, and generated relatively impoverished, static assemblages dominated by fewer species. Species responses ranged from extirpation to irruption, with smaller short-lived taxa (≤ 1 generation per year), notably chironomids and worms, replacing larger taxa with longer life cycles (> 1 generation per year). This research provides one of the first experimental tests of resilience to drought in aquatic ecosystems.

1. INTRODUCTION

1.1. Disturbance, community structure and climate change

Climate change is one of the most critical disturbances imposed on natural systems on a global scale. Its effects at the higher (multispecies) levels of organisation are still poorly understood, although this area has recently seen dramatic increases in research activity (e.g. Dossena et al., 2012; Mintenback et al., 2012; Perkins et al., 2010). It is in itself a compound stressor, associated with changes in temperature and atmospheric conditions, and it can also interact with the effects of other local or regional perturbations, including eutrophication, acidification and habitat fragmentation (Hagen et al., 2012; Layer et al., 2010, 2011; Meerhoff et al., 2012; Olesen et al., 2010).

Disturbance is integral to the organisation of the globe's freshwater, marine and terrestrial ecosystems (Sousa, 1979) and can be defined as 'any discrete event in time that disrupts the structure of an ecosystem, community, or population, and changes resource availability or the physical environment' (Pickett and White, 1985). The spatial and temporal occurrence of such events define a system's disturbance regime, in terms of frequency, intensity, timing, duration, extent and severity (Pickett and White, 1985).

In many regions, natural disturbance regimes have been modified by human activities (Turner, 2010). In the near future, disturbances as diverse as fire, floods, drought, hurricanes and landslides are expected to increase in frequency in many parts of the world due to human intervention (Turner, 2010). Since many disturbances are also associated with strong climate forcing, future intensification of weather extremes could have profound impacts on habitats, communities and ecosystems.

Extreme events can be characterised by their strength, timing and abruptness relative to the life cycles of affected organisms (Jentsch et al., 2007). In Europe, for example, climate models predict increased incidence of extreme events in future, including heat waves, heavy precipitation, drought, wind storms and storm surges (Beniston et al., 2007) that are likely to cause disturbances and trigger profound changes in local ecosystems that are disproportionate to their short duration. It is often challenging to assess the impact of extreme events on ecosystems because of their rarity and unpredictability, so controlled experiments have been advocated as a more logistically feasible alternative, which can also provide insight into underlying mechanisms (Jentsch et al., 2007).

1.2. Disturbance and diversity

Several theoretical models predict how increasing disturbance might reshape diversity in natural communities (e.g. Connell, 1978; Huston, 1979; Petraitis et al., 1989), yet despite decades of research, no clear patterns have emerged among empirical studies (Miller et al., 2011). Disturbance effects on species diversity are often expressed graphically as diversity–disturbance relationships, which plot a descriptor of species diversity (e.g. richness) against a descriptor of disturbance (e.g. frequency). The influential intermediate disturbance hypothesis (Connell, 1978) predicted a unimodal relationship between diversity and disturbance, whereby low disturbance leads to competitive exclusion and high disturbance filters out all but the most opportunistic disturbance-adapted species. Evidence in support of the intermediate disturbance hypothesis (IDH) came from communities characterised by high rates of competitive exclusion (e.g. Lubchenco, 1978; Sousa, 1979), but several recent studies revealed that unimodal diversity–disturbance relationships are far from ubiquitous (Mackey and Currie, 2001), and there are many examples of positive (e.g. Armesto and Pickett, 1985; Fox, 1985) and negative (e.g. Death and Winterbourn, 1995; Wilson and Tilman, 1991) relationships across different ecosystem types. Despite widespread and extensive research on disturbance dynamics, ecologists still lack a clear predictive understanding as to how changing disturbance regimes will affect multispecies systems (e.g. Death, 2002; Miller et al., 2011).

1.3. Climate change and drought disturbance in streams

Disturbance is expected to play an important role in governing ecosystem responses to future changes in climate (Easterling et al., 2003), and anticipated increases in the frequency and intensity of disturbance events (IPCC, 2007) may alter biodiversity and ecosystem functioning in the future (Daufresne et al., 2007; Ledger et al., 2008). For example, climate change is expected to alter precipitation patterns at global, regional and local scales (Beniston et al., 2007; IPCC, 2007), with consequences for the temporal and spatial distribution of water across and within ecosystems (Acuña and Tockner, 2010; Vörösmarty et al., 2010). Water scarcity is one of this century's most pressing environmental issues (Schindler and Donahue, 2006) and climate-induced shifts in the availability of water are likely to increase the incidence of drought which can be further exacerbated at supraseasonal scales by overabstraction of water for human use, as an additional indirect consequence of climate change (Chessman, 2009; Poff and Zimmerman, 2010).

Droughts, defined here as low-flow periods unusual in their frequency, duration, extent, severity or intensity, occur naturally in many rivers and streams during times of low rainfall (Boulton, 2003). In Mediterranean regions regular stream drying occurs as a natural part of the hydrologic cycle and stream biota possess traits adapted to tolerate or escape dry conditions (Bonada et al., 2007). By contrast, drought events occurring unpredictably may have devastating effects in river communities containing few drought-adapted species (Lake, 2003). Across Europe, drought has already increased dramatically in frequency over the past 30 years (IPCC, 2007). Despite this growing prevalence, research on the ecological impacts of drought in rivers and streams lags behind that of other forms of disturbance, especially floods (James et al., 2008; Sponseller et al., 2010). Flow is widely regarded as a master variable that shapes the ecological characteristics of rivers and streams (Poff and Zimmerman, 2010), and extreme events like drought are known to exert strong selective pressure on populations, influencing their relative success, and both the lethal and sub-lethal effects on stream biota have the potential to profoundly alter biodiversity, food web structure and ecosystem functioning (Daufresne and Boët, 2007; Ledger et al., 2011;

Woodward et al., 2012). Under climate change, uncharacteristically frequent or intense (Beche et al., 2009; Gurvich et al., 2002) droughts may drive species beyond their 'hydrological envelope'—just as global warming can drive species populations beyond their 'thermal envelope' (Ledger et al., 2008; Woodward et al., 2010). These novel disturbance regimes challenge the capacity of river systems to recover and could elicit unforeseen dynamics that erode the supply of ecological goods and services, including the maintenance of viable fisheries or waste-processing capabilities. Drought impacts are also likely to be contingent upon the many facets of the hydrologic regime itself, and the imperative now is to develop approaches to explore how the frequency, intensity and duration of low-flow periods influence aquatic systems (Ledger et al., 2011; Poff and Zimmerman, 2010).

Freshwater organisms possess traits that confer a degree of resistance (ability to withstand disturbances; Pimm, 1984) and resilience (tendency to return to a reference state after disturbance; Chapin et al., 2002) that together govern their ecological stability and associated biodiversity–ecosystem functioning relationships (Ptacnik et al., 2010). Here, we define ecological stability simply as the propensity for a system to maintain its species composition in the face of disturbance (Woodward et al., 2002). Several reviews indicate that where droughts occur unpredictably, biotic resistance is typically low whereas resilience is more varied, and may be related to life-history traits such as body size and voltinism, or rarity (Lake, 2003; Ledger et al., 2011). The rate of community recovery following drought episodes may be relatively rapid (days-months) but local species loss, strongly skewed abundance patterns, and other biological legacies may markedly delay it or divert its trajectory (Lake, 2003; Ledger et al., 2006). Drought disturbances may also skew community dynamics, either by accelerating species turnover or by decelerating or arresting successional processes in stream benthic habitats (Ledger et al., 2008) and recovery may be difficult to establish, especially given the inherently dynamic and patchy nature of many freshwater communities in both space and time (Hagen et al., 2012; Olesen et al., 2010).

1.4. Mesocosm experiments

Droughts occurring unpredictably in natural systems can be a challenge to study and research to establish the ecological effects of these events is inevitably opportunistic and often beset by confounding gradients and/or lacking adequate controls or pre-impact data (James et al., 2008; Lake, 2003).

Controlled manipulative experiments are required to overcome these confounding influences and to identify the mechanistic basis of cause-and-effect relationships that cannot be resolved via field survey approaches (e.g. Jentsch et al., 2007; Ledger and Winterbourn, 2000; Ledger et al., 2002, 2011; Woodward et al., 2012). However, laboratory-based flume or microcosm studies inevitably sacrifice realism for control and replication, undermining the validity of extrapolating to relevant spatiotemporal scales and levels of biological complexity needed to understand system-level responses (Harris et al., 2007; Ledger et al., 2009). In this study, we adopt a novel experimental approach using mesocosms that circumvent the shortcomings of correlational studies (see Dewson et al., 2007; Ledger et al., 2008) while maintaining greater realism than would be possible in smaller scale (e.g. laboratory flumes or microcosms) manipulations. The main objectives of this study were (1) to examine how drying disturbances induced by supraseasonal drought affects stream macroinvertebrate communities and (2) to characterise how the resilience of biota is challenged by disturbances of different frequencies. A series of stream mesocosms fed by a chalk stream were used to simulate drought disturbance episodes, via a direct manipulation of flows (see Ledger et al., 2008, 2011). Mesocosm-scale research provides the means to make direct comparisons among replicated communities under different, experimentally applied, flow regimes, and when conducted for at least one generation of the longest-lived organism, perturbation experiments can yield insights into the intergenerational responses within and across food webs (Woodward et al., 2012; Yodzis, 1988). Previous research has shown that the mesocosms used in this study are both replicable and realistic for water quality and biodiversity (algae and macroinvertebrates) (Harris et al., 2007; Ledger et al., 2009) and contain complex food webs with structural properties (e.g. connectance, path lengths, degree distributions) which shape ecological responses to stress that are consistent with those of natural systems (Brown et al., 2011; Ledger et al., 2011; Woodward et al., 2012).

The mesocosm experiment was used to test three hypotheses: first, that drying disturbance would restructure benthic macroinvertebrate communities, with the extent of the effect contingent upon the frequency of occurrence of disturbance events. We expected that small species with fast life histories would replace larger taxa with longer life cycles, consistent with theory (Brown et al., 2004; Pianka, 1983). Second, we proposed that species richness would peak at intermediate levels of disturbance

(Connell, 1978) and third, we proposed that disturbance would arrest the process of macroinvertebrate community development over time, with the extent of the constraining effect greatest in patches disturbed at high frequency (HF).

2. METHODS

2.1. Mesocosms

Research was conducted between February 2000 and February 2002 using a series of stream mesocosms at the Freshwater Biological Association River Laboratory, East Stoke, Dorset, UK (50°40′48″ N, 2°11′06″ W). The mesocosms were arranged in four spatial blocks next to a chalk stream (Fig. 1). Each block of mesocosms contained three stainless-steel linear channels (each width 0.33 m, length 12 m, depth 0.30 m) fed unfiltered water by gravity from the stream (containing invertebrates, algae and detritus) through a branching 110 mm diameter pipe (Harris et al., 2007). The upper end of each mesocosm channel was closed and fitted with a flow control valve whereas the lower end was open, allowing free drainage of water and suspended particles into an outlet stream. Each mesocosm was positioned 5 cm below the inlet and 10 cm above the outlet to avoid transfer of biota among the channels. Mesocosms were filled to 20 cm depth with clean substrate dominated by chert gravel (volumetric proportions of particle sizes, 85% 11–25 mm, 5% 2–11 mm, 5% 0.35–2 mm, 5% $<$0.35 mm), matching the source stream (Harris et al., 2007). Consistent with many chalk streams, mesocosms did not have extensive hyporheic zones (Trimmer et al., 2010), but substrata provided refugia for suitably adapted species during drying disturbances (Harris, 2006). Physicochemical conditions were highly congruent among mesocosms (Harris et al., 2007) and closely paralleled those of the source stream (Ledger et al., 2009). During the main study period, water temperature (mean 12.2 °C) varied seasonally, with summer maxima (18.7 °C in June 2000) and winter minima (6.0 °C in December 2001). Inflowing water was nutrient rich (mean PO_4: 0.16 mg L^{-1}; NO_3: 5.62 mg L^{-1} from March 2000–February 2002) with high pH (mean 8.1) and conductivity (mean 460 μS cm^{-1}) (Harris et al., 2007). Outside the experimentally simulated dewatering periods, discharge in the mesocosms was stable (cross treatment mean 0.005 m^3 s^{-1}), with mean water velocity (at two-thirds depth) and depth over the gravel of 0.20 m s^{-1} and 81 mm, respectively, and water residence times were short (mean 66 s) (Harris, 2006).

A

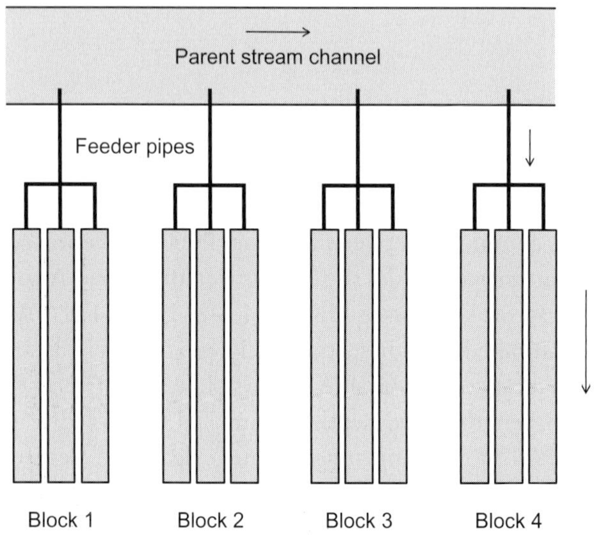

B

Figure 1 Schematic diagram (A) and photograph (B) of the stream mesocosm facility at the Freshwater Biological Association River Laboratory, Dorset, UK. Four blocks of three stream mesocosms (each channel 12 m length × 0.3 m width) were fed water through pipes (6 m length) from the parent stream. Water flow (direction indicated by arrows) in to each mesocosm was controlled by a valve. Each block contained one undisturbed control, one low-frequency (LF) and one high-frequency (HF) disturbed channel. (For colour version of this figure, the reader is referred to the online version of this chapter.)

2.2. Experimental design and application

Unfiltered water from the source stream was diverted into the mesocosms to initiate colonisation and community development (February-March 2000). Macroinvertebrate colonisation was either passive, in drift from the source stream, or by adult oviposition (Ledger et al., 2009). Two months immediately after the the initial colonization period an experiment was established consisting of two drought disturbance treatments—low-frequency (LF) and high-frequency (HF) flow cessation—and an undisturbed control (Table 1). Disturbance regimes were short periods (6-day) of flow cessation applied at either approximately quarterly (LF treatment, 99-day dry/wet cycles, 7 disturbance events) or monthly intervals (HF treatment, 33-day dry/wet cycles, 20 disturbance events). By contrast, flows in controls remained uninterrupted for the duration of the experiment (693 days, Table 1). The dewatering treatments mimicked severe drought conditions in which low fluctuating water levels repeatedly disturb patches of the river bed over a prolonged period, as might be expected in supraseasonal drought scenarios (see Ledger et al., 2008, 2011). These events are expected to become more frequent in the UK under Intergovernmental Panel on Climate Change (IPCC) scenarios (Vidal and Wade, 2009). Disturbances were applied by slowly closing inflow ducts and allowing water to drain from the channels over the course of several days. During dewatering, surface flows ceased and drying of exposed substrata occurred in patches, whereas the interstices beneath the bed surface remained wet, and small pools persisted at intervals along the length of the dewatered channels, providing refugia for suitably adapted species (Harris, 2006). Surfaces of

Table 1 Disturbance treatments applied in stream mesocosms

	Drying disturbance		
	High frequency	**Low frequency**	**Control**
Mean dry days per cycle	6	6	0
Mean wet days per cycle	27	93	693
Disturbance events (*n*)	20	7	0
Total duration (*d*)	693	693	693
% time disturbed	16.7	5.6	0

exposed substrata dried at ambient rates, such that the stress experienced by organisms stranded in the mesocosms was consistent with those in adjacent naturally drying stream reaches (Harris, 2006; Ledger et al., 2008). A blocked experimental design (Zar, 1999) was used in which each treatment was replicated four times, with each block of channels containing each drought treatment (i.e. HF and LF disturbance) and a control (4 blocks × 3 treatments = 12 channels in total; Fig. 1).

2.3. Sampling and processing

Benthic macroinvertebrates were sampled monthly (21 occasions) in each mesocosm, immediately before disturbances were applied. On each occasion, a small Surber sample (0.025 m^2, 300 μm mesh) was taken from the upper, central and lower section of each channel. Animals in samples were later sorted from debris, identified to the lowest practicable taxonomic unit (species or genus) and counted. Data from each of the three mesocosm sections were combined to provide a single estimate of macroinvertebrate species composition for each replicate mesocosm on each sampling occasion.

2.4. Statistical analysis

Repeated-measures analysis of variance (RM-ANOVA) was used to test the main effect of disturbance frequency (between-subject factor), time (within-subject factor) and their interaction, on (1) macroinvertebrate taxon richness, (2) abundances (numbers m^{-2}) of core taxa and (3) community similarity metrics (Jaccard's and Spearman's rank coefficients). For core taxa (> 1% total abundance), ANOVA summary statistics are reported in Table 3, with full tables provided in Appendix B. Data were log-transformed, and where necessary, departures from sphericity were corrected using the Huynh–Feldt adjustment. One-way ANOVA with Tukey HSD tests were used subsequently to examine differences between treatment means at endpoints. Sequential Bonferroni corrections were applied to groups of tests to preserve an alpha of 0.05 (Rice, 1989).

Resilience of community structure was assessed using two standard methods (Bradley and Ormerod, 2001). First, within each mesocosm block, Spearman's rank correlation coefficients were used to compare macroinvertebrate relative abundances in controls with those of each disturbed treatment, with high positive values (maximum +1) indicating high stability. Second, resilience of taxonomic composition was assessed using Jaccard's coefficient of similarity (*J*; Magurran, 2004): $J(AB) = j/(a+b-j)$ where *a* is the number of taxa in control assemblage A, *b* is the number of taxa in disturbed assemblage B within the same experimental block and *j* is the number of taxa common to both assemblages, with values of *J* ranging from 0 (no similarity, low resilience) to 1 (complete similarity, high resilience). For both of these measures of community resilience, pairwise comparisons between controls and each disturbance treatment (LF, HF) within a each block were made at every endpoint (4 blocks × 21 endpoints).

A series of partial constrained ordinations (redundancy analysis, RDA) was performed using CANOCO 4.5 (ter Braak and Šmilauer, 2002) to establish the effect of the disturbance regimes on community structure. Accordingly, three ordinations incorporating various combinations of explanatory variables and covariables were used to test for terms analogous to univariate repeated-measures ANOVA (Lepš and Šmilauer, 2003), and specifically explored effects of disturbance treatment, time, and their interaction on macroinvertebrate community composition (see Table 2). Linear ordination was used because gradient lengths on a preliminary DCA were short (<2 SD). In the analyses, disturbance treatment, experimental block, mesocosm units and season were coded as binary dummy variables (0 or 1), and sampling occasions (endpoints) were coded as either dummy variables or as quantitative

Table 2 Results of partial redundancy analyses (1–6) examining macroinvertebrate community structure in stream mesocosms, based on numerical (1–3) or relative (4–6) abundance

Analysis	Explanatory variables	Covariables	% var	*r*	*F*	*P*
1	Treatment	Time, block	10.5	0.85	26.51	0.001
2	Time	Treatment, block	14.5	0.96	38.30	0.001
3	Treatment × time	Mesocosm, time	3.2	0.76	5.90	0.001
4	Treatment	Time, block	8.3	0.85	20.56	0.001
5	Time	Treatment, block	14.8	0.96	39.20	0.001
6	Treatment × time	Mesocosm, time	2.7	0.82	5.01	0.001

% var: percentage of species variability explained by the first ordination axis; *r*: species-environment correlation of the first axis; *F*, *P*: *F*-ratio and corresponding probability value of each Monte Carlo permutation test. 'Mesocosm' denotes use of dummy identifying variables for stream channels.

variables (experimental time, days), to test temporal effects as both discrete and continuous variables. Macroinvertebrate abundance data were *ln*-transformed densities (numbers m^{-2}). In the first analysis, the effect of drought on the macroinvertebrate assemblage was determined using the three treatments as dummy explanatory variables (0 or 1) and the four blocks and 21 times as dummy (0 or 1) covariables. Thus, variance attributable to time, related to shifts in community composition among seasons and endpoints, together with progressive change during the experiment, was partialled out to reveal underlying differences among the three treatments. In a second analysis, temporal variation in community composition was determined with time coded as 21 dummy explanatory variables and mesocosm identifiers used as covariables, to partial out variance among channels related to treatment or block effects. A third ordination examined the interaction between treatment and time, with each combination of treatment × time entered as an explanatory variable. In this analysis, mesocosm units and experimental time were entered as covariables, removing the 'main effect' of each treatment in each mesocosm, thereby revealing variance specific to particular treatments through time (Lepš and Šmilauer, 2003). The ordinations above were performed on macroinvertebrate species data as both absolute densities (in the CANOCO analysis 'non-standardised', analyses 1–3 in Table 2) and as proportions of total abundance (i.e. 'centred and standardized', analyses 4–6 in Table 2). In each case, a restricted Monte Carlo permutation test (999 permutations) to account for non-independence within the data was used to test the significance of the model.

Redundancy analysis was also used to determine the degree of community development (succession) over time in each treatment (Woodward et al., 2002). Again, this method was most appropriate because a preliminary DCA revealed short time-constrained gradient lengths (<2 SD) (ter Braak and Šmilauer, 2002). Community development was analysed for each individual replicate mesocosm community ($n=12$ analyses) and overall for each treatment ($n=3$ analyses) (see Table 4). In the partial RDA analysis, elapsed experimental time (days) was used as a single constraining explanatory variable and mesocosm units and seasons were entered as covariables. The percentage of variation explained by the first RDA axis quantified the strength of any linear time trend in community structure, and a Monte Carlo permutation test (999 permutations) was used to assess the statistical significance ($P<0.05$) of the trend.

3. RESULTS

3.1. Disturbance effects on community descriptors

A total of 127 macroinvertebrate taxa were collected from the mesocosms during the experimental period, with the overall number of taxa encountered within individual treatments decreasing with increasing disturbance frequency (control 114 taxa, LF disturbance 106, HF disturbance 100, Appendix A). In total, 26 taxa present in controls were absent from drought-disturbed channels, most notably rare (< 1% total numbers) Trichoptera (8 taxa), Coleoptera (5), Ephemeroptera (4), Diptera (3) and Gastropoda (3). Overall, HF disturbance excluded more taxa (20 control taxa absent) than the LF treatment (17 control taxa absent). A further 12 taxa, including rare semi-aquatic Diptera (6 taxa), were found only in the drought-disturbed channels (Appendix A). RM-ANOVA revealed that macroinvertebrate taxon richness varied significantly with disturbance treatment ($F_{2,6}=29.87$, $P=0.001$), experimental time ($F_{20,120}=40.18$, $P<0.0005$) and their interaction ($F_{40,120}=2.46$, $P<0.0005$), but there was no effect of mesocosm block ($F_{3,6}=3.57$, $P=0.086$). One-way ANOVAs with multiple comparisons showed that taxon richness under LF disturbance (endpoint mean 30.2 ± 1.5 taxa) was not significantly different from undisturbed controls (mean 32.6 ± 1.6 taxa, Tukey HSD, $P<0.05$) at any endpoint, whereas by contrast, richness in HF disturbance treatments (mean 24.1 ± 1.2 taxa) was significantly lower than in controls (by a mean of 26%) at 13 endpoints ($P<0.05$, Fig. 2A).

Mean total abundance (numbers m^{-2}) of macroinvertebrates varied among treatments and sampling endpoints, with peaks in summer and troughs in winter (Fig. 2B). RM-ANOVA revealed non-significant main effects of treatment ($F_{2,6}=0.55$, $P=0.606$) and block ($F_{3,6}=2.74$, $P=0.131$), whereas time effects ($F_{20,120}=51.46$, $P<0.0005$) and the interaction between treatment and time ($F_{40,120}=2.99$, $P<0.0005$) were statistically significant, the latter reflecting varied effects of the drying among endpoints. Although mean abundances in LF ($12{,}235\pm1972$ m^{-2}) and HF treatments ($12{,}958\pm2434$ m^{-2}) were similar to controls ($12{,}093\pm1618$ m^{-2}) ANOVA with multiple comparisons revealed HF drought significantly (Tukey $P<0.05$) increased abundances in summer and reduced them in autumn and winter (Fig. 2B).

Stability in rank abundance of the fauna between control and disturbed assemblages depended on drying frequency (RM-ANOVA, $F_{1,3}=600.67$, $P<0.0005$), time ($F_{20,60}=5.16$, $P<0.0005$) and their interaction ($F_{20,60}=2.76$, $P=0.001$). Declining resilience was evident with increasing

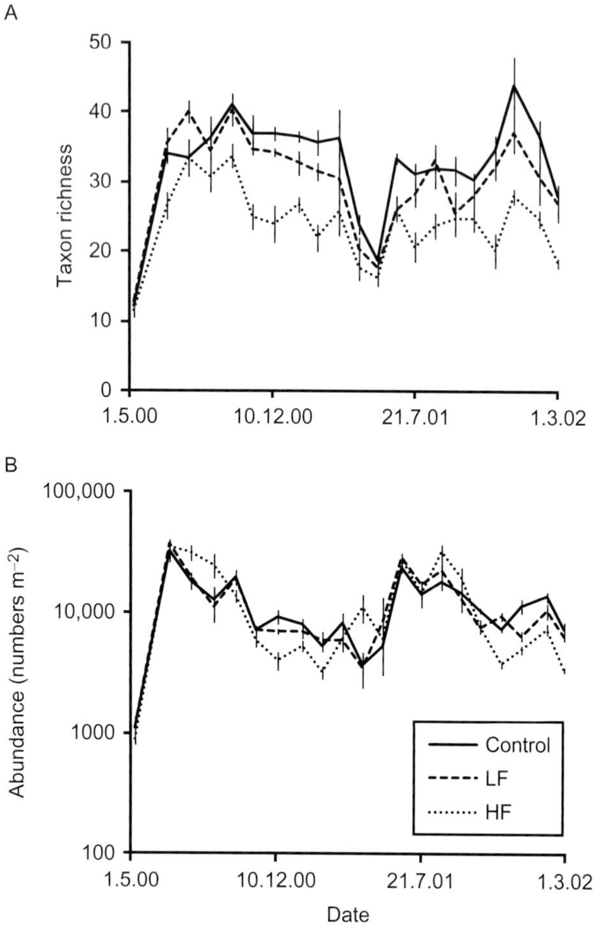

Figure 2 Mean (±1 SE) taxon richness (A) and total abundance (numbers m^{-2}) (B) of macroinvertebrates in mesocosms disturbed at high frequency (HF), low frequency (LF) and in undisturbed controls (C) over 21 months.

disturbance frequency, as revealed by Spearman's rank correlation coefficients (LF mean $=0.80\pm0.02$, HF mean $=0.45\pm0.03$; Fig. 3A), and was statistically significant on most sampling occasions (*t*-tests, $P<0.05$). Similarly, Jaccard's coefficients were consistently lower in HF disturbance treatment (LF mean $=0.55\pm0.01$, HF mean $=0.47\pm0.01$; Fig. 3B), indicating greatest community dissimilarity from controls, as revealed by a RM-ANOVA with statistically significant effects of treatment ($F_{1,3}=76.87$, $P=0.003$), time ($F_{20,60}=4.73$, $P<0.0005$) and a non-significant interaction term ($F_{20,60}=1.44$, $P=0.140$).

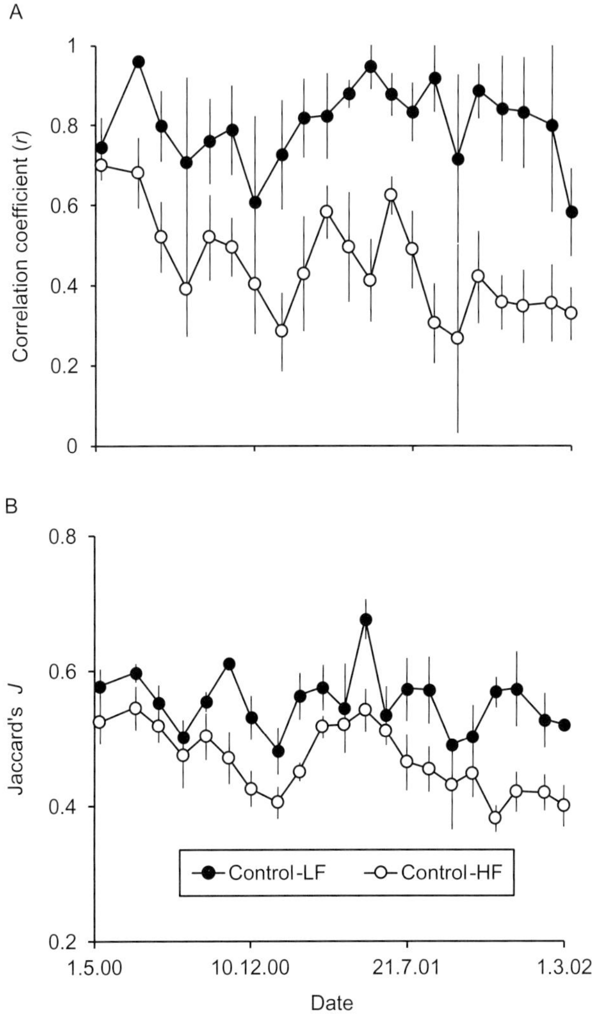

Figure 3 Resilience of macroinvertebrate communities to drought disturbance, as revealed by Spearman's rank correlation coefficients (A) and Jaccard's similarity coefficients (B) comparing LF disturbance and HF disturbance treatments with undisturbed controls.

3.2. Disturbance effects on community structure

Drought disturbance had significant effects on macroinvertebrate community structure (RDA, Table 2) and on the abundances of the main constituent species (RM-ANOVA, Table 3). In undisturbed controls, communities

Table 3 Summary of repeated-measures analysis of variance (RM-ANOVA) testing the main effects of drought treatment, mesocosm block (between-subject factors) and time (within-subject factors), and the interaction between treatment and time, on abundance (numbers m^{-2}) of 12 core taxa (full RM-ANOVA tables in Appendix B)

Taxon	Treatment	Time	Treatment × time	*Post hoc*
G. pulex	**0.003**	**<0.0005**	0.043	C, LF > HF
Pisidium sp.	0.035	**<0.0005**	**<0.0005**	
Chironomini	0.018	**<0.0005**	**<0.0005**	
Tanytarsini	0.571	**<0.0005**	**0.001**	
Diamesinae	0.005	**<0.0005**	**0.001**	
Orthocladiinae	0.031	**<0.0005**	0.012	
Ceratopogonidae	**<0.0005**	**<0.0005**	0.196	C > LF, HF
P. antipodarum	0.243	**<0.0005**	0.011	
R. peregra	0.010	**<0.0005**	**<0.0005**	
Tubificidae	0.608	**<0.0005**	**<0.0005**	
Naididae	0.032	**<0.0005**	0.046	
P. flavomaculatus	0.003	**<0.0005**	**<0.0005**	

Bold denotes statistically significant *P* values following sequential Bonferroni correction.

were dominated numerically by 17 core taxa (each >1% total numbers) which collectively accounted for 89% of total numbers, together with 97 rare taxa that contributed a further 11% (Fig. 4; Appendix A). Core taxa in controls were Tanytarsini midge larvae (1917±545 individuals m^{-2}), Tubificidae worms (mean 1843±378 m^{-2}), Orthocladiinae midge larvae (1435±746 m^{-2}), *Gammarus pulex* L. amphipods (1165±319 mm^{-2}), Naididae worms (976±429 m^{-2}), Chironomini midge larvae (677±120 m^{-2}), *Potamopyrgus antipodarum* J.E. Grey gastropods (562±154 m^{-2}), *Radix balthica* (L.) gastropods (363±155 m^{-2}), *Pisidium* sp. bivalves (360±116 m^{-2}), Diamesinae midge larvae (253±152 m^{-2}), Ceratopogonidae midge larvae (218±55 m^{-2}), *Polycentropus flavomaculatus* (Pictet) caddisfly larvae (196±65 m^{-2}), Tanypodinae midge larvae (178±48 m^{-2}), *Caenis luctuosa* (Bürmister) mayfly larvae (148±51 m^{-2}), *Oulimnius* sp. beetles (132±31 m^{-2}), *Asellus aquaticus* (L.) isopods (116±49 m^{-2}) and *Tinodes waeneri* (L.) caddisfly larvae (115±51 m^{-2}) (Fig. 5, Appendix A).

The shape of the rank-abundance curve for LF disturbed communities was similar to controls (16 core taxa accounted for 92% total numbers) but the upper portion for HF disturbance was elevated, indicating that frequently disturbed communities were less even, being dominated by relatively

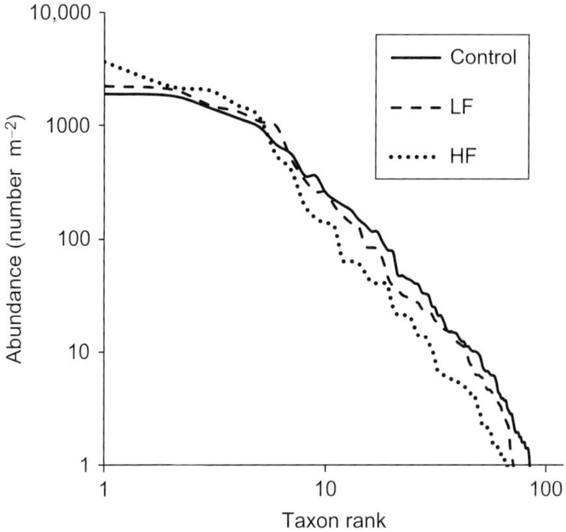

Figure 4 Mean abundance (number m^{-2}) of macroinvertebrates in undisturbed controls (C) and treatments disturbed at high (HF) and low (LF) frequency. For each treatment, taxa were ranked from left to right in order of decreasing abundance.

few taxa (11 core taxa accounted for 93% total numbers), most notably worms and chironomids (Fig. 4, Appendix A). Partial redundancy analyses based on both the numerical (analyses 1–3, Table 2) and relative abundance (analyses 4–6) of constituent taxa showed that community structure varied significantly ($P < 0.001$) with drought treatment, time and their interaction. Ordination revealed that some mayflies (*Caenis luctuosa*, *Ephemera danica* Müller), snails (*Ancylus fluviatilis* Müller, *Potamopyrgus antipodarum*, *Valvata piscinalis* Müller), caddisflies (*Sericostoma personatum* (Spense), *Hydropsyche contubernalis* McLachlan) and beetles (*Limnius volckmari* (Panzer)) were highly susceptible to drought, being significantly less abundant in drought treatments, irrespective of frequency, than in undisturbed controls (Fig. 6A). Resilience varied with disturbance frequency for many other taxa; densities of amphipods (*Gammarus pulex*) and leeches (*Erpobdella octoculata* L.), together with several caddisflies (*Athripsodes cinereus* Curtis, *Lepidostoma hirtum* (Fabricius), *Polycentropus flavomaculatus*, *Tinodes weaneri* L.) and beetles (*Oulimnius tuberculatus* Müller, *Elmis aenea* (Müller)), were sustained under LF disturbance but markedly reduced under HF drought (Fig. 6A).

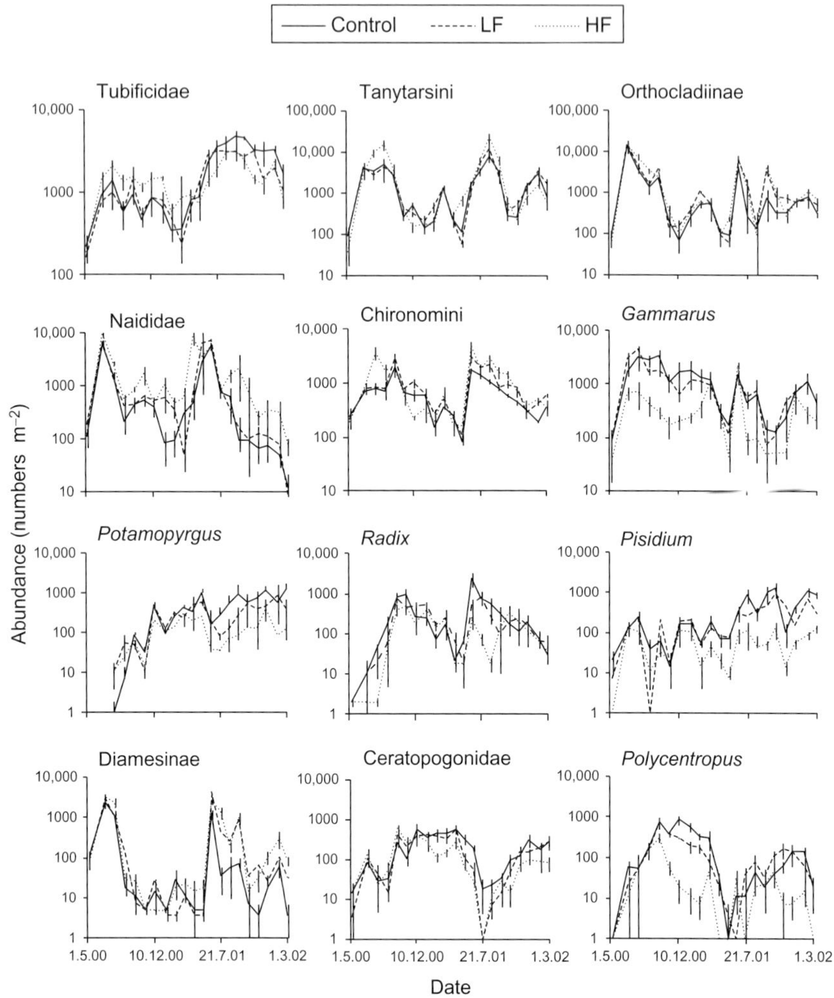

Figure 5 Mean (±1 SE) densities of 12 core taxa in mesocosms disturbed at high frequency (HF), low frequency (LF) and in undisturbed controls (C) over 21 months. Note *y*-axes are scaled to data.

Chironomini midges exploited drying irrespective of frequency, whereas Oligochaeta (*Eiseniella tetredra* (Savigny), Lumbriculidae, Naididae) worms only increased in HF disturbance treatments. Repeated-measures ANOVA revealed that effects on the majority of common taxa depended on the time of sampling (significant treatment × time effect, Table 3). One-way ANOVA with multiple comparisons revealed that for *Pisidium* sp., *R.*

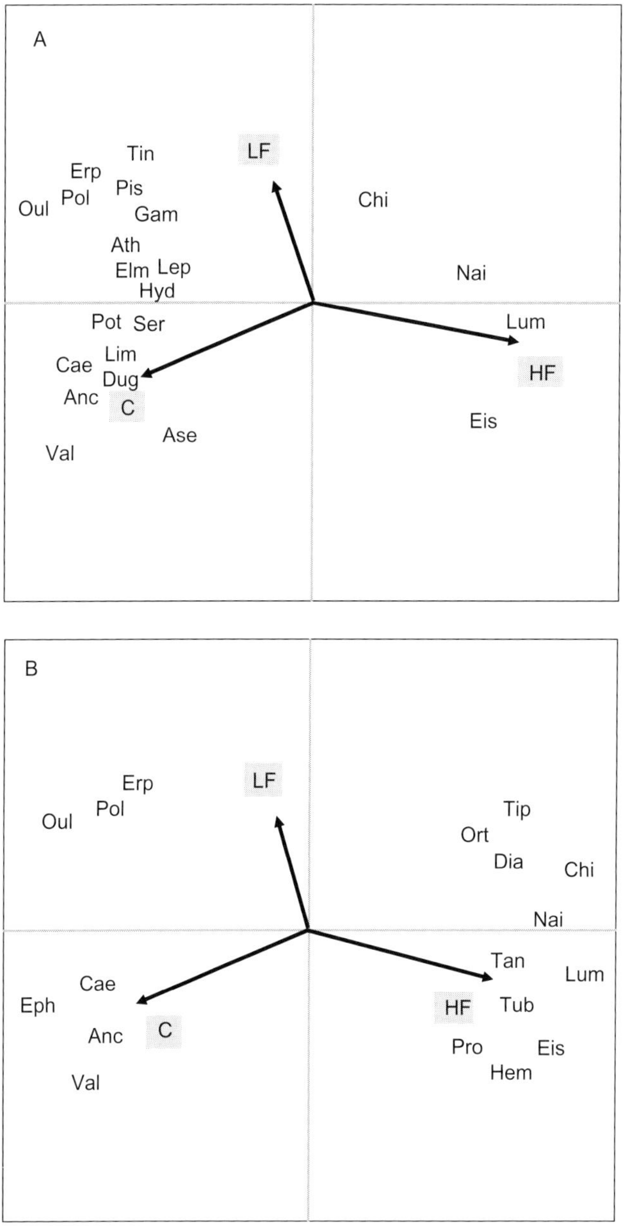

Figure 6 Partial redundancy analysis (RDA) diagram illustrating overall differences in macroinvertebrate community structure among patches disturbed at low (LF) and high (HF) frequency, and undisturbed controls (C). Ordinations were based on either relative (A) or numerical abundance (B) of component species. The direction, and length, of taxa vectors relative to the origin is indicated by abbreviated taxa labels, and reflects the

peregra and *P. flavomaculatus*, densities were lower in HF treatments than in LF treatments or controls at most endpoints (Tukey HSD, $P<0.05$; Fig. 5).

Contrasts in the numerical responses of component species to disturbance led to marked shifts in relative abundance, as revealed by RDA (analyses 4–6, Table 2; Fig. 6B). Thus, LF treatments contained proportionally more fly larvae (some chironomid midges and tipulid craneflies) and fewer mayflies and gastropods than controls, but were otherwise similar to controls in taxonomic composition, whereas HF treatments deviated markedly from controls, being dominated by fly larvae (especially chironomids) and most notably oligochaetes (Naididae, Lumbriculidae Tubificidae) and heavily depleted of mayflies, snails and caddisflies (Fig. 6B).

3.3. Disturbance effects on temporal dynamics

Macroinvertebrate community structure differed markedly among sampling endpoints and seasons (RDA analyses 2 and 5, Table 2) but the occurrence and extent of statistically significant directional change (community development) depended on disturbance frequency (RDA analyses 1–15, Table 4). Linear time-trends reflecting strong turnover in community composition during the experiment were statistically significant ($P=0.001$) for controls and LF disturbance treatments (RDA analyses 1–10, Table 4), where densities of gastropods (*Potamopyrgus antipodarum*), bivalves (*Pisidium* sp.) and oligochaetes (Lumbriculidae, Tubificidae) increased with time as amphipods (*Gammarus pulex*), isopods (*Asellus aquaticus* L.), mayflies (*Caenis luctuosa*) and caddisflies (*Polycentropus flavomaculatus*) declined (Figs. 5 and 7A, B). The percentage of variation attributable to a linear time trend in the two treatments (mean $C=28.6\%$, $LF=20.7\%$) was similar (*t*-test, $P>0.05$). By contrast, RDAs were not statistically significant for HF

---

trend, and extent of increase in, (relative or numerical) abundance among treatments. Taxa in diagrams were those in which >15% species variance was explained by the RDA model. Abbreviations as follows: Anc, *Ancylus fluviatilis*; Ase, *Asellus aquaticus*; Ath, *Athripsodes cinereus*; Cae, *Caenis luctuosa*; Chi, Chironomini; Dia, Diamesinae; Dug, *Dugesia polychroa*; Eis, *Eiseniella tetradra*; Elm, *Elmis aenea*; Eph, *Ephemera danica*; Erp, *Erpobdella octoculata*; Gam, *Gammarus pulex*; Hem, *Hemerodromia*; Hyd c, *Hydropsyche contubernalis*; Lep, *Lepidostoma hirtum*; Lim, *Limnius* sp.; Lum, Lumbriculidae; Nai, Naididae; Ort, Orthocladiinae; Oul, *Oulimnius* sp.; Pis, *Pisidium* sp.; Pol, *Polycentropus flavomaculatus*; Pot, *Potamopyrgus antipodarum*; Pro, *Procleon bifidum*; Ser, *Sericostoma personatum*; Tan, Tanytarsini; Tin, *Tinodes waeneri*; Tip, *Tipula montium*; Tub, Tubificidae; Val, *Valvata piscinalis*.

Table 4 Results of partial redundancy analyses (1–15) examining linear time trends in macroinvertebrate community structure in stream mesocosms, based on the relative abundance of component taxa

Analysis	Treatment	Block	% Var	*r*	*F*	*P*
1	Control	1	34.1	0.90	8.27	0.001
2		2	21.7	0.82	4.44	0.001
3		3	22.7	0.88	4.70	0.001
4		4	35.6	0.85	8.85	0.001
5		1-4	21.8	0.82	21.21	0.001
6	LF	1	23.6	0.84	4.94	0.001
7		2	17.2	0.87	3.33	0.001
8		3	21.8	0.86	4.45	0.001
9		4	20.0	0.83	3.91	0.001
10		1-4	14.6	0.76	13.01	0.001
11	HF	1	8.0	0.56	1.39	0.225
12		2	4.2	0.42	0.70	0.553
13		3	7.3	0.48	1.26	0.271
14		4	5.4	0.37	0.91	0.426
15		1–4	4.2	0.37	3.37	0.016

% Var, percentage of species variability explained by the first ordination axis; *r*, species-environment correlation of the first axis; *F*, *P*, *F*-ratio and corresponding probability value of each Monte Carlo permutation test.

disturbance communities ($P > 0.05$, RDA analyses 11–14), indicating that monthly substratum drying arrested community development (Fig. 7C), and the percentage of variation attributable to a linear time trend in HF channels (mean HF = 6.2%) was significantly lower than in controls or LF treatments (*t*-test, $P > 0.05$).

4. DISCUSSION

Climate change is expected to increase the occurrence of extreme events and change the nature of disturbance regimes across a variety of ecosystems, but direct evidence as to how these effects might be manifested remains scarce (Durance and Ormerod, 2007; Woodward et al., 2010). In the present study, we simulated supraseasonal drought by repeatedly dewatering stream sediments. Such droughts occur across the globe (e.g. Lake, 2008; Lind et al., 2006; Schlief and Mutz, 2011) and are expected to increase in frequency and intensity with climate change in many areas (IPCC, 2007). The results of our mesocosm experiment reveal how disturbance regimes

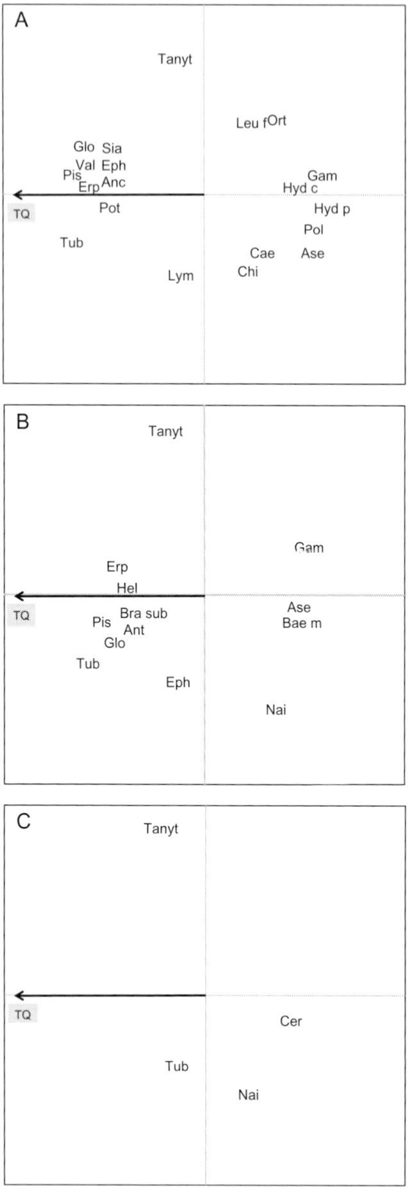

Figure 7 Partial RDA diagram illustrating directional change in macroinvertebrate community structure in undisturbed controls (A) and in patches disturbed at low (B) and high (C) frequency. Axis 1 was constrained by experimental time (days) as a linear trend. The direction, and length, of taxa vectors relative to the origin is indicated by abbreviated taxa labels, and reflects the trend, and extent of increase in, (relative or numerical) abundance through time. Abbreviations as in Figure 6 plus: Ant, *Antocha vitripennis*; Bae m, *Baetis muticus*; Bra sub, *Brachycentrus subnubilis*; Cer, Ceratopogonidae; Glo, *Glossiphonia complanata*; Hel, *Helobdella stagnalis*; Hyd p, *Hydropsyche pellucidula*; Leu, *Leuctra fusca*; Rad, *Radix balthica*; Sia, *Sialis lutaria*.

associated with drought shape ecological assemblages. Macroinvertebrates were relatively resilient to drying disturbances occurring at low frequency (quarterly), sustaining abundant and diverse assemblages which developed over time despite repeated dewatering. These data provide further evidence of the strong capacity for biotic recovery in streams following modest levels of disturbance (Feld et al., 2011; Hladyz et al., 2011; Townsend and Hildrew, 1994), including drought (e.g. Lake, 2003; Ledger and Hildrew, 2001). However, more frequent (monthly) drying eroded this innate resilience and generated relatively impoverished, compositionally static assemblages dominated by a few opportunistic species. Such severe impacts on community structure are likely to alter process rates (e.g. decomposition, herbivory, primary and secondary production), with wider effects on the amenity and conservation value of streams, and their provision of ecological goods (production of fisheries) and services (water quality) (Ledger et al., 2011), but much research remains to be done to elucidate many of these effects (Feld et al., 2011; Friberg et al., 2011).

4.1. Disturbance and diversity

Drying disturbances reduced macroinvertebrate taxon richness at high, but not low, frequency. Our results are thus broadly consistent with the prevailing view that disturbance in streams can reduce diversity (e.g. Death, 2002; Death and Winterbourn, 1995; Matthaei et al., 1996), but that macroinvertebrate taxon richness is rapidly restored by recolonisation (e.g. McCabe and Gotelli, 2000; Townsend and Hildrew, 1994). Although our study tested only two levels of disturbance, the results were not consistent with the unimodal diversity–disturbance relationship predicted by the IDH (Connell, 1978). The most-often cited explanation for deviation from IDH predictions is a lack of trade-off between competitive ability and disturbance sensitivity among the biota (Chesson and Huntly, 1997; dos Santos et al., 2011). Consistent with this, whilst we observed that HF disturbance excluded large rare species and favoured good colonists, the latter were still were present in undisturbed controls and LF disturbance habitats, albeit at lower density.

4.2. Resilience and disturbance frequency

Flow cessation led to patchy dewatering of mesocosm habitats, and resistance to drying among the biota was generally low, with high mortality of macroinvertebrates stranded on desiccating sediments (Harris, 2006).

However, wet interstices and small pools provided refugia for suitably adapted species (Harris, 2006; Ledger et al., 2011), and high resilience to LF disturbance restored endpoint community structure. Although some fauna exploited LF disturbance, especially some chironomids, overall increases were less marked than in HF treatments. Chironomid proliferations may have been constrained by the high resilience of algae to infrequent drought (Ledger et al., 2006): crustose green algae recovered from dewatering, displacing diatoms mats that provide favourable habitat and food (Ledger et al., 2006, 2008). The re-establishment of the community occurred via several routes, including in-channel reproduction by survivors in interstitial sediments (see Burrell and Ledger, 2003), immigration by drift from upstream (Mill Stream) and oviposition by winged adults (Harris, 2006). The year-round presence of colonists above and below the surface water in the channel is thought to buffer stream communities against disturbance (Townsend and Hildrew, 1994). In our experiment, undisturbed upstream source communities (i.e. the parent river) provided a ready supply of recolonists as may be the case in many natural droughts. However, our results could be considered conservative, since drought typically causes extensive drying across the riverscape and recolonist sources are more remote (Ledger et al., 2011).

HF disturbance led to a marked shift in faunal composition, with substantial irruptions in the abundance of chironomid midges and oligochaete worms and declines in amphipods, mayflies and caddisflies, among others. Chironomids are characterised by *r*-selected traits (relative to the more *K*-selected larger macroinvertebrates), including high fecundity, small size and short generation times, which enable rapid exploitation of food, habitat or enemy-free space created by disturbances (Pinder, 1992; Townsend and Hildrew, 1994). Post-disturbance irruptions of chironomid larvae are often observed in streams (e.g. Power et al., 2008) and can be particularly marked in the warm nutrient-rich waters of chalk streams after drought (Wright et al., 2004). Many chironomids live in soft sediments that provide refuge from high temperatures and low oxygen concentrations associated with drought (Armitage et al., 1995), and the proliferation of diatoms under HF drying may have triggered irruptions of these primary consumers (Ledger et al., 2008). Together with chironomids, oligochaetes dominated communities subject to HF disturbance. Faced with disturbance, many oligochaetes increase their reproductive output, some by adopting semelparity to maximize resources invested in reproduction (Bird, 1982). For naidid worms, an immediate (within a day) increase in numbers following disturbances was often observed that we attribute to asexual reproduction, consistent with observations elsewhere (Brinkhurst and Jamieson, 1971), although it is not clear whether this is a response to favourable (e.g. warm temperature, abundant resources) or deleterious (e.g. dessication, low oxygen) environmental cues.

4.3. Resilience and ecosystem functioning

Our study revealed significant variation in species responses to drying, ranging from irruption to extirpation, including those within the same functional groups (Fig. 8), attributable to body size and life-history traits (Ledger et al., 2011). Such variability in responses to environmental change among species that contribute to the same processes is recognised as a key determinant of ecosystem resilience (Elmqvist et al., 2003). Theoretically, where assemblages contain species with a diversity of responses to specific local conditions, as shown here, turnover in species composition when faced with environmental stress may lead to maintenance of ecosystem processes, and in such communities high taxonomic biodiversity can be viewed as conferring a degree of ecological insurance against change (Elmqvist et al., 2003; Mulder et al., 2012). On the other hand, widespread uniformity in responses to change among species that perform the same ecological roles should curtail redundancy, leading to potentially dramatic shifts in key processes as the environment changes (Folke et al., 2004). To date, however, drought research in streams has centred on impacts on community structure, with uncertain consequences for functioning (but see Chadwick and Huryn, 2007; Dewson et al., 2007). As part of the present study, however, Ledger et al. (2011) showed how drought affected the process of macroinvertebrate secondary production and discovered that, while there was clear evidence of contrasting responses in production to drought among biota within some functional groups (grazers and collectors), substantial reductions were far more frequent than no change or increased production, both within and among functional groups, and overall secondary production was more than halved (58% reduction) by frequent (monthly) drying (Ledger et al., 2011). This substantial erosion of macroinvertebrate biomass in the face of drought stress is likely to have wider effects on ecosystem processes such as leaf litter decomposition and herbivory.

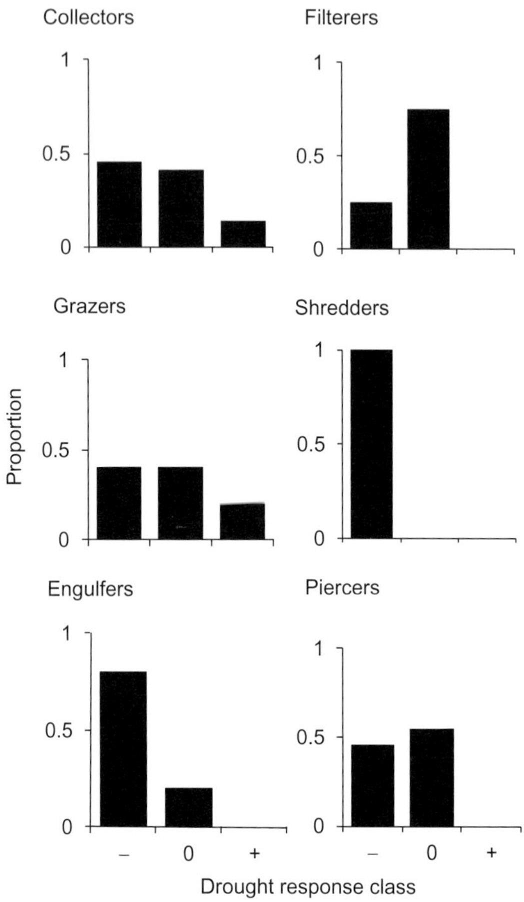

Figure 8 Distribution of high-frequency (monthly) dewatering effects on secondary production of macroinvertebrate taxa in six functional feeding groups (figure redrawn after Ledger et al., 2011). Taxa were classified according to their statistically significant positive (+) negative (–) or lack of (0) response to HF drought, as revealed by one-sample *t*-tests.

Our mesoscosm experiment revealed strong reductions among core taxa known to perform functionally important roles in stream communities. The suppression of potent shredders, most notably amphipods (*Gammarus pulex*) and caddis larvae (*Sericostoma personatum*, Limnephilidae) (see Jonsson and Malmqvist, 2000; Woodward et al., 2008), could reduce leaf litter decomposition rates and increase the accumulation of detritus in drought affected streams (Lake, 2003), potentially causing wider indirect effects on secondary production and the dynamic stability of food webs under donor control (Rooney et al., 2006; Walters and Post, 2011). Many of the most effective grazers, particularly gastropods, declined under drought (e.g. *P. antipodarum*), raising the possibility that diatom proliferations in HF disturbed treatments (see Ledger et al., 2008) was caused by herbivore release, notwithstanding irruptions in herbivorous chironomids. Drought also reduced populations of macroinvertebrate predators, especially those of large size with limited access to refugia such as caddis (e.g. *P. flavomaculatus*), alderflies (*S. lutaria*) and leeches (e.g. *E. octoculata*), and we speculate this may lessen top-down pressure on prey assemblages, such as chironomids, with unknown indirect effects. However, further experiments are needed to establish the wider effects of drought on ecosystem functioning.

4.4. Disturbance and community development

The experiment revealed dynamic, progressive change in community structure in the absence of disturbance. Theoretically, turnover in community composition can be driven by a range of biotic processes including competition and ecosystem engineering, and/or by abiotic processes, such as shifts in climate, flow or habitat (Milner, 1994). In the mesocosms, benthic habitats changed subtly with time, with patches of sand and fine organic sediment developing among coarse gravels and macrophytes (Harris, 2006). Changes in substratum particle size distributions may therefore account for some of the observed turnover, with taxa including caddis (*P. flavomaculatus*) and crustaceans (*G. pulex, A. aquaticus*) associated with gravels replaced gradually by those tolerant of finer sediments, such as molluscs (*P. antipodarum*, *Pisidium* sp.), worms (Lumbriculidae, Tubificidae) and burrowing mayflies (*E. danica*) (Harris, 2006). However, there was also a gradual fall in diatom biomass in the absence of disturbance (Ledger et al., 2008) which we suggest could explain declines in herbivorous mayflies (*C. luctuosa*).

The extent of community development under LF disturbance was similar to that in undisturbed controls, indicating, first, that component taxa were largely resilient to infrequent dewatering, and second, that key habitat features were not permanently changed by the disturbance. By contrast, HF disturbance arrested community development completely by recurrently eroding population abundances and/or by constraining habitat development itself. HF disturbances acted to tightly constrain communities to relatively few tolerant taxa, with traits which enable survival in harsh conditions, as has been reported elsewhere (e.g. Chase, 2007; Lepori and Malmqvist, 2009). However, we found no evidence of progressive erosion of biodiversity or numerical abundance as experimental time elapsed, consistent with the notion that immigration and extinction rates either reached equilibrium or declined to zero, during the experiment (see Death, 2002).

4.5. Drought as an environmental filter

Much debate has centred upon the extent to which community composition is determined by stochastic processes of colonisation and extinction (neutral theory) or by deterministic processes in which species associate with specific ecological conditions (niche theory) (see e.g. Chase, 2007; Thompson and Townsend, 2006). Our data suggest that drought acted as a harsh environmental filter that stripped species lacking resistance and/or resilience traits from the regional pool, consistent with the notion of niche selection in communities (see Chase, 2007; Poff, 1997). We also observed high compositional similarity among the disturbed channels—similar habitats hosted similar communities—with species–habitat associations shaped by traits (Chase, 2007). These largely predictable drought-adapted assemblages consisted of small short-lived taxa (>1 life cycle per year, e.g. chironomids), or those with refuge-seeking or interstitial habits (e.g. ceratopogonids, oligochaetes), whereas many larger taxa with longer life cycles (≤ 1 life cycle per year), including some caddis, mayflies and beetles, were eliminated by frequent dewatering (Ledger et al., 2011; Fig. 9). Community development was also canalised, with turnover in controls corresponding closely to changes in the physical nature of the substrate, and we observed little evidence of stochastic ecological drift at the scale of the channel (Harris, 2006). However, we have shown that stochastic arrival in a colonisation sequence can shape community composition at smaller (patch) spatial scales (Ledger et al., 2006), and it seems likely that both deterministic and stochastic processes influence community development in stream habitats, with species composition shaped by the physical nature of the environment and dispersal (Lepori and Malmqvist, 2009).

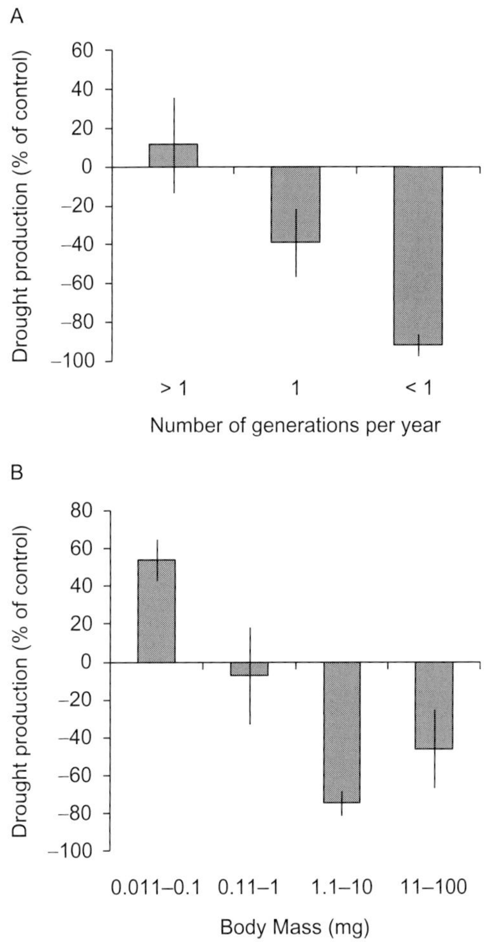

Figure 9 Mean ($\pm$1 SE) effect of high-frequency drought on secondary production of macroinvertebrate taxa in relation to (A) the potential number of life cycles per year and (B) mean individual body mass *(figure redrawn after Ledger et al., 2011).*

5. CONCLUSIONS

Hydrologic drought is a natural phenomenon and an important component of the flow regime in many riverine ecosystems (Boulton, 2003; Lake, 2003). Future shifts in climate are expected to increase the frequency, intensity and extent of drought events in river systems, with potentially devastating effects on benthic ecosystems (Sponseller et al., 2010).

The ecological effects of drought are still relatively poorly understood when compared with other environmental stressors such as flooding or acid episodicity, despite a recent intensification of research effort (e.g. James et al., 2008). In our view, controlled manipulative field experiments are needed to develop a strong mechanistic understanding of the many structural and functional impacts of drought, particularly in relation to biogeochemical processes and dynamics at higher levels of biological organisation (Ledger et al., 2011). In this study, we simulated prolonged drought conditions that caused repeated patchy dewatering of benthic habitat and found that the effects of the stress depend on the frequency with which benthic habitats are exposed to stream bed dewatering. The effects of these harsh low-flow disturbances are likely to be very different from those that do not lead to dewatering, and research is needed to explore the relationships between drought regimes and ecological impacts in rivers to identify resilience thresholds, non-linear responses (i.e. tipping points) and the potential for system recovery, before a more predictive science can emerge (Friberg et al., 2011).

ACKNOWLEDGMENTS

The Freshwater Biological Association (FBA) and the Centre for Ecology and Hydrology Dorset generously supported this research. The project was funded by a FBA/Natural Environmental Research Council (NERC) postdoctoral fellowship to MEL, a tied studentship to RMLH, and NERC grant NER/B/S/2002/00215. Dr Iwan Jones kindly supplied the photograph for Fig. 1. We are grateful to the many people who kindly assisted in the field, with special thanks to Mr Brian Godfrey, Dr Bethan Ledger and Dr John Murphy.

APPENDIX A

Summary (experimental mean and standard error [SE]) of benthic macroinvertebrate numerical (numbers m^{-2}) and relative (% total numbers) abundance in mesocosm patches disturbed at high (LF) and low (HF) frequency, and in undisturbed controls. Taxa within orders are ranked alphabetically.

Taxon	Control				LF drought				HF drought			
	m^{-2}		%		m^{-2}		%		m^{-2}		%	
	Mean	SE	Mean	SE	Mean	SE	Mean	SE	Mean	SE	Mean	SE
Tricladida												
Dendrocoelum lacteum (Müller)	2.8	1.8	< 0.1	< 0.1	3.2	1.9	< 0.1	< 0.1	0.5	0.8	< 0.1	< 0.1
Dugesia polychroa Schmidt	10.0	5.9	< 0.1	< 0.1	3.5	2.5	< 0.1	< 0.1	0.2	0.4	< 0.1	< 0.1
Dugesia tigrina (Girard)	1.9	1.4	< 0.1	< 0.1	0.5	0.6	< 0.1	< 0.1	0	0	0	0
Glossoscolecidae	0	0	0	0	0	0	0	0	0.2	0.4	< 0.1	< 0.1
Planaria torva (Müller)	1.6	1.2	< 0.1	< 0.1	0.7	0.9	< 0.1	< 0.1	0.4	0.5	< 0.1	< 0.1
Polycelis nigra Müller	6.2	3.8	< 0.1	< 0.1	6.2	3.3	< 0.1	< 0.1	2.1	1.8	< 0.1	< 0.1
Oligochaeta												
Eiseniella tetraedra (Savigny)	0.9	0.8	< 0.1	< 0.1	0.9	0.9	< 0.1	< 0.1	4.9	2.1	< 0.1	< 0.1
Lumbriculidae	24.7	13.4	0.3	0.2	27.5	8.1	0.3	0.1	60.8	14.2	0.9	0.3
Naididae	975.9	429.1	6.8	2.4	1318.3	653.3	9.0	2.9	2034.1	680.1	15.8	4.2
Tubificidae	1843.5	378.3	17.6	3.2	1492.5	269.9	15.5	2.5	1505.3	203.3	18.2	2.6
Gastropoda												
Acroloxus lacustris (L.)	0.2	0.4	< 0.1	< 0.1	0	0	0	0	0	0	0	0

Continued

Summary (experimental mean and standard error [SE]) of benthic macroinvertebrate numerical (numbers m^{-2}) and relative (% total numbers) abundance in mesocosm patches disturbed at high (LF) and low (HF) frequency, and in undisturbed controls. Taxa within orders are ranked alphabetically.—cont'd

Taxon	Control				LF drought				HF drought			
	m^{-2}		%		m^{-2}		%		m^{-2}		%	
	Mean	SE	Mean	SE	Mean	SE	Mean	SE	Mean	SE	Mean	SE
Ancylus fluviatilis (Müller)	43.0	25.6	0.4	0.2	12.2	8.8	0.1	<0.1	0.5	0.8	<0.1	<0.1
Bithynia leachii (Sheppard)	0.5	0.6	<0.1	<0.1	0	0	0	0	0	0	0	0
Bithynia tentaculata (L.)	0.4	0.5	<0.1	<0.1	0.4	0.7	<0.1	<0.1	0	0	0	0
Physa fontinalis (L.)	3.7	3.0	<0.1	<0.1	4.2	2.9	<0.1	<0.1	0.2	0.4	<0.1	<0.1
Potamopyrgus antipodarum (J.E. Grey)	561.5	154.6	5.6	1.4	329.7	117.7	4.0	1.4	127.3	40.6	2.0	0.8
Radix balthica (L.)	362.8	154.7	2.9	1.0	254.1	72.8	2.6	0.7	150.9	48.1	2.4	0.9
Theodoxus fluviatilis (L.)	3.9	2.8	<0.1	<0.1	0.7	0.9	<0.1	<0.1	0.2	0.4	<0.1	<0.1
Valvata piscinalis (Müller)	47.6	17.2	0.5	0.2	11.6	8.7	0.1	<0.1	0.9	0.9	<0.1	<0.1
Bivalvia												
Pisidium sp.	360.2	116.3	3.3	1.0	257.1	70.2	2.7	0.8	63.8	17.0	0.8	0.3
Sphaerium sp.	1.9	1.8	<0.1	<0.1	0.4	0.7	<0.1	<0.1	0	0	0	0
Hirudinea												
Erpobdella octoculata (L.)	44.6	13.3	0.4	0.1	31.6	8.6	0.3	<0.1	6.5	2.8	0.1	<0.1
Glossiphonia complanata (L.)	7.4	3.3	<0.1	<0.1	4.6	2.4	<0.1	<0.1	1.9	1.3	<0.1	<0.1
Helobdella stagnalis (L.)	2.8	2.4	<0.1	<0.1	2.6	2.1	<0.1	<0.1	1.4	1.4	<0.1	<0.1
Piscicola geometra (L.)	0.4	0.7	<0.1	<0.1	0.4	0.5	<0.1	<0.1	0.2	0.4	<0.1	<0.1
Trocheta subviridis (Dutrochet)	0	0	0	0	0.2	0.4	<0.1	<0.1	0	0	0	0

Isopoda												
Asellus aquaticus (L.)	115.7	49.0	1.1	0.3	85.5	35.4	1.1	0.5	39.1	11.5	0.5	0.2
Amphipoda												
Crangonyx pseudogracilis Bousfield	15.0	8.1	0.2	<0.1	15.0	6.1	0.3	0.1	16.9	8.5	0.3	0.2
Gammarus pulex (L.)	1165.1	318.9	10.2	2.3	1076.0	268.5	9.7	1.9	406.0	133.9	4.5	1.3
Ephemeroptera												
Alainites muticus (L.)	14.5	7.3	0.1	<0.1	30.3	13.1	0.3	0.2	10.4	4.1	0.2	<0.1
Baetis buceratus Eaton	21.0	13.7	0.1	<0.1	22.9	11.8	0.2	<0.1	25.6	26.3	0.2	0.1
Baetis rhodani (Pictet)	10.4	7.1	<0.1	<0.1	4.6	3.2	<0.1	<0.1	4.2	2.7	<0.1	<0.1
Baetis scambus Eaton	47.3	28.3	0.3	0.2	64.7	28.2	0.4	0.2	50.6	26.3	0.2	0.1
Brachycercus harrisellus Curtis	6.7	5.2	<0.1	<0.1	2.5	2.7	<0.1	<0.1	1.4	1.3	<0.1	<0.1
Caenis horaria (L.)	0.7	0.7	<0.1	<0.1	0	0	0	0	0	0	0	0
Caenis luctuosa Bürmeister	147.9	50.6	1.6	0.5	45.0	13.9	0.5	0.1	21.2	9.0	0.2	<0.1
Caenis rivulorum Eaton	1.4	1.3	<0.1	<0.1	0.2	0.4	<0.1	<0.1	0	0	0	0
Centroptilum luteolum (Müller)	15.0	7.3	0.2	<0.1	10.4	6.6	0.1	<0.1	5.3	3.6	<0.1	<0.1
Ecdyonurus sp.	0.2	0.4	<0.1	<0.1	0	0	0	0	0	0	0	0
Ephemera danica Müller	92.6	21.5	0.9	0.2	28.7	7.4	0.3	<0.1	5.6	2.3	<0.1	<0.1
Heptagenia sulphurea (Müller)	6.3	3.3	<0.1	<0.1	2.6	1.7	<0.1	<0.1	0.5	0.6	<0.1	<0.1
Lasiobaetis atrebatinus Eaton	0.4	0.7	<0.1	<0.1	0.4	0.7	<0.1	<0.1	0	0	0	0
Nigrobaetis niger (L.)	0.5	0.8	<0.1	<0.1	0.4	0.5	<0.1	<0.1	0.5	0.6	<0.1	<0.1
Paraleptophlebia submarginata (Stephens)	1.9	1.5	<0.1	<0.1	0.7	0.7	<0.1	<0.1	0.2	0.4	<0.1	<0.1

Continued

Summary (experimental mean and standard error [SE]) of benthic macroinvertebrate numerical (numbers m^{-2}) and relative (% total numbers) abundance in mesocosm patches disturbed at high (LF) and low (HF) frequency, and in undisturbed controls. Taxa within orders are ranked alphabetically.—cont'd

Taxon	Control				LF drought				HF drought			
	m^{-2}		%		m^{-2}		%		m^{-2}		%	
	Mean	SE	Mean	SE	Mean	SE	Mean	SE	Mean	SE	Mean	SE
Procloeon bifidum Bengtsson	1.4	1.2	<0.1	<0.1	0.5	0.6	<0.1	<0.1	2.3	1.6	<0.1	<0.1
Procloeon pennulatum Eaton	0.2	0.4	<0.1	<0.1	1.1	1.8	<0.1	<0.1	0.7	0.7	<0.1	<0.1
Serratella ignita (Poda)	76.5	52.0	0.3	0.2	80.4	58.1	0.3	0.2	62.2	50.8	0.2	0.2
Plecoptera												
Isoperla grammatica (Poda)	2.1	1.6	<0.1	<0.1	0.5	0.8	<0.1	<0.1	0.7	0.7	<0.1	<0.1
Leuctra fusca (L.)	9.9	5.7	<0.1	<0.1	7.6	4.5	<0.1	<0.1	0.5	0.6	<0.1	<0.1
Leuctra geniculata (Stephens)	21.0	9.9	0.2	0.1	13.4	5.7	0.2	<0.1	4.8	3.4	<0.1	<0.1
Perlodes microcephalus (Pictet)	0	0	0	0	0.5	0.8	<0.1	<0.1	0	0	0	0
Odonata												
Calopteryx splendens (Harris)	0.2	0.4	<0.1	<0.1	0.2	0.4	<0.1	<0.1	0	0	0	0
Cordulegaster boltonii (Donovan)	0.5	0.6	<0.1	<0.1	0.2	0.4	<0.1	<0.1	0.2	0.4	<0.1	<0.1
Hemiptera												
Aphelocheirus aestivalis (Fabricius)	6.9	3.8	<0.1	<0.1	6.3	3.6	<0.1	<0.1	2.3	1.4	<0.1	<0.1

Coleoptera												
Brychius elevatus (Panzer)	13.0	9.9	<0.1	<0.1	4.8	3.4	<0.1	<0.1	3.7	3.4	<0.1	<0.1
Deronectes sp.	0.2	0.4	<0.1	<0.1	0	0	0	0	0.4	0.7	<0.1	<0.1
Elmis aenea (Müller)	33.5	14.8	0.3	0.1	18.9	5.6	0.2	<0.1	6.2	2.4	0.1	<0.1
Haliplus haliplus sp.	0.4	0.5	<0.1	<0.1	0.2	0.4	<0.1	<0.1	0	0	0	0
Haliplus lineatocollis (Marsham)	78.8	43.3	0.6	0.3	23.3	9.9	0.2	<0.1	20.3	12.0	0.1	<0.1
Hydraena riparia Kugelann	0.2	0.4	<0.1	<0.1	0	0	0	0	0	0	0	0
Hydroporinae	0.7	1.4	<0.1	<0.1	0	0	0	0	0	0	0	0
Laccobius sp.	15.9	16.2	<0.1	<0.1	2.1	2.6	<0.1	<0.1	3.2	3.1	<0.1	<0.1
Limnius sp.	40.4	17.8	0.5	0.2	16.7	5.7	0.2	<0.1	6.0	2.6	<0.1	<0.1
Limnius volckmari (Panzer)	4.4	2.5	<0.1	<0.1	5.8	3.7	<0.1	<0.1	1.2	0.9	<0.1	<0.1
Nebrioporus depressus elegans (Fabricius)	6.0	4.9	<0.1	<0.1	4.1	2.9	<0.1	<0.1	9.2	6.0	<0.1	<0.1
Orectochilus villosus (Müller)	3.9	1.8	<0.1	<0.1	2.3	2.0	<0.1	<0.1	1.1	1.2	<0.1	<0.1
Oulimnius sp.	132.2	31.4	1.3	0.4	82.3	19.1	0.9	0.2	20.8	6.8	0.3	<0.1
Oulimnius troglodytes (Gyllenhal)	0.4	0.7	<0.1	<0.1	0	0	0	0	0	0	0	0
Oulimnius tuberculatus (Müller)	1.4	1.2	<0.1	<0.1	0.9	0.9	<0.1	<0.1	0.2	0.4	<0.1	<0.1
Platambus maculatus (L.)	11.1	6.0	<0.1	<0.1	6.0	2.8	<0.1	<0.1	5.8	3.2	<0.1	<0.1
Potamonectes d. elegans (Fabricius)	9.5	6.8	<0.1	<0.1	3.3	2.3	<0.1	<0.1	6.9	3.7	0.1	<0.1
Riolus cupreus (Müller)	0	0	0	0	0.2	0.4	<0.1	<0.1	0	0	0	0
Megaloptera												
Sialis lutaria (L.)	10.0	3.9	<0.1	<0.1	7.6	3.3	<0.1	<0.1	3.0	1.9	<0.1	<0.1

Continued

Summary (experimental mean and standard error [SE]) of benthic macroinvertebrate numerical (numbers m^{-2}) and relative (% total numbers) abundance in mesocosm patches disturbed at high (LF) and low (HF) frequency, and in undisturbed controls. Taxa within orders are ranked alphabetically.—cont'd

Taxon	Control				LF drought				HF drought			
	m^{-2}		%		m^{-2}		%		m^{-2}		%	
	Mean	SE	Mean	SE	Mean	SE	Mean	SE	Mean	SE	Mean	SE
Trichoptera												
Athripsodes albifrons (L.)	0.4	0.5	<0.1	<0.1	0.7	0.7	<0.1	<0.1	0.2	0.4	<0.1	<0.1
Athripsodes cinereus (Curtis)	11.1	4.5	0.1	<0.1	8.5	3.9	<0.1	<0.1	1.2	1.0	<0.1	<0.1
Brachycentrus subnubilus Curtis	3.0	2.0	<0.1	<0.1	5.8	3.5	<0.1	<0.1	5.6	4.6	<0.1	<0.1
Ceraclea dissimilis (Stephens)	0	0	0	0	0.2	0.4	<0.1	<0.1	0.5	0.8	<0.1	<0.1
Drusus annulatus (Stephens)	0.2	0.4	<0.1	<0.1	0	0	0	0	0	0	0	0
Goera pilosa (Fabricius)	1.6	1.6	<0.1	<0.1	0.4	0.5	<0.1	<0.1	0.5	0.6	<0.1	<0.1
Holocentropus sp.	0	0	0	0	0	0	0	0	0.4	0.7	<0.1	<0.1
Hydropsyche contubernalis Mclachlan	32.6	13.8	0.3	0.1	37.6	25.2	0.2	<0.1	12.3	8.1	<0.1	<0.1
Hydropsyche pellucidula (Curtis)	25.4	9.6	0.3	0.1	18.0	6.6	0.2	<0.1	5.1	2.6	<0.1	<0.1
Hydropsyche siltalai Dohler	0.5	0.6	<0.1	<0.1	1.2	1.1	<0.1	<0.1	0.4	0.5	<0.1	<0.1
Hydroptila sp.	8.8	6.1	<0.1	<0.1	15.9	10.8	0.1	<0.1	13.8	10.4	<0.1	<0.1
Ithytrichia sp.	12.3	6.3	0.1	<0.1	15.0	6.4	0.2	<0.1	13.6	7.4	0.2	0.1
Lepidostoma hirtum (Fabricius)	18.7	9.6	0.2	0.1	21.3	21.4	0.2	0.1	1.6	1.1	<0.1	<0.1

Limnephilus lunatus Curtis	0	0	0	0	0.2	0.4	<0.1	<0.1	0	0	0	0
Lype sp.	0.9	1.4	<0.1	<0.1	0	0	0	0	0	0	0	0
Molanna angustata Curtis	0.2	0.4	<0.1	<0.1	0	0	0	0	0.2	0.4	<0.1	<0.1
Mystacides azurea (L.)	0.5	0.8	<0.1	<0.1	0.2	0.4	<0.1	<0.1	0	0	0	0
Oecetis testacea (Curtis)	0.2	0.4	<0.1	<0.1	0	0	0	0	0	0	0	0
Plectrocnemia sp.	1.8	2.9	<0.1	<0.1	0	0	0	0	2.1	3.6	<0.1	<0.1
Polycentropus flavomaculatus (Pictet)	196.1	64.7	2.1	0.7	135.2	46.3	1.5	0.4	39.7	19.0	0.3	0.1
Potamophylax latipennis (Curtis)	0.5	0.6	<0.1	<0.1	0.4	0.5	<0.1	<0.1	0.5	0.6	<0.1	<0.1
Psychomyia pusilla (Fabricius)	4.2	2.7	<0.1	<0.1	6.2	3.7	<0.1	<0.1	1.4	1.1	<0.1	<0.1
Rhyacophila dorsalis (Curtis)	3.2	2.9	<0.1	<0.1	0.9	0.9	<0.1	<0.1	0	0	0	0
Sericostoma personatum (Spence)	14.8	6.7	0.2	<0.1	10.9	8.0	0.1	<0.1	1.9	1.4	<0.1	<0.1
Tinodes waeneri (L.)	115.1	51.0	1.3	0.6	128.2	58.7	1.4	0.6	13.9	6.0	0.2	<0.1
Ylodes conspersus (Rambur)	0.2	0.4	<0.1	<0.1	0	0	0	0	0.2	0.4	<0.1	<0.1
Diptera												
Antocha vitripennis (Meigen)	6.7	3.7	<0.1	<0.1	11.3	5.4	0.1	<0.1	1.4	1.9	<0.1	<0.1
Atrichopogon sp.	0	0	0	0	0.2	0.4	<0.1	<0.1	0	0	0	0
Ceratopogonidae	218.4	54.5	3.1	1.0	187.9	47.5	2.9	1.0	135.4	39.7	2.1	0.6
Chelifera sp.	0.5	0.6	<0.1	<0.1	0	0	0	0	1.2	1.2	<0.1	<0.1
Chironomini	677.2	120.4	6.4	1.0	971.5	207.4	9.0	1.3	1209.0	316.1	10.1	1.6
Chrysops sp.	0	0	0	0	0.4	0.5	<0.1	<0.1	0	0	0	0
Clinocera sp.	3.7	3.6	<0.1	<0.1	4.4	2.5	<0.1	<0.1	4.2	2.6	<0.1	<0.1
Diamesinae (*Potthastia*)	253.2	152.4	1.6	0.7	493.8	272.2	2.3	0.9	550.8	234.2	2.9	0.9
Dicranota sp.	1.8	1.5	<0.1	<0.1	0.9	0.8	<0.1	<0.1	0.5	0.6	<0.1	<0.1

Continued

Summary (experimental mean and standard error [SE]) of benthic macroinvertebrate numerical (numbers m^{-2}) and relative (% total numbers) abundance in mesocosm patches disturbed at high (LF) and low (HF) frequency, and in undisturbed controls. Taxa within orders are ranked alphabetically.—cont'd

Taxon	Control				LF drought				HF drought			
	m^{-2}		%		m^{-2}		%		m^{-2}		%	
	Mean	SE	Mean	SE	Mean	SE	Mean	SE	Mean	SE	Mean	SE
Dolichopodidae	0.2	0.4	<0.1	<0.1	0.2	0.4	<0.1	<0.1	0.5	0.8	<0.1	<0.1
Elaeophila sp.	12.2	5.9	0.1	<0.1	15.3	6.8	0.1	<0.1	5.3	2.9	<0.1	<0.1
Ephydridae	0	0	0	0	0.4	0.7	<0.1	<0.1	0	0	0	0
Gonomyia sp.	0	0	0	0	0	0	0	0	0.2	0.4	<0.1	<0.1
Hemerodromia sp.	2.5	1.7	<0.1	<0.1	1.4	1.3	<0.1	<0.1	4.2	2.0	<0.1	<0.1
Ibrisia marginata (Fabricius)	0.2	0.4	<0.1	<0.1	0.2	0.4	<0.1	<0.1	0.5	0.6	<0.1	<0.1
Limnophora sp.	1.6	1.3	<0.1	<0.1	0.7	0.9	<0.1	<0.1	0.9	0.9	<0.1	<0.1
Limonia sp.	0.2	0.4	<0.1	<0.1	0	0	0	0	0	0	0	0
Orthocladiinae	1435.0	746.0	7.8	2.0	2069.3	815.4	11.7	2.3	2206.0	821.2	12.9	2.4
Oxycera trilineata (Fabricius)	0	0	0	0	0	0	0	0	0.2	0.4	<0.1	<0.1
Pericoma trivialis Eaton	1.2	1.1	<0.1	<0.1	0.7	0.9	<0.1	<0.1	0.7	0.9	<0.1	<0.1
Pilaria sp.	0	0	0	0	0.2	0.4	<0.1	<0.1	1.1	1.0	<0.1	<0.1
Prodiamesinae	39.3	22.5	0.3	0.1	34.9	17.0	0.2	<0.1	41.4	16.3	0.3	0.1
Simuliidae	31.4	15.0	0.3	0.2	26.8	8.3	0.4	0.2	21.0	7.2	0.4	0.3
Tabanus sp.	5.3	2.2	<0.1	<0.1	3.7	2.7	<0.1	<0.1	4.6	3.3	<0.1	<0.1
Tanypodinae	177.9	47.6	1.5	0.4	155.9	43.7	1.3	0.3	199.1	59.7	1.5	0.3
Tanytarsini	1917.4	545.2	13.0	2.7	2212.2	796.7	15.4	3.3	3639.2	1422.1	19.0	4.3
Tipula montium Egger	6.2	2.7	<0.1	<0.1	23.8	9.6	0.2	<0.1	39.5	19.5	0.4	0.2
Tipula sp.	0.5	1.1	<0.1	<0.1	0	0	0	0	0.5	0.8	<0.1	<0.1
Total abundance (m^{-2})	12,093	1618	100		12,235	1972	100		12,958	2434	100	

APPENDIX B

Tables for repeated-measures ANOVA testing the main effects of drought treatment, experimental block (between-subject factors) and time (within-subject factor), and their interactions, on numerical abundance (m^{-2}) of the most numerically abundant core taxa (a–l).

(a) *Gammarus pulex*

	SS	df	MS	F	P	Post hoc
Within-subjects						
Time	433.30	16	27.27	21.07	<0.0005	
Time × drought	65.64	32	2.07	1.60	0.043	
Time × block	89.71	48	1.88	1.45	0.061	
Error (time)	123.38	96	1.29			
Between-subjects						
Drought	64.21	2	32.10	17.01	0.003	C, LF > HF
Block	85.83	3	28.61	15.16	0.003	
Error	11.33	6	1.89			

(b) Tubificidae

	SS	df	MS	F	P
Within-subjects					
Time	159.12	20	7.96	37.97	<0.0005
Time × drought	21.68	40	0.54	2.59	<0.0005
Time × block	18.76	60	0.31	1.49	0.195
Error (time)	25.15	120	0.21		
Between-subjects					
Drought	1.18	2	0.59	0.54	0.608
Block	12.91	3	4.30	3.96	0.072
Error	6.53	6	1.09		

(c) Chironomini

	SS	df	MS	F	P
Within-subjects					
Time	186.38	20	9.32	44.94	<0.0005
Time × drought	22.97	40	0.57	2.77	<0.0005
Time × block	18.30	60	0.31	1.47	0.038
Error (time)	24.88	120	0.21		

Continued

Tables for repeated-measures ANOVA testing the main effects of drought treatment, experimental block (between-subject factors) and time (within-subject factor), and their interactions, on numerical abundance (m^{-2}) of the most numerically abundant core taxa (a–l).—cont'd

Between-subjects					
Drought	4.56	2	2.28	8.54	0.018
Block	3.98	3	1.33	4.97	0.046
Error		6			

(d) Tanytarsini

	SS	df	MS	*F*	*P*
Within-subjects					
Time	551.51	20	27.58	52.04	<0.0005
Time × drought	43.85	40	1.10	2.07	0.001
Time × block	27.30	60	0.46	0.86	0.742
Error (time)	63.59	120	0.53		
Between-subjects					
Drought	1.08	2	0.54	0.62	0.571
Block	5.46	3	1.82	2.08	0.204
Error					

(e) Orthocladiinae

	SS	df	MS	*F*	*P*
Within-subjects					
Time	859.70	12	71.64	30.72	<0.0005
Time × drought	113.16	24	4.72	2.02	0.012
Time × block	115.51	36	3.21	1.38	0.125
Error (time)	167.92	72	2.33		
Between-subjects					
Drought	21.68	2	10.84	6.60	0.031
Block	19.64	3	6.55	3.98	0.071
Error					

(f) Naididae

	SS	df	MS	*F*	*P*
Within-subjects					
Time	743.91	20	37.20	27.15	<0.0005
Time × drought	82.73	40	2.07	1.51	0.046
Time × block	159.69	60	2.66	1.94	0.001
Error (time)	164.41	120	1.37		

Tables for repeated-measures ANOVA testing the main effects of drought treatment, experimental block (between-subject factors) and time (within-subject factor), and their interactions, on numerical abundance (m^{-2}) of the most numerically abundant core taxa (a–l).—cont'd

Between-subjects					
Drought	77.23	2	38.61	6.48	0.032
Block	12.87	3	4.29	0.72	0.575
Error					

(g) *Potamopyrgus antipodarum*

	SS	df	MS	*F*	*P*
Within-subjects					
Time	430.30	19	22.78	15.49	<0.0005
Time × drought	98.19	38	2.60	1.77	0.011
Time × block	120.62	57	2.13	1.45	0.049
Error (time)	166.65	113	1.47		
Between-subjects					
Drought	112.18	2	56.09	1.81	0.243
Block	16.06	3	5.35	0.17	0.911
Error	185.89	6	30.98		

(h) *Radix balthica*

	SS	df	MS	*F*	*P*
Within-subjects					
Time	728.65	20	36.43	27.27	<0.0005
Time × drought	122.06	40	3.05	2.28	<0.0005
Time × block	138.34	60	2.31	1.73	0.006
Error (time)	160.32	120	1.34		
Between-subjects					
Drought	41.60	2	20.80	10.96	0.010
Block	101.96	3	33.99	17.90	0.002
Error	11.39	6	1.90		

(i) *Pisidium* sp.

	SS	df	MS	*F*	*P*
Within-subjects					
Time	483.63	20	24.18	16.86	<0.0005
Time × drought	127.09	40	3.18	2.22	<0.0005

Continued

Tables for repeated-measures ANOVA testing the main effects of drought treatment, experimental block (between-subject factors) and time (within-subject factor), and their interactions, on numerical abundance (m^{-2}) of the most numerically abundant core taxa (a–l).—cont'd

Time × block	83.06	60	1.38	0.97	0.553
Error (time)	172.13	120	1.43		
Between-subjects					
Drought	117.38	2	58.67	6.19	0.035
Block	105.01	3	35.00	3.69	0.081
Error		6			

(j) Diamesinae

	SS	df	MS	*F*	*P*
Within-subjects					
Time	1367.90	20	68.40	40.67	<0.0005
Time × drought	141.94	40	3.55	2.11	0.001
Time × block	182.58	60	3.04	1.81	0.003
Error (time)	201.83	120	1.68		
Between-subjects					
Drought	73.84	2	36.92	14.73	0.005
Block	27.47	3	9.16	3.65	0.083
Error	15.04	6	2.51		

(k) Ceratopogonidae

	SS	df	MS	*F*	*P*	*Post hoc*
Within-subjects						
Time	639.25	20	31.96	23.40	<0.0005	
Time × drought	67.21	40	1.68	1.23	0.196	
Time × block	144.76	60	2.41	1.77	0.004	
Error (time)	163.92	120	1.37			
Between-subjects						
Drought	21.33	2	10.67	41.81	<0.0005	C > LF, HF
Block	7.76	3	2.59	10.14	0.009	
Error	1.53	6	0.26			

(l) *Polycentropus flavomaculatus*

	SS	df	MS	*F*	*P*
Within-subjects					
Time	651.44	20	32.57	17.03	<0.0005
Time × drought	171.88	40	4.30	2.25	<0.0005
Time × block	113.82	60	1.90	0.99	0.504

Error (time)	229.49	120	1.91		
Between-subjects					
Drought	178.33	2	89.17	19.01	0.003
Block	11.69	3	3.90	0.83	0.524
Error	28.14	6	4.69		

REFERENCES

Acuña, V., Tockner, K., 2010. The effects of alterations in temperature and flow regime on organic carbon dynamics in Mediterranean river networks. Global Change Biol. 16, 2638–2650.

Armesto, J.J., Pickett, S.T.A., 1985. Experiments on disturbance in old-field plant communities: impacts on species richness and abundance. Ecology 66, 230–240.

Armitage, P.D., Cranston, P.S., Pinder, L.C.V. (Eds.), 1995. The Chironomidae—Biology and Ecology of Non-Biting Midges. Chapman and Hall, London.

Beche, L.A., Connors, P.G., Resh, V.H., Merenlender, A.M., 2009. Resilience of fishes and invertebrates to prolonged drought in two California streams. Ecography 32, 778–788.

Beniston, M., Stephenson, D.B., Christensen, O.B., Ferro, C.A.T., Frei, C., Goyette, S., Halsnaes, K., Holt, T., Jylhä, K., Koffi, B., Palutikof, J., Schöll, R., Semmler, T., Woth, K., 2007. Future extreme events in European climate: an exploration of regional climate model projections. Climate Change 81, 71–95.

Bird, G.J., 1982. The Ecology of Oligochaetes in Dorset Chalk Streams. PhD thesis.University of Reading, UK.

Bonada, N., Dolédec, S., Statzner, B., 2007. Taxonomic and biological trait differences of stream macroinvertebrate communities between mediterranean and temperate regions: implications for future climatic scenarios. Global Change Biol. 13, 1658–1671.

Boulton, A.J., 2003. Parallels and contrasts in the effects of drought on stream macroinvertebrate assemblages. Freshw. Biol. 48, 1173–1185.

Bradley, D.C., Ormerod, S.J., 2001. Community persistence among stream invertebrates tracks the North Atlantic Oscillation. J. Anim. Ecol. 70, 987–996.

Brinkhurst, R.O., Jamieson, B.G.M., 1971. Aquatic Oligochaeta of the World. Oliver and Boyd, Edinburgh.

Brown, J.H., Gilooly, J.F., Allen, A.P., Savage, V.M., West, G.B., 2004. Toward a metabolic theory of ecology. Ecol. Lett. 85, 1173–1185.

Brown, L.E., Edwards, F.K., Milner, A.M., Woodward, G., Ledger, M.E., 2011. Food web complexity and allometric scaling relationships in stream mesocosms: implications for experimentation. J. Anim. Ecol. 80, 884–895.

Burrell, G.P., Ledger, M.E., 2003. Growth of a stream-dwelling caddisfly (*Olinga feredayi*: Conosucidae) on surface and hyporheic food resources. J. N. Am. Benthol. Soc. 22, 92–104.

Chadwick, M.A., Huryn, A.D., 2007. Role of habitat in determining macroinvertebrate production in an intermittent-stream system. Freshw. Biol. 52, 240–251.

Chapin III, F.S., Lovecraft, A.L., Zavaleta, E.S., Nelson, J., Robards, M.D., Kofinas, G.P., Trainor, S.F., Peterson, G.D., Huntingdon, H.P., Naylor, R.L., 2002. Principles of Terrestrial Ecosystem Ecology. Springer-Verlag, New York, USA.

Chase, J.M., 2007. Drought mediates the importance of stochastic community assembly. Proc. Natl. Acad. Sci. USA 104, 17430–17434.

Chessman, B.C., 2009. Climatic changes and 13-year trends in stream macroinvertebrate assemblages in New South Wales, Australia. Global Change Biol. 15, 2791–2802.

Chesson, P., Huntly, N., 1997. The roles of harsh and fluctuating conditions in the dynamics of ecological communities. Am. Nat. 150, 519–553.

Connell, J.H., 1978. Diversity of tropical rainforests and coral reefs. Science 199, 1302–1310.

Daufresne, M., Boët, P., 2007. Climate change impacts on structure and diversity of fish communities in rivers. Global Change Biol. 13, 2467–2478.

Daufresne, M., Bady, P., Fruget, J.F., 2007. Impacts of global changes and extreme hydroclimatic events on macroinvertebrate community structures in the French Rhône River. Oecologia 151, 544–559.

Death, R.G., 2002. Predicting invertebrate diversity from disturbance regimes in forest streams. Oikos 97, 18–30.

Death, R.G., Winterbourn, M.J., 1995. Diversity patterns in stream benthic invertebrate communities: the influence of habitat stability. Ecology 76, 1446–1460.

Dewson, Z.S., James, A.B.W., Death, R.G., 2007. Stream ecosystem functioning under reduced flow conditions. Ecol. Appl. 17, 1797–1808.

dos Santos, F.A.S., Johst, K., Grimm, V., 2011. Neutral communities may lead to decreasing diversity-disturbance relationships: insights from a generic simulation model. Ecol. Lett. 14, 653–660.

Dossena, M., Grey, J., Montoya, J.M., Perkins, D.M., Trimmer, M., Woodward, G., Yvon-Durocher, G., 2012. Warming alters community size structure and ecosystem functioning. Proc. R. Soc. Lond. B. http://dx.doi.org/10.1098/rspb.2012.0394.

Durance, I., Ormerod, S.J., 2007. Climate change effects on upland stream macroinvertebrates over a 25-year period. Global Change Biol. 13, 942–957.

Easterling, D.R., Meehl, G.A., Parmesan, C., Changnon, S.A., Karl, T.R., Mearns, L.O., 2003. Climate extremes: observations, modelling, and impacts. Science 289, 2068–2074.

Elmqvist, T., Folke, C., Nyström, M., Peterson, G., Bengtsson, J., Walker, B., Norberg, J., 2003. Response diversity and ecosystem resilience. Front. Ecol. Environ. 1, 488–494.

Feld, C.K., Birk, S., Bradley, D.C., Hering, D., Kail, J., Marzin, A., Melcher, A., Nemitz, D., Pedersen, M.L., Pletterbauer, F., Pont, D., Verdonschot, P.F.M., et al., 2011. From natural to degraded rivers and back again: a test of restoration ecology theory and practice. Adv. Ecol. Res. 44, 119–210.

Folke, C., Carpenter, S., Walker, B., Scheffer, M., Elmqvist, T., Gunderson, L., Holling, C.S., 2004. Regime shifts, resilience, and biodiversity in ecosystem management. Ann. Rev. Ecol. Syst. 35, 557–581.

Fox, J.W., 1985. Plant diversity in relation to plant production and disturbance by voles in Alaskan tundra communities. Arctic Alpine Res. 17, 199–204.

Friberg, N., Bonada, N., Bradley, D.C., Dunbar, M.J., Edwards, F.K., Grey, J., Hayes, R.B., Hildrew, A.G., Lamouroux, N., Trimmer, M., Woodward, G., 2011. Biomonitoring of human impacts in freshwater ecosystems: the good, the bad, and the ugly. Adv. Ecol. Res. 44, 1–68.

Gurvich, D.E., Diaz, S., Falczuk, V., Perez-Harguindeguy, N., Cabido, M., Christopher Thorpe, P., 2002. Foliar resistance to simulated extreme temperature events in contrasting plant functional and chorological types. Global Change Biol. 8, 1139–1145.

Hagen, M., Kissling, W.D., Rasmussen, C., de Aguiar, M.A.M., Brown, L.E., Carstensen, D.W., Alves-dos-Santos, I., Dupont, Y.L., Edwards, F.K., Genini, J., Guimarães, P.R., Jenkins, G.B., Jordano, P., Kaiser-Bunbury, C.N., Ledger, M., Maia, K.M., Marquitti, F.M.D., McLaughlin, O., Morellato, L.P.C., O'Gorman, E., Trøjelsgaard, K., Tylianakis, J.M., Vidal, M.M., Woodward, G., Olesen, J.M., 2012. Biodiversity, species interactions and ecological networks in a fragmented world. Adv. Ecol. Res. 46, 89–210.

Harris, R.M.L., 2006. The Effect of Experimental Drought Disturbance on Macroinvertebrate Assemblages in Stream Mesocosms.. Ph.D. Thesis. University of Birmingham, UK.

Harris, R.M.L., Milner, A.M., Armitage, P.D., Ledger, M.E., 2007. Replicability of physicochemistry and macroinvertebrate assemblages in stream mesocosms: implications for experimental research. Freshw. Biol. 52, 2434–2443.

Hladyz, S., Åbjörnsson, K., Cariss, H., Chauvet, E., Dobson, M., Elosegi, A., Ferreira, V., Fleituch, T., Gessner, M.O., Giller, P.S., Graça, M.A.S., Gulis, V., Hutton, S.A., Lacoursière, J., Lamothe, S., Lecerf, A., Malmqvist, B., McKie, B.G., Nistorescu, M., Preda, E., Riipinen, M.P., Rîşnoveanu, G., Schindler, M., Tiegs, S.D., Vought, L.B.M., Woodward, G., 2011. Stream ecosystem functioning in an agricultural landscape: the importance of terrestrial-aquatic linkages. Adv. Ecol. Res. 44, 211–276.

Huston, M.A., 1979. A general hypothesis of species diversity. Am. Nat. 113, 81–99.

IPCC, 2007. Contribution of Working Group II to the Fourth Assessment. Report of the Intergovernmental Panel on Climate Change. In: Parry, M.L., Canziani, O.F., Palutikof, J. P., van der Linden, P.J., Hanson, C.E. (Eds.), Climate Change 2007: Impacts, Adaptation and Vulnerability. Cambridge University Press, Cambridge, UK, p. 976.

James, A.B.W., Dewson, Z.S., Death, R.G., 2008. Do stream macroinvertebrates use instream refugia in response to severe short-term flow reduction in New Zealand streams? Freshw. Biol. 53, 1316–1334.

Jentsch, A., Kreyling, J., Beierkuhnlein, C., 2007. A new generation of climate-change experiments: events, not trends. Front. Ecol. Environ. 5, 365–374.

Jonsson, M., Malmqvist, B., 2000. Ecosystem process rate increases with animal species richness: evidence from leaf-eating, aquatic insects. Oikos 89, 519–523.

Lake, P.S., 2003. Ecological effects of perturbation by drought in flowing water. Freshw. Biol. 48, 1161–1172.

Lake, P.S., 2008. Drought, the "creeping disaster". Effects on Aquatic Ecosystems. Land and Water Australia, Canberra, Australia.

Layer, K., Riede, J.O., Hildrew, A.G., Woodward, G., 2010. Food web structure and stability in 20 streams across a wide pH gradient. Adv. Ecol. Res. 42, 265–301.

Layer, K., Hildrew, A.G., Jenkins, G.B., Riede, J., Rossiter, S.J., Townsend, C.R., Woodward, G., 2011. Long-term dynamics of a well-characterised food web: four decades of acidification and recovery in the Broadstone Stream model system. Adv. Ecol. Res. 44, 69–117.

Ledger, M.E., Hildrew, A.G., 2001. Recolonization by the benthos of an acid stream following a drought. Arch. Hydrobiol. 152, 1–17.

Ledger, M.E., Winterbourn, M.J., 2000. Growth of New Zealand stream insect larvae in relation to food type. Arch. Hydrobiol. 149, 353–364.

Ledger, M.E., Crowe, A.L.M., Woodward, G., Winterbourn, M.J., 2002. Is the mobility of stream insects related to their diet? Arch. Hydrobiol. 154, 41–59.

Ledger, M.E., Harris, R.M.L., Milner, A.M., Armitage, P.D., 2006. Disturbance, biological legacies and community development in stream mesocosms. Oecologia 148, 682–691.

Ledger, M.E., Harris, R.M.L., Armitage, P.D., Milner, A.M., 2008. Disturbance frequency influences patch dynamics in stream benthic algal communities. Oecologia 155, 809–819.

Ledger, M.E., Harris, R.M.L., Armitage, P.D., Milner, A.M., 2009. Realism of model ecosystems: an evaluation of physicochemistry and macroinvertebrate assemblages in artificial streams. Hydrobiologia 617, 91–99.

Ledger, M.E., Edwards, F.K., Brown, L.E., Milner, A.M., Woodward, G., 2011. Impact of simulated drought on ecosystem biomass production: an experimental test in stream mesocosms. Global Change Biol. 17, 2288–2297.

Lepori, F., Malmqvist, B., 2009. Deterministic control on community assembly peaks at intermediate levels of disturbance. Oikos 118, 471–479.
Lepš, J., Šmilauer, P., 2003. Multivariate Analysis of Ecological Data Using CANOCO. Cambridge University Press, Cambridge.
Lind, P.R., Robson, B.J., Mitchell, B.D., 2006. The influence of reduced flow during a drought on patterns of variation in macroinvertebrate assemblages across a spatial hierarchy in two lowland rivers. Freshw. Biol. 51, 2282–2295.
Lubchenco, J., 1978. Plant species diversity in a marine intertidal community: important of herbivore food preference and algal competitive abilities. Am. Nat. 112, 23–39.
Mackey, R., Currie, D., 2001. The diversity-disturbance relationship: is it generally strong and peaked? Ecology 82, 3479–3492.
Magurran, A.E., 2004. Measuring Biological Diversity. Blackwell Science, Oxford, UK 256 pp.
Matthaei, C.D., Uehlinger, U., Meyer, E.I., Frutiger, A., 1996. Recolonization by benthic invertebrates after experimental disturbance in a Swiss prealpine river. Freshw. Biol. 35, 233–248.
McCabe, D.J., Gotelli, N.J., 2000. Effects of disturbance frequency, intensity, and area on assemblages of stream macroinvertebrates. Oecologia 124, 270–279.
Meerhoff, M., Mello, F.T-d., Kruk, C., Alonso, C., González-Bergonzoni, I., Pablo Pacheco, J., Lacerot, G., Arim, M., Beklioğlu, M., Brucet, S., Goyenola, G., Iglesias, C., Mazzeo, N., Kosten, S., Jeppesen, E., 2012. Environmental warming in shallow lakes: a review of potential changes in community structure as evidenced from space-for-time substitution approaches. Adv. Ecol. Res. 46, 259–350.
Miller, A.D., Roxburgh, S.H., Shea, K., 2011. How frequency and intensity shape diversity-disturbance relationships. Proc. Natl. Acad. Sci. USA 108, 5643–5648.
Milner, A.M., 1994. System recovery. In: Calow, P., Petts, G.E. (Eds.), The Rivers Handbook: Hydrological and Ecological Principles, vol. 2. Blackwell, Oxford, UK, pp. 76–98.
Mintenbeck, K., Barrera-Oro, E.R., Brey, T., Jacob, U., Knust, R., Mark, F.C., Moreira, E., Strobel, A., Arntz, W.E., 2012. Impact of Climate Change on Fishes in Complex Antarctic Ecosystems. Adv. Ecol. Res. 46, 351–426.
Mulder, Christian, Boit, Alice, Shigeta Mori, J., Vonk, Arie, Dyer, Scott D., Faggiano, Leslie, Geisen, Stefan, González, Angélica L., Kaspari, Michael, Lavorel, Sandra, Marquet, Pablo A., Rossberg, Axel G., Sterner, Robert W., Voigt, Winfried, Wall, Diana H., 2012. Distributional (in)congruence of biodiversity-ecosystem functioning. Adv. Ecol. Res. 46, 1–88.
Olesen, J.M., Dupont, Y.L., O'Gorman, E., Ings, T.C., Layer, K., Melian, C.J., Troejelsgaard, K., Pichler, D.E., Rasmussen, C., Woodward, G., 2010. From Broadstone to Zackenberg: space, time and hierarchies in ecological networks. Adv. Ecol. Res. 42, 1–71.
Perkins, D.M., McKie, B.G., Malmqvist, B., Gilmour, S.G., Reiss, J., Woodward, G., 2010. Environmental warming and biodiversity-ecosystem functioning in freshwater microcosms: partitioning the effects of species identity, richness and metabolism. Adv. Ecol. Res. 43, 177–209.
Petraitis, P.S., Latham, R.E., Niesenbaum, R.A., 1989. The maintenance of species diversity by disturbance. Q. Rev. Biol. 64, 393–418.
Pianka, E.R., 1983. On r and K selection. Am. Nat. 104, 592–597.
Pickett, S.T.A., White, P.S. (Eds.), 1985. The Ecology of Natural Disturbance and Patch Dynamics. Academic Press, London.
Pimm, S.L., 1984. The complexity and stability of ecosystems. Nature 307, 321–326.
Pinder, L.C.V., 1992. Biology on epiphytic Chironomidae (Diptera: Nematocera) in chalk streams. Hydrobiologia 248, 39–51.

Poff, N.L., 1997. Landscape filters an species traits: towards mechanistic understanding and prediction in stream ecology. J. N. Am. Benthol. Soc. 16, 391–409.
Poff, N.L., Zimmerman, J.K.H., 2010. Ecological responses to altered flow regimes: a literature review to inform the science and management of environmental flows. Freshw. Biol. 55, 194–205.
Power, M.E., Parker, M.S., Dietrich, W.E., 2008. Seasonal reassembly of a river food web: floods, droughts, and impacts of fish. Ecol. Monogr. 78, 263–282.
Ptacnik, R., Moorthi, S.D., Hillebrand, H., 2010. Hutchinson reversed, or why there need to be so many species. Adv. Ecol. Res. 43, 1–44.
Rice, W.R., 1989. Analyzing tables of statistical tests. Evolution 43, 223–225.
Rooney, N., McKann, K., Gellner, G., Moore, J.C., 2006. Structural asymmetry and the stability of diverse food webs. Nature 442, 265–269.
Schindler, D.W., Donahue, W.F., 2006. An impending water crisis in Canada's western prairie provinces. Proc. Natl. Acad. Sci. USA 103, 7210–7216.
Schlief, J., Mutz, M., 2011. Leaf decay processes during and after a supra-seasonal hydrological drought in a temperate lowland stream. Int. Rev. Hydrobiol. 96, 633–655.
Sousa, W.P., 1979. Disturbance in marine intertidal boulder fields: the nonequilibrium maintenance of species diversity. Ecology 60, 1225–1239.
Sponseller, R.A., Grimm, N.B., Boulton, A.J., Sabo, J.L., 2010. Responses of macroinvertebrate communities to long-term flow variability in a Sonoran Desert stream. Global Change Biol. 16, 2891–2900.
Ter Braak, C.J.F., Šmilauer, P., 2002. CANOCO Reference Manual and CanoDraw for Windows User Guide: Software for Canonical Community Ordination (version 4.5). Microcomputer Power, Ithaca, NY.
Thompson, R., Townsend, C., 2006. A truce with neutral theory: local deterministic factors, species traits and dispersal limitation together determine patterns of diversity in stream invertebrates. J. Anim. Ecol. 75, 476–484.
Townsend, C.R., Hildrew, A.G., 1994. Species traits in relation to a habitat templet for river systems. Freshw. Biol. 31, 265–275.
Trimmer, M., Maanoja, S., Hildrew, A.G., Pretty, J.L., Grey, J., 2010. Potential carbon fixation by methane oxidation in well-oxygenated riverbed gravels. Limnol. Oceanogr. 55, 560–568.
Turner, M.G., 2010. Disturbance and landscape dynamics in a changing world. Ecology 91, 2833–2849.
Vidal, S., Wade, A., 2009. Multimodel assessment of future climatological droughts in the United Kingdom. Int. J. Climatol. 29, 2056–2071.
Vörösmarty, C.J., McIntyre, P.B., Gessner, M.O., Dudgeon, D., Prusevich, A., Green, P., Glidden, S., Bunn, S.E., Sullivan, C.A., Liermann, C.R., Davies, P.M., 2010. Global threats to human water security and river biodiversity. Nature 467, 555–561.
Walters, A.W., Post, D.M., 2011. How low can you go? Impacts of a low-flow disturbance on aquatic insect communities. Ecol. Appl. 21, 163–174.
Wilson, S.D., Tilman, D., 1991. Interactive effects of fertilisation and disturbance on community structure and resource availability in an old-field plant community. Oecologia 88, 61–71.
Woodward, G., Jones, J.I., Hildrew, A.G., 2002. Community persistence in Broadstone Stream (U.K.) over three decades. Freshw. Biol. 47, 1419–1435.
Woodward, G., Papantoniou, G., Edwards, F.E., Lauridsen, R., 2008. Trophic trickles and cascades in a complex food web: impacts of a keystone predator on stream community structure and ecosystem processes. Oikos 117, 683–692.
Woodward, G., Benstead, J.P., Beveridge, O.S., Blanchard, J., Brey, T., Brown, L.E.L., Cross, W.F., Friberg, N., Ings, T.C., Jacob, U., Jennings, S., Ledger, M.E., Milner, A.M., Montoya, J.M., O'Gorman, E.O., Olesen, J.M., Petchey, O.L.,

Pichler, D.E., Reuman, D.C., Thompson, M.S., Van Veen, F.J.F., Yvon-Durocher, G., 2010. Ecological networks in a changing climate. Adv. Ecol. Res. 42, 71–138.

Woodward, G., Brown, L.E., Edwards, F.K., Hudson, L.N., Milner, A.M., Reuman, D.C., Ledger, M.E., 2012. Climate change impacts in multispecies systems: drought alters food web size-structure in a field experiment. Philos. Trans. R. Soc. Lond. B (in press).

Wright, J.F., Clarke, R.T., Gunn, R.J.M., Kneebone, N.T., Davey-Bowker, J., 2004. Impact of major changes in flow regime on the macroinvertebrate assemblages of four chalk stream sites, 1997-2001. Riv. Res. Appl. 20, 775–794.

Yodzis, P., 1988. The indeterminacy of ecological interactions as perceived through perturbation experiments. Ecology 69, 508–515.

Zar, J.H., 1999. Biostatistical Analysis, fourth ed. Prentice Hall, London.

Environmental Warming in Shallow Lakes: A Review of Potential Changes in Community Structure as Evidenced from Space-for-Time Substitution Approaches

Mariana Meerhoff[*,†,‡,1], Franco Teixeira-de Mello[*], Carla Kruk[§,¶], Cecilia Alonso[||], Iván González-Bergonzoni[*,†], Juan Pablo Pacheco[*], Gissell Lacerot[||], Matías Arim[*,#,], Meryem Beklioğlu[††], Sandra Brucet[‡‡], Guillermo Goyenola[*], Carlos Iglesias[*], Néstor Mazzeo[*,‡], Sarian Kosten[§§,¶¶], Erik Jeppesen[†,||||,##]**

*Departamento de Ecología y Evolución, Centro Universitario Regional Este (CURE), Facultad de Ciencias, Universidad de la República, Burnett s/n, Maldonado, Uruguay
†Department of Bioscience, Aarhus University, Vejlsøvej, Silkeborg, Denmark
‡South American Institute for Resilience and Sustainability Studies (SARAS), Maldonado, Uruguay
§Laboratory of Ethology, Ecology and Evolution, Instituto de Investigaciones Biológicas Clemente Estable, Italia, CP 11600, Montevideo, Uruguay
¶Ecología Funcional de Sistemas Acuáticos, Limnología, IECA, Facultad de Ciencias, Universidad de la República, Iguá, CP 11400, Montevideo, Uruguay
||Ecología Funcional de Sistemas Acuáticos, Centro Universitario Regional Este (CURE), Universidad de la República, Ruta 9, km 204, Rocha, Uruguay
#Facultad de Ciencias, Universidad de la República, Iguá, CP 11400, Montevideo, Uruguay
**Center for Advanced Studies in Ecology and Biodiversity (CASEB), Depto. de Ecología, Facultad de Ciencias Biológicas, Pontificia Universidad Católica, CP 6513677, Santiago, Chile
††Department of Biology, Limnology Laboratory, Middle East Technical University, Üniversiteliler Mahallesi, Dumlupınar Bulvarı, Çankaya, Ankara, Turkey
‡‡European Commission, Joint Research Centre, Institute for Environment and Sustainability, Ispra, Italy
§§Department of Aquatic Ecology and Water Quality Management, Wageningen University, Wageningen, The Netherlands
¶¶Leibniz-Institute of Freshwater Ecology and Inland Fisheries (IGB), Berlin/Neuglobsow, Germany
||||Greenland Climate Research Centre (GCRC), Greenland Institute of Natural Resources, Kivioq, P.O. Box 570 3900, Nuuk, Greenland
##Sino-Danish Centre for Education and Research (SDC), Beijing, China
[1]Corresponding author: e-mail address: mm@dmu.dk; merluz@fcien.edu.uy

Advances in Ecological Research, Volume 46
ISSN 0065-2504
http://dx.doi.org/10.1016/B978-0-12-396992-7.00004-6

Contents

Abstract

Shallow lakes, one of the most widespread water bodies in the world landscape, are very sensitive to climate change. Several theories predict changes in community traits, relevant for ecosystem functioning, with higher temperature. The space-for-time substitution approach (SFTS) provides one of the most plausible empirical evaluations for these theories, helping to elucidate the long-term consequences of changes in climate.

Here, we reviewed the changes at the community level for the main freshwater taxa and assemblages (i.e. fishes, macroinvertebrates, zooplankton, macrophytes, phytoplankton, periphyton and bacterioplankton), under different climates. We analyzed data obtained from latitudinal and altitudinal gradients and cross-comparison (i.e. SFTS) studies, supplemented by an analysis of published geographically dispersed data for those communities or traits not covered in the SFTS literature.

We found only partial empirical evidence supporting the theoretical predictions. The prediction of higher richness at warmer locations was supported for fishes, phytoplankton and periphyton, while the opposite was true for macroinvertebrates and zooplankton. With decreasing latitude, the biomass of cladoceran zooplankton and periphyton and the density of zooplankton and macroinvertebrates declined (opposite for fishes for both biomass and density variables). Fishes and cladoceran zooplankton showed the expected reduction in body size with higher temperature. Life history changes in fish and zooplankton and stronger trophic interactions at intermediate positions in the food web (fish predation on zooplankton and macroinvertebrates) were evident, but also a weaker grazing pressure of zooplankton on phytoplankton occurred with increasing temperatures. The potential impacts of lake productivity, fish predation and other factors, such as salinity, were often stronger than those of temperature itself. Additionally, shallow lakes may shift between alternative states, complicating theoretical predictions of warming effects. SFTS and meta-analyses approaches have their shortcomings, but in combination with experimental and model studies that help reveal mechanisms, the "field situation" is indispensable to understand the potential effects of warming.

1. INTRODUCTION

1.1. Global change and freshwater communities

Anthropogenic impacts on natural ecosystems are increasing apace in both the terrestrial and aquatic (both freshwater and marine) realms, and environmental stressors, such as climate change, threaten to alter community structure and ecosystem functioning from local to global scales (Hagen et al., 2012; Ledger et al., 2012; Mintenbeck et al., 2012; Mulder et al., 2012). The Millennium Ecosystem Assessment (2005) has quantified existing and projected deterioration or loss of natural ecosystems through the intensification of agriculture, urbanization and other anthropogenic impacts that are likely to have significant impacts on most of the terrestrial ecosystems on Earth by the year 2070. The key global drivers include climate warming, changes in precipitation patterns, land use changes (Vitousek, 1994), increasing atmospheric CO_2 concentrations (Rockström et al., 2009), and alterations in the global nitrogen cycle and global fertilization of ecosystems (Galloway et al., 2008; Gruber and Galloway, 2008). Invasive species (Walther et al., 2009) and decreasing biodiversity due to habitat loss and rising water demands (Vörösmarty et al., 2000) are among the most widely reported biological responses to these changes (Parmesan and Yohe, 2003).

The impacts of environmental warming, although increasingly recognized as a key component of climate change following the recent reports from the Intergovernmental Panel on Climate Change (IPCC, 2007), and a growing number of ecological studies, are still poorly understood at the higher (multispecies) levels of biological organization. It is recognized, however, that its impacts are expected to be strongest at high altitude and high latitudes (Phoenix and Lee, 2004; Rouse et al., 1997; Smol et al., 2005; Woodward et al., 2010a,b). Large parts of the polar regions, particularly the Arctic, are expected to show a much faster increase in the mean annual temperature than lower latitudes (Howard-Williams et al., 2006).

Temperature affects a myriad of biological processes, including individual growth and respiration rates (potentially affecting primary production and community respiration, e.g. Yvon-Durocher et al., 2010), changes in life history traits, changes in phenology and trophic dynamics, with potential temporal or spatial mismatches arising between prey availability and consumer demands (e.g. Winder and Schindler, 2004). Species may not respond with the same strength or synchronously in time, since they are affected not only by changes in temperature but also by other environmental factors, such as changes in photoperiod (Winder and Schindler, 2004). However, after accounting for size dependence, temperature explains the largest amount of variation in almost all biological rates (Brown et al., 2004; Peters, 1983). Given sufficient resource availability, increasing temperatures generally accelerate growth and development rates of individual organisms (Forster et al., 2011b), although changes in absolute abundances may be species specific or ecosystem specific (Adrian et al., 2006; Blenckner et al., 2007; Reist et al., 2006).

Warming may also contribute to changes in the latitudinal or altitudinal distributional range of some species, thus likely affecting diversity and community structure. Stenothermal species (narrow thermal range) will most probably shift range or become locally extinct, whereas eurythermal species (wide thermal range) will likely be able to adapt to new thermal regimes (Lappalainen and Lehtonen, 1997; Woodward et al., 2010a,b). The already observed and the predicted changes in global and regional temperatures make understanding warming effects on ecological communities a priority (Moss et al., 2009; Petchey et al., 2010). However, the effects of temperature on many aspects of community structure, such as the distribution of diversity and biomass across trophic levels, or the extent and distribution of specialism and generalism across species, are still poorly understood (Petchey et al., 2010; Woodward et al., 2010b).

1.2. Shallow lakes and ecosystem responses to changes in temperature

Shallow lakes (typically polymictic, maximum depth ca. <5 m) are particularly strongly affected by human activity, as they are key providers of critical ecosystem services that underpin aquaculture, crop irrigation and drinking water supply. Besides being one of the most common and widespread inland water bodies in the world landscape (Downing et al., 2006), these lakes represent one of the ecosystem types most rapidly affected by external perturbations, including climate change (Jeppesen et al., 2009; Moss et al., 2009).

Shallow lakes are typically isolated and fragmented in the landscape and are extremely sensitive to climate variability such as evaporation and precipitation balances, since they have a very large surface:volume ratio (Coops et al., 2003), which affects the persistence of climatic signatures. Lake depth and hydrologic residence time, together with stratification and mixing regime, are also physical features that interact to control the duration and strength of climate signals, for instance, in water temperature (Blenckner, 2005). In deep lakes, the winter climate signal (as described, for instance, by the winter North Atlantic Oscillation index, NAO, in the Northern Hemisphere) can persist until late summer, whereas in shallow lakes, the winter temperature signal lasts for much shorter time (Gerten and Adrian, 2001). Identification of climate signals in current measurements may thus be difficult due to time-lags arising from the action of other processes at different scales.

Most lakes are also commonly affected by multiple interacting stressors (Christensen et al., 2006; Yan et al., 2008), potentially confounding the detection of signals due to climate change. Freshwater systems, not least shallow lakes, are subject to increasing deterioration processes in many parts of the world (Feld et al., 2011; Friberg et al., 2011; Hladyz et al., 2011; Moss, 1998; Moss et al., 2011). These include wetland area loss, local extinction of native species and introduction of exotic species, acidification, water level changes due to water extraction or canalization and, especially, eutrophication (nutrient enrichment in the water bodies), among the major drivers of global change that aquatic ecosystems are currently facing (reviewed in, e.g. Carpenter et al., 1998; Dodds, 2007; Friberg et al., 2011; Schindler, 2006; Smith, 2003). While some shallow lakes and ponds might simply dry out (Beklioğlu et al., 2007), the lower water level will often concentrate pollutants, enhance resuspension and, together with the higher temperature, amplify the sediment release of nutrients, especially of phosphorus (McKee et al., 2003; Özen et al., 2010). Under increasing precipitation, in contrast, the runoff of nutrients from the catchments may increase (Jeppesen et al., 2009, 2011; Özen et al., 2010). An amplification of eutrophication symptoms by climate warming seems to occur also by changes in trophic dynamics and interactions (Meerhoff et al., 2007a) and shifts in fish community structure (Jeppesen et al., 2010a) and population size structure (Daufresne et al., 2009). As a consequence, the likelihood of dominance of nuisance phytoplankton taxa, such as cyanobacteria and filamentous algae, may rise (Jeppesen et al., 2009; Kosten et al., 2011a; Mooij et al., 2005; Paerl and Huisman, 2009; Trochine et al., 2011).

The interaction of climate signals with nutrient enrichment symptoms has been addressed by several studies based on palaeolimnological records (Bjerring et al., 2009), mesocosm experiments (e.g. Moss et al., 2004 and references herein), latitudinal gradient analysis (Kosten et al., 2009b and references herein) and synthesis papers (Jeppesen et al., 2010a, 2011; Moss, 2010; Moss et al., 2011). However, disentangling the specific effects of environmental warming from nutrient enrichment (as well as from other global changes) is crucial if we are to further our understanding of the responses of freshwater ecosystems to a changing climate and contribute to their conservation with adequate mitigation measures. The interactive effects of warming and other global changes and the indirect impacts on community structure, food web functioning and ecosystem processes are even far less well known (Montoya and Raffaelli, 2010; Woodward et al., 2010b).

The influence of the catchment on within-lake processes can vary depending on land use, the regional climate and geographical location. However, shallow lakes are especially complex due to their numerous internal feedbacks. Shallow lakes may respond differently from other aquatic systems to external perturbations or stressors, given the recognized potential for them to shift between alternative states over an intermediate range of nutrient concentrations, in a typical hysteresis process (Moss et al., 1996; Scheffer et al., 1993, 2001): such extreme regime shifts seem far less prevalent in many running waters, which appear to be far more stable in the face of perturbations or stressors in general (Layer et al., 2010, 2011; Ledger et al., 2012). Biological and physicochemical mechanisms related to the presence or absence of macrophytes and water turbidity maintain either state via a range of positive feedbacks. The stability of such alternative states has been challenged by recent studies conducted in other regions of the globe than the temperate zone (Jeppesen et al., 2007; Kosten et al., 2011a; Meerhoff et al., 2007a), which was the birth place of this theoretical framework, suggesting that they might be modulated by climatic influences. Similar to enhanced nutrient loading, changes in water level, enhanced temperature and longer growing season have been identified as potential weakening factors that may trigger a change from a clear water, macrophyte-dominated state to undesirable states (Beklioğlu et al., 2006) involving the dominance by free-floating plants (Scheffer et al., 2003) or filamentous green algae (Trochine et al., 2011). The hysteresis phenomenon complicates our ability to make predictions for the response of a specific community, not least at ecosystem level, to changing temperatures, as several community and food web configurations may occur under similar environmental conditions.

1.3. Indirect effects of climate on community structure through availability of nutrients

Trends extracted from latitudinal gradient or cross-comparison analyses are based on correlative data. This may render it difficult to disentangle strict climate signals from potential indirect effects of climate and human-related impacts (such as nutrient enrichment and salinization through cultivation, e.g. Bjerring et al., 2009). An increase in nutrient loading may determine a series of important changes in community traits in shallow lake communities. For instance, changes in fish community composition, biomass and body size structure, with non-random loss of predatory species (due to increased turbidity and lower concentrations or larger variations in dissolved oxygen), occur worldwide with eutrophication (Jeppesen et al., 2005). Large increases in nutrient concentrations due to eutrophication thus affect lake trophic webs, from the basal resources to the top predators.

Both the absolute and relative concentrations of nutrients in the water column are essential for the development of phytoplankton communities, as they affect both total biomass and composition. A warmer climate can affect the identity of the limiting nutrient and the availability of nutrients to primary producers: for instance, higher denitrification under warmer conditions could lead to nitrogen limitation (Lewis, 1996, 2000). There are conflicting observations as to which is the main limiting nutrient across different climates. In cross-comparison mesocosm experiments on nutrient and fish addition under different temperatures, the nature of the limiting nutrient varied between temperate and tropical systems (Danger et al., 2009). Addition of phosphorus (P) favoured phytoplankton in temperate lakes, while nitrogen (N) had positive effects in tropical mesocosms, and water N:P ratios tended to be higher in the tropical than in the temperate experiment. However, several latitudinal gradient studies show no clear consistent differences in N or P limitation across large spatial scales in terms of nutrient ratios and chlorophyll-a:nutrient relationships, among other indicators (Huszar et al., 2006; Kosten et al., 2009b; Mazumder and Havens, 1998). Kosten et al. (2011b) showed that while total phosphorus (TP) explained most of the variance in phytoplankton chlorophyll-a (Chl-a) in the cool region of South America, total nitrogen (TN) explained most of the variance in the intermediate and warm regions. A recent work based on an extensive meta-analysis of various ecosystems has shown equivalence in N and P limitation in freshwater ecosystems and synergistic effects of N and P enrichment (i.e. co-limitation) (Elser et al., 2007). These authors found only a weak negative correlation between latitude and the positive response of primary producers to enrichment in N, but not in P. In summary, our review of nutrient limitation among different climates in the space-for-time substitution (SFTS) approach literature shows no clear or consistent pattern.

1.4. Theoretical predictions

Changes in temperature, due to both anthropogenic activities and natural temperature variations, are a main determinant of community structure and ecosystem processes, via the direct effects on the metabolic demands of individuals and the attendant changes in the distribution of body sizes of organisms (Arim et al., 2007; Charnov and Gillooly, 2004; Forster et al., 2011a,b; Gillooly, 2000; Yvon-Durocher et al., 2011a). The metabolic theory of ecology (hereafter, MTE) considers the effects of both body size and environmental temperature on relevant ecological rates and patterns (Brown et al., 2004). MTE, and other physiologically based theories relating environmental temperature with organisms' performance and community function, provides the basis for predictive analysis of the effect of climate warming on ecosystems (e.g. Angilletta, 2009; Karasov and Martínez del Rio, 2007; McNab, 2002; Perkins et al., 2010; Yvon-Durocher et al., 2010). The MTE is based on a set of empirical generalizations about scaling of biological rates with body size (Brown et al., 2004; Peters, 1983), a mechanistic explanation for this scaling (West et al., 1997) and its response to changes in temperature (Dell et al., 2011; Gillooly et al., 2001; Huey and Kingsolver, 2011). This theory remains at the centre of a heated debate (Brown et al., 2005; Etienne et al., 2006; Forster et al., 2011b; Kozlowski and Konarzewski, 2004; Reiss et al., 2010). Nonetheless, despite some deviations from general expectations (e.g. Algar et al., 2007; Caruso et al., 2010), the predictive potential of this ecological theory, which is based on basic principles of biology and kinetics, is notable (e.g. Weber et al., 2011).

MTE quantitatively predicts changes in species (or higher taxa) richness due to changes in environmental temperature and in the mean body size of the individuals composing local communities (Allen et al., 2002; Brown et al., 2004). This theory predicts an increase in richness with temperature (specifically, a linear relationship between the logarithm of species richness and the inverse of temperature, Allen et al., 2002) (Fig. 1, Table 1). The detection of some degree of curvilinear association between these variables in invertebrate and vertebrate ectotherms could be considered as a limitation of the theory or as an area in which further theoretical advances are needed (Algar et al., 2007). Considering the possible change in species body size and in metabolic rates due to changes in temperature, several predictions can be made.

From a population perspective, the carrying capacity, the growth rates, the incidence among local populations, and the geographic range of species could be potentially affected by temperature. A basic determinant of a population's carrying capacity is the relationship between the demands of individuals and the availability of resources (Damuth, 1981). The effect of temperature on the demands of an individual predicts a displacement of the density–mass relationship, as well as of the biomass distribution (e.g. Brown et al., 2004). These displacements take place because the rise in energetic demands with increasing temperature determines that the same amount of resources is divided among individuals that require more from the environment; consequently, fewer individuals can satisfy their demands and a decrease in density and total biomass is expected (Fig. 1, Table 1). However, if the amounts of available resources and/or predation strength depend on body size and temperature, changes in the slope and modes of the size distribution could also be expected (see Arim et al., 2011; Brown et al., 2004). In addition to these changes in the carrying capacity of species, population dynamics could also be affected since the maximum growth rate is expected to increase with temperature (Brown et al., 2004). In this sense, the rise in temperature could lead to less abundant and more variable populations (e.g. Beisner et al., 1997) with more frequent local extinction, affecting metapopulation dynamics and the fraction of local habitats occupied by the species (Hanski, 1999).

A causal connection between temperature and body size has long been recognized (reviewed by Angilletta, 2009; Kingsolver and Huey, 2008), and Bergmann's rule, now more than 160 years old, is one of the first empirical generalizations about ecological patterns (Gaston and Blackburn, 2000) and still plays a central role in biogeography (Olalla-Tárraga, 2011). It states that organisms from higher latitudes have large body sizes in comparison with individuals of the same species, related species or assemblages inhabiting lower latitudes (Olalla-Tárraga et al., 2010) (Fig. 1, solid line). This trend was conceived for endotherms that obtain a

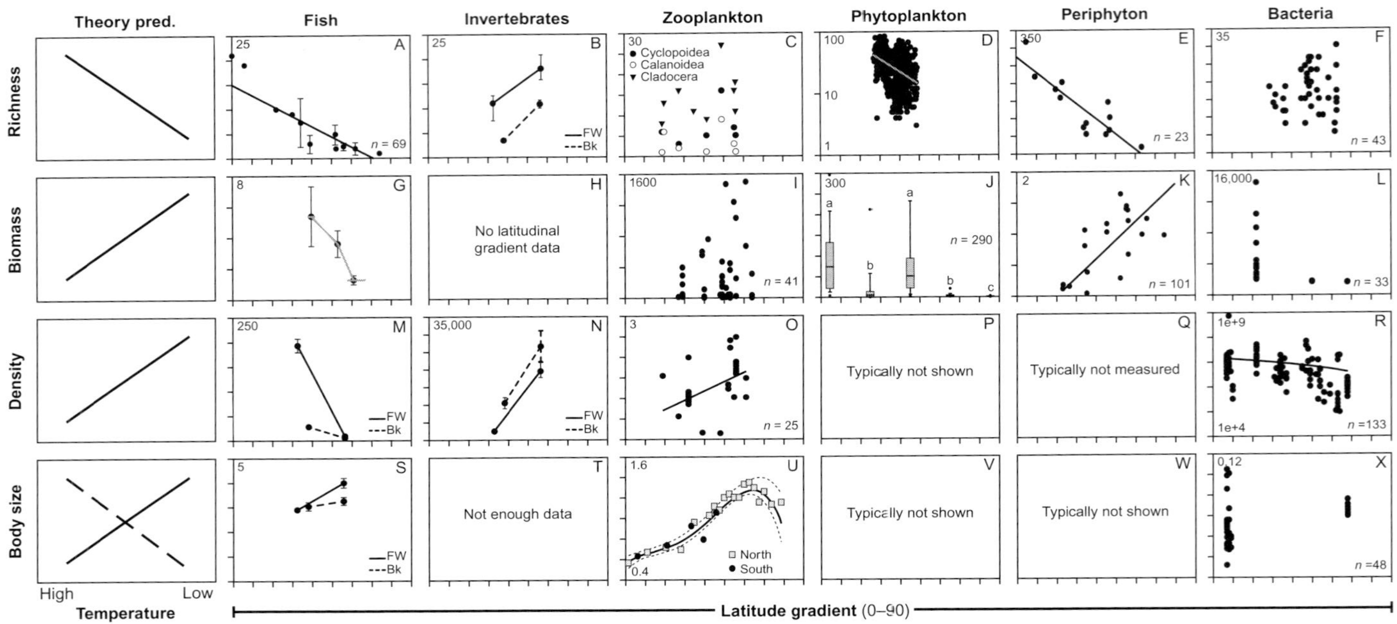

Figure 1 Theoretical expectations and most representative empirical trends in main traits of different communities of shallow lakes along a latitude gradient, as found in space-for-time substitution studies or built on meta-analysis. See Table 1 for a summary of the respective units and references from which data were extracted, and respective Appendixes for the meta-analyses. From right to left: theoretically predicted relationships (solid lines: most supported trend, dashed line: alternative expected trend, see Section 1), fish, macroinvertebrates, zooplankton, phytoplankton, periphyton and bacterioplankton, and from top to bottom: richness, biomass, density and body size, indicating the maximum scale value on each *y*-axis, the significant relationships (indicated by a line or an * where appropriate) and the reasons for empty panels. Macrophytes were excluded due to lack of data on all traits. Data on proximate latitudes were averaged and shown as mean value for that latitude range, in the case of fish biomass data (extracted from Gyllström et al., 2005) after removing the lowest phosphorus values from the analysis in order to have a comparable range among climate zones. To calculate zooplankton dry weight, we used a conversion factor of 48% to transform μg C L^{-1} to μg DW L^{-1} (Hessen, 1989). Phytoplankton biomass box-plots show mean, mean ± 1 SE, mean ± 1 SD and extreme values. The letters in the right corner of the panels are used for identification in the main text. When shown, error bars correspond to 1 standard error. Codes: Bk = brackish lakes, FW = freshwater lakes.

Table 1 Summary of the data (variables and units) and respective references describing patterns of changes in main community traits along a latitudinal gradient, shown in Fig. 1

	Richness	Biomass	Density	Body size
Theoretical predictions	$\log(s) = (-E/1000k)(1000/T) + C_1$ C_1: constant, k: Boltzmann's constant, E: activation energy	$W \sim [R]\ M^{1/4} e^{E/kT}$ [R]: Supply rate of limiting resource	$K \sim [R]\ M^{3/4}\ e^{E/kT}$	See text for explanation
	Allen et al. (2002)	Brown et al. (2004)	Brown et al. (2004)	Gardner et al. (2011), Marquet and Taper (1998), and Olalla-Tárraga and Rodríguez (2007)
Fish	Species number	CPUE (kg net^{-1} night^{-1})	ind m^{-2}	cm (standard length)
	González-Bergonzoni et al. (2012)	Gyllström et al. (2005)	Brucet et al. (2010) and Teixeira-de Mello et al. (2009)	Brucet et al. (2010) and Teixeira-de Mello et al. (2009)
Macroinvertebrates	Taxa number	No data	ind m^{-2}	No data
	Brucet et al. (2012) and Meerhoff et al. (2007a)		Brucet et al. (2012) and Meerhoff et al. (2007a)	

Continued

Table 1 Summary of the data (variables and units) and respective references describing patterns of changes in main community traits along a latitudinal gradient, shown in Fig. 1—cont'd

	Richness	Biomass	Density	Body size
Zooplankton	Genera number (cladocerans and copepods)	μg dry weight L^{-1} (total zooplankton)	ind L^{-1} (cladocerans)	mm (cladocerans)
	Brucet et al. (2010), Meerhoff et al. (2007b), and Pinto-Coelho et al. (2005)	Gyllström et al. (2005), Havens et al. (2007), Jackson et al. (2007), Pinto-Coelho et al. (2005), and Vakkilainen et al. (2004)	Hansson et al. (2007), Meerhoff et al. (2007a), Meerhoff et al. (2007b), and Pinto-Coelho et al. (2005)	Gillooly and Dodson (2000)
Phytoplankton	Species number	μg Chl-a L^{-1}	No data	No data
	Stomp et al. (2011)	Meta-analysis: Danger et al. (2009), Ganf (1974), Kalff and Watson (1986), Kruk et al. (2011), and Melack (1976, 1979) plus all references in Table 3		
Periphyton	Species number	log μg Chl-a cm^{-2}	No data	No data
	Analysis of metadata. Appendix A	Analysis of metadata. Appendix A		
Bacterioplankton	DGGE bands	μg C L^{-1}	cells ml^{-1}	μm^3
	Analysis of metadata. Appendix B	Analysis of metadata. Appendix B	Analysis of metadata. Appendix B	Analysis of metadata. Appendix B

The equations for the theoretical predictions and their respective references are also shown, while the references used to extract data for meta-analyses are presented in Appendixes. Most data were recalculated or redrawn from the original literature, except the figure from Gillooly and Dodson (2000), republished with kind permission from the Association for the Sciences of Limnology and Oceanography (copyright 2000) and the figure from Stomp et al. (2011), republished with kind permission from the authors (copyright 2011). Abbreviation DGGE, denaturing gradient gel electrophoresis thermoregulatory benefit by the reduction in the area:volume ratio at larger masses (at high latitudes and thus cold conditions). However, several mechanisms were proposed to account for latitudinal trends in the body size of endo and ectothermic species, focusing on the effects of resource availability and thermoregulation (reviewed by Watt et al., 2010).

The "heat balance hypothesis" predicts either an increase or a reduction in body size depending on the original size and thermal biology of a given species (Olalla-Tárraga, 2011). Individuals are typically active at a range of body temperatures that could be achieved by different mechanisms (McNab, 2002). Thermoconformers, for instance, are those animals that match to their environmental temperature, in contrast to thermoregulators that maintain a nearly constant body temperature via behavioural (ectotherms) or physiological (endotherms) mechanisms (see Angilletta, 2009; McNab, 2002). Among thermoconformers, a reverse Bergmann's pattern is expected to reduce heating times in smaller organisms (Olalla-Tárraga and Rodríguez, 2007). The body temperature of thermoconformers closely fluctuates with ambient temperature with a heating time proportional to body size, being unable to accelerate the heating process. As consequence, large individuals in colder environments have a short activity window (e.g. to forage) while the reduced available energy compromises their viability and, therefore, small individuals might be favoured (see Olalla-Tárraga and Rodríguez, 2007). Small ectothermic thermoregulators, however, would become larger at lower temperatures to increase heat conservation (from Olalla-Tárraga, 2011) (Fig. 1, dashed vs. solid lines, respectively). The existence of size thresholds and potential range limits of body size for these processes to occur are not yet defined. While originally focused on amphibians, this hypothesis has the potential to explain observed patterns in fishes (Belk and Houston, 2002; Blanck and Lamouroux, 2007), invertebrates (Blanckenhorn and Demont, 2004) and ectotherms in general (Olalla-Tárraga and Rodríguez, 2007), all of which dominate the fauna of shallow lakes and freshwaters in general.

A common empirical generalization is the "temperature-size rule", a widely recognized phenomenon, according to which ectothermic individuals growing at lower temperature reach larger body sizes (Angilletta, 2009; Forster et al., 2011a,b; Kingsolver and Huey, 2008). This process prevails independently of thermoregulatory behaviour and has been reported for unicellular and multicellular organisms, and may arise due to differential thermal responses of development and growth rates, which are not addressed by the MTE (Forster et al., 2011a,b). From Bergmann's rule, MTE and the "temperature-size rule", it is thus expected that for several, not necessarily exclusive reasons, warming will promote changes in body size (Gardner et al., 2011) although the direction of change will vary depending, among other factors, on the thermal biology of the organisms.

Some macroecological approaches also support a change in body size not only within, but also among, species with global warming. Individual fitness may be summarized, from an energetic perspective, by two processes (Brown et al., 1993), namely, the acquisition of energy from the environment and the transformation of this energy into progeny. While energy acquisition increases with an increase in body size, the individual potential to produce offspring decreases. From the balance between these two processes, an optimum body size requiring less energy to survive in a given environment is expected (Marquet and Taper, 1998; Marquet et al., 2008). In this sense, population viability is determined by the balance between individual energetic demands and the availability of energy in the environment, determining constraints to the maximum and minimum body sizes in local communities (see Marquet and Taper, 1998; Marquet et al., 2008). Based on this energetic context, small species could experience a selection for larger sizes under a relative reduction in the supply–demand balance, whereas the opposite is true for large-bodied species (Marquet and Taper, 1998; Marquet et al., 2008). Specifically, when the resource availability: demands ratio is reduced (e.g. after an increase in temperature), the maximum and minimum size constraints become stronger and the smaller species experience a selection pressure towards larger sizes, while the opposite occurs for larger species (Brown et al., 1993; Marquet and Taper, 1998; Marquet et al., 2008). Within this framework, the optimum body sizes represent a threshold between positive and negative trends in response to changes in environmental conditions (Brown, 1995; Brown et al., 1993; Marquet and Taper, 1998; Marquet et al., 2008). The body size distributions expected from this theory have been evaluated in mammals, birds (e.g. Marquet and Taper, 1998) and fishes (Fu et al., 2004; Knouft, 2004; Knouft and Page, 2003; Rosenfield, 2002).

Trophic interactions may also be affected by temperature in several other ways. The rise in energetic demands with temperature could involve more consumption events (McNab, 2002), an increase in the range of prey consumed and an increase in spatial movements to capture the prey (Petchey et al., 2010; Pike, 1984; Schoener, 1974). In addition, it has been proposed that community food chain length and species trophic position can be negatively associated with environmental temperature (Arim et al., 2007). Temperature might also indirectly affect biotic interactions by altering the final body size of adults and the time elapsed to achieve this size (Forster et al., 2011a,b). Size refugia from predation is a common phenomenon in aquatic environments (Chase, 1999), where gape-limited predation is prevalent (Gilljam et al., 2011; Woodward et al., 2010c). Smaller prey will likely be more frequently preyed upon at higher temperatures, while smaller predators might face a reduction in the total amount of available resources (Arim et al., 2010, 2011). The relative rate of variation in predator and prey sizes will determine the strength of this effect. These changes in species richness, food chain length, frequency and intensity of trophic interactions, and consequently in connectivity, have a significant potential to alter food webs and create changes in community structure and stability, and ecosystem functioning (Dunne et al., 2002; Fox and McGrady-Steed, 2002; May, 1972; McCann, 2000; Petchey et al., 2010). Contradictory results of models on the responses of community traits and food web metrics to increasing temperatures (such as connectance, Petchey et al., 2010) highlight the need for more and better empirical data to parameterize models, test predictions and produce models of mechanistic effects of temperature change on interspecific interactions, especially in freshwater ecosystems.

1.5. Space-for-time substitution approach

Various methods have been used to investigate the potential effects of global change in natural systems. In particular, a large battery of approaches is used to study the effects of climate warming on shallow lakes, including warming experiments at different scales, from individual-species responses to increased temperature to larger field scale experiments where temperature and associated climate factors are drivers (e.g. Feuchtmayr et al., 2009; Liboriussen et al., 2005; McKee et al., 2002a,b; Netten et al., 2010; Yvon-Durocher et al., 2011a), palaeolimnological surveys (e.g. Smol et al., 2005), long-term contemporary surveys (e.g. Adrian et al., 2006; Winder and Schindler, 2004) and mathematical models (e.g. Mooij et al., 2007; Trolle et al., 2011).

Another research approach to study the potential impacts of climate warming on ecosystems is the so-called SFTS (Pickett, 1989). The origins of this approach, although not meant to analyze biological responses to climate change, can be traced back to the end of the XVIII century with the studies of von Humboldt and Bonpland, who after latitudinal and altitudinal gradient studies in South America suggested that temperature and precipitation played a key role in the spatial distribution of vegetation

(Von Humboldt and Bonpland, 1805, 1807). It has been widely used to gauge community and ecosystem level responses to major environmental gradients in freshwaters, including pH (Layer et al., 2010) and nutrients (Rawcliffe et al., 2010), and many other systems (Olesen et al., 2010).

Within the current context of climate change, this approach is now used to identify key differences between systems located in different climates (slightly warmer or cooler) but otherwise similar (such as in area, trophic state or productivity level, etc.). Comparable ecosystems under different natural climatic conditions showing different ecosystem processes help elucidate the potential long-term consequences of changes in climate, as long-term data are often not available beyond a few years of sampling in most systems (but see Durance and Ormerod, 2007; Jeppesen et al., submitted for publication; Layer et al., 2011). Thus, this approach highlights the structure and interactions of communities arguably at, or close to, equilibrium conditions rather than the responses to short-term perturbations (i.e. experimental increase in temperature) (Woodward et al., 2010a,b). SFTS is a common method used in the attempt to understand change when it is not possible or desirable to wait for years or decades to detect changes (Jackson, 2011). The SFTS approach includes cross-comparison studies, latitudinal gradients (with mean temperature generally increasing towards lower latitudes) and altitudinal (with temperature generally increasing towards lowlands) gradients and has been applied to a series of environments, such as marine and inland ecosystems in Antarctica (e.g. Howard-Williams et al., 2006), Icelandic geothermal streams (Friberg et al., 2009; Woodward et al., 2010a), temperate and subtropical streams (Teixeira-de Mello et al., 2012) and shallow lakes in Europe (Gyllström et al., 2005; Moss et al., 2004; Stephen et al., 2004; and references herein) and South America (Kosten et al., 2009c, 2011a and references herein). SFTS represents one of the most widely used approaches in ecological climate change research, although its potentially weak aspects should be considered before application, including the need for a trait- or size-based approach so as to avoid the potentially confounding effects of geology, biogeography and land use, and the findings should be analyzed carefully since equilibrial or non-equilibrial conditions may be hard to distinguish (Woodward et al., 2010b). We argue that the SFTS approach provides one of the most plausible empirical evaluations for these theories using real-world data and, by the generalization of these theories, provides the potential for inductive advances and the proposition of new mechanisms.

Our aim with this work is primarily to review and summarize the changes in community traits (richness, biomass, density, body size) and in growth, reproduction and strength of trophic interactions in shallow lakes in relation to climate regimes, as shown by studies applying the SFTS approach, either in the form of cross-comparisons studies or investigations along latitudinal and altitudinal gradients and as shown by analyses of published metadata. Extraction of spatial trends that can be associated with climate signals may be done confidently when the sites chosen for comparison are similar as to trophic state, area and other relevant limnological characteristic (e.g. Brucet et al., 2010, 2012, submitted for publication; Meerhoff et al., 2007a,b; Teixeira-de Mello et al., 2009), or when many sites are sampled with a random distribution of trophic states (e.g. Kosten et al., 2011b and references herein), so that physicochemical gradients are nested within climate regions (e.g. Declerck et al., 2005). Thereby, we aimed to detect patterns of change that can be associated with differences in ambient temperature and thus to elucidate changes with warming.

We also analyzed metadata generated in different parts of the world in order to construct gradients for the traits or communities not covered by the current SFTS literature. We searched for shallow lake systems (maximum depth < 5 m, polymictic, or identified by the authors as such) and checked for records on characteristics such as area, trophic state, pH, salinity, turbidity and the presence of exotic species (if identified in the work). The introduction or extraction of species, particularly invertebrates and fish (Villéger et al., 2011), could blur latitudinal patterns of native fauna and should thus be considered in the analysis. We analyzed trends at different levels of organization, mostly focusing on community structure parameters and compared these findings with theoretical expectations basically from MTE (*sensu* Brown et al., 2004) and with findings from other research approaches. A significant challenge is to predict the effects of climate warming on lake community structure, since the effects of temperature on a particular community depend on the feedbacks with other communities that are also affected by warming and other aspects of global change (Moss et al., 2011). It is beyond the scope of this chapter to address all observed or potential climate-driven changes in freshwaters, as many have been reviewed recently for lakes (e.g. Adrian et al., 2009; Moss, 2010, 2011; Moss et al., 2009) or freshwaters in general (Perkins et al., 2010; Woodward et al., 2010b). Instead, our target is to highlight the trends in climate-associated changes for community traits that may have profound implications for the functioning of these ecosystems and to identify areas that require further research and experimental testing of mechanisms, while stressing the limitations and advantages of this research approach.

Being a comprehensive work, this chapter can be read either as a single monograph or in part by considering the self-contained sections, following the axis of community traits (e.g. richness, biomass, density, body size) or the axis of specific shallow lake assemblages (i.e. fishes, macroinvertebrates, zooplankton, macrophytes, phytoplankton, periphyton and bacterioplankton). The articles most frequently cited in the following subsections are briefly described when first mentioned and thereafter referred to for a summary of their main characteristics (Table 2).

Table 2 Summary of the most frequently quoted articles used in the Results section

References	Approach	Type of study	Latitude range	*n* lakes	Climate
Bécares et al. (2008)	LG	Mesocosm-exp	39–61°N	6	Med–Temp–Cold
Brucet et al. (2010)	CC	Semi exp lit.	38–56°N	8	Med–Temp
Brucet et al. (submitted for publication)	LG	Whole lake	28–69°N	1632	Med–Temp–Cold
Brucet et al. (2012)	CC	Semi exp lit.	38–56°N	8	Med–Temp
Danger et al. (2009)	CC	Mesocosm-exp	14–48°N	2	Trop–Temp
Declerck et al. (2005)	LG	Whole lake	36–56°N	98	Med–Temp–Cold
Griffiths (1997)	LG	Whole lake	29-–74°N	474	Subtrop–Cold
Gyllström et al. (2005)	LG	Whole lake	38–68°N	81	Warm–Temp–Cold
Havens et al. (2009)	CC	Whole lake	29–49°N	2	Subtrop–Temp
Heino (2009)	Review		World wide		

Table 2 Summary of the most frequently quoted articles used in the Results section—cont'd

References	Approach	Type of study	Latitude range	*n* lakes	Climate
Jackson et al. (2007)	CC	Whole lake	51–56°N	252	Temp–Cold
Kosten et al. (2009c)	LG	Whole lake	5–55°S	83	Trop–Subtrop–Cold
Kosten et al. (2011a)	LG	Whole lake	5–55°S, 38–68°N	143	Trop–Cold
Kosten et al. (2011b)	LG	Whole lake	5–55°S	83	Trop–Subtrop–Cold
Lacerot (2010)	LG	Whole lake	5–55°S	83	Trop–Subtrop–Cold
Lewis (1996)	CC	Whole lake	0–70°N	26	Trop–Temp
Meerhoff et al. (2007a)	CC	Semi exp lit.	32.5°S–56°N	10	Subtrop–Temp
Meerhoff et al. (2007b)	CC	Semi exp lit.	32.5°S–56°N	10	Subtrop–Temp
Muylaert et al. (2010)	LG	Whole lake	36–56°N	98	Med–Temp–Cold
Moss et al. (2004)	LG	Mesocosm-exp	36–56°N	6	Med–Temp–Cold
Pinto-Coelho et al. (2005)	LG	Whole lake	19–20°S, 27–55°N	64	Temp–Trop
Schiaffino et al. (2011)	LG	Whole lake	45–63°S	45	Cold–SubPolar
Stephen et al. (2004)	LG	Mesocosm-exp	36–56°N	6	Med–Temp–Cold
Teixeira-de Mello et al. (2009)	CC	Semi exp lit.	32.5°S–56°N	19	Subtrop–Temp

Abbreviations indicate the approach and type of study: LG, latitudinal gradient; CC, cross-comparison; exp, experimental; lit, littoral (i.e. artificial plants were introduced). Broad climate regimes were assigned based on the information given by authors or based on geographic locations, abbreviations: Med, mediterranean; Subtrop, subtropical; Temp, temperate. References are ordered alphabetically.

2. FINDINGS IN SPACE-FOR-TIME STUDIES

2.1. Richness changes with climate

The strong latitudinal relationship of increasing diversity towards low latitudes is one of the most familiar and widespread patterns in ecology and has puzzled many generations of scientists, since Darwin and von Humboldt (e.g. Allen and Gillooly, 2006; Hillebrand, 2004). Lakes are good model systems for the study of taxon richness in relation to environmental gradients, because they are well-delineated ecological entities in the landscape (Dodson et al., 2000), yet compared with terrestrial systems studies on diversity patterns in lakes are underrepresented (Waide et al., 1999), especially in shallow lakes (e.g. Declerck et al., 2005; Jeppesen et al., 2000; Kruk et al., 2009; Scheffer et al., 2006). In this section, we will briefly cover the findings for shallow lakes, stressing that we do not attempt to explain the patterns in taxonomic richness solely (or mostly) by climate factors, due to the fundamental effects of evolutionary history and biogeography on this particular variable.

2.1.1 Fishes

At a global scale, a meta-analysis detected a decreasing latitudinal pattern in fish assemblage species richness in shallow lakes (linear regression, $n=69$, $r^2=0.88$, $p<0.0001$) (from González-Bergonzoni et al., 2012) (Fig. 1A). Greater fish biodiversity in warm climates has also been found in a worldwide latitudinal gradient investigation and in lowland versus high altitude lakes (Amarasinghe and Welcomme, 2002). This recurrent pattern is also consistent at latitudinal continental scales in Europe (Brucet et al., submitted for publication) and North America (Griffiths, 1997; Mandrak, 1995) (Table 2). Particularly, the broad-scale patterns of fish species richness and diversity in European lakes were mainly explained by environmental temperature together with morphometric variables (Brucet et al., submitted for publication; based on $n=1632$ lakes), while in a wide spatial range of Chinese lakes ($n=109$ lakes, Zhao et al., 2006), climatic variables (temperature and potential evapotranspiration) were among the most important explanatory variables. Specifically in shallow lakes, cross-comparison studies of littoral communities have found higher fish richness at the warmer locations in a subtropical–temperate investigation (Meerhoff et al., 2007a; Teixeira-de Mello et al., 2009) and in a temperate versus cold climate comparison (Jackson et al., 2007).

In brackish lakes, the higher richness of fish at warmer locations seems, however, less pronounced, as found in a similar cross-comparison study of temperate and Mediterranean shallow lakes (Brucet et al., 2010). Overall, there is an increase in fish richness from cold to warm climates, modulated by hydrology-related factors (such as drought intensity and salinity) that may promote lower relative richness in warm lakes with high hydrological stress (Jeppesen et al., 2010a; M. Beklioğlu et al., unpublished data).

The functional diversity of fish seems also to increase with decreasing latitude, as found in a review of North American freshwaters (Moss, 2010) and in cross-comparison studies (Teixeira-de Mello et al., 2009). In particular, the richness of omnivorous fish species follows a strong and negative worldwide latitudinal gradient in all aquatic ecosystem types, this being associated with the same pattern followed by taxonomic richness (González-Bergonzoni et al., 2012), not least in shallow lakes (Jeppesen et al., 2010a).

2.1.2 Macroinvertebrates

Studies on invertebrate diversity patterns in freshwaters are clearly unbalanced in that they mainly focus on aquatic insects and primarily in running waters. Overall, the few available studies suggest that differences in macroinvertebrate richness between climate regions are ambiguous and depend on taxonomic groups (Heino, 2009). At a global scale, the diversity of some invertebrate groups, such as the orders Ephemeroptera and Trichoptera, seems to show a negative relationship with latitude (Barber-James et al., 2008; de Moor and Ivanov, 2008; data not available for our meta-analysis), while other groups (e.g. Plecoptera, Decapoda, Simulidae dipterans) do not show any clear trend (e.g. Plecoptera species richness, linear regression $r^2=0.001$, $p>0.05$; with data from Heino, 2009 and Palma and Figueroa, 2008) or seem to increase their diversity towards high latitudes, as in some studies in large biogeographic areas (Crandall and Buhay, 2008; McCreadie et al., 2005). At family level, a latitudinal gradient in assemblage composition of chironomids has revealed the important role of ambient temperature as a key factor for the distribution and relative abundance of different taxa (Larocque et al., 2006). Unfortunately, in most studies, the original data are not shown, the taxonomic resolution achieved differs from paper to paper, or data on lakes and streams are pooled, preventing us from conducting a formal meta-analysis of all published works.

In the few SFTS studies including only shallow lakes, plant-associated macroinvertebrates had much lower richness (at different taxonomic resolution, mostly order level) in the subtropics than in comparable temperate lakes (Meerhoff et al., 2007a), a climate pattern also found in brackish lakes in a similar cross-comparison study between Mediterranean and temperate regions (Brucet et al., 2012) (Fig. 1B, Table 2). Similarly, in a study comparing 16 tropical and 10 temperate lakes, Lewis (1996) found lower diversity and abundance of benthic invertebrates in the tropics, especially among Nematoda and Chironomidae (although data are not available for our meta-analysis). Higher risk of drought and large variations in salinity at the southernmost locations may reduce the species richness in arid or Mediterranean shallow lakes (Boix et al., 2008).

2.1.3 Zooplankton

In contrast to theoretical predictions, the literature shows evidence for a negative or no relationship between mesozooplankton (i.e. cladocerans and copepoda) richness and increasing temperature (in our review, decreasing latitude) (Fig. 1C). Cladoceran species richness exhibited no significant relationship with latitude (from 58.1° to 70.7°N) and altitude (from 0 to 1000 m) over a set of 336 Norwegian lakes, although high latitude and high altitude lakes had relatively low richness (Hessen et al., 2006): these differences were attributed to constraints imposed during colonization and to temperature, although the main explanatory factor was, in fact, productivity. Lewis (1996) reported a similar number of zooplankton species (in total and for cladocerans, copepods and rotifers), with slightly more cladoceran species in temperate than in tropical lakes. A decline in cladoceran and copepod species richness, in this case with increasing surface water temperature, was also observed with a data set of 1042 lake-years collected from 53 lakes in the northern hemisphere (Shurin et al., 2010; showing no latitudinal data and thus unavailable for our meta-analysis). Temperate shallow lakes often host a richer local assemblage of cladocerans (at genus level) than warm shallow lakes (Fig. 1C) as found in different studies (LSmeans $p < 0.001$ in Meerhoff et al., 2007b; no statistics reported in Pinto-Coelho et al., 2005; see details in Table 2). Typically, larger-bodied genera, such as *Daphnia*, *Sida*, *Eurycercus*, *Leptodora* and *Polyphemus*, were missing from both pelagic and littoral habitats in warm shallow lakes (Meerhoff et al., 2007b), which is in accordance with previous studies showing that large zooplankters were rare in tropical lakes (Fernando, 2002; Fernando et al., 1987; Lewis, 1996). In brackish lakes, however, cladoceran species richness was similar in temperate and Mediterranean regions and salinity emerged as the key controlling factor (Brucet et al., 2009). Overall, local factors such as productivity, salinity and fish predation pressure, despite being factors likely ultimately related to climate often seem better proximate predictors of cladoceran richness than temperature itself.

2.1.4 Macrophytes

The few latitudinal gradient studies on macrophytes diversity are not specific to shallow lakes; they refer to freshwaters in general and often show contrasting results and unclear or no latitudinal pattern. A comparison between temperate and tropical freshwaters revealed higher macrophytes richness in temperate regions, a pattern attributed to the frequent dominance of a particular species within a given water body in the tropics (Crow, 1993). A Finnish study, however, reported a general decline in macrophyte richness in lakes towards the north (Linkola, 1933; reported in Rorslett, 1991), related to the duration of the ice-free period since short growing seasons may not only limit macrophyte growth but their distribution as well (Heino, 2001). Latitudinal changes in macrophyte richness have been recorded in the Great Lakes basin (Canada, USA), largely associated with latitudinal changes in geology and related factors such as sediment composition (Lougheed et al., 2001) rather than climate. Yet another study in North European lakes demonstrated no or only marginal latitudinal changes in macrophyte diversity (Rorslett, 1991), coinciding with the lack of latitudinal pattern in submerged macrophyte species richness in 83 South American shallow lakes (S. Kosten et al., unpublished data). Although regional and continental studies seemed to be inconclusive, a global analysis of the numbers and distribution of vascular macrophytes in freshwaters (not discriminating shallow lakes) showed that their diversity is highest in the tropics (Afrotropics, Neotropics and Orient) and lowest in the Nearctic, Palaeoarctic and Australasia (Chambers et al., 2008), suggesting a marked divergence between alpha and gamma diversity.

2.1.5 Phytoplankton

Reports about phytoplankton composition and richness across latitudes, especially those including warm shallow lakes, are very scarce. Altitudinal gradient studies of phytoplankton richness are also limited (but see Wang et al., 2011). However, Stomp et al. (2011) analyzed patterns of species richness of freshwater phytoplankton in 540 lakes and reservoirs across the United States and found strong latitudinal (range: 26–51°N; linear regression: $r^2=0.16$,

$p < 0.001$; Fig. 1D), longitudinal and altitudinal gradients, overall showing an increase in richness with rising water temperature. A comparison between temperate and subtropical climates reported a higher richness of species comprising at least 5% of total biovolume in the subtropical lakes (Kruskal–Wallis test $H = 14.31$, $p < 0.002$, see Appendix C).

In terms of composition, warmer phytoplankton communities are often described as overlapping with temperate communities and little endemism (Lewis, 1996). Phytoplankton composition (considering both phylogenetic groups and dominant species) seemed to be affected more by local conditions and nutrient concentrations than by factors associated with latitude *per se*, as found in 27 lakes situated at cold-temperate (56–60°N) and subarctic (67–68°N) latitudes (Trifonova, 1998). Local factors such as the presence of submerged vegetation cover may also override latitudinal differences, as described in a European study revealing a negative association between phytoplankton richness and macrophyte cover in 98 shallow lakes (Declerck et al., 2005), although the opposite pattern was traced in subtropical shallow lakes (Kruk et al., 2009).

2.1.6 Periphyton

That periphyton, including all substrata-attached microalgae and associated bacteria and fungi, plays a fundamental role in shallow lake food webs has been demonstrated by several studies during the past decade (Liboriussen and Jeppesen, 2003; Vadeboncoeur et al., 2003, and references herein). However, we did not find any studies on the latitudinal variation on periphytic algae richness. To remedy this, we conducted a meta-analysis of patterns in species richness and biomass (the only community traits measured or shown for this assemblage, as far as we are aware) of periphyton algae over a latitudinal gradient ranging from 68°S to 83°N, including data from a total of 23 publications and 14 countries (see Appendix A for further information). In our meta-analysis, we detected a strong and significant linear decrease in periphyton algae richness with increasing latitude (linear regression: $r^2 = 0.755$, $p < 0.001$, Fig. 1E).

2.1.7 Bacterioplankton

Inclusion of bacteria in the SFTS literature is restricted to a few papers on investigations carried out along gradients in Europe (Declerck et al., 2005; Van der Gutch et al., 2006), Argentina and Antarctica (Schiaffino et al., 2011), and China (Wu et al., 2006). We therefore reviewed the literature and extracted crude data on community traits whenever available from both gradient and single site studies ($n=140$ lakes in 17 publications, see Appendix B for further information). We could not detect significant trends in bacterioplankton richness (usually estimated through molecular fingerprinting methods) along a latitudinal (or temperature) gradient from our dataset (linear regression $n=43$, $r^2=0.008$, $p=0.58$; Fig. 1F). Bacterioplankton richness is described in just a few studies, mostly conducted in temperate regions, potentially biasing our results. However, there is increasing evidence for the existence of an inverse relationship of richness with latitude for both continental (Schiaffino et al., 2011) and marine (Fuhrman et al., 2008; Pommier et al., 2007) water bodies. Despite the underlying mechanisms still being under debate (e.g. species sorting, history), there are indications that the bacterial community composition in the water column (including shallow systems) changes along latitudinal gradients (Schiaffino et al., 2011; Sommaruga and Casamayor, 2009; Van der Gucht et al., 2007; Yannarell and Triplett, 2005).

2.2. Climate effects on biomass

2.2.1 Fishes

Fish biomass seems to increase with decreasing latitude, or increasing ambient temperature (Fig. 1G). Pelagic fishes increased significantly in biomass in warm climates (as CPUE; g net^{-1} night^{-1}) in a series of shallow lakes of varying nutrient concentrations along a climate gradient in Europe (ANOVA $p<0.04$, for log-transformed biomass, recalculated from Gyllström et al., 2005) (Fig. 1G). Also, cross-comparison studies have found a significantly higher fish biomass (as g m^{-2}) in warmer climates, both in freshwater (ANOVA $F_{1,100}=40.7$, $p<0.0001$, Teixeira-de Mello et al., 2009) and in brackish shallow lakes (Brucet et al., 2010) at similar TP concentrations in each set under comparison (although brackish lakes seem to have a lower biomass than freshwaters). In a European-scale study (Brucet et al., submitted for publication), fish biomass was mainly related to lake productivity, but when differences in productivity were taken into account in the analysis, altitude was negatively correlated with biomass.

2.2.2 Macroinvertebrates

Information on changes in the biomass of macroinvertebrates along a climate gradient is extremely scarce in general and almost absent in shallow lake studies: most research records abundance and identity at local scales, rather than biomass. Therefore, we were not able to identify or test a trend of change in macroinvertebrate biomass under different climates or latitudes.

2.2.3 Zooplankton

Although our analysis of published metadata (Table 1) showed no significant relationship, zooplankton biomass exhibits an apparent increase towards colder regions of the globe (Fig. 1I). A latitudinal gradient analysis of 81 lakes in Europe (38–68 °N) documented lower zooplankton biomass levels in the warmer lakes of Mediterranean Spain, with climate being the second most important predictor of zooplankton biomass after TP ($F=19.7$, $r^2=0.49$, $p<0.001$, $n=65$) (Gyllström et al., 2005). Lower zooplankton biomass at lower latitudes was also evident in an outdoor experimental study where nutrient enrichment and fish predation were analyzed in six European lakes (Vakkilainen et al., 2004). A comparison of zooplankton biomass between two eutrophic lakes, one in temperate Italy and the other in subtropical Florida (USA), documented higher biomass in the temperate lake (Havens et al., 2009). The small eutrophic tropical reservoirs in the study by Pinto-Coelho et al. (2005) were the exception to this pattern, with higher zooplankton biomass, particularly of Cladocera, in the warm region. However, the subtropical Florida lakes included in this same study had a lower cladoceran total biomass compared to the colder lakes. Mediterranean brackish lakes also exhibited a remarkably lower biomass of large zooplankton than similar temperate lakes (Brucet et al., 2010). In all cases, colder lakes had a significantly higher proportion of *Daphnia* than similar warmer systems (Brucet et al., 2010; Gyllström et al., 2005; Jackson et al., 2007; Meerhoff et al., 2007b). Trends of change in zooplankton biomass with altitude appear to be less clear than along latitudinal gradients, according to a study of subarctic ponds (Karlsson et al., 2005).

The majority of the articles analyzed address the effect of lake trophic state (i.e. nutrient concentrations) on zooplankton biomass, showing ambiguous relationships between both variables and climate. TP seems the most important predictor of zooplankton biomass and the biomass of large pelagic crustaceans, irrespective of the climatic region considered, as found along a European latitudinal gradient (Gyllström et al., 2005), a latitudinal gradient in Turkey (Beklioğlu et al., in prep.) and in cross-comparisons (Jackson et al., 2007): all these studies indicated that increases in TP concentration have a positive effect on zooplankton biomass. However, the response of zooplankton, and particularly that of cladocerans, to increases in TP was more intense in tropical regions (measured as the slope in the regressions) (Pinto-Coelho et al., 2005). Interestingly, intermediate (subtropical) lakes in this study had the flattest slope in the relationship between zooplankton biomass and TP (Pinto-Coelho et al., 2005), as also found in subtropical Lake Apopka (Florida, USA) (Havens et al., 2009).

2.2.4 Macrophytes

Most evidence of effects of warming on macrophyte biomass comes from interannual comparisons rather than from SFTS studies, preventing us from assessing potential latitudinal trends. Early season warm temperatures seem to favour a strong increase in whole-lake submerged macrophyte biomass, as recorded in five lakes in the Eastern Townships of Quebec, Canada, despite increased turbidity in such period (Rooney and Kalff, 2000). However, this temperature effect seems to be overruled by the strong influence of underwater irradiance, which in turn is most often impacted by lake trophic state. In an analysis of data from 139 lakes between latitudes 46° and 69°, underwater irradiance and not latitude explained a large part of the variance in submerged macrophyte biomass (Duarte et al., 1986). For emergent macrophytes, however, lake morphometry is an important factor determining plant biomass; however, no significant positive relationship between the biomass per area colonized and latitude has been found (Duarte et al., 1986).

2.2.5 Phytoplankton

Most studies deal with biomass of phytoplankton in terms of chlorophyll-a (Chl-a) concentrations (indirect estimate of biomass), and, to a lower extent, biovolume (directly related to biomass). We analyzed published data from a large amount of shallow lakes in different climate zones (Tables 1 and 3) and discovered neither linear trends with latitude nor clear differences between climatic regions, except for higher total biomass (as Chl-a) in the tropics and temperate systems, followed by subtropical and subpolar regions (Fig. 1J; $n=90$ lakes, Kruskal–Wallis: $H=205$, $p<0.001$). Larger biomass of phytoplankton in the tropics has been reported before (Danger et al., 2009; Huszar et al., 2006; Lewis, 1996, 2000). Kruk et al. (2010) found that the largest values of Chl-a occurred in subtropical lakes ($n=40$) and the lowest in tropical lakes ($n=42$, over a total of 210 shallow lakes), while Kosten et al. (2011b) recorded a decrease in phytoplankton Chl-a from subpolar regions towards the tropics in South America (83 lakes in total). On the other hand, in a study of 143 shallow lakes along a latitudinal transect ranging from subarctic Europe to southern South America, lakes in warmer climates did not exhibit an overall higher phytoplankton biomass, although cyanobacteria biomass seemed favoured (Kosten et al., 2011a). Along a latitudinal gradient in Europe, both high and low Chl-a concentrations were found in southern and northern lakes (Nõges et al., 2003), while higher Chl-a

Table 3 Relationship between phytoplankton biomass (as chlorophyll-a, Chl-a, μg L^{-1}) and total phosphorus (TP, μg L^{-1}) in shallow lakes under different climate regimes, as described in published studies

References	Location	Climate	*n*	Model	R^2	Slope
Antoniades et al. (2003a)*	Canadian Arctic: Ellesmere Island	Polar	31	$y=0.16x-0.22$	0.36	0.16
Antoniades et al. (2003a)*	Canadian Arctic: Prince Patrick Island	Polar	29	$y=0.007x-0.40$	0.00	0.01
Antoniades et al. (2003b)*	Canadian Arctic: Ellef Ringnes Island	Polar	24	$y=0.75x-1.58$	0.36	0.75
Flanagan et al. (2003)	Canada	Cold	113	$y=0.33x-0.42$	0.07	0.33
Kosten et al. (2011b)	Argentina	Cold	11		0.94	
Mazumder and Havens (1998)	North America, Europe	Cold	126	$y=(0.87\pm0.05)x+0.60\pm0.07$	0.73	0.87
Ogbebeo et al. (2009)*	Canada	Cold (flooded Arctic-tundra)	10	$y=0.22x+0.25$	0.38	0.22
Ogbebeo et al. (2009)*	Canada	Cold (forest-tundra)	6	$y=0.94x-0.19$	0.84	0.94
Ogbebeo et al. (2009)*	Canada	Cold (non-flooded Arctic tundra)	9	$y=0.50x+0.07$	0.39	0.50
Flanagan et al. (2003)	Canada	Temperate	316	$y=0.92x-0.38$	0.28	0.92
Jackson et al. (2007)	Canadian prairie	Cold	30	$y=0.86x+1.72$		0.86
Jackson et al. (2007)	Denmark	Temperate	222	$y=0.82x+2.35$		0.82
Kruk et al. (2011)**	The Netherlands	Temperate	95	$y=0.91x-0.33$	0.34	0.91
Meerhoff et al. (2007a,b)**	Denmark	Temperate	9	$y=1.64x-1.50$	0.70	1.64

Prairie et al. (1989)	North America, Europe	Temperate	133		0.69	0.87
Mazumder (1994)	North America, Europe	Temperate LH	126	$y=(0.87\pm0.05)x+0.60\pm0.07$	0.73	0.87
Mazumder and Havens (1998)	North America, Europe	Temperate SH	235	$y=-(0.97\pm0.02)x+0.21\pm0.03$	0.87	0.97
Mazumder and Havens (1998)	Florida, USA	Temperate SH and LH	361	$y=-(0.94\pm0.03)x+0.35\pm0.05$	0.71	0.94
Kosten et al. (2011b)	Brazil, Uruguay, Argentina	Temperate–subtropical	34		0.62	
Brown et al. (2000)		Subtropical	359		0.67	0.42
Wang et al. (2008)	China	Subtropical	45	$y=1.04x-0.97$	0.80	1.04
Kruk et al. (2009)★★	Uruguay	Subtropical	18	$y=0.25x+0.28$	0.05	0.25
Kruk et al. (2011)★★	Uruguay	Subtropical	12	$y=-0.12x+0.99$	0.00	−0.12
Meerhoff et al. (2007a,b)★★	Uruguay	Subtropical	10	$y=0.71x-0.59$	0.34	0.71
Mazumder and Havens (1998)	Florida, USA	Subtropical SH	59	$y=-(1.06\pm0.07)x+0.36\pm0.09$	0.79	1.06
Huszar et al. (2006)	South America, Africa	Tropical–subtropical	192	$y=0.70x-0.15$	0.42	0.70
Kosten et al. (2011b)	Brazil	Tropical	38		0.20	
Parinet et al. (2004)	Ivory Coast	Tropical	9	$y=0.93x+1.66$	0.32	0.93

The explained variance (R^2), slope and linear models of the relationship between $\log_{10}$ Chl-a (y) and $\log_{10}$ TP (x) in the water column of polar, cold, temperate, subtropical and tropical shallow lakes, indicating the respective references, location, and number of cases (n) are displayed. LH, lakes with large herbivorous zooplankton; SH, lakes with small herbivorous zooplankton, as classified by the authors. ★ calculated from data available, ★★ data obtained in the cited references provided by authors. References are ordered by increasing temperature of the region of the data.

concentrations were found in the southern lakes along a latitudinal gradient in Turkey (M. Beklioğlu et al., unpublished data).

Interestingly, the other proxy for total phytoplankton biomass, that is, biovolume, often shows a different pattern to that of Chl-a. In a comparison of a large data set from subtropical and temperate regions, greater total biovolume values appeared in the subtropics than in similar temperate lakes over the same wide gradient in nutrient concentration (Kruskal–Wallis test $H=11.34$, $p<0.001$, $n=650$, see Appendix C). Some works argue the tropics to have a higher frequency of cyanobacteria (Paerl and Huisman, 2009). In the subtropical–temperate lakes comparison (see Appendix C for analysis information), cyanobacteria and Dinophyceae had a significantly higher relative biovolume in the subtropics, whereas Chlorophyceae, Cryptophyceae, Euglenophyceae, Zygnematophyceae, Chrysophyceae and Xantophyceae had a higher relative biovolume in the temperate lakes (Kruskal–Wallis tests, Fig. 2). Mesocosm experiments conducted in different European climate regions have also suggested that warmer climates may increase the relative contribution of cyanobacteria to total phytoplankton biovolume in shallow lakes (Stephen et al., 2004; Van de Bund et al., 2004; Table 2) as also found in a recent mesocosm warming experiment in southern England (Yvon-Durocher et al., 2011a,b). In contrast, other mesocosm experiments

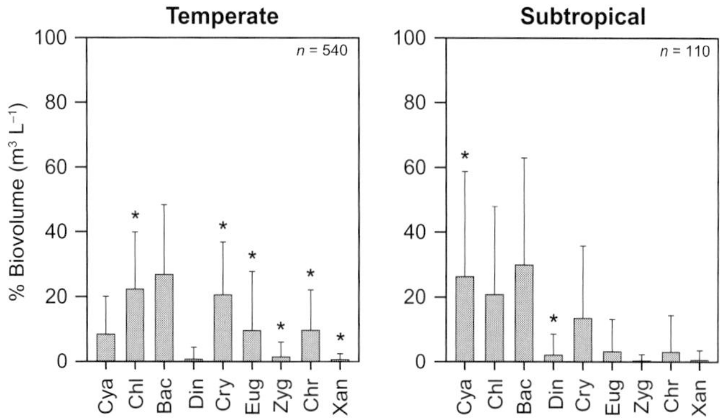

Figure 2 Relative biovolume (%) of phylogenetic classes of phytoplankton in shallow lakes located in temperate ($n=540$) and subtropical ($n=110$) climates. Abbreviations: Cyanobacteria (Cya), Chlorophyceae (Chl), Bacillariophyceae (Bac), Dinophyceae (Din), Cryptophyceae (Cry), Euglenophyceae (Eug), Zygnematophyceae (Zyg), Chrysophyceae (Chr) and Xanthophyceae (Xan). Error bars correspond to 1 standard error and * indicates classes with significantly higher biovolume according to the Kruskal–Wallis test ($p<0.05$). Data were extracted from Kruk *et al.* (2011), see Appendix C for a description of methodology.

conducted in tropical and temperate shallow lakes have not revealed any effect on cyanobacteria or any drastic changes in phytoplankton community structure (Danger et al., 2009).

As the relationship between Chl-a and nutrients is directly associated with the predictability of phytoplankton biomass, it is important to analyze its variability and strength under different climatic conditions. The relationship between phytoplankton Chl-a concentrations and nutrients (typically TP) along latitudinal gradients is one of the most commonly reported in the literature of gradients (Table 3). An increase in phytoplankton biomass with increases in TP occurs across climates (positive slope in the regressions) (Fig. 3). However, we found no clear patterns of change in the slope or intercepts across climates (Table 3), even with opposite results being evident in studies at the same latitudinal location (Brown et al., 2000; Gyllström et al., 2005; Huszar et al., 2006; Jeppesen et al., 2007; Muylaert et al., 2010). Although differences were not significant, high variability in the importance of nutrients for phytoplankton biomass (indicated by a wide range of r^2 values in the regressions) emerges in subtropical and also in cold regions (Huszar et al., 2006; Kosten et al., 2009b,c) (Fig. 3), likely suggesting that factors other than TP (or

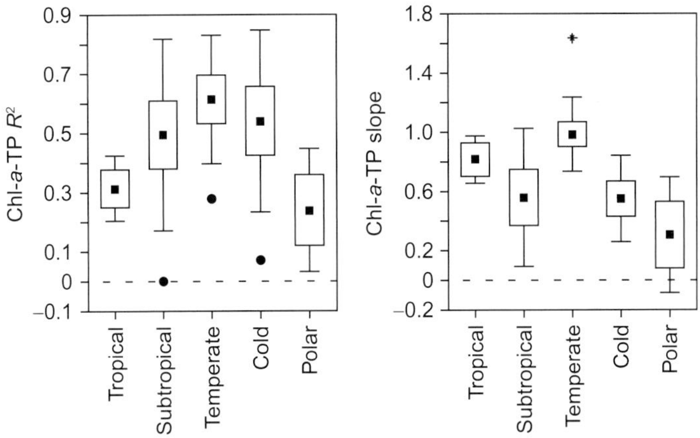

Figure 3 Relationship between phytoplankton biomass and nutrients along a climate gradient, showing the explained variance (R^2) and slope in the linear regressions between total phosphorus (TP) and phytoplankton chlorophyll-a (Chl-a) in shallow lakes from tropical ($n=3$), subtropical ($n=8$), temperate ($n=9$), cold ($n=7$) and polar ($n=3$) regions, according to published works applying the space-for-time approach. Box-plots show mean, mean ± SE, mean ± SD and extreme values. Respective references and full regression models are shown in Tables 1 and 3.

nutrients) explain a large proportion of the phytoplankton biomass variation. Besides, the amount of Chl-a per unit of TP may be higher in temperate and tropical lakes than in other climatic regions (marginally significant differences, Kruskal–Wallis test, $p=0.07$; Fig. 3).

2.2.6 Periphyton

Only a few published works applying the SFTS approach have included periphyton in their target communities, and those that have, exclusively focus on biomass (Bécares et al., 2008; Brucet et al., 2010, 2012; Meerhoff et al., 2007a). In our meta-analysis, we found a significant and positive linear response of periphyton algae biomass with increasing latitude ($n=16$ publications, $r^2=0.378$, $p<0.01$) (Fig. 1K). This pattern is consistent with findings in other studies based on both field sampling campaigns in temperate, subarctic and Antarctic regions (Hansson, 1992), mesocosm experiments along a latitudinal gradient (Bécares et al., 2008) and cross-comparisons (Brucet et al., 2012; Meerhoff et al., 2007a). Nonetheless, the effects of latitude or temperature might be very complex, because the periphyton response curve to increasing nutrients shows saturation at different points depending on climate regions and is clearly higher in the colder regions of northern Europe than in the Mediterranean zone (Bécares et al., 2008).

2.2.7 Bacterioplankton

In our constructed database, we could not detect any significant relationship between bacterioplankton biomass and latitude, possibly due to the relatively few geographic locations included (despite the fairly high number of systems, $n=33$ lakes, Fig. 1L).

2.3. Climate effects on density

Abundance of individuals is typically measured, but not always reported, for several communities in shallow lakes, particularly animal or single-cell communities. Bacterioplankton represent an exception here, since most works provide numerical counts of abundance rather than estimations of biomass.

2.3.1 Fishes

Published SFTS literature reporting absolute fish density data is rare, and, typically, density and biomass proxies such as CPUE are shown. However, the scarce findings support the trend of higher fish densities towards warmer climates, over a wide range in nutrient concentrations and system area. The evidence comes from cross-comparison studies in warm and temperate climates, both in freshwater (Teixeira-de Mello et al., 2009) and in brackish (Brucet et al., 2010) shallow lakes (Fig. 1), and from an extensive study including both latitudinal and altitudinal gradients (Brucet et al., submitted for publication). A comparison of fish abundance in cold Canadian lakes to comparatively warmer Danish shallow lakes showed greater abundances in Denmark, as a likely consequence of greater winter mortality in Canada (Jackson et al., 2007). On the other hand, drought-induced fish mortality can also explain the lower fish densities found in southern Mediterranean lakes (M. Beklioğlu et al., unpublished data). All these works, despite showing opposite results to general theoretical expectations based primarily on metabolism (Brown et al., 2004; Fig. 1M), discard potential effects of trophic state on the observed patterns, as the cross-comparison studies included lakes paired in terms of nutrient levels and the large spatial gradient study also covered the same gradient in TP for the different locations.

2.3.2 Macroinvertebrates

We found no latitudinal gradient study in shallow lakes including macroinvertebrates, and thus our climate-related data come only from cross-comparison studies. In contrast to fishes, but fitting predictions, the relatively few available data indicate greater densities at colder locations (Fig. 1N). Cross-comparison studies, both in freshwater (Meerhoff et al., 2007a) and in brackish (Brucet et al., 2012) shallow lakes, traced up to eight times lower densities of plant-associated macroinvertebrates in the warmer systems.

2.3.3 Zooplankton

Our meta-analysis of the SFTS studies showed a significant trend of lower microcrustacean abundance towards lower latitudes (linear regression: $r^2=0.16$, $p<0.05$, $n=25$, Fig. 1O), although variability in density patterns was high. The cross-comparison studies detected overall higher cladoceran densities in colder climates, both in freshwater (Meerhoff et al., 2007b) and in brackish shallow lakes (Brucet et al., 2010). However, temperature *per se* did not explain the low abundances of zooplankton in warm lakes in the latter, since low abundances also occurred in cold lakes with high fish densities (Brucet et al., 2010), suggesting a top-down food web effect. In contrast, both cladocerans and copepods were more abundant in the warmer lakes in a comparative study of diel migration patterns of microcrustaceans in similar shallow lakes between latitudes 39° and 61°N in Europe (Hansson et al., 2007). Both the highest and lowest crustacean densities occurred in the warmer lakes in a study comparing several lakes and reservoirs of different depth in temperate, subtropical and tropical countries (Pinto-Coelho et al., 2005). Reservoirs might exhibit different patterns than true lakes, since deeper sections are common, which might offer a hypolimnion refuge to large zooplankton, thereby masking the otherwise typical patterns in warm shallow lakes. Calanoid copepods, however, were more abundant in the colder lakes (Pinto-Coelho et al., 2005). The patterns followed by other zooplankton groups, such as rotifers, remain unexplained due to the scarce number of works including this fraction. Typically, however, rotifers seem the dominant group in warm lakes (Brucet et al., 2010; Fernando, 2002).

2.3.4 Macrophytes

Different variables that may be used as proxies for density are regularly reported in shallow lakes, such as the percent area covered and the volume of the lake taken up by submerged macrophytes. All these variables are strongly influenced by the underwater light conditions and therefore by lake trophic state but are influenced by temperature as well. High coverage of submerged macrophytes has been related to high spring temperatures in temperate systems (Rooney and Kalff, 2000; Scheffer et al., 1992). In contrast, both space-for-time (Duarte et al., 1986; Kosten et al., 2009a) and inter-annual comparative studies (Hargeby et al., 2004) found that cold winters—typically quantified by the cumulative number of frost—days generally lead to a higher coverage of submerged macrophytes. Even after accounting for the effect of underwater light conditions, however, the variance in coverage among lakes located at similar latitudes remains high (Kosten et al., 2009a), again suggesting action of different processes at a local scale. This likely explains why no latitudinal gradient in submerged macrophyte coverage emerged from the literature, even when variations in lake nutrient status were taken into account, as done for 83 shallow lakes along a latitudinal study in South America (S. Kosten et al., unpublished data) (Fig. 4). The great subjectivity and variety associated with plant cover estimation methods (typically visual, for a brief overview of different methods, see Kosten et al., 2009a) may also prevent the identification of clear patterns.

2.3.5 Phytoplankton

We found only one study reporting microalgal density values in connection to environmental gradients and particularly with latitude (Søballe and Kimmel, 1987). These authors found a negative relationship between algal

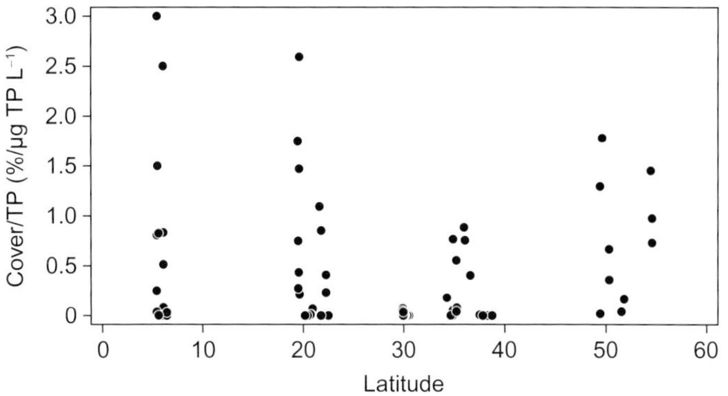

Figure 4 Coverage (%) of submerged macrophytes corrected by the lake total phosphorus concentration in 83 shallow lakes along a latitudinal gradient in South America. S. Kosten et al. (unpublished data), see Kosten et al. (2009a) for details on methodology.

counts (log-transformed) and latitude (correlation $r = -0.45$, $p < 0.01$) for natural lakes all over United Sates. Due to the lack of raw data and the deeper nature of the lakes including in this study (ca. 8 m *versus* ca. 5 m in our review), we could not include this finding in our analysis (Fig. 1P).

2.3.6 Bacterioplankton

In our constructed database (Appendix B), we found a significant increase in bacterial abundance in the water column with decreasing latitude (linear regression: $n = 133$, $r^2 = 0.08$, $p = 0.001$; Fig. 1R), which is in agreement with a previous latitudinal gradient study specifically including shallow lakes (Schiaffino et al., 2011). This correlation was also clear when we grouped the lakes according to climatic zones (which do not necessarily overlap with latitude), following the revision by Leemans and Cramer (1991) of the Köppen climate system (1936) (Fig. 5). We attempted to identify the main factors behind the emerging pattern by applying Spearman's correlations and stepwise multiple regressions with the following parameters: latitude, altitude, lake area, water temperature, conductivity, TP, TN, phytoplankton Chl-a, and dissolved organic carbon (DOC) (see Appendix B for a summary of results). In particular, the concentration of DOC is a major controlling factor of bacterial production in shallow polar lakes (Karlsson et al., 2001). Increasing temperatures may stimulate photosynthate extracellular release, thus indirectly contributing to increase DOC concentrations (Morán et al., 2006). However, in our analysis, temperature and, secondly, Chl-a were the two factors explaining most of the latitudinal variation in

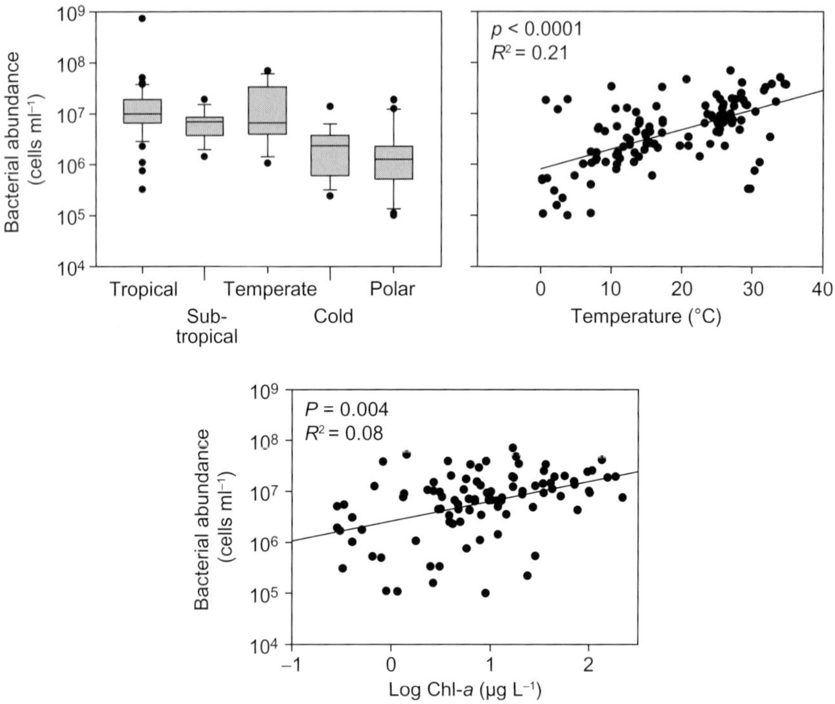

Figure 5 Density of bacterioplankton in lakes along a climate gradient as found in a meta-analysis, showing densities in different climate regions and the relationship between bacterioplankton and temperature and bacterioplankton and phytoplankton biomass (measured as Chl-a). Box-plots show median values (lines), while lower and upper boundaries depict the 25th and 75th percentiles, respectively; whiskers indicate the 10th and 90th percentiles and dots represent outliers. For both regressions, explained variance (R^2) and probability (p) are shown. Respective references are shown in Appendix B.

bacterioplankton abundance (adj. $r^2=0.66$, $p=0.005$) (Fig. 5). Latitude and DOC, however, co-varied strongly with temperature ($r=-0.83$, $p<0.0001$, $n=112$; and $r=-0.94$, $p<0.0001$, $n=26$, respectively) and were, therefore, not included in the regression model due to multicollinearity.

Our regression model agrees with results obtained by White et al. (1991), who found strong dependence of bacterial growth rate on temperature and also that a substantial portion of the residual variation was explained by phytoplankton biomass in a broad range of aquatic systems, although no shallow lakes were included in their investigation.

2.4. Climate effects on body size and size structure

The reduction in body size of aquatic organisms seems to be one of the clearest responses to environmental warming, and it has even been described as the third universal response to warming together with changes in phenology and distribution (Daufresne et al., 2009; Gardner et al., 2011). In this review, we sought to test for general trends in body size and community size structure across the different trophic levels that typically constitute shallow lake food webs.

2.4.1 Fishes

Mean body size of fish at the assemblage level tends to decrease with increasing temperature (Fig. 1S), as shown in several latitudinal and cross-comparison studies of lake fish assemblages (e.g. Griffiths, 1997; Meerhoff et al., 2007a; Teixeira-de Mello et al., 2009). Littoral subtropical fishes were significantly smaller than their temperate counterparts in cross-comparison studies (LSmeans $p<0.0001$ in Meerhoff et al., 2007a; ANOVA $F_{1,94}=32.5$ $p<0.0001$ in Teixeira-de Mello et al., 2009). A similar climate pattern occurred in the brackish shallow lakes, although the differences were not statistically significant, most likely due to an overriding effect of salinity over temperature (Brucet et al., 2010). An extensive fish analysis of North American lake fish showed that the proportion of large fish species (>20 cm TL) within the assemblage increased with latitude and that small fish species (<20 cm TL) were more common at lower latitudes (Griffiths, 1997). The same trends of an increase in the relative proportion of small fishes (<10 cm standard length) at lower latitudes occurred along a latitudinal gradient in South America (Lacerot, 2010), and in a review of lake studies from 11 European countries (Brucet et al., submitted for publication). Fish size has been mainly related to altitude, air temperature and amplitude of temperature, whereas trophic state was of little importance in a large study of European lakes (Brucet et al., submitted for publication). Lack of crude data on body sizes (instead of proportions of size classes) prevents us from incorporating several of these studies in our analysis, but the findings support the pattern of a reduction in the mean body size of fish communities with decreasing latitude and increasing temperature.

The body size difference associated with latitude seems clear at the assemblage or community level, whereas the pattern at population level is more ambiguous (Jeppesen et al., 2010a), probably as a result of contingent effects of species-specific biology and biotic interactions. Typical temperate fish species, such as roach (*Rutilus rutilus*) and perch (*Perca fluviatilis*), show different responses with latitude; perch having the strongest decrease in mean body size with decreasing latitude or increasing temperature, while the mean body size of roach does not exhibit any such changes (Jeppesen et al., 2010a).

2.4.2 Macroinvertebrates

Overall, data on changes in the body size of macroinvertebrates along spatial gradients or changing temperatures are still too scarce and fragmented to permit the drawing of firm conclusions (Fig. 1T), highlighting another important research gap that merits further work.

Using species lists and occurrence data on diving beetles (Coleoptera, Dysticidae) in more than 400 Canadian lakes and ponds, Vamosi et al. (2007) found that the proportion of large species tended to increase at higher latitudes (along a narrow gradient: 49–55°N). Along an elevation gradient, both the proportion of large species and total species richness peaked at mid-altitudes, being exceptionally low at high altitudes (above 2000 m). The authors attributed this to oxygen limitation and low productivity at high altitude locations, overriding the expected temperature effect for body sizes and species richness. In the cross-comparison study of subtropical and temperate shallow lakes, the taxa occurring in the former were characterized by either very small or very large body sizes (such as shrimps, applesnails, crayfish, etc.), whereas the taxa in the set of temperate lakes covered a far wider range in body sizes (Meerhoff et al., 2007a).

2.4.3 Zooplankton

A comprehensive review has shown that cladoceran mean body size decreases towards lower latitudes ($n = 1100$ water bodies, latitudinal gradient between 81°N and 77°S, Gillooly and Dodson, 2000) (Fig. 1U), a pattern mostly driven by differences in body size of individual species of three large and dominant pelagic genera (*Daphnia*, *Ceriodaphnia* and *Diaphanosoma*). In South America, zooplankton mean body size (length), as well as the body size of different taxonomic and functional groups, decreased from temperate to tropical lakes (Lacerot, 2010). This pattern was generated by large differences in the body length of pelagic cladocerans, particularly *Ceriodaphnia*, *Moina*, *Daphnia* and *Diaphanosoma* genera (Lacerot, 2010). Along a latitudinal gradient in Turkey, the mean body size of Cladocera was larger in the higher latitude lakes (M.Beklioğlu et al., unpublished data). In contrast, despite large differences in biomass, cladoceran zooplankton body size did not vary along climate regions in the European latitudinal gradient analysis by Gyllström et al. (2005), although the large-bodied *Daphnia* spp. were scarce in the warmer lakes. Lack of large pelagic or littoral genera occurred in freshwater subtropical lakes, whereas these taxa were common in similar temperate lakes (Meerhoff et al., 2007b). Similarly, in European brackish shallow lakes, the normalized biomass size spectra of zooplankton exhibited a unimodal pattern with dominance of rotifers in the Mediterranean lakes, whereas cold Danish lakes showed a bimodal distribution, with the second dome corresponding to cladoceran and copepod species (Brucet et al., 2010) (Fig. 6).

2.4.4 Macrophytes

Being a variable most often ignored in the plant community, we found no data reporting changes in plant body size or data allowing us to indirectly estimate body size (e.g. ratio biomass:cover) in the SFTS or in geographically isolated literature.

2.4.5 Phytoplankton

Phytoplankton body size variation with latitude is seldom reported in empirical literature, despite the fact that cells and colonies are routinely measured to estimate biovolume, one of the most typical proxies for phytoplankton biomass (Kruk et al., 2010). Although individual sizes are not available in the literature to construct a latitudinal gradient (Fig. 1V), we compared the biovolume distribution of different size classes of phytoplankton species in shallow lakes in two contrasting climates (subtropical Uruguay and temperate The Netherlands, see Appendix C for a description of the methodology). We found no significant differences in the biovolume of the different size classes, despite a higher total biovolume of phytoplankton in the subtropical lakes (see Biomass). In mesocosm experiments studying the effects of fish addition and nutrient loading on shallow lakes distributed from Finland to southern Spain (Moss et al., 2004), fish shifted the phytoplankton assemblage composition towards smaller phytoplankton, especially chlorophytes and cyanobacteria, presumably by removing large-bodied grazers; however, this effect was not affected by latitude (Moss et al., 2004; Stephen et al., 2004).

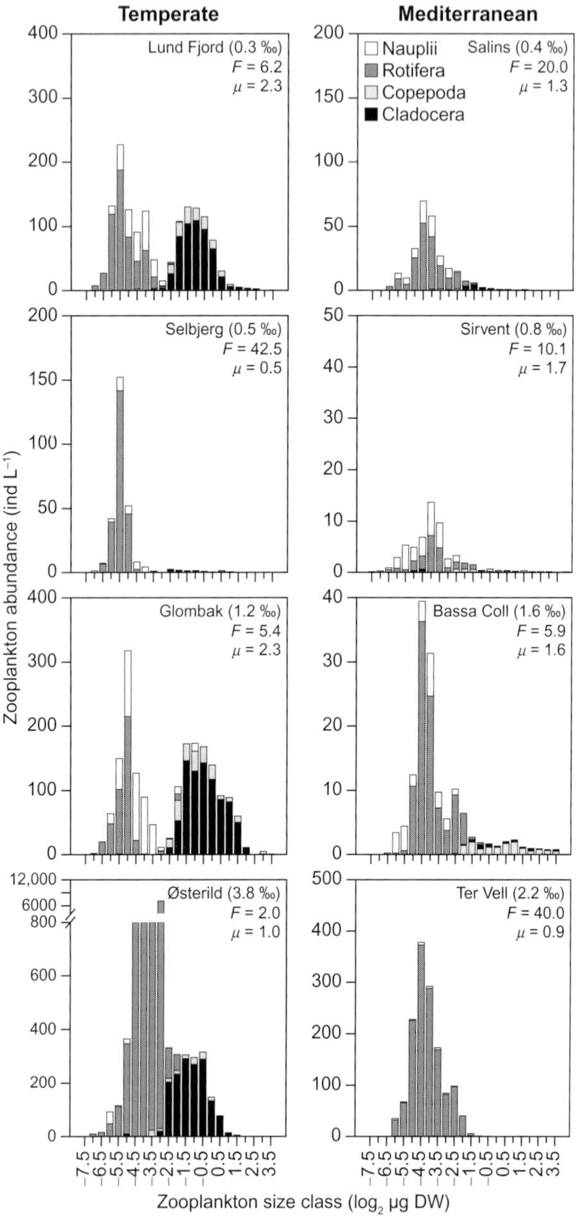

Figure 6 Zooplankton size distribution in temperate (left column) and Mediterranean (right column) brackish shallow lakes, showing the abundance contribution of nauplii, Rotifera, Copepoda and Cladocera (each group shown with a different grey tone). Each panel represents a lake, indicating the respective salinity (‰), mean zooplankton size diversity (μ) and mean fish density (F, ind m^{-2}). *Modified from Brucet et al. (2010).*

2.4.6 Periphyton

Neither in the SFTS studies nor in the geographically isolated studies used in our previous meta-analysis have we found published data on periphyton size structure in shallow lakes, preventing us from constructing our own gradient.

2.4.7 Bacterioplankton

We found a relatively large number of lakes with data available on body size (cell biovolume) of bacterioplankton ($n=48$), allowing us to construct a database for the meta-analysis. However, all are derived from two extreme geographic locations and showed strong variability, and no trend in body size could be associated with latitude (Fig. 1X). Nevertheless, a wide range of variation for the tropical locations was observed, indicating either that this trait in bacterioplankton may be more linked to local conditions than to latitude or that variations in this trait are constrained along latitudinal gradients.

2.5. Climate effects on reproduction and growth

2.5.1 Fishes

Animals with indeterminate growth experience a life history trade-off in resource allocation between onset and duration of reproduction, fecundity, growth and survival (Gunderson and Dygert, 1988; Heino and Kaitala, 1999). Three primary life history strategies representing essential trade-offs among those demographic parameters, namely, opportunistic, equilibrium and periodic, have been proposed for freshwater fishes (see Winemiller and Rose, 1992) and interpreted as being adaptive to the relative predictability and variability of the environment (Winemiller, 2005).

Most of the large-scale studies of intraspecific variation in the life history traits of fishes focus on single species, grouping populations from lentic and lotic habitats (e.g. Heibo et al., 2005; Lappalainen and Tarkan, 2007; Lappalainen et al., 2003, 2008), with very few studies on life history traits of whole fish assemblages at a continental level (Mims et al., 2010). Different hydrological characteristics of rivers and lakes could potentially influence life history traits in a variety of ways (Lytle and Poff, 2003); however, the work by Blanck and Lamouroux (2007) noted only a slight effect of habitat type on life history traits within European freshwater fishes. In shallow lakes, responses of life history traits of fishes to warming, among other responses, have been reviewed by Jeppesen et al. (2010a). With variations among communities and species-specific responses, empirical evidence indicates that higher-latitude populations grow more slowly, mature later and have longer life and reproductive spans and a greater maximal size than populations at lower latitudes or higher temperatures (Jeppesen et al., 2010a). Reproductive seasons are limited to periods of the year when limiting physical factors, typically temperature, permit the survival of offspring (Conover, 1992). In temperate systems, most freshwater fish species have seasonal reproduction, with peaks in spring or summer (Wootton, 1984). Breeding and growing seasons, growth rate, number of reproductive events per year and annual reproductive investment are often negatively related to latitude (Fig. 7, Table 4). Smaller, short-lived species with generally opportunistic strategies are more abundant at low latitudes, whereas large, longer-lived species with generally periodic or equilibrium strategies predominate at high latitudes, according to a very comprehensive study of North American fishes (Mims et al., 2010). The general pattern of variation in life history traits with latitude (Fig. 7, Table 4) seems to be common for fishes of inland waters (Conover, 1992; Gotelli and Pyron, 1991; Hubbs, 1985; Morin et al., 1982;

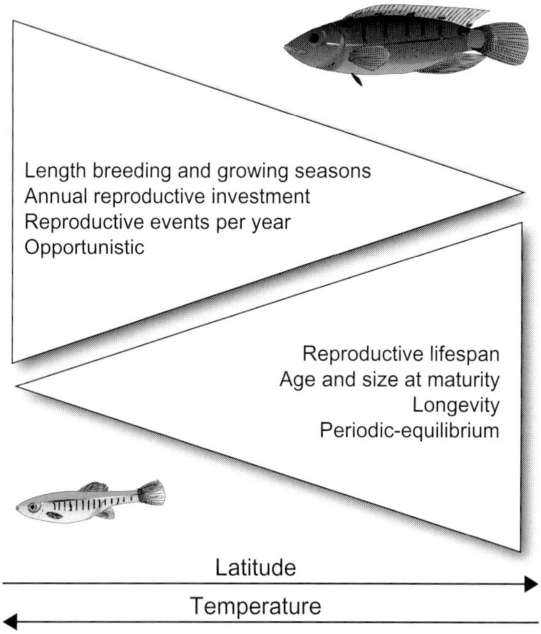

Figure 7 Expected changes in the main life history traits of fish along a temperature and latitudinal gradient based on our review of literature from inland water systems. With increases in temperature, longer breeding and growing seasons, annual reproductive investment and more reproduction events are predicted; a longer reproductive lifespan, greater age and size at maturity and longevity being negatively related with temperature. See text for details and supporting literature.

Table 4 Evidence of variation in reproductive life history traits of fishes with latitude, indicating example species and respective publications

Life history traits	Example species	Respective reference
Length of breeding and growing seasons	*Jenynsia multidentata* *Sander lucioperca* *Micropterus salmoides*	Goyenola et al. (2011), Lappalainen et al. (2003), and Rogers and Allen (2009)
Annual reproductive investment	*Jenynsia multidentata* *Perca fluviatilis*	Goyenola et al. (2011) and Heibo et al. (2005)
Reproductive life span	*Rutilus rutilus*	Lappalainen and Tarkan (2007) and Lappalainen et al. (2008)
Age and size at maturity	*Salvelinus alpinus*	Malmquist (2004) and Venne and Magnan (1989)
Longevity	*Perca fluviatilis*	Heibo et al. (2005)

Please see conceptual Fig. 7 for the expected changes.

Paine, 1990; Reznick et al., 2006) as well as marine habitats (Vila-Gispert et al., 2002).

2.5.2 Macroinvertebrates

Patterns on latitudinal changes in reproduction traits of invertebrates in diverse water bodies along a latitudinal gradient (12°N to 30°S) show that life history traits are highly variable and very likely influenced by regional and local factors (Hart, 1985). Despite this, recruitment seems to be more connected with seasonality in temperate climates, as evidenced by the greater timing flexibility in reproduction observed at low latitudes. Most of the evidence for this pattern comes from lotic environments and supports the same general trends as observed for fish, where larger body sizes, later maturity, fewer broods and larger embryos of some macroinvertebrates seem more common towards higher latitudes (reviewed by Sainte-Marie, 1991).

2.5.3 Zooplankton

We found few works on changes in zooplankton reproduction associated with a latitude or altitude gradient. An increase in autumn temperatures, as projected by different climate models, can switch the reproduction of zooplankton from sexual to asexual, resulting in a lower genetic diversity of these organisms (Chen and Folt, 1996). Likewise, studies of sediments from shallow lakes from Greenland to Spain have shown a 100-times decline in the ratio of resting eggs to carapaces of the pelagic microcrustacean *Bosmina* spp., strongly correlated with changes in summer air temperature, while Chl-a (food resources) and fish density (predation) explained a comparatively minor proportion of the variation (Jeppesen et al., 2003). Studies of chydorids have also shown higher egg:carapace ratios in cold lakes than in warmer waters (Sarmaja-Korjonen, 2003).

2.5.4 Macrophytes

Macrophytes display a range of overwintering strategies (e.g. either in the water column or as seeds, tubers or turions in the sediments), which are differently affected by changes in ambient temperature, particularly in winter. In comparison to other lake communities, there is relatively little information published on phenological changes in macrophytes as a consequence of warming. Both longer (Jeppesen et al., 2010b) and shorter growing seasons (Barko and Smart, 1981) have been reported. At higher temperatures, an enhanced growth rate of submerged macrophytes allows a quick access to the water surface (Barko and Smart, 1981), leading to deeper colonization depths and higher biomass (Rooney and Kalff, 2000). Various studies have indeed found deeper colonization depths towards the equator (Duarte and Kalff, 1987; Kosten et al., 2011b). Duarte and Kalff (1987) attributed the difference in colonization depth to higher irradiance in the tropics. Maximum daily irradiance (which is a combination of light intensity and day length), however, does not vary substantially between lower and higher latitudes (Lewis, 1986). Another factor in favour of the temperature effect is the finding in Canadian lakes where colonization depth increased during a relatively warm year, in spite of increased turbidity (Rooney and Kalff, 2000). Mesocosm investigations have also shown enhanced macrophyte growth as a result of high temperature-mediated reduction in water level (Bucak et al., 2012). Nevertheless, changes in periphyton densities, likely affected by alternate trophic web configurations under different climates (Meerhoff et al., 2007a), may also partially explain changes in colonization depth (Kosten et al., 2011b and references therein).

2.5.5 Phytoplankton, periphyton and bacterioplankton

Life history traits of other assemblages, such as of phytoplankton, periphyton and bacterioplankton, appear less frequently in the SFTS literature. For instance, studies on phytoplankton growth rates along latitudinal transects mainly belong to the marine realm (e.g. Marañón et al.,

2000). One of the reasons lies in the microscopic and unicellular type of life of these organisms, in which sexual reproduction is typically absent or rare. Most of the studies on phytoplankton reproduction deal with particular species (e.g. Hickel, 1988; Jewson, 1992), without covering many species or their change in different regions, despite the circumstance that this might be important for understanding some community processes (Eilertsen and Wyatt, 2000). Furthermore, it is relatively easy to develop laboratory experiments to evaluate the effects of temperature on microorganism growth (e.g. Butterwick et al., 2005), thus making perturbation experiments a more common approach than the SFTS.

2.6. Climate effects on intensity of trophic interactions

Climate warming may affect not only species performances and community traits but also species interactions. Interspecific interactions within food webs or other multi-species networks are highly underrepresented within the context of climate change research (Woodward et al., 2010b), not least in shallow lakes (but see Meerhoff et al., 2007a). Indication of changes in the intensity of trophic interactions may be obtained from certain indexes that typically include ratios of biomass or densities between predator–prey pairs relevant to shallow lake functioning (e.g. piscivorous fish:planktivorous fish, zooplankton:phytoplankton biomass ratio) (e.g. Jeppesen et al., 2000), or between groups of known different sensitivity to predation (such as motile organisms) and the total amount of potential prey (e.g. Meerhoff et al., 2007a). Also, the changes in size distribution of potential prey are used as indicators of predation pressure (e.g. Brucet et al., 2010).

Evidence from SFTS studies indicates a clear difference in the strength of trophic interactions with increasing temperature in shallow lakes. A stronger predation pressure by fish under warm conditions was supported by cross-comparison studies where littoral trophic interactions in subtropical and temperate lakes were compared (Meerhoff et al., 2007a). Several predation indexes constructed based on different sensitivities to predation by several groups, such as the proportion of small fish, the proportion of small snails, the proportion of chironomids, and the proportion of *Daphnia* relative to small cladocerans (Meerhoff et al., 2007a), suggest stronger effects of fish predation in warm lakes than in similar temperate lakes. The predation pressure exerted by planktivorous fish, particularly on zooplankton (and also on macroinvertebrates), increases with increasing temperature, as also found in European (Gyllström et al., 2005) and South American (Lacerot, 2010) latitudinal gradient analyses, and in cross-comparison studies (Brucet et al.,

2010, 2012; Meerhoff et al., 2007a, b) (Fig. 8). Mesocosm experiments conducted from Finland to southern Spain (Moss et al., 2004) have revealed significant effects of fish on the community structure of zooplankton and phytoplankton (leading towards dominance of smaller algae) along this latitudinal gradient. However, this effect of fish was not influenced by latitude (Moss et al., 2004; Stephen et al., 2004), possibly due to the similar density of fish stocked in all experimental locations.

Cross-comparison studies on *Daphnia* ephippia in sediments of subtropical and temperate shallow lakes indicated that they are also present in warm lakes despite being almost absent from contemporary water samples (Iglesias et al., 2011). These findings support the hypothesis that the typical lack of *Daphnia* in warm climates is more a consequence of high predation rather than of metabolic constraints (Havens and Beaver, 2011; Iglesias et al., 2011; Meerhoff et al., 2007a). Anti-predator behavioural responses by zooplankton also change with climate and the differential associated risk of predation by fish (Brucet et al., 2010; Meerhoff et al., 2007b; Tavsanoğlu et al., 2012).

Most likely as a consequence of increased predation by fish, the size distribution and composition of zooplankton in warm lakes typically change towards dominance of smaller taxa and individuals (Brucet et al., 2010; Gillooly and Dodson, 2000; Iglesias et al., 2011; Lacerot, 2010; Meerhoff et al., 2007b); consequently, the capacity of zooplankton to control phytoplankton crops (estimated as the zooplankton:phytoplankton biomass ratio) may decrease with increasing temperature or decreasing latitude (Gyllström et al., 2005) (Fig. 8). This last pattern occurred in all freshwater studies we examined, regardless of approach (i.e. latitudinal gradient, cross-comparisons), geographic areas, and use of different biomass estimators (Fig. 8). In brackish lakes, the zooplankton: phytoplankton biomass ratio remained low throughout the year in the temperate regions, whereas the ratio was markedly higher in winter than in summer in the Mediterranean lakes, likely reflecting shifts in zooplankton and fish structure related to a fluctuating hydrology (Brucet et al., 2009). The occurrence of macrophytes in the lakes seems to enhance the grazing pressure of zooplankton in intermediate and cold regions (most likely indirectly via the provision of refugia from fish predation), while no or negligible effect of plants appears at low latitudes (Meerhoff et al., 2006, 2007b; Muylaert et al., 2010; Tavsanoğlu et al., 2012). Similar patterns were found in the gradient analysis in South

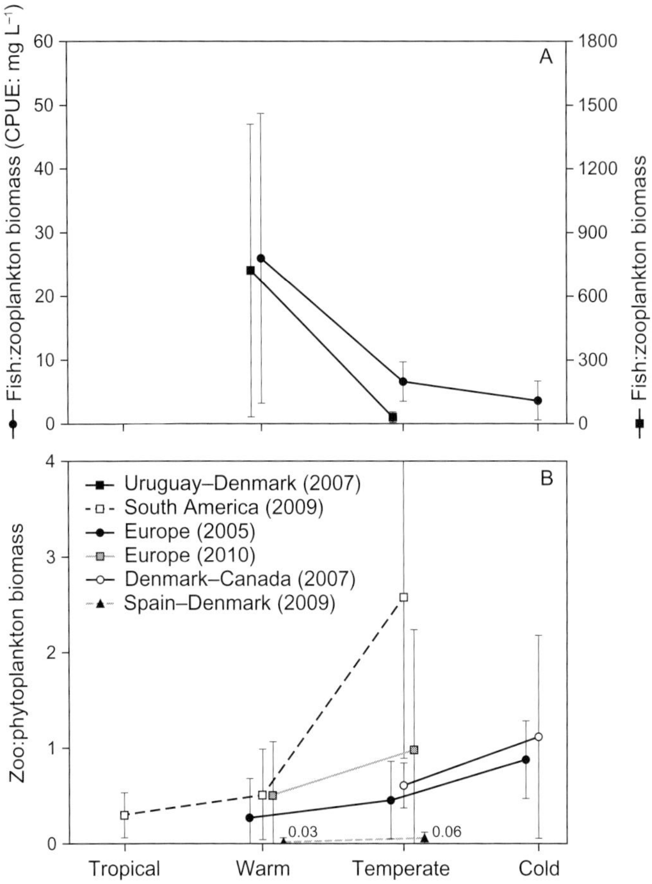

Figure 8 Variation in consumption pressure along a climate gradient, estimated as biomass ratios. (A) predation pressure of planktivorous fish on zooplankton (fish:zooplankton biomass) and (B) grazing pressure of zooplankton on phytoplankton (zoo: phytoplankton biomass). The latitudinal gradient corresponds to ranges from 5° to 55°S and from 38° to 64°N in Europe and America. Data were grouped in wide climate types: Tropical (Brazil), Warm (intermediate Argentina, Uruguay and Spain), Temperate (Denmark, Belgium, The Netherlands, UK and Poland) and Cold (Estonia, Sweden, Finland, Canada and southern Argentina). Zooplankton (either densities or biomass) median values were extracted from the literature and converted into biomass when necessary, according to Jeppesen et al. (1994). Full lines connect the points belonging to the same paper (see symbols). Values were extracted from (A) Fig. 6 in Gyllström et al. (2005) and Table 2 and Fig. 2 in Meerhoff et al. (2007a) and (B) Fig. 6 in Gyllström et al. (2005), Fig. 6 in Brucet et al. (2009), Fig. 2B in Jackson et al. (2007), Fig. 4 in Muylaert et al. (2010) and Fig. 7A in Kosten et al. (2009a). Error bars are 1 SE.

America (Kosten et al., 2009c), where the zooplankton grazing pressure decreased from cold to warm regions; the patterns were, however, influenced by macrophytes and the abundance of fish.

Despite lower total abundances, the relative abundance of herbivorous plant-associated macroinvertebrates was higher in the subtropical climate than in the temperate communities (Meerhoff et al., 2007a). This could imply that a larger proportion of the macroinvertebrates, besides the large numbers of small omnivorous fish, is feeding on periphyton and macrophytes under warmer conditions. On an evolutionary timescale, this could explain the increase in chemical defences of macrophytes at low latitudes, as suggested by Morrison and Hay (2011). On the other hand, grazing can affect periphyton by diminishing its biomass and changing taxonomic or group composition: grazing-mediated changes in composition have been revealed in a study showing a shift to dominance of fast-growing or resistant species (Sumner and McIntire, 1982).

Fully controlled experiments testing fish-mediated trophic cascading effects in shallow lakes under different climates are limited to a few mesocosm investigations. Trophic cascades seem to be weaker in tropical mesocosms, with stronger fish effects on nutrient cycling promoting a more pronounced bottom-up effect than in temperate mesocosms (Danger et al., 2009). Another conclusion of the cross-comparison of littoral food webs was that trophic cascades (concomitant with stronger fish predation pressure) from fish to the bottom of the webs were weaker and with relatively direct effects (perhaps reflecting shorter path lengths through the more reticulate food webs) on most prey communities (Meerhoff et al., 2007a) (Fig. 9).

3. DISCUSSION

3.1. Can we predict changes in community traits with warming?

So far, we have attempted to detect whether common patterns of change in community traits as a response to changes in temperature emerge from the empirical SFTS literature: some traits of a few communities changed in accordance to theoretical predictions, but for most organizational levels, responses were often unclear or non-linear, and only partly supported predictions. Below, we contrast our findings with theoretical expectations and compare them with findings from other research approaches.

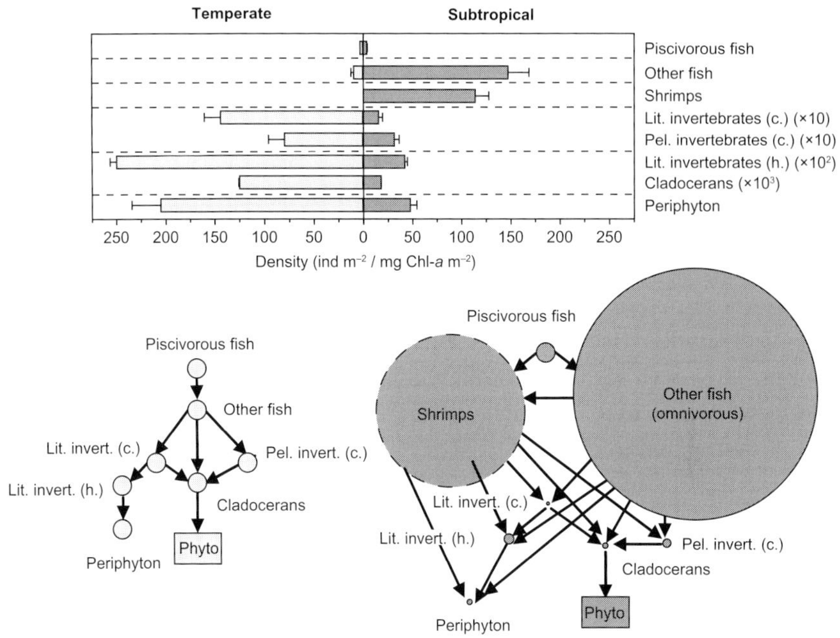

Figure 9 Structure of main communities in temperate (left) and subtropical (right) shallow lakes. (Above) Density of potentially piscivorous fish, all other fish, shrimps, littoral carnivorous invertebrates (Lit. invert. (c)), pelagic carnivorous invertebrates (Pel. Invert (c)), littoral herbivorous macroinvertebrates (Lit. invert. (h)), cladocerans and biomass of periphyton. (Below) Simplified scheme of trophic interactions among the same trophic groups. The densities in the subtropics are expressed relative to those in the temperate lakes (considered as the unit). Except fish, the same taxa share the same trophic classification in both climate zones. Shrimp relative density is dashed due to shrimp absence in the temperate lakes. Phytoplankton (Phyto) box has the same size as it was fixed in the study (surrogate of turbidity) by pairing the lakes in both climates. Data are sample means of five selected lakes and error bars (above) show the standard error. *Modified from Meerhoff et al. (2007a).*

3.1.1 Congruence of SFTS findings and theoretical expectations

Some of the shallow lake communities surveyed here clearly followed the predictions of an increase in richness with an increase in ambient temperature (Allen et al., 2002), namely, fish, phytoplankton, and periphyton; whereas macroinvertebrates and cladoceran zooplankton (at genera level) typically showed the opposite trend, and no clear response could be identified for macrophytes and bacterioplankton. Fish taxonomic richness exhibited the clearest pattern (meta-analysis by González-Bergonzoni et al., 2012; Fig. 1A), with a significant positive relationship with decreasing latitude. Macroinvertebrate richness data are often not available at species level and latitudinal studies show ambiguous trends, besides being commonly based on pooled data from lakes and streams (e.g. Heino, 2009). In studies including only shallow lakes, however, we found an opposite trend to the expected response (at genus and family levels) (Fig. 1B), with lower local taxon richness in warm lakes (e.g. Brucet et al., 2012; Lewis, 1996; Meerhoff et al., 2007a). Most of the reviewed articles also described a decrease in local cladoceran zooplankton richness (at genus level) at warmer locations (e.g. Brucet et al., 2010; Meerhoff et al., 2007b, Pinto-Coelho et al., 2005) (Fig. 1C). Whether similar or contrasting latitudinal richness patterns occur at species level remains unresolved for these groups. However, richness at different taxonomic resolutions (e.g. species and genera) tends to be highly correlated, as found for, for example, terrestrial vegetation (Enquist et al., 2002), macroinvertebrates (Heino and Soininen, 2007) and phytoplankton (Gallego et al., 2012) and as predicted by theoretical models (Allen et al., 2002).

Microorganisms, ranging from phytoplankton to bacteria, however, are usually considered to either lack or display only weak latitudinal gradients in richness, because of their high abundances and small sizes, enabling them to disperse easily over few geographical barriers and over large distances (Fenchel and Finlay, 2004; Finlay, 2002; Hillebrand and Azovsky, 2001, but see Van der Gucht et al., 2007). The distribution of these groups is suggested to be more strongly driven by local environmental factors (Baas-Becking, 1934; Beisner et al., 2006; Kruk et al., 2009; Ptacnik et al., 2010a,b) than by latitudinal gradients. In our review, however, phytoplankton and periphyton richness increased with decreasing latitude (or rising temperature) (Fig. 1D and E). It has been suggested that most microbial biodiversity is hidden from our observation, because most species will occur at densities below current limits of detection (de Wit and Bouvier, 2006); a limitation increasingly overcome by the application of massive tag-sequencing, which allows for the detection of the microbial components of the so-called rare biosphere (Purdy et al., 2010; Sogin et al., 2006). In our analysis, bacterioplankton richness did not show any clear trend, maybe reflecting the narrow geographic range of the available data (i.e. temperate regions) or, alternatively, the suggested stronger effect of local factors on this diverse group.

MTE predicts a decrease in community biomass and density with an increase in ambient temperature, due to the enhanced metabolic demands and consequent limitation of resources for individuals (Brown et al., 2004; Fig. 1, Table 1). In our review, only periphyton biomass and macroinvertebrate and zooplankton densities showed a significant decline with decreasing latitudes (Fig. 1K, N and R).

Fish biomass and density showed the opposite pattern (Fig. 1G and M), with a clear increase towards low latitude locations as found in several studies applying different approaches (e.g. Brucet et al., 2010; Gyllström et al., 2005; Meerhoff et al., 2007a; Teixeira-de Mello et al., 2009; see Table 2 for a summary). This is an intriguing phenomenon, since the potential co-variation of lake productivity with latitude was considered in the respective studies (i.e. comparing lakes with similar TP ranges in Brucet et al., 2010; Gyllström et al., 2005, Teixeira-de Mello et al., 2009) and in the data we extracted from the literature. Besides, we found no evidence for limitations of specific nutrients or productivity gradients associated with latitude (reviewed in Section 1.3). Increases in fish biomass and abundance with temperature have also been reported recently from Icelandic geothermal streams, which span a broad thermal gradient within a single catchment (Woodward et al., 2010a).

The meta-analyses of zooplankton (in this case, cladocerans and copepods), however, yielded no significant relationship between biomass and latitude despite that the expected trend of reduction occurred in most reviewed studies (e.g. Gyllström et al., 2005; Havens et al., 2007; Jackson et al., 2007; Meerhoff et al., 2007b; Pinto-Coelho et al., 2005; Vakkilainen et al., 2004).

Contrary to the predictions by MTE, warmer conditions have been argued to lead to higher phytoplankton biomass and yield (Lewis, 1996), not least in shallow lakes. This has been explained by longer growing seasons (Lewis, 1996, 2000), increased availability of phosphorus from internal loading (Jeppesen et al., 2009; Søndergaard et al., 2003) and smaller size of zooplankton grazers (Meerhoff et al., 2007b), resulting in lower grazing pressure as suggested in our review (e.g. Gyllström et al., 2005; Kosten et al., 2009c; Meerhoff et al., 2007a; Fig. 8). In our analysis of large databases from two contrasting climates, we did find higher total phytoplankton biovolume and higher total biovolume of cyanobacteria in subtropical climates than in similar temperate lakes (Fig. 2), as reported in studies following other approaches (Paerl and Huisman, 2009). The general pattern was, however, not maintained for other phytoplankton biomass proxies, such as Chl-a concentration (Fig. 1J), as also reported by Kosten et al. (2012). The typical absence of counts of phytoplankton and periphyton in the shallow lake literature prevented us from testing density predictions on these communities.

The estimation of bacterial biomass relies on the application of conversion factors to translate the volume assessments into carbon values, and these factors are not yet fully established, leading to widely varying estimations when applying the different proposed values to the same dataset (e.g. Posch et al., 2001). Therefore, bacterioplankton does not seem to be the most appropriate portion of the community with to test predictions of changes in biomass with warming at present, at least not with the information found in this review. Nonetheless, bacterioplankton still showed a significant increase in density with decreasing latitudes (i.e. increasing temperature) (Fig. 1R). According to our findings, higher temperatures could promote higher bacterioplankton density both through a direct effect (via acceleration of metabolic rates leading to enhanced multiplication) and indirect effects (e.g. via increasing substrate availability, DOC) (Fig. 5). Higher temperatures could also promote competitive advantages for bacteria over algae (Carey et al., 2012; Yvon-Durocher et al., 2011a). The fact that temperature and Chl-a appeared in our regression model as separate factors supports the hypothesis that the effects of temperature and resource availability on bacterial biomass production are independent (Lopez-Urrutia et al., 2006).

Freshwater communities, and particularly shallow lake communities, seem to be strongly size structured (Hildrew et al., 2007), as recognized in the pioneering works on the effect of fish predation on the size structure of zooplankton in ponds (Brooks and Dodson, 1965; Hrbacek et al., 1961). Different trends of changes in body size with temperature could be expected according to several theories, which are not necessarily mutually exclusive (see Section 1, Fig. 1, Table 1 and Daufresne et al., 2009). In our review, fish and zooplankton communities showed a reduction in mean body size with increasing temperatures (Fig. 1S and U, for fish and zooplankton, respectively), supporting the predictions by MTE (Brown et al., 2004) and the "temperature-size" rule (Angilletta, 2009). Bacterioplankton body size, however, revealed no relationship with latitude or temperature, but the data were geographically too restricted and the variability too large to allow detection of latitudinal gradients. The lack of sufficient data for four of our target communities (i.e. macroinvertebrates, macrophytes, phytoplankton and periphyton) prevented us from generalizing a trend for real lake communities. Moreover, at the population level, different mechanisms may occur across latitudinal gradients as predicted by several theories, leading to a decrease in the mean body size of populations with a concomitant reduction of mean body size across the community as a whole (Daufresne et al., 2009). At the community level, a decrease in average body size with increasing temperature may also reflect competitive exclusion of larger species, size-selective predation (e.g. Brooks and Dodson, 1965) or a reduction of the density of larger species due to inefficient energy transfer along the food web with increasing temperatures (Arim et al., 2007). These mechanisms, however, would not necessarily appear across latitudinal gradients due to local variations in the amount of resources available and the thermal setting in which the communities develop. The trends we found in the SFTS literature mostly refer to changes at assemblage or community level and may thus reflect the effect of temperature through a variety of mechanisms or likely the combined effect of temperature and inter-specific interactions.

3.1.2 Evidence of warming effects obtained by other approaches

Other approaches (e.g. perturbation experiments) to study the potential effects of climate change have shown no effects of increased temperature on richness of some freshwater communities. In outdoor heated mesocosms, cladoceran diversity (McKee et al., 2002a) and phytoplankton richness (Moss et al., 2003) were not significantly affected by experimental warming, although measures of community evenness increased (McKee et al., 2002a). Warming, rather than affecting macrophyte richness, may promote a switch from submerged to free-floating macrophyte dominance in shallow eutrophic lakes, as reported mainly from perturbation experiments (Feuchtmayr et al., 2009; Netten et al., 2010) and previously suggested by models (Scheffer et al., 2003). Warming may thus promote a polewards expansion of the geographical distribution of this type of plants (Scheffer et al., 2003), with potential strong and negative consequences for local biodiversity.

In passively warmed mesocosms in alpine regions, zooplankton biomass was suppressed due to the decline in large cladocerans, even in the absence of fish (Strecker et al., 2004). At the same time, warming did not affect total phytoplankton biomass but significantly altered the assemblage composition by favouring fast-growing phytoflagellates over larger filamentous green algae, supporting the hypothesis that moderate warming can destabilize plankton dynamics in shallow cold-water ecosystems (Strecker et al., 2004). An experimental increase in temperature in mesocosms had also very minor effects on phytoplankton Chl-a and total biovolume, although it significantly decreased the biovolume of Cryptophyceae and Dinophyceae and did not affect the abundance of cyanobacteria as might be expected (Moss et al., 2003). No obvious effects on the densities of zooplankton (McKee et al., 2002a) and macroinvertebrates (Feuchtmayr et al., 2007; McKee et al., 2003) were observed, suggesting that the overall abundances of most invertebrate taxa will not be severely affected by the predicted temperature rise (Feuchtmayr et al., 2007; but see Dossena et al., 2012). In contrast, in another series of outdoor heating experiments, warming shifted the composition and distribution of phytoplankton size and biomass from assemblages of large individuals with high standing biomass to assemblages with low standing biomass and many smaller-bodied species, including small cyanobacteria (Yvon-Durocher et al., 2011a). In this case, however, such an effect did not occur in the distribution of size and biomass of zooplankton. Warming thus seems to lead to reorganization of the biomass structure in both the benthic and planktonic assemblages (Dossena et al., 2012; Yvon-Durocher et al., 2011a), at least in experimental, relatively simple, food webs with fixed fish densities.

In contrast to our findings (Fig. 1R), in experimental outdoor mesocosms, the abundances of picoalgae, bacteria and heterotrophic nanoflagellates changed in a similar manner over time, with no direct effect of experimental warming (Christoffersen et al., 2006). However, experimental warming modified the effects of nutrient additions (Christoffersen et al., 2006; Özen et al., 2012), indicating that interactive effects may be significant in the future given the current and expected increase in nutrient loading to many shallow lakes worldwide (Jeppesen et al., 2009, 2010b; Moss et al., 2011). The difference between our findings and these experimental studies may reflect the co-variation in nutrient availability (i.e. DOC) and temperature found in our data (Appendix B) rather than different direct effects of temperature on bacterioplankton populations.

A meta-analysis of the effects of warming on body size of ectothermic aquatic organisms, from the individual to the community level, revealed a significant increase in the proportion of small species and organisms at warmer temperatures (Daufresne et al., 2009), as we have found for fish and cladoceran zooplankton (Fig. 1S and U). A study of Cladocera subfossil distribution in 54 European lakes located in different climates showed that small species dominated warm water lakes with high conductivity, macrophyte coverage and fish abundance. In contrast, large pelagic species were more common in low-conductivity high-latitude lakes (Bjerring et al., 2009). Higher summer temperatures have promoted a reduction in the mean body size of temperate zooplankton (Moore et al., 1996), as also found in heating experiments for phytoplankton communities (e.g. Sommer and Lewandowska, 2011; Yvon-Durocher et al., 2011a). This effect was not found in otherwise similar outdoor experiments, however (Moss et al., 2003). The pieces of evidence showing a reduction in the body size of organisms with increasing temperature suggest that the underlying size structure of aquatic ecosystems might not be robust to global warming (Yvon-Durocher et al., 2011a). The magnitude and sometimes direction of change, however, seem to vary with taxonomic groups and environmental conditions (Gardner et al., 2011). It should be also noted that the spatial scale and range of body sizes under analysis may also affect the generation and testing of congruent predictions (Tilman et al., 2004).

Our review of the SFTS literature revealed trends of a reduction in size at maturity, life-span and the occurrence of more reproductive events in fish towards warmer locations, and although the information on other communities is more scarce, it also suggests comparable changes in the reproduction mode of zooplankton and macroinvertebrates (Section 2.5). In line with these findings, experimental warming can influence the reproduction of fish by affecting embryonic development time and offspring numbers (unimodally linked to temperature) (e.g. García-Ulloa et al., 2011). It can also lead to a reduction in the body size at maturity, as reported for the cladoceran *Simocephalus vetulus* in heated mesocosms (McKee et al., 2002a). Differential effects of temperature on life history traits such as fecundity, development and survivorship will affect the overall fitness of organisms. Mathematical models suggest, for instance, that warming will cause ectothermic species with a high temperature sensitivity of development to reach the age at first reproduction earlier in the year, potentially leading to mismatches between consumers and resources if the latter have a lower temperature sensitivity of development (Amarasekare and Savage, 2012). Studies in other heating experiments (outdoor tanks, Liboriussen et al., 2005) have demonstrated the occurrence of a rapid microevolutionary response of the cladoceran *S. vetulus*, in both survival and subcomponents of individual performance (i.e. age at reproduction and number of offspring), suggesting that populations may persist locally under the predicted scenarios of global warming (Van Doorslaer et al., 2007). According to those results, such microevolutionary responses may buffer changes in community structure under climate warming and help explain the apparent lack of effects in some experiments and field data.

3.1.3 The importance of local factors and trophic interactions

Local characteristics have a large potential of blurring latitudinal gradients in traits of, at least, several communities. Some basic relationships, such as the density–mass relationship, might be robust at the level of biomes or at a very large range of body sizes but could exhibit large variations among local communities (e.g. Arim et al., 2011; Hechinger et al., 2011; White et al., 2007; but see, Brown et al., 2011; Layer et al., 2010; McLaughlin et al., 2010; O'Gorman and Emmerson, 2010).

In particular, the strong imprint of biogeography and local factors may override temperature-related effects on richness and could alter the predicted trends. In many ecosystems, including shallow lakes, explanatory variables that are potentially important as main drivers of local diversity tend to be highly inter-correlated, such as productivity level and the abundance of key structuring organisms, such as aquatic plants (Declerck et al., 2005) and fish. Besides the potential influence of climatic factors such as the length of the growing season, the annual input of solar radiation and the length of the hydroperiod (Della Bella et al., 2008), local factors such as lake area (e.g. Rorslett, 1991), sediment characteristics (Barko et al., 1991), maximum depth, TP (Jeppesen et al., 2000) and nitrate concentration (James et al., 2005) also have strong influences on plant species richness at local scales. Trophic interactions may also indirectly affect submerged macrophytes along latitudinal gradients: the positive effect of long ice cover on macrophyte coverage (e.g. Kosten et al., 2009a) may thus be due to a cascading effect through anoxia causing fish kills (Jackson et al., 2007) and subsequent stronger zooplankton and snail control on phytoplankton and periphyton competitors (Jones and Sayer, 2003). The effect of warming on phytoplankton and periphyton densities, however, also depends on the relative importance of trophic cascading effects, nutrient status and water residence time (Søballe and Kimmel, 1987).

Our review highlights that several properties of the potential prey communities, from richness to biomass and body size, indicate enhanced predation pressure with increased temperatures. Temperature influences the predator–prey interactions both directly and indirectly, either via the alteration of predator feeding rates and feeding modes (Beisner et al., 1996) or through changes in both prey and predator body size (Moore et al., 1996; Woodward et al., 2010b). Experimental warming studies measuring feeding intake rates at different temperatures have demonstrated higher consumption of macroinvertebrates, both on other invertebrates and on periphyton in warmer scenarios (Kishi et al., 2005).

Warming experiments have also shown that trophic cascade effects may vary with temperature, due to physiological depression of predators (Kishi et al., 2005).

Higher consumption by vertebrate and invertebrate grazers (Lazzaro, 1997; Meerhoff et al., 2007a) could thus also explain our observed local patterns of richness of prey communities in warmer regions, the lower macroinvertebrate densities and lower zooplankton density and mean body size of zooplankton (Meerhoff et al., 2007a) as also suggested by mathematical models (van Leeuwen et al., 2007) and confirmed in our review (Fig. 8). The grazing pressure by zooplankton on phytoplankton showed the opposite trend, being weaker at lower latitudes (Fig. 8). The lack of a relationship or relatively weak latitude effect on phytoplankton biomass could reflect local variations in contrasting mechanisms. The final outcome may be the result at local scales of the balance between temperature effects on metabolism, which should reduce phytoplankton biomass, and the effects of longer growing seasons and lower zooplankton grazing pressure, which would increase it. Ultimately, our results indicate that overall the link between zooplankton and phytoplankton weakens with increasing water temperature. Weak interactions may dampen biomass oscillations between consumers and resources (McCann et al., 1998), implying that not all responses at a specific trophic level are propagated to lower trophic levels or have significant impacts on ecosystem processes (Pace et al., 1999). An increase in the duration of climatic variability effects due to food web interactions seems possible, as the signal of winter climate can be detected in the clear water phase (i.e. a reduction in phytoplankton biomass) in early summer (Straile and Adrian, 2000) or in summer phytoplankton composition and biomass (Blenckner et al., 2002; Weyhenmeyer, 2001) in temperate lakes.

Species higher in the food web are often especially sensitive to changes in ambient temperature (Petchey et al., 1999; Woodward et al., 2010a,b). The decline and maybe the extinction of a top predator, as a potential consequence of warming, may significantly influence the effects of size-structured interactions in food webs and ecosystem functioning. Other temperature-related changes in fish structure and predation pressure might also be expected. After conducting feeding experiments with the omnivore fish species opalaye (*Girella nigricans*) at different ambient temperatures, Behrens and Lafferty (2007) suggested that warming may lead to changes in the diet of fish towards greater consumption and assimilation of plant material. An increase in the relative proportion of herbivorous and omnivorous species within the fish assemblage with decreasing latitude has been found in meta-analyses of aquatic systems worldwide, not least shallow lakes (González-Bergonzoni et al., 2012; Jeppesen et al., 2010a), giving further support to the previous hypothesis. This change in diet would result in food webs being more truncated, as predicted both theoretically (Arim et al., 2007) and empirically (Meerhoff et al., 2007a). Altogether, the evidence suggests that food webs in temperate or cold shallow lakes will be affected by warming, also due to the current lower number of feeding links (Fig. 9; Meerhoff et al., 2007a) and the lower richness of top species (typically fish) (Fig. 1A). As found in our review, however, the direction of changes seems difficult to predict for the specific communities given the different relative importance of direct or indirect effects of higher temperature under specific local scenarios and the interactions with other environmental stressors.

3.2. Advantages and disadvantages of the SFTS approach

The actual test of theoretical predictions with empirical data from shallow lake literature represents a practical challenge relative to the heterogeneous nature of currently available data. The description of different entities (e.g. populations, communities) and the differences in taxonomic resolution (e.g. species, genera, classes) make comparisons, in our case across studies and different communities, a challenging task (Yvon-Durocher et al., 2011b). Like all other approaches, the SFTS approach has a series of advantages and disadvantages (many already summarized in Woodward et al., 2010b), as it studies communities that have long been exposed to different temperatures, rather than in a transient state in response to warming *per se*. Latitude is probably a better predictor of long-term regimes of climatic variables than current climate values registered by local weather stations (Naya et al., 2008). However, latitude may be correlated with other climatic (e.g. wind speed, seasonality), ecological (e.g. day length and length of growing season), and potentially also historical factors (e.g. biogeographical boundaries) that may affect organisms' physiology (Naya et al., 2008) and, with it, the observed latitudinal patterns. Furthermore, climate is much more than temperature alone, and other potential confounding processes interact with climate to shape communities in lakes. Lake trophic state, as reflected by nutrient levels (more commonly TP), affects community structure by augmenting the fish densities and biomass in increasingly productive systems around the world

(Brucet et al., submitted for publication; Jeppesen et al., 2000, 2005; Mehner et al, 2007; Teixeira-de Mello et al., 2009). Lake productivity thus often overrides the effects of temperature on fish assemblage densities (e.g. Brucet et al., submitted for publication; Mehner et al., 2007). Experimental warming in outdoor mesocosms promoted an increase in TP, total alkalinity, and conductivity, decreased pH and oxygen saturation and increased the frequency of severe deoxygenation (McKee et al., 2003). Several pieces of evidence also support the idea that increasing temperature acts synergistically with eutrophication (e.g. Feuchtmayr et al., 2007; Jeppesen et al., 2010b; McKee et al., 2003; Moss et al., 2011), so these two key drivers could produce outcomes that are not predictable from studying either in isolation.

Climate warming may lead to a higher frequency of dramatic changes in water level, transforming permanent water bodies into temporary systems (Beklioğlu et al., 2007) and leading to changes for the whole food web. For example, strong differences in zooplankton assemblage structure appeared between permanent and temporary shallow lakes of the same wetland system, with higher diversity and lower density of large zooplankton in permanent shallow lakes, which was attributed to a higher fish predation pressure compared to temporary lakes (Brucet et al., 2005). Warming may also promote an increase in salinity in shallow lakes due to enhanced evaporation, especially in arid and semi arid climates (Beklioğlu et al., 2007), saltwater intrusions into freshwater bodies (Schallenberg et al., 2003) and enhanced use of freshwater for irrigation and industry under a likely increased consumption pressure (Williams, 2001). Trophic structure in shallow lakes typically changes along a salinity gradient (Brucet et al., 2009, 2010; Jeppesen et al., 1994, 2007), as also shown in this review. Indirect effects of climate warming, such as changes in salinity and hydrology, could thus have larger consequences for the diversity, size distribution and abundance of grazers of shallow lake ecosystems than an increase in nutrients or temperature *per se* (Brucet et al., 2009, 2010). Such shifts with increasing salinity and the subsequent decrease in the top-down control on phytoplankton may negatively affect the resilience of brackish lakes to the enhanced nutrient loading (Brucet et al., 2010, 2012).

A further factor that potentially masks natural responses of aquatic communities to temperature variations is the accidental or purposeful introduction of alien species (mostly fishes in freshwaters) by humans. The SFTS approach has clear limitations as to predicting the impact of invasive species associated with climate warming because a native species may become invasive when expanded beyond its natural distributional range (as expected for large free-floating plants, Scheffer et al., 2003), and we cannot presume that ecological interactions among species will necessarily be similar in the native and the foreign ecosystem. Climate warming also has the potential to modify the impacts of invasive alien species by affecting the whole process of invasion and further increasing ecosystem vulnerability (Rahel and Olden, 2008). For instance, climate change may affect aquatic invasive species by altering thermal regimes, reducing ice cover duration, altering stream flow regimes, increasing salinity, and augmenting the demand for water storage and conveyance structures (Rahel and Olden, 2008). Furthermore, potentially synergistic effects of climate warming and non-native species make forecasting even more difficult (Woodward et al., 2010b). For example, as pointed out by Rahel and Olden (2008), global climate change will lead to an increase in water temperatures in northern-latitude lakes, which may result in seasonally stressful conditions for coldwater-adapted fish species. On the other hand, however, it may provide suitable thermal conditions for non-native warm water fish species (Sharma et al., 2007). Such species could prey on or compete with native fishes and this competition may result in a decline or loss of native populations (Jackson and Mandrak, 2002).

However, potentially confounding factors such as eutrophication, acidification, and the equivalent reverse phenomena generally fluctuate less rapidly than climatic variables. This difference in behaviour and signal-response times can be used to distinguish between climatic forcing and confounding factors, by removing trends and low-frequency fluctuations from relevant time-series by detrending or high-pass filtering, and analyzing only the remaining, higher-frequency fluctuations that are driven mainly by external physical forcing (Adrian et al., 2009).

Notwithstanding the current shortage of empirical data on all relevant communities or traits in the literature at a truly global scale, a clearer picture of what we should expect with climate warming is starting to emerge, especially when we combine trends from SFTS studies with modelling studies, *in situ* warming experiments and long-term empirical studies (e.g. Jeppesen et al., 2010a; Jeppesen et al., submitted for publication; Sarmento et al., 2010). In some cases, however, apparently contradictory patterns emerge, as highlighted here. Experimental warming studies are often of short-term duration, thus showing transient phenomena (i.e. warming as a stressor), whereas SFTS studies in natural systems are more likely to reveal equilibrial conditions with respect to temperature rather than responses to changing conditions *per se*. Moreover, warming experiments most often simplify food webs and, in particular, the role of predators by using fixed densities of a single predatory fish species (e.g. Liboriussen et al., 2005; McKee et al., 2002a; Yvon-Durocher et al., 2011a), thereby not allowing for the assessment of temperature-driven effects on prey communities through temperature impacts on the assemblage of predators. For instance, the majority of the reviewed SFTS studies suggest that fish predation, rather than temperature alone, is also responsible for the changes in the structure of several communities, mainly as a result of cascading effects of the temperature-driven decrease in fish size (Lacerot, 2010; Meerhoff et al., 2007a) leading to a stronger predation pressure on macroinvertebrates and zooplankton (Gyllström et al., 2005; Meerhoff et al., 2007a,b).

The SFTS approach thus highlights empirical trends that allow researchers to generate hypotheses regarding climate-driven future changes in community and ecosystem processes, thus potentially stimulating further research and the design of appropriate experiments to disentangle effects of potential drivers and to test mechanisms (e.g. Friberg et al., 2009; Liboriussen et al., 2005 and other references mentioned along the text). The combination of empirical data, mechanistic approaches and the framework given by food web and metabolic theories to climate change research in shallow lakes seems also fundamental, as suggested for other ecosystems or in theoretical works (e.g. Montoya and Raffaelli, 2010; Van der Putten et al., 2010; Woodward et al., 2010b).

3.3. Topics for further research

The occurrence of processes acting at different scales renders it difficult to disentangle their specific effects and the relative importance of local and regional characteristics on multispecies richness. In particular, both the experimental evidence and our review of whole lake data might suggest that the effects of nutrient enrichment and predation by fish on zooplankton and macroinvertebrate communities might be stronger than the direct effects of warming. It is thus open for further analysis whether some of our seemingly contradictory findings are a result of temperature-enhanced fish predation in warm locations, and therefore with consequences for a climate warming context, or a result of biogeographic processes or of geographically restricted databases. SFTS studies analyzing changes in community traits in fishless systems could provide some answers to these questions.

Our target community traits (i.e. richness, biomass, density and body size) are very seldom combined in the literature within a given study, and information on the responses of key communities to warming is therefore fragmentary. The common use of proxies for some traits (such CPUE, with different units in different works, as a proxy of fish density and biomass per unit area) also renders it difficult to compare results from different publications and to estimate predation pressure indexes. The lack of reports in the literature of traits typically measured (such as the body size and density of phytoplankton and periphyton) represents information gaps that should be relatively easy to solve. On the other hand, some traits are typically not measured, such as the body size and density of macrophytes or traits of macroinvertebrates at low levels of aggregation (i.e. species and genera), while some assemblages are typically underrepresented in shallow lake research. In particular, and despite the importance of bacteria for geochemical cycles, biomass and nutrient and energy transport within aquatic trophic webs (Azam, 1998), data on microbial communities in shallow lakes are scarce. When present, bacterioplankton community traits are typically not described, reflecting the very different methodological skills and facilities required to evaluate all community traits, which will hopefully become less of a hindrance in the future with the advent of more sophisticated molecular approaches (Purdy et al., 2010). Besides, microbial ecologists have modified their focus, from evaluating bacterioplankton communities in terms of taxa-independent approaches (e.g. measurements of biomass) to elucidating factors controlling the community composition with the aim to understand the role and fate of specific populations (Pernthaler and Amann, 2005), and a move towards more *in situ* characterizations of multispecies systems (Purdy et al., 2010).

Possibly, more serious, though, than the scarcity of data on bacterioplankton is the lack of data on their predators (i.e. heterotrophic nanoflagellates, or HNF). The absence of this link from the literature prevents us from estimating potential changes in energy and nutrient fluxes within the food web of shallow lakes under warming, since these organisms link the microbial loop with the classic food web (i.e. phytoplankton–zooplankton–fish) (Reiss et al., 2010). Changes in lake trophic state could promote a decoupling between bacterioplankton and their potential predators (Gasol, 1994), potentially leading to changes in the efficiency of energy transfer and the occurrence and strength of trophic links (Zingel et al., 2007) in the whole system. The scarcity or lack of data (and large uncertainties associated with laboratory techniques) on the microscopic components of shallow lakes also renders it very difficult to search for body size thresholds on which to test the predictions of the different theories that relate temperature with body size.

More data on changes in trophic position, diet breadth and feeding rates are needed to test the predicted trends of a reduction in food web length with increasing temperature (Arim et al., 2007) and of changes in key measures of interaction density, such as connectance (Petchey et al., 2010). Climate effects on life history traits other than reproductive characteristics, and seasonal dynamics of lake communities other than plankton (Adrian et al., 2006), are areas that could be usefully investigated applying the space-for-time framework. Another important caveat that deserves further investigation is the effects of climate on host–parasite interactions: several studies suggest that climate strongly influences parasite infection in lake communities (e.g. phytoplankton, Ibelings et al., 2011; fish, Macnab and Barber, 2012), yet this field of trophic ecology is still surprisingly poorly understood (Thompson, Dunne and Woodward, in press). Despite the fact that the overall effects still have yet to be fully elucidated, temperature-mediated changes in parasite infection can potentially alter the expected outcomes of trophic cascades or predator–prey dynamics between the main lake communities. The likely important effects of changes in temperature on the production and action of allelopathic substances represent another research area of great potential.

Arguably, the most familiar characteristic of shallow lake ecosystems is their capacity to shift between alternative, relatively stable, states (Moss, 1990; Scheffer et al., 1993, 2001). This implies path dependency or hysteresis, in which a return to the environmental condition (e.g. a particular nutrient concentration) before the shift is not sufficient to recover the previous ecosystem state. Several pieces of evidence suggest that the differences in the biotic interactions between cold and warmer regions (Meerhoff et al., 2007a) might result in a weaker clearing effect of submerged plants under warmer conditions (Jeppesen et al., 2007), and thus in a weaker stability of the clear water state. Kosten et al. (2009b) assessed whether the effect of submerged plants changes with climate in South America and found that the competitive balance between submerged macrophytes and phytoplankton is not straightforward. However, the effect of submerged macrophytes on water clarity in the lakes (outside the plants beds) was generally lower in South America (Kosten et al., 2009b) than in temperate Northern Hemisphere lakes (e.g. Jeppesen et al., 2007), suggesting that other processes may have been overlooked. How global changes, including here the changes in C and N cycles and land use, besides climate warming, affect the critical thresholds over which the ecosystems experience such dramatic shifts is thus one of the most relevant questions for ecosystem management in the future, particularly for shallow lakes.

ACKNOWLEDGEMENTS

We are very grateful to Anne Mette Poulsen for chapter editing and to Juana Jacobsen and Tinna Christensen for superb graphical assistance. We thank R. Schiaffino for kindly sharing data on bacteria for our analysis, J. Huisman and M. Stomp for kindly permitting reproduction of the panel of phytoplankton richness in Fig. 1D; J. Gorga for the nice drawings of fish in Fig. 8. We thank two anonymous reviewers and Editor G. Woodward for constructive comments that greatly improved the chapter. C. A., C. K., C. I., F. T. M., G. G., G. L., M. A., M. M. and N. M. receive support from the S. N. I. (Agencia Nacional de Investigación e Innovación, ANII, Uruguay). C. A., C. K., M. A., M. M. and N. M. are supported by PEDECIBA, M. M. also by ANII-FCE (2009–2749) and the national award by L'Oréal-UNESCO 2011 for Women in Science (Uruguay, with support of DICyT). C. A., C. K. and G. L. are also supported by CSIC (Programa Grupos I+D—1037, Ecología Funcional de Sistemas Acuáticos), C. K. and G. L. by Wetenschappelijk Onderzoek van de Tropen en Ontwikkelingslanden (WOTRO, Foundation for the Advance of Tropical Research), The Netherlands, and C. A. also by the Max Planck Society. F. T. M. and I. G. B. are supported by SNB-ANII and IGB also by ANII-FCE (2009–2530). E. J., M. B. and S. B. are supported by FP7/ENV-2009-1 under grant agreement 244121 (REFRESH Project), and E. J. also by the Research Council for Nature and Universe (272-08-0406), the STF project CRES, CIRCE, CLEAR2 and FNU (16-7745). M. B. is also supported by TUBİTAK ÇAYDAĞ 105Y332.

APPENDIX A. PERIPHYTON LATITUDINAL GRADIENT

We analyzed papers covering a latitudinal gradient range from 68°S to 83°N. All the data come from field sampling campaigns of periphyton on different substrates, either natural or artificial, in freshwater shallow lakes along a nutrient concentration range of 8.0–4050 μg TN L^{-1}, 0.4–2750 μg TP L^{-1}, and a planktonic chlorophyll-a concentration gradient from 0.4 to 400 μg Chl-a L^{-1}. We performed linear regressions between periphyton data and latitude using the statistical package STATISTICA.

Literature analyzed:

Ács, E., Borsodi, A.K., Makk, J., Molnár, P., Mózes, A., Rusznyak, A., Reskone, M.N., and Kiss, K.T. (2003). Algological and bacteriological investigations on reed periphyton in Lake Velencei, Hungary. *Hydrobiologia* **506/509**, 549–557.

Bonaventura, S.M., Vinocur, A., Allende, L., and Pizarro H. (2006). Algal structure of the littoral epilithon in lentic water bodies at Hope Bay, Antarctic Peninsula. *Polar Biol.* **29**, 668–680.

Bonilla, S., Villeneuve, V., and Vincent, W.F. (2005). Benthic and planktonic algal communities in a high arctic lake: pigment structure and contrasting responses to nutrient enrichment. *J. Phycol.* **41**, 1120–1130.

Conde, D., Bonilla, S., Aubriot, L., de León, R., and Pintos, W. (1999). Comparison of the areal amount of chlorophyll *a* of planktonic and attached microalgae in a shallow coastal lagoon. *Hydrobiologia* **408/409**, 285–291.

Fermino, F.S., Bicudo, C.E.M., and Bicudo, D.C. (2011). Seasonal influence of nitrogen and phosphorus enrichment on the floristic composition of the algal periphytic community in a shallow tropical, mesotrophic reservoir (São Paulo, Brazil). *Oecol. Aust.* **15**, 476–493.

Ferragut, C., Rodello, A.F., and Bicudo, C.E.M. (2010). Seasonal variability of periphyton nutrient status and biomass on artificial and natural substrates in a tropical mesotrophic reservoir. *Acta Limnol. Brasil.* **22**, 397–409.

Franca, R.C.S., Lopes, M.R.M., and Ferragut, C. (2009). Temporal variation of biomass and status nutrient of periphyton in shallow Amazonian Lake (Rio Branco, Brazil). *Acta Limnol. Bras.* **21**, 175–183.

Hansson, L.-A. (1992). The role of food chain composition and nutrient availability in shaping algal biomass development. *Ecology* **73**, 241–247.

Harrison, S.S.C., and Hildrew, A.G. (1998). Patterns in the epilithic community of a lake littoral. *Freshwater Biol.* **39**, 477–492.

Havens, K.E., East, T.L., Rodusky, A.J., and Sharfstein, B. (1999). Littoral periphyton responses to nitrogen and phosphorus: an experimental study in a subtropical lake. *Aquat. Bot.* **63**, 267–290.

Kiss, M.K., Lakatos, G., Borics, G., Gido, Z., and Deak, C. (2003). Littoral macrophyte-periphyton complexes in two Hungarian shallow waters. *Hydrobiologia* **506/509**, 541–548.

Lambert, D., Cattaneo, A., and Carignan, R. (2008). Periphyton as an early indicator of perturbation in recreational lakes. *Can. J. Fish Aquat. Sci.* **65**, 258–265.

Liboriussen, L., and Jeppesen, E. (2003). Temporal dynamics in epipelic, pelagic and epiphytic algal production in a clear and a turbid shallow lake. *Freshwater Biol.* **48**, 418–431.

Liboriussen, L., and Jeppesen, E. (2006). Structure, biomass, production and depth distribution of periphyton on artificial substratum in shallow lakes with contrasting nutrient concentrations. *Freshwater Biol.* **51**, 95–109.

Luttenton, M.R., and Lowe, R.L. (2006). Response of a lentic periphyton community to nutrient enrichment at low N:P ratios. *J. Phycol.* 42, 1007–1015.

Moschini-Carlos, V., Henry, R., and Pompêo, M.L.M. (2000). Seasonal variation of biomass and productivity of the periphytic community on artificial substrata in the Jurumirim Reservoir (São Paulo, Brazil). *Hydrobiologia* **434**, 35–40.

Murakami, E.A., Bicudo, D.C., and Rodrigues, L. (2009). Periphytic algae of the Garças Lake, Upper Paraná River floodplain: comparing the years 1994 and 2004. *Braz. J. Biol.* **69**, 459–468.

Ortega, M.R., Alvarado, R., Hernández, R., Israde, I., Sánchez, J.D., Arredondo, M., and Martínez, I. (2009). El perifiton de un lago hiposalino hipereutrófico en Michoacán, México. *Biológicas* **11**, 56–63.

Pizarro, H., Allende, L., and Bonaventura, S.M. (2004). Littoral epilithon of lentic water bodies at Hope Bay, Antarctic Peninsula: biomass variables in relation to environmental conditions. *Hydrobiologia* **529**, 237–250.

Rautio, M., and Vincent, W.F. (2006). Benthic and pelagic food resources for zooplankton in shallow high-latitude lakes and ponds. *Freshwater Biol.* **51**, 1038–1052.

Roberts, E., Kroker, J., Körner, S., and Nicklisch, A. (2003). The role of periphyton during the re-colonization of a shallow lake with submerged macrophytes. *Hydrobiologia* **506/509**, 525–530.

Rodríguez, P., Tell, G., and Pizarro, H. (2011). Epiphytic algal biodiversity in humic shallow lakes from the lower Paraná River basin (Argentina). *Wetlands* **31**, 53–63.

Rodusky, A.J., Steinman, A.D., East, T.L., Sharfstein, B., and Meeker, R.H. (2001). Periphyton nutrient limitation and other potential growth-controlling factors in Lake Okeechobee, U.S.A. *Hydrobiologia* **448**, 27–39.

Rouf, A.J.M., Ambak, M.A., Lokman, S., Siew-Moi, P., and Sinn, C.H. (2008). Temporal changes in the periphytic algal communities in a drowned tropical forest reservoir in Malaysia: Lake Kenyir. *Lakes Reservoirs: Res. Manage.* **13**, 271–287.

Sánchez, M.L., Pizarro, H., Tell, G., and Izaguirre, I. (2010). Relative importance of periphyton and phytoplankton in turbid and clear vegetated shallow lakes from the Pampa Plain (Argentina): a comparative experimental study. *Hydrobiologia* **646**, 271–280.

APPENDIX B. BACTERIOPLANKTON LATITUDINAL GRADIENT

We analyzed 17 publications including information on 140 lakes along a latitudinal gradient from 7° to 64°S and from 29.1° to 68.5°N. The lakes covered a gradient in trophic status from 1.1 to 506 μg TP L^{-1} and 0.29 to 221 μg phytoplankton Chl-a L^{-1}. We grouped the lakes according to climatic zones following the revision by Leemans and Cramer (1991) of the Köppen climate system (1936). We performed Spearman's correlations and stepwise multiple linear regressions (after log-transforming relevant data) between bacterioplankton data and relevant environmental variables, latitude, altitude, lake area, water temperature, conductivity, nutrients, such as phytoplankton Chl-a, dissolved organic carbon concentration, etc. We used the statistical packages STATISTICA and SigmaPlot.

The Spearman's correlations indicated that bacterioplankton abundance significantly declined with increasing latitude ($r=0.55$, $p<0.0001$, $n=133$) and increased with DOC ($r=0.90$, $p<0.0001$, $n=34$), temperature ($r=0.67$, $p<0.0001$, $n=112$), conductivity ($r=0.58$, $p<0.0001$, $n=98$) and Chl-a ($r=0.46$, $p<0.0001$, $n=95$). No significant relationships were found for nutrients (TP and TN).

Literature cited:

Alonso, C., Zeder, M., Piccini, C., Conde, D., and Pernthaler, J. (2009). Ecophysiological differences of betaproteobacterial populations in two hydrochemically distinct compartments of a subtropical lagoon. Environ. Microbiol. **11**, 867–876.

Anesio, A.M., Abreu, P.C., and de Assis Esteves, F. (1997). Influence of the hydrological cycle on the bacterioplankton of an impacted clear water Amazonian lake. Microb. Ecol. **34**, 66–73.

Bouvy, M., Falcão, D., Marinho, M., Pagano, M., and Moura, A. (2000). Occurrence of *Cylindrospermopsis* (Cyanobacteria) in 39 Brazilian tropical reservoirs during the 1998 drought. Aquat. Microb. Ecol. **23**, 13–27.

Farjalla, V.F., Faria, B.M., Esteves, F.A., and Bozelli, R.L. (2001). Bacterial density and biomass, and relations with abiotic factors, in 14 coastal lagoons of Rio de Janeiro State. In Aquatic Microbial Ecology in Brazil, Serie Oecologia Brasiliensis (Ed. by B.M. Faria, V.F. Farjalla and F.A. Esteves), pp. 65–76. PPGE-UFRJ, Rio de Janeiro, Brazil.

Farjalla, V.F., Laque, T., Suhett, A., Amado, A.M., and Esteves, F.A. (2005). Diel variations of bacterial abundance and productivity in tropical coastal lagoons: the importance of bottom up factors in a short time scale. Acta Limno. Bras. **17**, 373–383.

Farjalla, V.F., Enrich-Prast, A., Esteves, F.A., and Cimbleris, A.C.P. (2006). Bacterial growth and DOC consumption in a tropical coastal lagoon. Braz. J. Biol. **66**, 383–392.

Glöckner, F.-O., Zaichikov, E., Belkova, N., Denissova, L., Pernthaler, J., Pernthaler, A., and Amann, R. (2000). Comparative 16S rRNA analysis of lake bacterioplankton reveals globally distributed phylogenetic clusters including an abundant group of Actinobacteria. Appl. Environ. Microbiol. **6**, 5053–5065.

Izaguirre, I., Allende, L., and Marinone, M.C. (2003). Comparative study of the planktonic communities of two lakes of contrasting trophic status at Hope Bay (Antarctic Peninsula). J. Plankton Res. **25**, 1079–1097.

Izaguirre, I., Mataloni, G., Allende, L., and Vinocur, A. (2001). Summer fluctuations of microbial planktonic communities in an eutrophic lake—Cierva Point, Antarctica. J. Plankton Res. **23**, 1095–1109.

Karlsson, J., Jonsson, A., and Jansson, M. (2001). Bacterioplankton production in lakes along an altitude gradient in the subarctic north of Sweden. Microb. Ecol. **42**, 372–382.

Muylaert, K., Van der Gucht, K., Vloemans, N., Meester, L.D., Gillis, M., and Vyverman, W. (2002). Relationship between bacterial community composition and bottom-up versus top-down variables in four eutrophic shallow lakes. Appl. Environ. Microbiol. **68**, 4740–4750.

Piccini, C., Conde, D., Alonso, C., Sommaruga, R., and Pernthaler, J. (2006). Blooms of single bacterial species in a coastal lagoon of the southwestern Atlantic Ocean. Appl. Environ. Microbiol. **72**, 6560–6568.

Schauer, M., and Hahn, M. (2005). Diversity and phylogenetic affiliations of morphologically conspicuous large filamentous bacteria occurring in the pelagic zones of a broad spectrum of freshwater habitats. Appl. Environ. Microbiol. **71**, 1931–1940.

Schiaffino, R.M., Unrein, F., Gasol, J.M., Farías, M.E., Estévez, C., Balagué, V., and Izaguirre, I. (2009). Comparative analysis of bacterioplankton assemblages from maritime Antarctic freshwater lakes with contrasting trophic status. Polar Biol. **32**, 923–936.
Schiaffino, R.M., Unrein, F., Gasol, J.M., Massana, R., Balagué, V., and Izaguirre, I. (2011). Bacterial community structure in a latitudinal gradient of lakes: the roles of spatial versus environmental factors. Freshwater Biol. **56**, 1973–1991.
Sommaruga, R., and Casamayor, E. (2009). Bacterial 'cosmopolitanism' and importance of local environmental factors for community composition in remote high-altitude lakes. Freshwater Biol. **54**, 994–1005.
Wu, Q.L., Zwart, G., Schauer, M., Kamst-van Agterveld, M.P., and Hahn, M.W. (2006). Bacterioplankton community composition along a salinity gradient of sixteen high-mountain lakes located on the Tibetan Plateau, China. Appl. Environ. Microbiol. **72**, 5478–5485.

APPENDIX C. PHYTOPLANKTON UNPUBLISHED DATA AND LATITUDINAL GRADIENT

C.1. Previously unpublished data: The Netherlands–Uruguay comparison

We compiled a database from 650 shallow lakes located within two climate zones in the Netherlands (temperate) and Uruguay (subtropical), covering a wide range of environmental characteristics (Kruk et al., 2011). The sampling and sample analyses protocols were similar in both locations. Lakes were sampled at random points integrating the water column and covering the whole lake area. Water samples for nutrients and plankton were taken integrating the water column with a plastic tube (20-cm diameter) and combining from 3 to 20 random replicates in each lake. Phytoplankton samples were fixed in Lugol's solution and total phosphorus (TP, mg L^{-1}) was estimated. Details on sample analysis are provided in Kruk et al. (2009).

Phytoplankton populations were counted in random fields from fixed Lugol samples using the settling technique (Utermöhl, 1958). We examined the samples at multiple magnifications and counted until we reached at least 100 individuals of the most frequent species (Lund et al., 1958). Organisms between 5 and 100 μm were counted at 400×, larger organisms were counted at 200×, and organisms between 5 and 2 μm were counted at 1000×. We did not include picoplanktonic species (<2 mm) or species strongly associated with periphytic communities. For all lakes, we considered the organism as the unit (unicell, colony or filament). Cell numbers per colony as well as organism dimensions, including maximum linear dimension (MLD, μm) were estimated. Individual volume (*V*, μm^3) and surface (*S*, μm^3) were calculated according to geometric equations (Hillebrand et al., 1999). For colonial organisms with mucilage, *V* and *S* calculations were made for whole colonies including mucilage. Population biovolume was estimated ($mm^3\ L^{-1}$) and calculated as the individual volume of the species multiplied by the abundance of individuals. Most of the samples were analyzed by the same group of scientists using the same identification keys and a common protocol. However, to diminish the potential taxonomical discrepancies between different data sets, we did not include the organisms that were not identified at least at the class level. Also, organisms from the same genus but not identified to species level were grouped together. We also excluded the species with a contribution of less than 5% to the total community biomass in any individual lake. Richness was estimated as the sum of species per lake.

We classified species into phylogenetic classes following Van Den Hoeck et al. (1997), except cyanobacteria (Pérez et al., 1999; Komárek and Anagnostidis, 2005) and Bacillariophyceae (Round, 1992). We also classified the species into size classes, selected to represent the main growing strategies of phytoplankton. Following Reynolds (1988), we plotted the mean values per species of all lakes of log *S*/*V* versus log MLD and selected five MLD classes corresponding to <3, 3–10, 10–30, 30–100 and >100 μm that have a more homogeneous distribution of the species. For each class and each size class, the biovolumes were summed per sample. In the case of classes, the biovolume was transformed into percentage to compare the relative contribution in each climatic region.

Total richness of 5% species, total biovolume and the percentage of each phylogenetic group biovolume were tested for differences in their median value between subtropical and temperate regions using Kruskal–Wallis tests.

C.2. Latitudinal gradient meta-analysis

A meta-analysis of data was done searching in published and un-published sources for total chlorophyll-a (Chl-a) and its relationship with total phosphorus (TP), determined using simple linear regression with $\log_{10}$-transformed data in all studies. The considered studies had comparable sampling procedure and sample analysis. A total of 290 cases were obtained for Chl-a including tropical, subtropical, temperate, cold and polar lakes.

A total of 28 models were calculated from the data available with Statistica software or obtained from the cited references (see Table 3).

Chlorophyll concentration as well as the slope and r^2 of the relation between ($\log_{10}$-transformed) chl-*a* and TP were tested for differences in their median value among climatic regions (tropical, subtropical, temperate, cold and polar lakes) using Kruskal–Wallis tests and *post hoc* multiple comparisons (Z value). We used this type of analysis due to its widespread use in the relevant literature, thus allowing us to increase the power of our comparison. KW test was selected due to the uneven number of cases in each climate region for all categorical variables.

Data on phytoplankton species richness and latitude from Stomp et al. (2011) (Ecological Archives E092-183-S1 in http://esapubs.org/archive/ecol/E092/183/suppl-1.htm) were used to estimate the r^2 of the linear relation between richness and latitude.

Literature cited:

Komárek, J., and Anagnostidis, K. (2005). Cyanoprokaryota II. Teil Oscillatoriales. Spektrum Akademischer Verlag, München.

Kruk, C., Peeters, E.T.H.M., Van Nes, E.H., Huszar, V.L.M., Costa, L. S., and Scheffer, M. (2011). Phytoplankton community composition can be predicted best in terms of morphological groups. *Limnol. Oceanogr.* **56**, 110–118.

Kruk, C., Rodríguez-Gallego, L., Meerhoff, M., Quintans, F., Lacerot, G., Mazzeo, N., Scasso, F., Paggi, J., Peeters, E.T.H.M. and Scheffer, M. (2009). Determinants of biodiversity in subtropical shallow lakes (Atlantic coast, Uruguay). *Freshwater Biol.* **54**, 2628–2641.

Lund, J.W.G., Kipling, C., and Le Cren, E.D. (1958). The inverted microscope method of estimating algal numbers and the statistical basis of estimations by counting. Hydrobiologia **11**, 143–170.

Pérez, M.C., Bonilla, S., De León, L., Smarda, J., and Komárek, J. (1999). A bloom of *Nodularia baltica-spumigena* group (Cyanobacteria) in a shallow coastal lagoon of Uruguay, South America. *Algol. Studies* **93**, 91–101.

Reynolds, C.S. (1988). Functional morphology and the adaptive strategies of freshwater phytoplankton, pp. 388–433. In *Growth and reproductive strategies of freshwater phytoplankton* (Ed. C.D. Sandgren) Cambridge University Press.

Round, F.E., Crawford, R. M., and Mann, D.G. (1992). The Diatoms. Biology and morphology of the genera. Cambridge University Press, Cambridge.

Stomp, M., Huisman, J., Mittelbach, G.G., Litchman, E., and Klausmeier, C.A. (2011). Large-scale biodiversity patterns in freshwater phytoplankton. *Ecology* **92**, 2096–2107.

Utermöhl, H. (1958). Zur Vervollkommnung der quantitativen Phytoplankton-Methodik. *Mitt int.ver. Limnol.* **9**, 1–38.

Van Den Hoeck, D., Mann, G., and Jahns, H.M. (1997). Algae: an introduction of phycology. Cambridge University Press.

REFERENCES

Adrian, R., Wilhelm, S., Gerten, D., 2006. Life history traits of lake plankton species may govern their phenological response to climate warming. Glob. Change Biol. 12, 652–661.

Adrian, R., O'Reilly, C.M., Zagarese, H., Baines, S.B., Hessen, D.O., Keller, W., Livingstone, D.M., Sommaruga, R., Straile, D., Van Donk, E., Weyhenmeyer, G.A., Winder, M., 2009. Lakes as sentinels of climate change. Limnol. Oceanogr. 54, 2283–2297.

Algar, A.C., Kerr, J.T., Currie, D.J., 2007. A test of Metabolic Theory as the mechanism underlying broad-scale species-richness gradients. Glob. Ecol. Biogeogr. 16, 170–178.

Allen, A.P., Gillooly, J.F., 2006. Assessing latitudinal gradients in speciation rates and biodiversity at the global scale. Ecol. Lett. 9, 947–954.

Allen, A.P., Brown, J.H., Gillooly, J.F., 2002. Global biodiversity, biochemical kinetics, and the energetic-equivalence rule. Science 297, 1545–1548.

Amarasekare, P., Savage, V., 2012. A framework for elucidating the temperature dependence of fitness. Am. Nat. 179, 178–191.

Amarasinghe, U.S., Welcomme, R.L., 2002. An analysis of fish species richness in natural lakes. Environ. Biol. Fish. 65, 327–339.

Angilletta, J.M.J., 2009. Thermal Adaptation: a theoretical and empirical synthesis. Oxford Univeristy Press, Oxford.

Antoniades, D., Douglas, M.S.V., Smol, J.P., 2003a. Comparative physical and chemical limnology of two Canadian high arctic regions: Alert (Ellesmere Island, NU) and Mould Bay (Prince Patrick Island, NWT). Arch. Hydrobiol. 158, 485–516.

Antoniades, D., Douglas, M.S.V., Smol, J.P., 2003b. The physical and chemical limnology of 24 ponds and one lake from Isachsen, Ellef Ringnes Island, Canadian High Arctic. Int. Rev. Hydrobiol. 88, 519–538.

Arim, M., Bozinovic, F., Marquet, P.A., 2007. On the relationship between trophic position, body mass and temperature: reformulating the energy limitation hypothesis. Oikos 116, 1524–1530.

Arim, M., Abades, S., Laufer, G., Loureiro, M., Marquet, P.A., 2010. Food web structure and body size: trophic position and resource acquisition. Oikos 119, 147–153.

Arim, M., Berazategui, M., Barreneche, J.M., Ziegler, L., Zaruki, M., Abades, S.R., 2011. Determinants of density–body size scaling within food webs and tools for their detection. Adv. Ecol. Res. 45, 1–40.

Azam, F., 1998. Microbial control of oceanic carbon flux: the plot thickens. Science 280, 694–696.

Baas-Becking, I.G.M., 1934. Geobiologie of Inleiding Tot de Milieukunde. Serie 18 /19. Van Stockum and Zoon, The Hague, The Netherlands.

Barber-James, H., Gattolliat, J.-L., Sartori, M., Hubbard, M., 2008. Global diversity of mayflies (Ephemeroptera, Insecta) in freshwater. Hydrobiologia 595, 339–350.

Barko, J.W., Smart, R.M., 1981. Comparative influences of light and temperature on the growth and metabolism of selected submersed freshwater macrophytes. Ecol. Monogr. 51, 219–236.

Bécares, E., Goma, J., Fernández-Aláez, M., Fernández-Aláez, C., Romo, S., Miracle, M.R., Stahl-Delbanco, A., Hansson, L.A., Gyllstrom, M., Van de Bund, W.J., Van Donk, E., Kairesalo, T., Hietala, J., Stephen, D., Balayla, D., Moss, B., 2008. Effects of nutrients and fish on periphyton and plant biomass across a European latitudinal gradient. Aquat. Ecol. 42, 561–574.

Behrens, M.D., Lafferty, K.D., 2007. Temperature and diet effects on omnivorous fish performance: implications for the latitudinal diversity gradient in herbivorous fishes. Can. J. Fish. Aquat. Sci. 64, 867–873.

Beisner, B.E., McCauley, E., Wrona, F.J., 1996. Temperature- mediated dynamics of planktonic food chains: the effect of an invertebrate carnivore. Freshw. Biol. 35, 219–232.

Beisner, B.E., McCauley, E., Wrona, F.J., 1997. The influence of temperature and food chain length on plankton predator-prey dynamics. Can. J. Fish. Aquat. Sci. 54, 586–595.

Beisner, B.E., Peres-Neto, P.R., Lindström, E.S., Barnett, A., Longhi, A.L., 2006. The role of environmental and spatial processes in structuring lake communities from bacteria to fish. Ecology 87, 2985–2991.

Beklioğlu, M., Altinayar, G., Tan Can, O., 2006. Water level control over submerged macrophyte development in five shallow lakes of Mediterranean Turkey. Arch. Hydrobiol. 166, 535–556.

Beklioğlu, M., Romo, S., Kagalou, I., Quintana, X., Bécares, E., 2007. State of the art in the functioning of shallow Mediterranean lakes: workshop conclusions. Hydrobiologia 584, 317–326.

Belk, M.C., Houston, D.D., 2002. Bergmann's rule in ectotherms: a test using freshwater fishes. Am. Nat. 160, 803–808.

Bjerring, R., Bécares, E., Declerck, S., Gross, E.M., Hansson, L.-A., Kairesalo, T., Nykänen, M., Halkiewicz, A., Kornijów, R., Conde-Porcuna, J.M., Seferlis, M., Nõges, T., Moss, B., Amsinck, S.L., Odgaard, B.V., Jeppesen, E., 2009. Subfossil Cladocera in relation to contemporary environmental variables in 54 Pan-European lakes. Freshw. Biol. 54, 2401–2417.

Blanck, A., Lamouroux, N., 2007. Large-scale intraspecific variation in life-history traits of European freshwater fish. J. Biogeogr. 34, 862–875.

Blanckenhorn, W.U., Demont, M., 2004. Bergmann and converse Bergmann latitudinal clines in arthropods: two ends of a continuum? Integr. Comp. Biol. 44, 413–424.

Blenckner, T., 2005. A conceptual model of climate-related effects on lake ecosystems. Hydrobiologia 533, 1–14.

Blenckner, T., Adrian, R., Livingstone, D.M., Jennings, E., Weyhenmeyer, G.A., George, D.G., Jankowski, T., Jarvinen, M., Aonghusa, C.N., Nõges, T., Straile, D., Teubner, K., 2007. Large-scale climatic signatures in lakes across Europe: a meta-analysis. Glob. Change Biol. 13, 1314–1326.

Boix, D., Gascón, S., Sala, J., Badosa, A., Brucet, S., López-Flores, R., Martinoy, M., Gifre, J., Quintana, X.D., 2008. Patterns of composition and species richness of crustaceans and aquatic insects along environmental gradients in Mediterranean water bodies. Hydrobiologia 597, 53–69.

Brooks, L., Dodson, I., 1965. Predation, body size and composition of the plankton. Science 50, 28–35.

Brown, J.H., Marquet, P.A., Taper, M.L., 1993. Evolution of body size: Consequences of an energetic deinition of fitness. Am. Nat. 142, 573–584.

Brown, C.D., Hoyer, M.V., Bachmann, R.W., Canfield Jr., D.E., 2000. Nutrient-chlorophyll relationships: an evaluation of empirical nutrient-chlorophyll models using Florida and north-temperate lake data. Can. J. Fish. Aquat. Sci. 57, 1574–1583.

Brown, J.H., Gilloly, J.F., Allen, A.P., Savage, V.M., West, G.B., 2004. Toward a metabolic theory of ecology. Ecology 85, 1771–1789.

Brown, J.H., West, G.B., Enquist, B.J., 2005. Yes, West, Brown and Enquist's model of allometric scaling is both mathematically correct and biologically relevant. Funct. Ecol. 19, 735–738.

Brown, L., Edwards, F., Milner, A., Woodward, G., Ledger, M., 2011. Food web complexity and allometric-scaling relationships in stream mesocosms: implications for experimentation. J. Anim. Ecol. 80, 884–895.

Brucet, S., Boix, D., López-Flores, R., Badosa, A., Moreno-Amich, R., Quintana, X.D., 2005. Zooplankton structure and dynamics in permanent and temporary Mediterranean salt marshes: taxon-based and size-based approaches. Arch. Hydrobiol. 162, 535–555.

Brucet, S., Boix, D., Gascón, S., Sala, J., Quintana, Z.D., Badosa, A., Søndergaard, M., Lauridsen, T.L., Jeppesen, E., 2009. Species richness of crustacean zooplankton and trophic structure of brackish lagoons in contrasting climate zones: north temperate Denmark and Mediterranean Catalonia (Spain). Ecography 32, 692–702.

Brucet, S., Boix, D., Quintana, X.D., Jensen, E., Nathansen, L.W., Trochine, C., Meerhoff, M., Gascón, S., Jeppesen, E., 2010. Factors influencing zooplankton size structure at contrasting temperatures in coastal shallow lakes: Implications for effects of climate change. Limnol. Oceanogr. 55, 1697–1711.

Brucet, S., Boix, D., Nathansen, L.W., Quintana, X.D., Jensen, E., Balayla, D., Meerhoff, M., Jeppesen, E., 2012. Effects of temperature, salinity and fish in structuring the macroinvertebrate community in shallow lakes: Implications for effects of climate change. PLoS One 7 (2), e30877. http://dx.doi.org/10.1371/journal.pone.0030877.

Brucet, S., Stephanie, P., Lauridsen, T., Mehner, T., Argillier, C., Winfield, I., Volta, P., Emmrich, M., Hesthagen, T., Holmgren, K., Krause, T., Palm, A., Rask, M. and Jeppesen, E. (submitted) Fish diversity in European lakes: climate, productivity and anthropogenic pressures.

Bucak, T., Saraoğlu, E., Jeppesen, E. and Beklioğlu, M., 2012. The role of water level for macrophyte growth and trophic interactions in eutrophic Mediterranean shallow lakes: a mesocosm experiment with and without fish. *Freshwater Biol.* 57, 1631–1642.

Butterwick, C., Heaney, S.I., Talling, J.F., 2005. Diversity in the influence of temperature on the growth rates of freshwater algae, and its ecological relevance. Freshw. Biol. 50, 291–300.

Carey, C.C., Ibelings, B.W., Hoffmann, E.P., Hamilton, D.P., Brookes, J.D., 2012. Eco-physiological adaptations that favour freshwater cyanobacteria in a changing climate. Water Res. 46, 1394–1407.

Carpenter, S.R., Caraco, N.F., Correll, D.L., Howarth, R.W., Sharpley, A.N., Smith, V.H., 1998. Nonpoint pollution of surface waters with phosphorus and nitrogen. Ecol. Appl. 8, 559–568.

Caruso, T., Garlaschelli, D., Bargagli, R., Convey, P., 2010. Testing metabolic scaling theory using intraspecific allometries in Antartic microarthropods. Oikos 119, 935–945.

Chambers, P., Lacoul, P., Murphy, K., Thomaz, S., 2008. Global diversity of aquatic macrophytes in freshwater. Hydrobiologia 595, 9–26.

Charnov, E.L., Gillooly, J.F., 2004. Size and temperature in the evolution of fish life histories. Integr. Comp. Biol. 44, 494–497.

Chase, J.M., 1999. Food web effects of prey size refugia: variable interactions and alternative stable equilibria. Am. Nat. 154, 559–570.

Chen, C.Y., Folt, C.L., 1996. Consequences of fall warming for zooplankton overwintering success. Limnol. Oceanogr. 41, 1077–1086.

Christensen, M.R., Graham, M.D., Vinebrooke, R.D., Findlay, D.L., Paterson, M.J., Turner, M.A., 2006. Multiple anthropogenic stressors cause ecological surprises in boreal lakes. Glob. Change Biol. 12, 2316–2322.
Christoffersen, K., Andersen, N., Søndergaard, M., Liboriussen, L., Jeppesen, E., 2006. Implications of climate-enforced temperature increases on freshwater pico- and nanoplankton populations studied in artificial ponds during 16 months. Hydrobiologia 560, 259–266.
Conover, D.O., 1992. Seasonality and the scheduling of life history at different latitudes. J. Fish Biol. 41, 161–178.
Coops, H., Beklioğlu, M., Chrisman, T.L., 2003. The role of water-level fluctuations in shallow lake ecosystems - workshop conclusions. Hydrobiologia 506, 23–27.
Crandall, K., Buhay, J., 2008. Global diversity of crayfish (Astacidae, Cambaridae, and Parastacidae-Decapoda) in freshwater. Hydrobiologia 595, 295–301.
Crow, G.E., 1993. Species-diversity in aquatic angiosperms - latitudinal patterns. Aquat. Bot. 44, 229–258.
Damuth, J., 1981. Population density and body size in mammals. Nature 290, 699–700.
Danger, M., Lacroix, G., Kâ, S., Ndour, H., Corbin, D., Lazzaro, X., 2009. Food-web structure and functioning of temperate and tropical lakes: A stoichiometric viewpoint. Ann. Limnol. Int. J. Lim. 45, 11–21.
Daufresne, M., Lengfellner, K., Sommer, U., 2009. Global warming benefits the small in aquatic ecosystems. Proc. Natl. Acad. Sci. USA 106, 12788–12793.
de Moor, F., Ivanov, V., 2008. Global diversity of caddisflies (Trichoptera: Insecta) in freshwater. Hydrobiologia 595, 393–407.
de Wit, R., Bouvier, T., 2006. "Everything is everywhere, but, the environment selects"; what did Baas Becking and Beijerinck really say? Environ. Microb. 8, 755–758.
Declerck, S., Vandekerkhove, J., Johansson, L., Muylaert, K., Conde-Porcuna, J.M., Van der Gucht, K., Pérez-Martínez, C., Lauridsen, T., Schwenk, K., Zwart, G., Rommens, W., López-Ramos, J., Jeppesen, E., Vyverman, W., Brendonck, L., De Meester, L., 2005. Multi-group biodiversity in shallow lakes along gradients of phosphorus and water plant cover. Ecology 86, 1905–1915.
Dell, A.L., Pawar, S., Savage, V.M., 2011. Systematic variation in the temperature dependence of physiological and ecological traits. Proc. Natl. Acad. Sci. USA 108, 10591–10596.
Della Bella, V., Bazzanti, M., Dowgiallo, M., Iberite, M., 2008. Macrophyte diversity and physico-chemical characteristics of Tyrrhenian coast ponds in central Italy: implications for conservation. Hydrobiologia 597, 85–95.
Dodds, W.K., 2007. Trophic state, eutrophication, and nutrient criteria in streams. Trends Ecol. Evol. 22, 669–676.
Dodson, S.I., Arnott, S.E., Cottingham, K.L., 2000. The relationship in lake communities between primary productivity and species richness. Ecology 81, 2662–2679.
Dossena, M., Grey, J., Montoya, J.M., Perkins, D.M., Trimmer, M., Woodward, G., Yvon-Durocher, G., 2012. Warming alters community size structure and ecosystem functioning. Proc. R. Soc. Lond. B 279, 1740, 3011–3019.
Downing, J.A., Prairie, Y.T., Cole, J.J., Duarte, C.M., Tranvik, L.J., Striegl, R.G., McDowell, W.H., Kortelainen, P., Caraco, N.F., Melack, J.M., Middelburg, J.J., 2006. The global abundance and size distribution of lakes, ponds, and impoundments. Limnol. Oceanogr. 51, 2388–2397.
Duarte, C.M., Kalff, J., 1987. Latitudinal influences on the depths of maximum colonization and maximum fiomass of submerged angiosperms in lakes. Can. J. Fish. Aquat. Sci. 44, 1759–1764.
Duarte, C.M., Kalff, J., Peters, R.H., 1986. Patterns in biomass and cover of aquatic macrophytes in lakes. Can. J. Fish. Aquat. Sci. 43, 1900–1908.

Dunne, J.A., Williams, R.J., Martinez, N.D., 2002. Network structure and biodiversity loss in food webs: robustness increases with connectance. Ecol. Lett. 5, 558–567.
Durance, I., Ormerod, S.J., 2007. Climate change effects on upland stream macroinvertebrates over a 25-year period. Glob. Change Biol. 13, 942–957.
Eilertsen, H.C., Wyatt, T., 2000. Phytoplankton models and life history strategies. South Afr. J. Mar. Sci. 22, 323–338.
Elser, J.J., Bracken, M.E.S., Cleland, E.E., Gruner, D.S., Stanley Harpole, W., Hillebrand, H., Ngai, J.T., Seabloom, E.W., Shurin, J.B., Smith, J.E., 2007. Global analysis of nitrogen and phosphorus limitation of primary producers in freshwater, marine and terrestrial ecosystems. Ecol. Lett. 10, 1135–1142.
Enquist, B.J., Haskell, J.P., Tiffney, B.H., 2002. General patterns of taxonomic and biomass partitioning in extant and fossil plant communities. Nature 419, 610–613.
Etienne, R.S., Apol, M.E.F., Olff, H., 2006. Demystifying the West, Brown & Enquist model of the allometry of metabolism. Funct. Ecol. 20, 394–399.
Feld, C.K., Birk, S., Bradley, D.C., Hering, D., Kail, J., Marzin, A., Melcher, A., Nemitz, D., Pedersen, M.L., Pletterbauer, F., Pont, D., Verdonschot, P.F.M., et al., 2011. From natural to degraded rivers and back again: a test of restoration ecology theory and practice. Adv. Ecol. Res. 44, 119–210.
Fernando, C.H., 2002. Zooplankton and tropical freshwater fisheries. In: Fernando, C.H. (Ed.), A Guide to Tropical Freshwater Zooplankton. Identification, Ecology and Impact on Fisheries. Backhuys Publishers, Leiden, pp. 255–280.
Fernando, C.H., Paggi, J.C., Rajapaksa, R., 1987. *Daphnia* in tropical lowlands. Mem. Ist. Ital. Idrobiol. 45, 107–141.
Feuchtmayr, H., McKee, D., Harvey, I.F., Atkinson, D., Moss, B., 2007. Response of macroinvertebrates to warming, nutrient addition and predation in large-scale mesocosm tanks. Hydrobiologia 584, 425–432.
Feuchtmayr, H., Moran, R., Hatton, K., Connor, L., Heyes, T., Moss, B., Harvey, I., Atkinson, D., 2009. Global warming and eutrophication: effects on water chemistry and autotrophic communities in experimental, hypertrophic, shallow lake mesocosms. J. Appl. Ecol. 46, 713–723.
Flanagan, K.M., McCauley, E., Wrona, F., Prowse, T., 2003. Climate change: the potential for latitudinal effects on algal biomass in aquatic ecosystems. Can. J. Fish. Aquat. Sci. 60, 635–639.
Forster, J., Hirst, A.G., Atkinson, D., 2011a. How do organisms change size with changing temperature? The importance of reproductive method and ontogenetic timing. Funct. Ecol. 25, 1024–1031.
Forster, J., Hirst, A.G., Woodward, G., 2011b. Growth and development rates have different thermal responses. Am. Nat. 178, 668–678.
Fox, J.W., McGrady-Steed, J., 2002. Stability and complexity in microcosm communities. J. Anim. Ecol. 71, 749–756.
Friberg, N., Dybkjær, J.B., Olafsson, J.S., Mar Gislason, G., Larsen, S.E., Lauridsen, T.L., 2009. Relationships between structure and function in streams contrasting in temperature. Freshw. Biol. 54, 2051–2068.
Friberg, N., Bonada, N., Bradley, D.C., Dunbar, M.J., Edwards, F.K., Grey, J., Hayes, R.B., Hildrew, A.G., Lamouroux, N., Trimmer, M., Woodward, G., 2011. Biomonitoring of human impacts in freshwater ecosystems: the good, the bad, and the ugly. Adv. Ecol. Res. 44, 2–68.
Fu, C., Wu, J., Wang, X., Lei, G., Chen, J., 2004. Patterns of diversity, altitudinal range and body size among freshwater fishes in the Yangtze River basin, China. Glob. Ecol. Biogeogr. 13, 543–552.

Fuhrman, J.A., Steele, J.A., Hewson, I., Schwalbach, M.S., Brown, M.V., Green, J.L., Brown, J.H., 2008. A latitudinal diversity gradient in planktonic marine bacteria. Proc. Natl. Acad. Sci. USA 105, 7774–7778.

Gallego, I., Davidson, T.A., Jeppesen, E., Pérez-Martínez, C., Sánchez, P., Juan, M., Fuentes, F., León, D., Peñalver, P., Toja, J., Casas, J.J., 2012. Taxonomic or ecological approaches? Searching for phytoplankton surrogates in the determination of richness and assemblage composition in ponds. Ecol. Indicators 18, 575–585.

Galloway, J.N., Townsend, A.R., Erisman, J.W., Bekunda, M., Cai, Z., Freney, J.R., Martinelli, L.A., Seitzinger, S.P., Sutton, M.A., 2008. Transformation of the nitrogen cycle: recent trends, questions, and potential solutions. Science 320, 889–892.

Ganf, G.G., 1974. Phytoplankton biomass and distribution in a shallow eutrophic lake (Lake George, Uganda). Oecologia 16, 9–29.

García-Ulloa, M., Álvarez-Gallardo, M.P., Torres-Bugarín, O., Buelna-Osben, H.R., Zavala-Aguirre, J.L., 2011. Temperature effects on the reproduction of *Xenoteca variata* Bean, 1887 (Pisces, Goodeidae). Avances Invest. Agropec. 15, 61–67.

Gardner, J.L., Peters, A., Kearney, M.R., Joseph, L., Heinsohn, R., 2011. Declining body size: a third universal response to warming? Trends Ecol. Evol. 26, 285–291.

Gasol, J.M.A., 1994. Framework for the assessment of top-down vs bottom-up control of heterotrophic nanoflagellate abundance. Mar. Ecol. Prog. Ser. 113, 291–300.

Gaston, K.J., Blackburn, T.M., 2000. Patterns and Process in Macroecology. Blackwell Science, Oxford.

Gerten, D., Adrian, R., 2001. Differences in the persistency of the North Atlantic Oscillation signal among lakes. Limnol. Oceanogr. 46, 448–455.

Gilljam, D., Thierry, A., Edwards, F.K., Figueroa, D., Ibbotson, A.T., Jones, J.I., Lauridsen, R.B., Petchey, O.L., Woodward, G., Ebenman, B., 2011. Seeing double: size-based and taxonomic views of food web structure. Adv. Ecol. Res. 45, 67–133.

Gillooly, J.F., 2000. Effect of body size and temperature on generation time in zooplankton. J. Plankt. Res. 22, 241–251.

Gillooly, J.F., Dodson, S.I., 2000. Latitudinal patterns in the size distribution and seasonal dynamics of new world, freshwater cladocerans. Limnol. Oceanogr. 45, 22–30.

Gillooly, J.F., Brown, J.H., West, G.B., Savage, V.M., Charnov, E.L., 2001. Effects of size and temperature on metabolic rate. Science 293, 2248–2251.

González-Bergonzoni, I., Meerhoff, M., Davidson, T.A., Teixeira-de Mello, F., Baattrup-Pedersen, A., Jeppesen, E., 2012. Meta-analysis shows a consistent and strong latitudinal pattern in fish omnivory across ecosystems. Ecosystems 15, 492–503.

Gotelli, N.J., Pyron, M., 1991. Life history variation in North American freshwater minnows: effects of latitude and phylogeny. Oikos 62, 30–40.

Goyenola, G., Iglesias, C., Mazzeo, N., Jeppesen, E., 2011. Analysis of the reproductive strategy of *Jenynsia multidentata* (Cyprinodontiformes, Anablepidae) with focus on sexual differences in growth, size, and abundance. Hydrobiolgia 673, 245–257.

Griffiths, D., 1997. Local and regional species richness in North American lacustrine fish. J. Anim. Ecol. 66, 49–56.

Gruber, N., Galloway, J.N., 2008. An Earth-system perspective of the global nitrogen cycle. Nature 451, 293–296.

Gunderson, D.R., Dygert, P.H., 1988. Reproductive effort as a predictor of natural mortality rate. ICES J. Mar. Sci. 44, 200–209.

Gyllström, M., Hansson, L.A., Jeppesen, E., García-Criado, F., Gross, E., Irvine, K., Kairesalo, T., Kornijow, R., Miracle, M.R., Nykänen, M., Nõges, T., Romo, S., Stephen, D., van Donk, E., Moss, B., 2005. The role of climate in shaping zooplankton communities of shallow lakes. Limnol. Oceanogr. 50, 2008–2021.

Hagen, M., Kissling, W.D., Rasmussen, C., De Aguiar, M.A.M., Brown, L., Carstensen, D.W., Alves-Dos-Santos, I., Dupont, Y.L., Edwards, F.K., Genini, J., Guimarães, P.K., Jenkins Jr., G.B., et al., 2012. Biodiversity, species interactions and ecological networks in a fragmented world. Adv. Ecol. Res. 46, 89–210.
Hanski, I., 1999. Metapopulation Ecology. Oxford University Press, New York.
Hansson, L.-A., 1992. The role of food chain composition and nutrient availability in shaping algal biomass development. Ecology 73, 241–247.
Hansson, L.-R., Bécares, E., Fernández-Aláez, M., Fernández-Aláez, C., Kairesalo, T., Miracle, M.R., Romo, S., Stephen, D., Vakkilainen, K., Van de Bund, W., Van Donk, E., Balayla, D., Moss, B., 2007. Relaxed circadian rhythm in zooplankton along a latitudinal gradient. Oikos 116, 585–591.
Hargeby, A., Blindow, I., Hansson, L.A., 2004. Shifts between clear and turbid states in a shallow lake: multi-causal stress from climate, nutrients and biotic interactions. Arch. Hydrobiol. 161, 433–454.
Hart, R.C., 1985. Seasonality of aquatic invertebrates in low-latitude and Southern Hemisphere inland waters. Hydrobiologia 125, 151–178.
Havens, K.E., Beaver, J.R., 2011. Composition, size, and biomass of zooplankton in large productive Florida lakes. Hydrobiologia 668, 49–60.
Havens, K.E., East, T.L., Beaver, J.R., 2007. Zooplankton response to extreme drought in a large subtropical lake. Hydrobiologia 589, 187–198.
Havens, K.E., Elia, A.C., Taticchi, M.I., Fulton, R.S., 2009. Zooplankton–phytoplankton relationships in shallow subtropical *versus* temperate lakes Apopka (Florida, USA) and Trasimeno (Umbria, Italy). Hydrobiologia 628, 165–175.
Hechinger, R.F., Lafferty, K.D., Dobson, A.P., Brown, J.H., Kuris, A.M., 2011. A common scaling rule for abundance, energetics, and production of parasitic and free-living species. Science 333, 445–448.
Heibo, E., Magnhagen, C., Vøllestad, L.A., 2005. Latitudinal variation in life-history traits in Eurasian perch. Ecology 86, 3377–3386.
Heino, J., 2001. Regional gradient analysis of freshwater biota: do similar biogeographic patterns exist among multiple taxonomic groups? J. Biogeogr. 28, 69–76.
Heino, J., 2009. Biodiversity of aquatic insects: spatial gradients and environmental correlates of assemblage-level measures at large scales. Freshw. Rev. 2, 1–29.
Heino, M., Kaitala, V., 1999. Evolution of resource allocation between growth and reproduction in animals with indeterminate growth. J. Evol. Biol. 12, 423–429.
Heino, J., Soininen, J., 2007. Are higher taxa adequate surrogates for species-level assemblage patterns and species richness in stream organisms? Biol. Conserv. 137, 78–89.
Hessen, D.O., 1989. Factors determining the nutritive status and production of zooplankton in a humic lake. J. Plankton Res. 11, 649–664.
Hessen, D.O., Faafeng, B.A., Smith, V.H., Bakkestuen, V., Walseng, B., 2006. Extrinsic and intrinsic controls of zooplankton diversity in lakes. Ecology 87, 433–443.
Hickel, B., 1988. Sexual reproduction and life cycle of *Ceratium furcoides* (Dinophyceae) in situ in the lake Plußsee (F.R.). Hydrobiologia 161, 41–48.
Hildrew, A., Rafaelli, D., Edmonds-Brown, R. (Eds.), 2007. Body Size: The Structure and Function of Aquatic Ecosystems. Cambridge University Press Cambridge, New York.
Hillebrand, H., 2004. On the generality of the latitudinal diversity gradient. Am. Nat. 163, 192–211.
Hillebrand, H., Azovsky, A.I., 2001. Body size determines the strength of the latitudinal diversity gradient. Ecography 24, 251–256.
Hladyz, S., Åbjörnsson, K., Cariss, H., Chauvet, E., Dobson, M., Elosegi, A., Ferreira, V., Fleituch, T., Gessner, M.O., Giller, P.S., Graça, M.A.S., Gulis, V., et al., 2011. Stream ecosystem functioning in an agricultural landscape: the importance of terrestrial-aquatic linkages. Adv. Ecol. Res. 44, 211–276.

Howard-Williams, C., Peterson, D., Lyons, W.B., Cattaneo-Vietti, R., Gordon, S., 2006. Measuring ecosystem response in a rapidly changing environment: the Latitudinal Gradient Project. Antarct. Sci. 18, 465–471.

Hrbacek, J., Dvorakova, M., Korinek, V., Prochazkova, L., 1961. Demonstration of the effect of the fish stock on the species composition of zooplankton and the intensity of metabolism of the whole plankton assemblage. Verh. Int. Ver. Theoret. Angew. Limnol. 14, 192–195.

Hubbs, C., 1985. Darter reproductive seasons. Copeia 1985, 56–68.

Huey, R.B., Kingsolver, J.G., 2011. Variation in universal temperature dependence of biological rates. Proc. Natl. Acad. Sci. USA 108, 10377–10378.

Huszar, V.L.M., Caraco, N., Roland, F., Cole, J., 2006. Nutrient–chlorophyll relationships in tropical–subtropical lakes: do temperate models fit? Biogeochemistry 79, 239–250.

Ibelings, B.W., Gsell, A.S., Mooij, W.M., Van Donk, E., Van Den Wyngaert, S., De Senerpont Domis, L.N., 2011. Chytrid infections and diatom spring blooms: paradoxical effects of climate warming on fungal epidemics in lakes. Freshw. Biol. 56, 754–766.

Iglesias, C., Mazzeo, N., Meerhoff, M., Lacerot, G., Clemente, J., Scasso, F., Kruk, C., Goyenola, G., García, J., Amsinck, S.L., Paggi, J.C., José de Paggi, S., Jeppesen, E., 2011. High predation is the key factor for dominance of small-bodied zooplankton in warm lakes—evidence from lakes, fish exclosures and surface sediment. Hydrobiologia 667, 133–147.

IPCC (2007). Climate Change 2007: Impacts, Adaptation and Vulnerability, IPCC, WGII

Jackson, L.J., 2011. Conservation of shallow lakes given an uncertain, changing climate: challenges and opportunities. Aquat. Conserv. 21, 219–223.

Jackson, D.A., Mandrak, N.E., 2002. Changing fish biodiversity: predicting the loss of cyprinid biodiversity due to global climate change. In: McGinn, N.A. (Ed.), Fisheries in a Changing Climate. Symposium 32. American Fisheries Society, Bethesda, Maryland, pp. 89–98.

Jackson, L.J., Lauridsen, T.L., Søndergaard, M., Jeppesen, E., 2007. A comparison of shallow Danish and Canadian lakes and implications of climate change. Freshw. Biol. 52, 1782–1792.

James, C., Fisher, J., Russell, V., Collings, S., Moss, B., 2005. Nitrate availability and hydrophyte species richness in shallow lakes. Freshw. Biol. 50, 1049–1063.

Jeppesen, E., Søndergaard, M., Kanstrup, E., Petersen, B., Henriksen, R.B., Hammershøj, M., Mortensen, E., Jensen, J.P., Have, A., 1994. Does the impact of nutrients on the biological structure and function of brackish and freshwater lakes differ? Hydrobiologia 275 (276), 15–30.

Jeppesen, E., Jensen, J.P., Sondergaard, M., Lauridsen, T., Landkildehus, F., 2000. Trophic structure, species richness and biodiversity in Danish lakes: changes along a phosphorus gradient. Freshw. Biol. 45, 201–218.

Jeppesen, E., Jensen, J.P., Lauridsen, T.L., Amsinck, S.L., Christoffersen, K., Mitchell, S.F., 2003. Sub-fossils in the surface sediment as proxies for community structure and dynamics of zooplankton in lakes: a study of 150 lakes from Denmark, New Zealand, the Faroe Islands and Greenland. Hydrobiologia 491, 321–330.

Jeppesen, E., Jensen, J.P., Søndergaard, M., Lauridsen, T.L., 2005. Response of fish and plankton to nutrient loading reduction in eight shallow Danish lakes with special emphasis on seasonal dynamics. Freshw. Biol. 50, 1365–2427.

Jeppesen, E., Meerhoff, M., Jakobsen, B.A., Hansen, R.S., Søndergaard, M., Jensen, J.P., Lauridsen, T.L., Mazzeo, N., Branco, C., 2007. Restoration of shallow lakes by nutrient control and biomanipulation: the successful strategy depends on lake size and climate. Hydrobiologia 581, 269–288.

Jeppesen, E., Kronvang, B., Meerhoff, M., Søndergaard, M., Hansen, K.M., Andersen, H.E., Lauridsen, T.L., Beklioğlu, M., Özen, A., Olesen, J.E., 2009. Climate change effects on

runoff, catchment phosphorus loading and lake ecological state, and potential adaptations. J. Environ. Qual. 38, 1030–1041.
Jeppesen, E., Meerhoff, M., Holmgren, K., González-Bergonzoni, I., Teixeira-de Mello, F., Declerck, S.A.J., De Meester, L., Søndergaard, M., Lauridsen, T.L., Bjerring, R., Conde-Porcuna, J.M., Mazzeo, N., Iglesias, C., Reizenstein, M., Malmquist, H.J., Liu, Z.W., Balayla, D., Lazzaro, X., 2010a. Impacts of climate warming on lake fish community structure and potential ecosystem effects. Hydrobiologia 646, 73–90.
Jeppesen, E., Moss, B., Bennion, H., Carvalho, L., DeMeester, L., Friberg, N., Gessner, M.O., Lauridsen, T.L., May, L., Meerhoff, M., Olafsson, J.S., Soons, M.B., Verhoeven, J.T.A., 2010b. Interaction of climate change and eutrophication. In: Kernan, M., Battarbee, R., Moss, B. (Eds.), Climate Change Impacts on Freshwater Ecosystems. Blackwell Publishing Ltd., London, pp. 119–151.
Jeppesen, E., Kronvang, B., Olesen, J.E., Audet, J., Søndergaard, M., Hoffmann, C.C., Andersen, H.E., Lauridsen, T.L., Liboriussen, L., Larsen, S.E., Beklioğlu, M., Meerhoff, M., Özen, A., Özkan, K., 2011. Climate change effects on nitrogen loading from cultivated catchments in Europe: implications for nitrogen retention, ecological state of lakes and adaptation. Hydrobiologia 663, 1–21.
Jeppesen, E., Mehner, T., Winfield, I.J., Kangur, K., Sarvala, J., Gerdeaux, D., Rask, M., Malmquist, H.J., Holmgren, K., Volta, P., Romo, S., Eckmann, R., Sandström, A., Blanco, S., Kangur, A., Ragnarsson Stabo, H., Tarvainen, M., Ventelä, A.-M., Søndergaard, M., Lauridsen, T.L., Meerhoff, M., 2012. Impacts of climate warming on the long-term dynamics of key fish species in 24 European lakes. *Hydrobiologia* 694, 1–39.
Jewson, D.H., 1992. Life cycle of a *Stephanodiscus* sp. (Bacillariophyta). J. Phycol. 28, 856–866.
Jones, J.I., Sayer, C.D., 2003. Does the fish-invertebrate-periphyton cascade precipitate plant loss in shallow lakes? Ecology 84, 2155–2167.
Kalff, J., Watson, S., 1986. Phytoplankton and its dynamics in two tropical lakes: a tropical and temperate zone comparison. Hydrobiologia 138, 161–176.
Karasov, W.H., Martínez del Rio, C., 2007. Physiological Ecology: How Animals Process Energy, Nutrients, and Toxins. Princeton University Press, Princeton, NJ.
Karlsson, J., Jonsson, A., Jansson, M., 2001. Bacterioplankton production in lakes along an altitude gradient in the subarctic north of Sweden. Microb. Ecol. 42, 372–382.
Karlsson, J., Jonsson, A., Jansson, M., 2005. Productivity of high-latitude lakes: climate effect inferred from altitude gradient. Glob. Change Biol. 11, 710–715.
Kingsolver, J.G., Huey, R.B., 2008. Size, temperature, and fitness: three rules. Evol. Ecol. Res. 10, 251–268.
Kishi, D., Murakami, M., Nakano, S., Maekawa, K., 2005. Water temperature determines strength of top-down control in a stream food web. Freshw. Biol. 50, 1315–1322.
Knouft, J.H., 2004. Latitudinal variation in the shape of the species body size distribution: an analysis using freshwater fishes. Oecologia 139, 408–417.
Knouft, J.H., Page, L.M., 2003. The evolution of body size in extant groups of North American freshwater fishes: speciation, size distributions, and Cope's rule. Am. Nat. 161, 413–421.
Kosten, S., Kamarainen, A., Jeppesen, E., Van Nes, E.H., Peeters, E.T.H.M., Mazzeo, N., Sass, L., Hauxwell, J., Hansel-Welch, N., Lauridsen, T.L., Søndergaard, M., Bachmann, R.W., Lacerot, G., Scheffer, M., 2009a. Climate-related differences in the dominance of submerged macrophytes in shallow lakes. Glob. Change Biol. 15, 2503–2517.
Kosten, S., Huszar, V.L.M., Mazzeo, N., Scheffer, M., Sternberg, L.S.L., Jeppesen, E., 2009b. Limitation of phytoplankton growth in South America: no evidence for increasing nitrogen limitation towards the tropics. Ecol. Appl. 19, 1791–1804.
Kosten, S., Huszar, V.L.M., Bécares, E., Costa, L.S., van Donk, E., Hansson, L.A., Jeppesen, E., Kruk, C., Lacerot, G., Mazzeo, N., De Meester, L., Moss, B.,

Lürling, M., Nõges, T., Romo, S., Scheffer, M., 2011a. Warmer climates boost cyanobacterial dominance in shallow lakes. Glob. Change Biol. 18, 118–126.
Kosten, S., Jeppesen, E., Huszar, V.L.M., Mazzeo, N., van Nes, E., Peeters, E.T.H.M., Scheffer, M., 2011b. Ambiguous climate impacts on competition between submerged macrophytes and phytoplankton in shallow lakes. Freshw. Biol. 56, 1540–1553.
Kozlowski, J., Konarzewski, M., 2004. Is West, Brown and Enquist's model of allometric scaling mathematically correct and biologically relevant? Ecology 18, 283–289.
Kruk, C., Rodríguez-Gallego, L., Meerhoff, M., Quintans, F., Scasso, F., Lacerot, G., Mazzeo, N., Paggi, J., Peeters, E., Scheffer, M., 2009. Determinants of diversity in subtropical shallow lakes (Atlantic coast, Uruguay). Freshw. Biol. 54, 2628–2641.
Kruk, C., Huszar, V.L.M., Peeters, E.T.H.M., Bonilla, S., Costa, L., Lürling, M., Reynolds, C.S., Scheffer, M., 2010. A morphological classification capturing functional variation in phytoplankton. Freshw. Biol. 55, 614–627.
Kruk, C., Peeters, E.T.H.M., Van Nes, E.H., Huszar, V.L.M., Costa, L.S., Scheffer, M., 2011. Phytoplankton community composition can be predicted best in terms of morphological groups. Limnol. Oceanogr. 56, 110–118.
Lacerot, G., 2010. Effects of climate on size structure and functioning of aquatic food webs. PhD Thesis. Wageningen University. 96pp.
Lappalainen, J., Lehtonen, H., 1997. Temperature habitats for freshwater fishes in a warming climate. Boreal Environ. Res. 2, 69–84.
Lappalainen, J., Tarkan, A.S., 2007. Latitudinal gradients in onset date, onset temperature and duration of spawning of roach. J. Fish Biol. 70, 441–450.
Lappalainen, J., Dörner, H., Wysujack, K., 2003. Reproduction biology of pikeperch (*Sander lucioperca* (L.))—a review. Ecol. Freshw. Fish 12, 95–106.
Lappalainen, J., Tarkan, A.S., Harrod, C., 2008. A meta-analysis of latitudinal variations in life-history traits of roach, *Rutilus rutilus*, over its geographical range: linear or non-linear relationships? Freshw. Biol. 53, 1491–1501.
Larocque, I., Pienitz, R., Rolland, N., 2006. Factors influencing the distribution of chironomids in lakes distributed along a latitudinal gradient in northwestern Quebec, Canada. Can. J. Fish. Aquat. Sci. 63, 1286–1297.
Layer, K., Riede, J.O., Hildrew, A.G., Woodward, G., 2010. Food web structure and stability in 20 streams across a wide pH gradient. Adv. Ecol. Res. 42, 265–299.
Layer, K., Hildrew, A.G., Jenkins, G.B., Riede, J., Rossiter, S.J., Townsend, C.R., Woodward, G., 2011. Long-term dynamics of a well-characterised food web: four decades of acidification and recovery in the Broadstone Stream model system. Adv. Ecol. Res. 44, 69–117.
Lazzaro, X., 1997. Do the trophic cascade hypothesis and classical biomanipulation approaches apply to tropical lakes and reservoirs? Verh. Int. Ver. Theor. Angew. Limnol. 26, 719–730.
Ledger, M.E., Harris, R.M.L., Armitage, P.D., Milner, A.M., 2012. Climate change impacts on community resilience: evidence from a drought disturbance experiment. Adv. Ecol. Res. 46, 211–258.
Leemans, R., Cramer, W., 1991. The IIASA database for mean monthly values of temperature, precipitation and cloudiness on a global terrestrial grid. Research Report RR-91-18. pp. 61. International Institute of Applied Systems Analyses, Laxenbur.
Lewis, W.M., 1996. Tropical lakes: how latitude makes a difference. In: Schiemer, F., Boland, K.T. (Eds.), Perspectives in Tropical Limnology. SPB Academic Publishing bv, Amsterdam, pp. 43–64.
Lewis, W.M., 2000. Basis for the protection and management of tropical lakes. Lakes Reservoir. Res. Manage. 5, 35–48.
Liboriussen, L., Jeppesen, E., 2003. Temporal dynamics in epipelic, pelagic and epiphytic algal production in a clear and a turbid shallow lake. Freshw. Biol. 48, 418–431.

Liboriussen, L., Landkildehus, F., Meerhoff, M., Bramm, M.E., Sondergaard, M., Christoffersen, K., Richardson, K., Sondergaard, M., Lauridsen, T.L., Jeppesen, E., 2005. Global warming: design of a flow-through shallow lake mesocosm climate experiment. Limnol. Oceanogr. Methods 3, 1–9.
Linkola, K., 1933. Regionale Artenstatistik der Süsswasserflora Finnlands. Ann. Bot. Soc. Zool. Bot. Fenn. "Vanamo" 3, 3–13.
Lopez-Urrutia, A., San Martin, E., Harris, R.P., Irigoien, X., 2006. Scaling the metabolic balance of the oceans. Proc. Natl. Acad. Sci. USA 103, 8739–8744.
Lougheed, V.L., Crosbie, B., Chow-Fraser, P., 2001. Primary determinants of macrophyte community structure in 62 marshes across the Great Lakes basin: latitude, land use, and water quality effects. Can. J. Fish. Aquat. Sci. 58, 1603–1612.
Lytle, D.A., Poff, N.L., 2003. Adaptation to natural flow regimes. Trends Ecol. Evol. 19, 94–100.
Macnab, V., Barber, I., 2012. Some (worms) like it hot: fish parasites grow faster in warmer water, and alter host thermal preferences. Glob. Change Biol. 18, 1540–1548.
Malmquist, H.J., 2004. Life history traits of Arctic charr and environmental factors: local variability and latitudinal gradients. In: The ACIA International Scientific Symposium on Climate Change in the Arctic: Extended Abstracts.
Mandrak, N.E., 1995. Biogeographic patterns of fish species richness in Ontario lakes in relation to historical and environmental factors. Can. J. Fish. Aquat. Sci. 52, 1462–1474.
Marquet, P.A., Taper, M.L., 1998. On size and area: patterns of mammalian body size extremes across landmasses. Evol. Ecol. 12, 127–139.
Marquet, P.A., Abades, S., Keymer, J.E., Zeballos, H., 2008. Discontinuities in body-size distributions: a view from the top. In: Allen, C.R., Holling, C.S. (Eds.), Discontinuities in ecosystems and other complex systems. Columbia University Press, New York, pp. 45–57.
May, R.M., 1972. Will a large complex system be stable? Nature 238, 413–414.
Mazumder, A., 1994. Phosphorus-chlorophyll relationships under contrasting herbivory and ther mal stratification: predictions and patterns. Can. J. Fish. Aquat. Sci. 51, 390–400.
Mazumder, A., Havens, K.E., 1998. Nutrient–chlorophyll–Secchi relationships under contrasting grazer communities of temperate versus subtropical lakes. Can. J. Fish. Aquat. Sci. 55, 1652–1662.
McCann, K.S., 2000. The diversity-stability debate. Nature 405, 228–233.
McCann, K., Hastings, A., Huxel, G.R., 1998. Weak trophic interactions and the balance of nature. Nature 395, 794–798.
McCreadie, J.W., Adler, P.H., Hamada, N., 2005. Patterns of species richness for blackflies (Diptera: Simuliidae) in the Nearctic and Neotropical regions. Ecol. Entomol. 30, 201–209.
McKee, D., Atkinson, D., Colling, S., Eaton, J.W., Harvey, I., Heyes, T., Hatton, K., Wilson, D., Moss, B., 2002a. Macro-zooplankter responses to simulated climate warming in experimental freshwater microcosms. Freshw. Biol. 47, 1557–1570.
McKee, D., Hatton, K., Eaton, J.W., Atkinson, D., Atherton, A., Harvey, I., Moss, B., 2002b. Effects of simulated climate warming on macrophytes in freshwater microcosm communities. Aquat. Bot. 74, 71–83.
McKee, D., Atkinson, D., Colling, S., Eaton, J.W., Gill, A.B., Harvey, I., Hatton, K., Heyes, T., Wilson, D., Moss, B., 2003. Response of freshwater microcosm communities to nutrients, fish, and elevated temperature during winter and summer. Limnol. Oceanogr. 48, 707–722.
McLaughlin, O., Jonsson, T., Emmerson, M., 2010. Temporal variability in predator-prey relationships of a forest floor food web. Adv. Ecol. Res. 42, 171–264.
McNab, B.K., 2002. The Physiological Ecology of Vertebrates. Cornell University Press, New York.

Meerhoff, M., Fosalba, C., Bruzzone, C., Mazzeo, N., Noordoven, W., Jeppesen, E., 2006. An experimental study of habitat choice by *Daphnia*: plants signal danger more than refuge in subtropical lakes. Freshw. Biol. 51, 1320–1330.

Meerhoff, M., Clemente, J.M., Teixeira-de Mello, F., Iglesias, C., Pedersen, A.R., Jeppesen, E., 2007a. Can warm climate-related structure of littoral predator assemblies weaken the clear water state in shallow lakes? Glob. Change Biol. 13, 1888–1897.

Meerhoff, M., Iglesias, C., Teixeira-de Mello, F., Clemente, J.M., Jensen, E., Lauridsen, T.L., Jeppesen, E., 2007b. Effects of habitat complexity on community structure and predator avoidance behaviour of littoral zooplankton in temperate versus subtropical shallow lakes. Freshw. Biol. 52, 1009–1021.

Mehner, T., Holmgren, K., Lauridsen, T.L., Jeppesen, E., Diekmann, M., 2007. Lake depth and geographical position modify lake fish assemblages of the European 'Central Plains' ecoregion. Freshw. Biol. 52, 2285–2297.

Melack, J.M., 1976. Primary productivity and fish yields in tropical lakes. Trans. Am. Fish. Soc. 105, 575–580.

Melack, J.M., 1979. Temporal variability of phytoplankton in tropical lakes. Oecologia 44, 1–7.

Millennium Ecosystem Assessment, 2005. Ecosystems and Human Well-being. Current State and Trends. Island Press, Washington, DC.

Mims, M.C., Olden, J.D., Shattuck, Z.R., Poff, N.L., 2010. Life history trait diversity of native freshwater fishes in North America. Ecol. Fresh. Fish 19, 390–400.

Mintenbeck et al., 2012. Impact of climate change on fish in complex Antarctic ecosystems. Adv. Ecol. Res. 46, 351–426.

Montoya, J., Raffaelli, D., 2010. Climate change, biotic interactions and ecosystem services Introduction. Philos. Trans. R. Soc. Lond. B 365, 2013–2018.

Mooij, W.M., Hulsmann, S., De Senerpont Domis, L.N., Nolet, B.A., Bodelier, P.L.E., Boers, P.C.M., Pires, M.D., Gons, H.J., Ibelings, B.W., Noordhuis, R., Portielje, R., Wolfstein, K., Lammens, E.H.R.R., 2005. The impact of climate change on lakes in the Netherlands: a review. Aquat. Ecol. 39, 381–400.

Mooij, W.M., Janse, J.H., De Senerpont Domis, L.N., Hülsmann, S., Ibelings, B.W., 2007. Predicting the effect of climate change on temperate shallow lakes with the ecosystem model PCLake. Hydrobiologia 584, 443–454.

Moore, M.V., Folt, C.L., Stemberger, R.S., 1996. Consequences of elevated temperatures for zooplankton assemblages in temperate lakes. Arch. Hydrobiol. 135, 289–319.

Morán, X.A.G., Sebastián, M., Pedrós-Alió, C., Estrada, M., 2006. Response of southern ocean phytoplankton and bacterioplankton production to short-term experimental warming. Limnol. Oceanogr. 51, 1791–1800.

Morin, R., Dodson, J.J., Power, G., 1982. Life history variations of anadromous Cisco (*Coregonus artedii*), lake whitefish (*C. clupeaformis*), and round whitefish (*Prosopium cylindraceum*) populations of Eastern James-Hudson Bay. Can. J. Fish. Aquat. Sci. 39, 958–967.

Morrison, W., Hay, M., 2011. Are lower latitude plants better defended? Palatability of freshwater macrophytes. Ecology 93, 65–74. http://dx.doi.org/10.1890/11-0725.1.

Moss, B., 1990. Engineering and biological approaches to the restoration from eutrophication of shallow lakes in which aquatic plant communities are important components. Hydrobiologia 200–2011, 367–377.

Moss, B., 1998. Ecology of Freshwaters. Man and Medium, Past to Future. Blackwell Science, London.

Moss, B., 2010. Climate change, nutrient pollution and the bargain of Dr Faustus. Freshw. Biol. 55, 175–187.

Moss, B., 2011. Cogs in the endless machine: lakes, climate change and nutrient cycles: a review. Sci. Total Environ. http://dx.doi.org/10.1016/j.scitotenv.2011.07.069.

Moss, B., Madgwick, J., Phillips, G.L., 1996. A Guide to the Restoration of Nutrient-Enriched Shallow Lakes. Broads Authority and Environment Agency (CE), Norwich.

Moss, B., McKee, D., Atkinson, D., Collings, S.E., Eaton, J.W., Gill, A.B., Harvey, I., Hatton, K., Heyes, T., Wilson, D., 2003. How important is climate? Effects of warming, nutrient addition and fish on phytoplankton in shallow lake microcosms. J. Appl. Ecol. 40, 782–792.

Moss, B., Stephen, D., Balayla, D.M., Bécares, E., Collings, S.E., Fernández Aláez, C., Fernández Aláez, M., Ferriol, C., García, P., Gomá, J., Gyllström, M., Hansson, L.A., Hietala, J., Kairesalo, T., Miracle, M.R., Romo, S., Rueda, J., Russell, V., Ståhl-Delbanco, A., Svensson, M., Vakkilainen, K., Valentín, M., Van de Bund, W.J., Van Donk, E., Vicente, E., Villena, M.J., 2004. Continental-scale patterns of nutrient and fish effects on shallow lakes: synthesis of a pan-European mesocosm experiment. Freshw. Biol. 49, 1633–1649.

Moss, B., Hering, D., Green, A.J., Adoud, A., Bécares, E., Beklioğlu, M., Bennion, H., Boix, D., Brucet, S., Carvalho, L., Clement, B., Davidson, T., Declerck, S., Dobson, M., Van Donk, E., Dudley, B., Feuchtmayr, H., Friberg, N., Grenouillet, G., Hering, D., Hillebrand, H., Hobaek, A., Irvine, K., Jeppesen, E., Johnson, R., Jones, I., Kernan, M., Lauridsen, T., Manca, M., Meerhoff, M., Olafson, J., Ormerod, S., Papastergiadou, E., Penning, E., Ptacnik, R., Quintana, X., Sandin, L., Seferlis, M., Simpson, G., Trigal, C., Verdonschot, P., Verschoor, A., Weyhenmeyer, G., 2009. Climate change and the future of freshwater biodiversity in Europe: a primer for policy-makers. Freshw. Rev. 2, 103–130.

Moss, B., Kosten, S., Meerhoff, M., Battarbee, R.W., Jeppesen, E., Mazzeo, N., Havens, K., Lacerot, G., Liu, Z., De Meester, L., Paerl, H., Scheffer, M., 2011. Allied attack: climate change and eutrophication. Inland Waters 1, 101–105.

Mulder et al., 2012. Distributional (in)congruence of Biodiversity-Ecosystem Functioning. Adv. Ecol. Res. 46, 1–88.

Muylaert, K., Pérez-Martínez, C., Sánchez-Castillo, P., Lauridsen, T., Vanderstukken, M., Declerck, S., Van der Gucht, K., Conde-Porcuna, J.-M., Jeppesen, E., De Meester, L., Vyverman, W., 2010. Influence of nutrients, submerged macrophytes and zooplankton grazing on phytoplankton biomass and diversity along a latitudinal gradient in Europe. Hydrobiologia 653, 79–90.

Naya, D.E., Bozinovic, F., Karasov, W.H., 2008. Latitudinal trends in digestive flexibility: testing the climatic variability hypothesis with data on the intestinal length of rodents. Am. Nat. 172, 122–134.

Netten, J.C., Arts, G.H.P., Gylstra, R., Van Nes, E.H., Scheffer, M., Roijackers, R.M.M., 2010. Effect of temperature and nutrients on the competition between free-floating *Salvinia natans* and submerged *Elodea nuttallii* in mesocosms. Fund. Appl. Limnol. 177, 125–132.

Nõges, P., Nõges, T., Tuvikene, L., Smal, H., Ligeza, S., Kornijów, R., Peczula, W., Bécares, E., Garcia-Criado, F., Alvarez-Carrera, C., Fernández-Aláez, C., Ferriol, C., Miracle, R.M., Vicente, E., Romo, S., van Donk, E., Van de Bund, W., Jensen, J.P., Gross, E.M., Hansson, L.-A., Gyllström, M., Nykänen, M., de Eyto, E., Irvine, K., Stephen, D., Collings, S., Moss, B., 2003. Factors controlling hydrochemical and trophic state variables in 86 shallow lakes in Europe. Hydrobiologia 506 (509), 51–58.

O'Gorman, E.J., Emmerson, M.C., 2010. Manipulating interaction strengths and the consequences for trivariate patterns in a marine food web. Adv. Ecol. Res. 42, 301–419.

Ogbebeo, F.E., Evans, M.S., Brua, R.B., Keating, J.J., 2009. Limnological features and models of chlorophyll-a in 30 lakes located in the lower Mackenzie River basin, Northwest Territories (Canada). J. Limnol. 68, 336–351.

Olalla-Tárraga, M.Á., 2010. (2011) "Nullius in Bergmann" or the pluralistic approach to ecogeographical rules: a reply to Watt *et al.* Oikos 120, 1441–1444.

Olalla-Tárraga, M.Á., Rodríguez, M.Á., 2007. Energy and interspecific body size patterns of amphibian faunas in Europe and North America: anurans follow Bergmann's rule, urodeles its converse. Glob. Ecol. Biogeogr. 16, 606–617.
Olalla-Tárraga, M.Á., Bini, L.M., Diniz-Filho, J.A.F., Rodríguez, M.Á., 2010. Cross-species and assemblage-based approaches to Bergmann's rule and the biogeography of body size in Plethodon salamanders of eastern North America. Ecography 33, 362–368.
Olesen, J.M., Dupont, Y.L., O'Gorman, E., Ings, T.C., Layer, K., Melian, C.J., Troejelsgaard, K., Pichler, D.E., Rasmussen, C., Woodward, G., 2010. From Broadstone to Zackenberg: space, time and hierarchies in ecological networks. Adv. Ecol. Res. 42, 1–71.
Özen, A., Karapınar, B., Kucuk, İ., Jeppesen, E., Beklioğlu, M., 2010. Drought-induced changes in nutrient concentrations and retention in two shallow Mediterranean lakes subjected to different degrees of management. Hydrobiologia 646, 61–72.
Özen, A., Šorf, M., Trochine, C., Liboriussen, L., Beklioğlu, M., Søndergaard, M., Lauridsen, T.L., Johansson, L.S. and Jeppesen, E., 2012. Temperature and nutrient effects on freshwater microbial and planktonic communities in climate warming experimental ponds after seven years of incubation. (in press)
Pace, M.L., Cole, J.J., Carpenter, S.R., Kitchell, J.F., 1999. Trophic cascades revealed in diverse ecosystems. Trends Ecol. Evol. 14, 483–488.
Paerl, H.W., Huisman, J., 2009. Climate change: a catalyst for global expansion of harmful cyanobacterial blooms. Environ. Microbiol. Reports 1, 27–37.
Paine, M.D., 1990. Life history tactics of darters (Percidae: Etheostomatiini) and their relationship with body size, reproductive behaviour, latitude and rarity. J. Fish Biol. 37, 473–488.
Palma, A., Figueroa, R., 2008. Latitudinal diversity of Plecoptera (insecta) on local and global scales. Illiesia 4, 81–90.
Parinet, B., Lhote, A., Legube, B., 2004. Principal component analysis: an appropriate tool for water quality evaluation and management-application to a tropical lake system. Ecol. Model. 178, 295–311.
Parmesan, C., Yohe, G., 2003. A globally coherent fingerprint of climate change impacts across natural systems. Nature 421, 37–42.
Perkins, D.M., McKie, B.G., Malmqvist, B., Gilmour, S.G., Reiss, J., Woodward, G., 2010. Environmental warming and biodiversity-ecosystem functioning in freshwater microcosms: partitioning the effects of species identity, richness and metabolism. Adv. Ecol. Res. 43, 177–209.
Pernthaler, J., Amann, R., 2005. Fate of heterotrophic microbes in pelagic habitats: focus on populations. Microbiol. Mol. Biol. Rev. 69, 440–461.
Petchey, O.L., McPhearson, P.T., Casey, T.M., Morin, P.J., 1999. Environmental warming alters food-web structure and ecosystem function. Nature 402, 69–72.
Petchey, O.L., Brose, U., Rall, B.C., 2010. Predicting the effects of temperature on food web connectance. Philos. Trans. R. Soc. B 365, 2081–2091.
Peters, R.H., 1983. The Ecological Implications of Body Size. Cambridge University Press, Cambridge.
Phoenix, G.K., Lee, J.A., 2004. Predicting impacts of Artic climate change: past lessons and future challenges. Ecol. Res. 19, 65–74.
Pickett, S.T.A., 1989. Space-for-time substitution as an alternative to long-term studies. In: Likens, G.E. (Ed.), Long-Term Studies in Ecology: Approaches and Alternatives. Springer-Verlag, New York, pp. 110–135.
Pike, G.H., 1984. Optimal foraging theory: a critical review. Annu. Rev. Ecol. Syst. 15, 523–575.

Pinto-Coelho, R., Pinel-Alloul, B., Methot, G., Havens, K.E., 2005. Crustacean zooplankton in lakes and reservoirs of temperate and tropical regions: variation with trophic status. Can. J. Fish. Aquat. Sci. 62, 348–361.
Pommier, T., Canbäck, B., Riemann, L., Boström, K.H., Simu, K., Lundberg, P., Tunlid, A., Hagström, Å., 2007. Global patterns of diversity and community structure in marine bacterioplankton. Mol. Ecol. 16, 867–880.
Posch, T., Loferer-Krössbacher, M., Gao, G., Alfreider, A., Pernthaler, J., Psenner, R., 2001. Precision of bacterioplankton biomass determination: a comparison of two fluorescent dyes, and of allometric and linear volume-to-carbon conversion factors. Aquat. Microb. Ecol. 25, 55–63.
Prairie, Y.T., Duarte, C.M., Kalff, J., 1989. Unifying nutrient-chlorophyll relationships in lakes. Can. J. Fish. Aquat. Sci. 46, 1176–1182.
Ptacnik, R., Andersen, T., Brettum, P., Lepistö, L., Willén, E., 2010a. Regional species pools control community saturation in lake phytoplankton. Proc. R. Soc. B 277, 3755–3764.
Ptacnik, R., Moorthi, S.D., Hillebrand, H., 2010b. Hutchinson reversed, or why there need to be so many species. Adv. Ecol. Res. 43, 1–44.
Purdy, K.J., Hurd, P.J., Moya-Larano, J., Trimmer, M., Woodward, G., 2010. Systems biology for ecology: from molecules to ecosystems. Adv. Ecol. Res. 43, 87–149.
Rawcliffe, R., Sayer, C.D., Woodward, G., Grey, J., Davidson, T.A., Jones, J.I., 2010. Back to the future: using palaeolimnology to infer long-term changes in shallow lake food webs. Freshw. Biol. 55, 600–613.
Reiss, J., Forster, J., Cássio, F., Pascoal, C., Stewart, R., Hirst, A.G., 2010. When microscopic organisms inform general ecological theory. Adv. Ecol. Res. 43, 45–85.
Reist, J.D., Wrona, F.J., Prowse, T.D., Power, M., Dempson, J.B., Beamish, R.J., King, J.R., Carmichael, T.J., Sawatzky, C.D., 2006. General effects of climate change on arctic fishes and fish populations. Ambio 35, 370–380.
Reznick, D., Schultz, E., Morey, S., Roff, D., 2006. On the virtue of being the first born: the influence of date of birth on fitness in the mosquitofish, *Gambusia affinis*. Oikos 114, 135–147.
Rockström, J., Steffen, W., Noone, K., Persson, Å., Chapin, F.S., Lambin, E.F., Lenton, T.M., Scheffer, M., Folke, C., Schellnhuber, H.J., Nykvist, B., de Wit, C.A., Hughes, T., van der Leeuw, S., Rodhe, H., Sörlin, S., Snyder, P.K., Costanza, R., Svedin, U., Falkenmark, M., Karlberg, L., Corell, R.W., Fabry, V.J., Hansen, J., Walker, B., Liverman, D., Richardson, K., Crutzen, P., Foley, J.A., 2009. A safe operating space for humanity. Nature 461, 472–475.
Rogers, M.W., Allen, M.S., 2009. Exploring the generality of recruitment hypotheses for largemouth bass along a latitudinal gradient of Florida lakes. Trans. Am. Fish. Soc. 138, 23–37.
Rooney, N., Kalff, J., 2000. Inter-annual variation in submerged macrophyte community biomass and distribution: the influence of temperature and lake morphometry. Aquat. Botany 68, 321–335.
Rorslett, B., 1991. Principal determinants of aquatic macrophyte richness in northern European lakes. Aquat. Botany 39, 173–193.
Rosenfield, J.A., 2002. Pattern and process in the geographical ranges of freshwater fishes. Glob. Ecol. Biogeogr. 11, 323–332.
Rouse, W., Douglas, M., Hecky, R., Hershey, A., Kling, G., Lesack, L., Marsh, P., McDonald, M., Nicholson, B., Roulet, N., Smol, J., 1997. Effects of climate change on the freshwaters of Arctic and Subarctic North America. Hydrol. Processes 11, 873–902.
Sainte-Marie, B., 1991. A review of the reproductive bionomics of aquatic gammaridean amphipods: variation of life history traits with latitude, depth, salinity and superfamily. Hydrobiologia 223, 189–227.

Sarmaja-Korjonen, K., 2003. Chydorid ephippia as indicators of environmental changes—biostratigraphical evidence from two lakes in southern Finland. Holocene 13, 691–700.
Sarmento, H., Montoya, J.M., Vázquez-Domínguez, E., Vaqué, D., Gasol, J.M., 2010. Warming effects on marine microbial food web processes: how far can we go when it comes to predictions? Philos. Trans. R. Soc. B 365, 2137–2149.
Schallenberg, M., Hall, C.J., Burns, C.W., 2003. Consequences of climate-induced salinity increases on zooplankton abundance and diversity in coastal lakes. Mar. Ecol. Prog. Ser. 251, 181–189.
Scheffer, M., Hosper, S.H., Meijer, M.L., Moss, B., Jeppesen, E., 1993. Alternative equilibria in shallow lakes. Trends Ecol. Evol. 8, 275–279.
Scheffer, M., Carpenter, S., Foley, J.A., Folke, C., Walker, B., 2001. Catastrophic shifts in ecosystems. Nature 413, 591–596.
Scheffer, M., Szabó, S., Gragnani, A., Van Nes, E.H., Rinaldi, S., Kautsky, N., Norberg, J., Oijackers, R., Franken, R., 2003. Floating plant dominance as an alternative stable state. Proc. Natl. Acad. Sci. USA 100, 4040–4045.
Scheffer, M., Van Geest, G.J., Zimmer, K., Jeppesen, E., Søndergaard, M., Butler, M.G., Hanson, M.A., Declerck, S., De Meester, L., 2006. Small habitat size and isolation can promote species richness: second-order effects on biodiversity in shallow lakes and ponds. Oikos 112, 227–231.
Schiaffino, R.M., Unrein, F., Gasol, J.M., Massana, R., Balagué, V., Izaguirre, I., 2011. Bacterial community structure in a latitudinal gradient of lakes: the roles of spatial versus environmental factors. Freshw. Biol. 56, 1973–1991.
Schindler, D.W., 2006. Recent advances in the understanding and management of eutrophication. Limnol. Oceanogr. 51, 356–363.
Schoener, T.W., 1974. Resource partitioning in ecological communities. Science 185, 27–39.
Sharma, S., Jackson, D.A., Minns, C.K., Shuter, B.J., 2007. Will northern fish populations be in hot water because of climate change? Glob. Change Biol. 13, 2052–2064.
Shurin, J.B., Winder, M., Adrian, R., Keller, W., Matthews, B., Paterson, A.M., Paterson, M.J., Pinel-Alloul, B., Rusak, J.A., Yan, N.D., 2010. Environmental stability and lake zooplankton diversity-contrasting effects of chemical and thermal variability. Ecol. Lett. 13, 453–463.
Smith, V., 2003. Eutrophication of freshwater and coastal marine ecosystems a global problem. Environ. Sci. Pollut. Res. 10, 126–139.
Smol, J.P., Wolfe, A.P., Birks, H.J.B., Douglas, M.S.V., Jones, V.J., Korhola, A., Pienitz, R., Ruhland, K., Sorvari, S., Antoniades, D., Brooks, S.J., Fallu, M.A., Hughes, M., Keatley, B.E., Laing, T.E., Michelutti, N., Nazarova, L., Nyman, M., Paterson, A.M., Perren, B., Quinlan, R., Rautio, M., Saulnier-Talbot, E., Siitoneni, S., Solovieva, N., Weckstrom, J., 2005. Climate-driven regime shifts in the biological communities of arctic lakes. Proc. Natl. Acad. Sci. USA 102, 4397–4402.
Søballe, D.M., Kimmel, B.L., 1987. A large-scale comparison of factors influencing phytoplankton abundance in rivers, lakes, and impoundments. Ecology 68, 1943–1954.
Sogin, M.L., Morrison, H.G., Huber, J.A., Welch, D.M., Huse, S.M., Neal, P.R., Arrieta, J.M., Herndl, G.J., 2006. Microbial diversity in the deep sea and the underexplored 'rare biosphere'. Proc. Natl. Acad. Sci. USA 103, 12115–12120.
Sommaruga, R., Casamayor, E., 2009. Bacterial 'cosmopolitanism' and importance of local environmental factors for community composition in remote high-altitude lakes. Freshw. Biol. 54, 994–1005.
Sommer, U., Lewandowska, A., 2011. Climate change and the phytoplankton spring bloom: warming and overwintering zooplankton have similar effects on phytoplankton. Glob. Change Biol. 17, 154–162.
Søndergaard, M., Jensen, J.P., Jeppesen, E., 2003. Role of sediments and internal loading of phosphorus in shallow lakes. Hydrobiologia 506/509, 135–145.

Stephen, D., Alfonso, T., Balayla, D., Bécares, E., Collings, S.E., Fernández-Aláez, M., Fernández-Aláez, C., Ferriol, C., García, P., Gomá, J., Gyllström, M., Hansson, L.A., Hietala, J., Kairesalo, T., Miracle, M.R., Romo, S., Rueda, J., Ståhldelbanco, A., Svensson, M., Vakkilainen, K., Valentín, M., Van de Bund, W.J., Van Donk, E., Vicente, E., Villena, M.J., Moss, B., 2004. Continental scale patterns of nutrient and fish effects on shallow lakes: introduction to a pan-European mesocosm experiment. Freshw. Biol. 49, 1517–1524.

Stomp, M., Huisman, J., Mittelbach, G.G., Litchman, E., Klausmeier, C.A., 2011. Large-scale biodiversity patterns in freshwater phytoplankton. Ecology 92, 2096–2107.

Strecker, A., Cobb, T., Vinebrooke, R., 2004. Effects of experimental greenhouse warming on phytoplankton and zooplankton communities in fishless alpine ponds. Limnol. Oceanogr. 49, 1182–1190.

Sumner, W.T., McIntire, C.D., 1982. Grazers-periphyton interactions in laboratory streams. Arch. Hydrobiol. 93, 135–157.

Tavsanoğlu, Ü.N., Çakıroğlu, A.Đ., Erdoğan, S., Meerhoff, M., Jeppesen, E., Beklioğlu, M., 2012. Sediment—not plants—is the preferred refuge for *Daphnia* against fish predation in Mediterranean shallow lakes: an experimental approach. Freshw. Biol. 57, 795–802.

Teixeira-de Mello, F., Meerhoff, M., Pekcan-Hekim, Z., Jeppesen, E., 2009. Substantial differences in littoral fish community structure and dynamics in subtropical and temperate shallow lakes. Freshw. Biol. 54, 1202–1215.

Teixeira-de Mello, F., Meerhoff, M., Baattrup-Pedersen, A., Maigaard, T., Kristensen, P., Andersen, T., Masdeu, M., Clemente, J.M., Fosalba, C., Mazzeo, N., Kristensen, E., Jeppesen, E., 2012. Community structure of fish in lowland streams differ substantially between subtropical and temperate climates. Hydrobiologia 684, 143–160.

Tilman, D., Hillerislambers, J., Harpole, S., Dybzinski, R., Fargione, J., Clark, C., Lehman, C., 2004. Does metabolic theory apply to community ecology? It's a matter of scale. Ecology 85, 1797–1799.

Trifonova, I.S., 1998. Phytoplankton composition and biomass structure in relation to trophic gradient in some temperate and subarctic lakes of northwestern Russia and the Prebaltic. Hydrobiologia 369/370, 99–108.

Trochine, C., Guerrieri, M., Liboriussen, L., Meerhoff, M., Lauridsen, T.L., Søndergaard, M., Jeppesen, E., 2011. Filamentous green algae inhibit phytoplankton with enhanced effects when lakes get warmer. Freshw. Biol. 56, 541–553.

Trolle, D., Hamilton, D.P., Pilditch, C., Duggan, I.C., Jeppesen, E., 2011. Predicting the effects of climate change on trophic status of three New Zealand lakes—implications for lake restoration. Environ. Model. Software 26, 354–370.

Vadeboncoeur, Y., Jeppesen, E., Vander Zanden, M.J., Schierup, H.-H., Christoffersen, K., Lodge, D.M., 2003. From Greenland to green lakes: cultural eutrophication and the loss of benthic energy pathways in lakes. Limnol. Oceanogr. 48, 1408–1418.

Vakkilainen, K., Kairesalo, T., Hietala, J., Balayla, D.M., Cares, E., Van de Bund, W.J., Van Donk, E., Fernández-Aláez, M., Gyllstrom, M., Hansson, L.-A., Miracle, M.R., Moss, B., Romo, S., Rueda, J., Stephen, D., 2004. Response of zooplankton to nutrient enrichment and fish in shallow lakes: a pan-European mesocosm experiment. Freshw. Biol. 49, 1619–1632.

Vamosi, J.C., Naydani, C.J., Vamosi, S.M., 2007. Body size and species richness along geographical gradients in Albertan diving beetle (Coleoptera, Dytiscidae) communities. Can. J. Zool. 85, 443–449.

Van de Bund, W.J., Romo, S., Villena, M.J., Valentín, M., van Donk, E., Vicente, E., Vakkilainen, K., Svensson, M., Stephen, D., Stahl-Delbanco, A., Rueda, J., Moss, B., Miracle, M.R., Kairesalo, T., Hansson, L.A., Hietala, J., Gyllström, M., Goma, J., Garcia, P., Fernández-Aláez, M., Fernández-Aláez, C., Ferriol, C., Collings, S.E., Bécares, E., Balayla, D.M., Alfonso, T., 2004. Responses of phytoplankton to fish predation and nutrient loading in shallow lakes: a pan-European mesocosm experiment. Freshw. Biol. 49, 1608–1618.

Van der Gucht, K., Cottenie, K., Muylaert, K., Vloemans, N., Cousin, S., Declerck, S., Jeppesen, E., Conde-Porcuna, J.-M., Schwenk, K., Zwart, G., Degans, H., Vyverman, W., De Meester, L., 2007. The power of species sorting: local factors drive bacterial community composition over a wide range of spatial scales. Proc. Natl. Acad. Sci. USA 104, 20404–20409.

Van der Putten, W.H., Macel, M., Visser, M.E., 2010. Predicting species distribution and abundance responses to climate change: why it is essential to include biotic interactions across trophic levels. Philos. Trans. R. Soc. B. 365, 2025–2034.

Van Doorslaer, W., Stoks, R., Jeppesen, E., De Meester, L., 2007. Adaptive microevolutionary responses to simulated global warming in *Simocephalus vetulus*: a mesocosm study. Glob. Change Biol. 4, 878–886.

Van Leeuwen, E., Lacerot, G., Van Nes, E.H., Hemerika, L., Scheffer, M., 2007. Reduced top-down control of phytoplankton in warmer climates can be explained by continuous fish reproduction. Ecol. Model. 206, 205–212.

Venne, H., Magnan, P., 1989. Life history tactics in landlocked Arctic charr (*Salvelinus alpinus*): a working hypothesis. Physiol. Ecol. Japan 1, 239–248.

Vila-Gispert, A., Moreno-Amich, R., García-Berhou, E., 2002. Gradients of life-history variation: an intercontinental comparison of fishes. Rev. Fish Biol. Fisheries 12, 417–427.

Villéger, S., Blanchet, S., Beauchard, O., Oberdorff, T., Brosse, S., 2011. Homogenization patterns of the world's freshwater fish faunas. Proc. Natl. Acad. Sci. USA 108, 18003–18008.

Vitousek, P.M., 1994. Beyond global warming: ecology and global change. Ecology 75, 1861–1976.

Von Humboldt, A., Bonpland, A., 1805. Essai sur la géographie des plantes; accompagne d'un tableau physique des régions equinoxiales. Levrault, Schoell et Compagnie, Paris 155pp.

Von Humboldt, A., Bonpland, A., 1807. Ideen zu einer Geographie der Pflanzen nebst einem Naturgemälde der Tropenländer. Nabu Press, Tübingen and Paris 182p.

Vörösmarty, C.J., Green, P., Salisbury, J., Lammers, R.B., 2000. Global water resources: vulnerability from climate change and population growth. Science 289, 284–288.

Waide, R.B., Willig, M.R., Steiner, C.E., Mittelbach, G., Gough, L., Dodson, S.I., Juday, G.P., Parmenter, R., 1999. The relationship between productivity and species richness. Annu. Rev. Ecol. Syst. 30, 257–300.

Walther, G.R., Roques, A., Hulme, P.E., Sykes, M.T., Pysek, P., Kuhn, I., Zobel, M., Bacher, S., Botta-Dukát, Z., Bugmann, H., Czucz, B., Dauber, J., Hickler, T., Jarosik, V., Kenis, M., Klotz, S., Minchin, D., Moora, M., Nentwig, W., Ott, J., Panov, V.E., Reineking, B., Robinet, C., Semenchenko, V., Solarz, W., Thuiller, W., Vilá, M., Vohland, K., Settele, J., 2009. Alien species in a warmer world: risks and opportunities. Trends Ecol. Evol. 24, 686–693.

Wang, H.-J., Liang, X.-M., Jiang, P.-H., Wang, J., Wu, S.-K., Wang, H.-Z., 2008. TN : TP ratio and planktivorous fish do not affect nutrient-chlorophyll relationships in shallow lakes. Freshw. Biol. 53, 935–944.

Wang, J., Soininen, J., Zhang, Y., Wang, B., Jang, X., Shen, J., 2011. Contrasting patterns in elevational diversity between microorganisms and macroorganisms. J. Biogeogr. 38, 595–603.

Watt, C., Mitchell, S., Salewski, V., 2010. Bergmann's rule: a concept cluster? Oikos 119, 89–100.

Weber, S.B., Blount, J.D., Godley, B.J., Witt, M.J., Broderick, A.C., 2011. Rate of egg maturation in marine turtles exhibits 'universal temperature dependence'. J. Anim. Ecol. 80, 1034–1041.

West, G.B., Brown, J.H., Enquist, B.J., 1997. A general model for the origin of allometric scaling laws in biology. Science 276, 122–126.

Weyhenmeyer, G.A., 2001. Warmer winters: are planktonic algal populations in Sweden's largest lakes affected? Ambio 30, 565–571.

White, P.A., Kalff, J., Rasmussen, J.B., Gasol, J.M., 1991. The effect of temperature and algal biomass on bacterial production and specific growth-rate in fresh-water and marine habitats. Microb. Ecol. 21, 99–118.

White, E.P., Ernest, S.K.M., Kerkhoff, A.J., Enquist, B.J., 2007. Relationships between body size and abundance in ecology. Trends Ecol. Evol. 22, 323–330.

Williams, W.D., 2001. Anthropogenic salinisation of inland waters. Hydrobiologia 466, 329–337.

Winder, M., Schindler, D.E., 2004. Climate change uncouples trophic interactions in an aquatic ecosystem. Ecology 85, 2100–2106.

Winemiller, K.O., 2005. Life history strategies, population regulation, and implications for fisheries management. Can. J. Fish. Aquat. Sci. 52, 872–885.

Winemiller, K.O., Rose, K.A., 1992. Patterns of life history diversification in North American fishes: implications for population regulation. Can. J. Fish. Aquat. Sci. 49, 2196–2218.

Woodward, G., Christensen, J.B., Olafsson, J.S., Gislason, G.M., Hannesdottir, E.R., Friberg, N., 2010a. Sentinel systems on the razor's edge: effects of warming on Arctic stream ecosystems. Glob. Change Biol. 16, 1979–1991.

Woodward, G., Benstead, J.P., Beveridge, O.S., Blanchard, J., Brey, T., Brown, L., Cross, W.F., Friberg, N., Ings, T.C., Jacob, U., Jennings, S., Ledger, M.E., Milner, A.M., Montoya, J.M., O'Gorman, E.O., Olesen, J.M., Petchey, O.L., Pichler, D.E., Reuman, D.C., Thompson, M.S., Van Veen, F.J.F., Yvon-Durocher, G., 2010b. Ecological networks in a changing climate. Adv. Ecol. Res. 42, 72–138.

Woodward, G., Blanchard, J., Lauridsen, R.B., Edwards, F.K., Jones, J.I., Figueroa, D., Warren, P.H., Petchey, O.L., 2010c. Individual-based food webs: species identity, body size and sampling effects. Adv. Ecol. Res. 43, 211–266.

Wootton, R.J., 1984. Introduction: tactics and strategies in fish reproduction. In: Potts, G. W., Wootton, R.J. (Eds.), Fish Reproduction: Strategies and Tactics. Academic Press, London, UK.

Wu, Q.L., Zwart, G., Schauer, M., Kamst-van Agterveld, M.P., Hahn, M.W., 2006. Bacterioplankton community composition along a salinity gradient of sixteen high-mountain lakes located on the Tibetan Plateau, China. Appl. Environ. Microbiol. 72, 5478–5485.

Yan, N.D., Somers, K.M., Girard, R.E., Paterson, A.M., Keller, W., Ramcharan, C.W., Rusak, J.A., Ingram, R., Morgan, G.E., Gunn, J.M., 2008. Long-term trends in zooplankton of Dorset, Ontario, lakes: the probable interactive effects of changes in pH, TP, DOC and predators. Can. J. Fish. Aquat. Sci. 65, 862–877.

Yannarell, A.C., Triplett, E.W., 2005. Geographic and environmental sources of variation in lake bacterial community composition. Appl. Environ. Microbiol. 71, 227–239.

Yvon-Durocher, G., Allen, A.P., Montoya, J.M., Trimmer, M., Woodward, G., 2010. The temperature dependence of the carbon cycle in aquatic systems. Adv. Ecol. Res. 43, 267–313.

Yvon-Durocher, G., Montoya, J.M., Trimmer, M., Woodward, G., 2011a. Warming alters the size spectrum and shifts the distribution of biomass in freshwater ecosystems. Glob. Change Biol. 17, 1681–1694.

Yvon-Durocher, G., Reiss, J., Blanchard, J., Ebenman, B., Perkins, D.M., Reuman, D.C., Thierry, A., Woodward, G., Petchey, O.L., 2011b. Across ecosystem comparisons of size structure: methods, approaches and prospects. Oikos 120, 550–563.
Zhao, S., Fang, J., Peng, C., Tang, Z., Piao, S., 2006. Patterns of fish species richness in China's lakes. Glob. Ecol. Biogeogr. 15, 386–394.
Zingel, P., Agasild, H., Nõges, T., Kisand, V., 2007. Ciliates are the dominant grazers on pico- and nanoplankton in a shallow, naturally highly eutrophic lake. Microb. Ecol. 53, 134–142.

Impact of Climate Change on Fishes in Complex Antarctic Ecosystems

Katja Mintenbeck*[,1], Esteban R. Barrera-Oro[†,‡], Thomas Brey*, Ute Jacob[§], Rainer Knust*, Felix C. Mark*, Eugenia Moreira[†], Anneli Strobel*, Wolf E. Arntz*

*Alfred Wegener Institute for Polar and Marine Research, Bremerhaven, Germany
[†]Instituto Antártico Argentino and CONICET, Buenos Aires, Argentina
[‡]Museo Argentino de Ciencias Naturales 'Bernardino Rivadavia', Buenos Aires, Argentina
[§]Institute for Hydrobiology and Fisheries Science, University of Hamburg, Hamburg, Germany
[1]Corresponding author: e-mail address: kmintenbeck@hotmail.com

Contents

Advances in Ecological Research, Volume 46
ISSN 0065-2504
http://dx.doi.org/10.1016/B978-0-12-396992-7.00006-X

Abstract

Antarctic marine ecosystems are increasingly threatened by climate change and are considered to be particularly sensitive because of the adaptation of most organisms to cold and stable environmental conditions. Fishes play a central role in the Antarctic marine food web and might be affected by climate change in different ways: (i) directly by increasing water temperatures, decreasing seawater salinity and/or increasing concentrations of CO_2; (ii) indirectly by alterations in the food web, in particular by changes in prey composition, and (iii) by alterations and loss of habitat due to sea ice retreat and increased ice scouring on the sea floor. Based on new data and data collected from the literature, we analyzed the vulnerability of the fish community to these threats.

The potential vulnerability and acting mechanisms differ among species, developmental stages and habitats. The icefishes (family Channichthyidae) are one group that are especially vulnerable to a changing South Polar Sea, as are the pelagic shoal fish species *Pleuragramma antarcticum*. Both will almost certainly be negatively affected by abiotic alterations and changes in food web structure associated with climate change, the latter additionally by habitat loss. The major bottleneck for the persistence of the majority of populations appears to be the survival of early developmental stages, which are apparently highly sensitive to many types of alterations. In the long term, if climate projections are realized, species loss seems inevitable: within the demersal fish community, the loss or decline of one species might be compensated by others, whereas the pelagic fish community in contrast is extremely poor in species and dominated by *P. antarcticum*. The loss of this key species could therefore have especially severe consequences for food web structure and the functioning of the entire ecosystem.

1. INTRODUCTION

Climate change in the Antarctic is not simply a future scenario but already a well-established fact (e.g. Curran et al., 2003; Gille, 2002; Jacob et al., 2011; Murphy et al., 2007, Rignot et al., 2008). Its impacts are most evident in the Antarctic Peninsula region (including the southern Bellingshausen and Amundsen seas), where average temperatures at the sea surface have increased by nearly 3 °C within just the past 50 years (~0.56 °C increase decade^{-1}; e.g. Domack et al., 2003; Turner et al., 2005), and winter temperatures have increased by 5–6 °C (Vaughan et al., 2003). This represents a dramatic increase in air temperature in this region far above the global mean and exceeds any other warming rate observed on Earth at comparable spatial scales; the causes, however, are still under discussion (Gille, 2008; Vaughan et al., 2003).

The corresponding warming of the seawater is less pronounced, though again highly significant: the upper water layers (down to about 50 m water depth) west off the Antarctic Peninsula have warmed by ~1 °C since 1955 (Meredith and King, 2005) and are predicted to rise by another 2 °C over the next century (Murphy and Mitchel, 1995). In Palmer Deep, West Antarctic Peninsula, bottom water temperature has been increasing at a rate of about 0.01 °C $year^{-1}$ since the 1980s (Smith et al., 2012). Moreover, there is some indication of a warming trend emerging in the deep waters of the Ross and Weddell seas (Ozaki et al., 2009; Robertson et al., 2002).

Increasing temperatures significantly affect ice dynamics: the warming has already resulted in a significant reduction in extent and duration of sea ice in the Antarctic Peninsula region, the Amundsen and Bellingshausen seas (Jacobs and Comiso, 1997; Loeb et al., 1997; Stammerjohn et al., 2008a,b) and has also contributed to disintegration and collapses of large ice shelves, such as the northern part of the Larsen ice shelf in the northwestern Weddell Sea (Domack et al., 2005; Marshall et al., 2006). In some shelf regions at the Antarctic Peninsula and in the Ross Sea, melting ice shelves, increased glacial meltwater runoff and reduced sea ice production have led to reduced seawater salinity, particularly in surface water layers (Jacobs et al., 2002; Moline et al., 2004). In Potter Cove, King George Island (South Shetland Islands, west Antarctic Peninsula), a glacier retreat of hundreds of metres and significant freshening of the upper water column have been observed within the past 15 years alone (Schloss et al., 2008).

Ocean acidification is considered a major threat for marine ecosystems that is concomitant with warming and atmospheric change (IPCC, 2007). Anthropogenic CO_2 emissions have increased atmospheric CO_2 concentrations since the industrialization in the 1850s, and about one-third of anthropogenic CO_2 from the atmosphere is absorbed by the world's oceans (Sabine et al., 2004). CO_2 is physically dissolved in seawater and this leads to progressive ocean acidification: several models predict a drop of seawater pH by 0.3–0.5 units by the year 2100 (atmospheric pCO_2 of 1000 μatm) and up to 0.77 units until the year 2300 (atmospheric pCO_2 of 1900–2300 μatm; Caldeira and Wickett, 2003, 2005; Feely et al., 2004; IPCC, 2007). So far, information on CO_2 changes in the Antarctic marine ecosystem is scarce, but local measurements of atmospheric CO_2 concentrations recorded at the permanent Argentinian station 'Carlini[1]', at the shoreline of Potter Cove, revealed a trend of increasing concentrations over a

[1] Formerly known as 'Jubany' (renamed in March 2012).

relatively short time (from 356 μatm in 1994 to 379 μatm in 2006; Ciattaglia et al., 2008).

Though all the mechanisms involved and their interactions are not yet fully understood, there is little doubt that many of these observed changes are beyond that associated with natural variability but caused at least in part by anthropogenic climate change. In the light of ongoing global climate change, it is most likely that those regions of the Antarctic where alterations are not yet evident will also be affected in the near future.

Extant Antarctic marine communities have already been significantly affected by these environmental changes. In Potter Cove, clear shifts in benthic community composition have been observed that appear to be related to increased sediment load in the water column and ice impact due to melting and disintegration of the glacier (Sahade et al., 2008). Changes in salinity alter seawater density and thus can affect stratification of the water column and the depth of the mixed layer: salinity and surface water stratification are two main factors determining phytoplankton composition (Arrigo et al., 1998; Moline et al., 2004). Off the west Antarctic Peninsula, Moline et al. (2004) observed a recurrent change in phytoplankton community structure, with a spatiotemporal shift from large diatoms towards small cryptophytes as salinity declined. Alterations in community structure are also evident in consumers higher in the food chain: since the 1970s, the abundance of krill (*Euphausia superba*) has declined in the southwestern Atlantic and salps have become more abundant (Atkinson et al., 2004). As life cycle and overwintering strategy of Antarctic krill are closely coupled to the sea ice, its accelerating retreat will suppress krill abundance (Atkinson et al., 2004; Loeb et al., 1997). Other factors contributing to krill decline might be water temperature itself, as krill prefers cooler water compared to salps (e.g. Pakhomov et al., 2002), and indirect (trophodynamic) effects, including predation of early krill stages by salps (Huntley et al., 1989) and the inability of krill to efficiently graze on small cryptophytes (see Moline et al., 2004 and references therein). Moreover, salps feed efficiently on a wide range of particles even when phytoplankton concentrations are low (Hopkins, 1985; Kremer and Madin, 1992; Madin, 1974) and are able to attain large population sizes and biomass rapidly (e.g. Mianzan et al., 2001). Under favourable environmental conditions, their efficient grazing and high ingestion rates (e.g. Perissinotto and Pakhomov, 1998a,b) could result in the competitive exclusion of other grazers, such as copepods.

Though there are no immediately obvious visible effects reported so far, Antarctic marine communities could be vulnerable to ocean acidification in the future. Recent experimental studies have shown that survival and development of early stages of calcifying invertebrates and Antarctic krill are significantly negatively affected by levels of predicted future CO_2 concentrations (Kawaguchi et al., 2011; Kurihara, 2008).

Increasing warming of the Southern Ocean will facilitate the invasion and/or introduction and colonization by species from adjacent oceans. Invasion by invertebrates via warming deep waters appears to be already underway. Lithodid crabs have been absent from Antarctic waters for millions of years (see e.g. Thatje et al., 2005), but reports of sightings in Antarctic waters have been accumulating in recent years (e.g. García Raso et al., 2005; Thatje et al., 2008). Recently, a large and reproductive population of king crabs (*Neolithodes yaldwyni*) was discovered in Palmer Deep, west of the Antarctic Peninsula (Smith et al., 2012). So far, the distribution of lithodids seems to be restricted to the deeper slope, but assuming a persistent warming of waters of the Antarctic Peninsula region, Smith et al. (2012) speculated that lithodids might migrate upwards and invade the shelf communities within the next 20 years. Invasion of such hitherto absent durophagous (i.e. shell- or skeleton-crushing) predators strongly affects benthic communities (Aronson et al., 2007). The presence of the king crabs in Palmer Deep has been associated with a decrease in diversity of the megabenthos, including an absence of echinoderms (Smith et al., 2012).

Another more direct impact of human activity is in the form of the increasing ship traffic by tourist cruise ships and research vessels, which further enhance the risk of introduction of exotic species to the South Polar Sea (Lee and Chown, 2007; Lewis et al., 2005, 2006). The North Atlantic spider crab *Hyas araneus*, for example, was found in benthic samples from the Antarctic Peninsula: a species usually only found in the North Atlantic and Arctic Ocean that was most likely introduced into the Southern Ocean via ships' sea chest or ballast water (Tavares and De Melo, 2004). Another passive, man-made pathway for invasion is the increasing amount of litter in the world's oceans. Non-indigenous species may be introduced into the Southern Ocean by transport on drifting plastic debris (Barnes, 2002; Lewis et al., 2005 and citations therein).

Invasion/introduction of alien species will most likely lead to strong alterations in food web structure owing to removal of prey for indigenous species, competition and predation (Woodward et al., 2010a). So far, only invasion by crustaceans has been detected, but if the warming trend continues, it is inevitable that further species, both benthic and pelagic, will invade the Antarctic marine ecosystem. As long as conditions in the South Polar Sea are appropriate for survival but still limit growth and in particular reproduction capacity of invasive species, the threat for indigenous species will remain low. However, once alien species become able to successfully reproduce and to build up populations (as it seems to be the case in the lithodids found in Palmer Deep; Smith et al., 2012), the threat for native Antarctic species significantly increases.

The direction and strength of ecosystem response to environmental change depend strongly upon responses of individual species and their interactions among each other. Fishes are an integral part of marine ecosystems and have been proposed to serve as useful bio-indicators of climate change (Dulvy et al., 2008; McFarlane et al., 2000). As organisms within an ecosystem are linked to each other directly or indirectly via trophodynamics, any kind of change affecting fishes will indirectly affect other members of the food web, with a huge range of potential indirect effects being triggered. For many decades, scientists retained a view that Antarctic food chains were relatively short and simple: essentially a connection from diatoms to krill to consumers. Krill, *E. superba*, in particular was regarded as an inexhaustible resource that underpinned the whole Antarctic food web, supporting fishes, penguins, seabirds, seals and whales (e.g. Murphy, 1962; Tranter, 1982). However, this paradigm has been challenged as being overly simplistic. Although krill does indeed seem to be a key species over large areas, many food chains are independent of it (e.g. Rodhouse and White, 1995), and high species numbers in the South Polar Sea (e.g. Gutt et al., 2004) suggest that the diatom–krill–consumer chain is only one component of a highly complex food web (Clarke, 1985; Jarre-Teichmann et al., 1995). Fishes take a central position in this ecological network: they occupy a variety of trophic niches, are the main consumers of benthos and plankton, and are an important food source for a multitude of species, including cephalopods, piscivorous fishes, penguins, flying birds, seals and whales (for review, see Barrera-Oro, 2002; Hureau, 1994; Kock, 1992; La Mesa et al., 2004). Fishes thus represent an important trophic link that connects small invertebrates and top predators of the Antarctic marine ecosystem, making their potential vulnerability to systemic shifts of particular interest.

In this chapter, we provide an overview of the potential effects of climate change on Antarctic fish species and communities. Based on our own data collected during several expeditions and data that were taken from the literature, we summarize the characteristics of the Antarctic marine ecosystem and the fish communities, and we evaluate the threats to fishes, the degree of endangerment of particular species and the potential consequences for overall ecosystem functioning.

2. THE ANTARCTIC MARINE ECOSYSTEM

2.1. Geographical and physical conditions

The ocean surrounding the Antarctic continent represents one of the most unique marine environments on Earth, an appreciation of which requires a brief review of the geological and climatological history and settings of the Antarctic and its adjacent waters and land masses (summarized in Clarke and Johnston, 1996; Eastman, 1991). In the Jurassic, Antarctica was still part of the supercontinent Gondwana, which subsequently broke up, leaving Antarctica connected to South America and Australia throughout the Cretaceous (about 65 Ma[2] ago), when the climate was temperate, with water temperatures above 10 °C. Between late Eocene and early Oligocene (about 38 Ma), separation of Antarctica from Australia was most likely completed, and seawater temperatures began to decrease sharply. The separation from South America and the formation of sea ice and the continental ice sheet began between 37 and 34 Ma (e.g. Ehrmann and Mackensen, 1992; Ivany et al., 2008; Pearson et al., 2009). The final separation from South America and the opening of the Drake Passage allowed for the development of the strongest current system in the world, the Antarctic Circumpolar Current (ACC), driven by strong westerly winds. The ACC encircles the whole continent and acts as a thermal barrier by effectively separating lower latitude warmer and higher latitude colder waters (see e.g. Orsi et al., 1995).

The Antarctic continent and shelf areas are now geographically isolated from other continents and shelves by great distances and the large abyssal basins of more than 4000 m water depths that surround it (Fig. 1): the only connection to other continents with water depths less than 2000 m is via the Scotia Ridge composed of numerous islands linking South America to the Antarctic Peninsula (Arntz et al., 2005; Tomczak and Godfrey, 1994). In addition, the continent and the surrounding ocean are thermally isolated by the ACC, which flows eastwards and connects the Atlantic, Indian

[2] Millions of years before present (Ma).

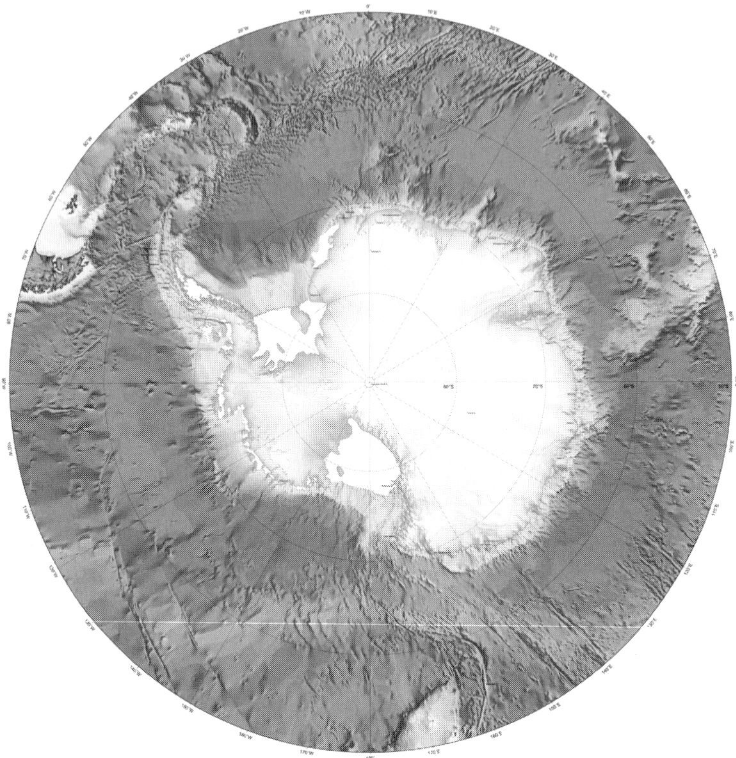

Figure 1 Map of Antarctica and the Southern Ocean *(Source: Centenary edition of the GEBCO Digital Atlas).*

and Pacific basins. This strong current system includes the Antarctic Polar Front, a region of downwelling and sharp temperature change of 3–4 °C (Knox, 1970). As a result, water temperatures in the South Polar Sea are consistently low (ranging from +1 to −1.86 °C close to the continent) with little seasonal variation (Deacon, 1984; Olbers et al., 1992). Close to the continent, the Antarctic Coastal Current (East Wind Drift) flows in the opposite direction and forms clockwise gyres in the Weddell Sea, Ross Sea and Bellingshausen Sea (Gordon and Goldberg, 1970). The region between both current systems is an area of wind- and density-driven upwelling of nutrient-rich circumpolar deep water (Antarctic Divergence), overlaid by Antarctic surface water in the upper layers (e.g. Eastman, 1993).

Beside the unique current system, the most important physical feature structuring the Antarctic marine ecosystem is the ice. The whole Antarctic shelf is narrow and depressed by the large continental ice sheet to depths of about 200 to >600 m. Shallow benthic habitats are thus extremely rare in the South Polar Sea and mostly restricted to the Islands of the Scotia Arc and along the Antarctic Peninsula. The continental ice sheet extends far beyond the coastline, forming large, floating ice shelves and a major source of calving icebergs (Nicol and Allison, 1997), which significantly affect vast areas of the shelf by grounding and seabed scouring (Gutt, 2001).

Sea ice is present year round but the overall coverage varies strongly with season, ranging from 4×10^6 km^2 in the austral summer to 20×10^6 km^2 in winter (Nicol and Allison, 1997; Zwally et al., 1983). Three major zones are distinguished based on prevailing sea ice conditions (Eicken, 1992): (i) the *high Antarctic zone* is almost permanently covered by ice and includes most areas close to the continent; (ii) the *seasonal sea ice zone* is characterized by open water in summer and ice coverage in winter; (iii) the *marginal ice zone* represents the transition from sea ice to the ice-free open ocean and is a region of enhanced ice drift, fragmentation and deformation. Sea ice dynamics significantly affect stratification of the underlying water column. During autumn, the depth of the mixed layer in the ice-free zone is mainly determined by the wind regime. During ice formation and growth, cold and highly saline (and thereby highly dense) seawater is ejected from the ice into the water below, resulting in thermohaline convection and a deepening of the mixed layer (and the pycnocline) to a depth of 50–200 m. In spring during sea ice melt, the entry of freshwater with low density lowers and stabilizes the pycnocline (Eicken, 1995; Gordon et al., 1984).

Light conditions in the Antarctic and in the upper layer of the South Polar Sea also undergo strong seasonal changes, ranging from 24 h of light in summer to complete darkness during the winter months.

Notwithstanding these strong seasonal fluctuations in ice coverage and light regime, the general physical conditions and cold climate in the South Polar Sea have been stable for more than 15 Ma (Dayton, 1990; Dayton et al., 1994).

2.2. Biological characteristics

The Antarctic marine biota are well adapted to the physical conditions in their environment, particularly in the high Antarctic where primary production, life cycles and strategies are closely coupled to seasonal sea ice dynamics. During winter, autotrophic primary production is low and mostly restricted to the sea ice (Arrigo et al., 1997; Lizotte, 2001). During spring and summer, when the sea ice is melting, the released ice algae fuel subsequent phytoplankton blooms in the shallow and stable mixed layer of the marginal ice edge (Lizotte, 2001; Smith and Nelson, 1986); these blooms are mainly formed by diatoms and *Phaeocystis* (Estrada and Delgado, 1990; Nöthig et al., 1991). In autumn, sea ice extends again and remaining algae are incorporated into newly formed ice (e.g. Melnikov, 1998). The large microphytoplankton (>20 µm) blooms account for most of the annual primary production (e.g. Scharek and Nöthig, 1995; Smith and Sakshaug, 1990) but their occurrence is limited in time and space. Pico- (0.2 to <2.0 µm) and nanoplankton (2.0 to <20 µm) are present in the water column throughout the whole year but these small size classes achieve much lower biomass and productivity compared with the bloom system (Detmer and Bathmann, 1997; Scharek and Nöthig, 1995).

Primary and secondary consumers in the water column are mainly represented by copepods, hyperiid amphipods, salps, fish larvae, chaetognaths and euphausiids; larger pelagic predators include squids and fishes (Hempel, 1985; Siegel et al., 1992). Antarctic krill, *E. superba*, is a dominate member of the community in the seasonal sea ice zone and the life history pattern of this species is closely linked to the seasonal sea ice cycle (Smetacek et al., 1990). In the high Antarctic zone, *E. superba* is replaced by a smaller congener *Euphausia crystallorophias*, the so-called ice krill (e.g. Hempel, 1985). Most zooplankton species are present and feeding in the upper water column or at the ice underside the whole year round (Bathmann et al., 1991; Marshall, 1988; Øresland, 1995; Smetacek et al., 1990).

Benthic shelf communities in the high Antarctic are characterized by extraordinarily high biomass and diversity (Brey and Gerdes, 1997; Dayton et al., 1994; Gutt et al., 2004) and are characterized by the dominance of suspension and deposit-feeding species such as sponges, ascidians and echinoderms (Dayton et al., 1974; Gutt and Starmans, 1998; Voss, 1988). In many regions, from shallow water coastal zones to deeper high Antarctic shelf areas, benthic community structure is shaped by physical disturbance, in particular by ice (Gutt, 2000, 2001; Sahade et al., 1998; Smale et al., 2008). In the eastern Weddell Sea, for example, the disturbance of the seafloor by grounding icebergs results in a patchy distribution of various successional stages, which increases between-habitat diversity (Gutt, 2000, 2001; Gutt and Piepenburg, 2003; Knust et al., 2003) which adds a spatiotemporal component to changes in the structure of the benthic food webs (Hagen et al., 2012).

In vast areas of the high Antarctic shelf, large sponges form a typical 3-dimensional habitat for a diverse invertebrate and fish community (Arntz et al., 1994; Gutt and Starmans, 1998). Sponges are often used by invertebrates and fishes as a refuge and/or nursery as well as an upper level substrate to benefit from enhanced access to the water column (Fig. 2A,B). In shallow, inshore areas (e.g. Potter Cove in King George Island), benthic macroalgae such as *Desmarestia* spp., *Himantothallus grandifolius* and *Palmaria decipiens* contribute significantly to primary production (Quartino and Boraso de Zaixso, 2008). As with the sponges on the deeper shelf, benthic macroalgae in shallow waters provide an analogously complex habitat and shelter for a multitude of species including fishes in coastal communities (e.g. Barrera-Oro and Casaux, 1990; Gambi et al., 1994; Moreno et al., 1982; Tada et al., 1996; Takeuchi and Watanabe, 2002) and are a major food source for secondary producers (Iken, 1996; Tatián et al., 2004). Below the depth zone of macroalgal presence, benthic consumers depend on pelagic production (e.g. Mincks et al., 2008). On the high Antarctic continental shelf, where benthic macroalgae are absent over vast areas, tight bentho-pelagic coupling plays an important role in the food web. The high benthic biomass found on the shelf indicates a highly efficient transfer of organic matter from surface waters towards the seafloor (Smith et al., 2006). The vertical export of energy is driven either passively, via sinking particulate organic matter (POM), or actively by migrating organisms.

POM flux on the shelves is dominated by faecal pellets and strings, and large diatoms (Bathmann et al., 1991; Bodungen et al., 1988; Fischer, 1989; Nöthig and Bodungen, 1989). Mass sedimentations of ice algae, *Phaeocystis*

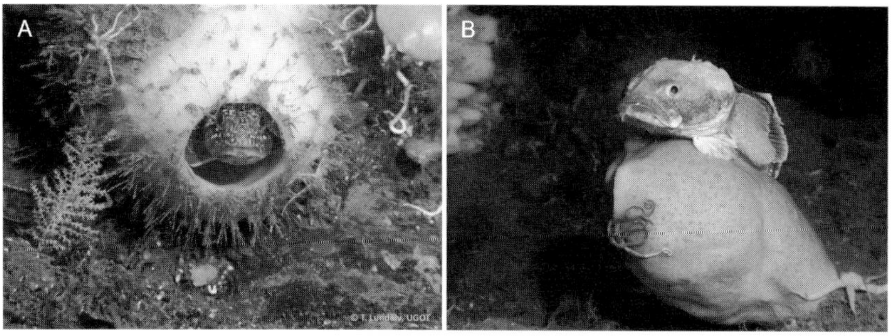

Figure 2 (A) *Trematomus* cf. *nicolai* hiding inside a sponge; (B) *Pogonophryne* sp. on top of a sponge (ANT XXVII-3 in 2011, western Weddell Sea). Photos: ©Tomas Lundälv, University of Gothenburg.

or diatoms after ice melt and termination of blooms are seasonally important export mechanisms (DiTullio et al., 2000; Riebesell et al., 1991; Scharek et al., 1999). Several zooplankton species including Antarctic krill, copepods and salps were observed to undertake extensive vertical migrations within the water column (Atkinson et al., 1992; Gili et al., 2006; Zhou and Dorland, 2004), thereby significantly contributing to the energy export from the euphotic zone towards the seafloor.

The marine living communities of the South Polar Sea are exploited by a multitude of endothermic animals. Whales and seabirds are seasonal visitors that forage in the seasonal sea ice zone and under the pack ice during summer (Boyd, 2002; Murase et al., 2002; Van Franeker et al., 1997). Penguins (mainly Emperor penguin, *Aptenodytes forsteri*, and Adélie penguin, *Pygoscelis adeliae*) and seals (Weddell seal, *Leptonychotes weddellii*; Ross seal, *Ommatophoca rossii*; Crabeater seal, *Lobodon carcinophagus*; Fur seal, *Arctocephalus gazella*; Elephant seal, *Mirounga leonina*) are permanent inhabitants of Antarctic coastal areas. Extensive cracks in the ice shelf covered by sea ice, such as the Drescher Inlet in the Riiser-Larsen Shelf ice (eastern Weddell Sea), represent particularly important breeding and foraging grounds for Weddell Seals and large Emperor Penguin colonies (Plötz et al., 1987).

3. ANTARCTIC FISH COMMUNITIES

3.1. Composition of the modern fauna

Despite the large area covered by the South Polar Sea (>20 million km^2), the modern fish fauna is composed of only about 320 species, belonging to 50 families (Eastman, 2005). This Antarctic ichthyofauna is unique for two reasons: (i) the modern fish fauna is highly endemic, with 88% of all species being confined to the South Polar Sea (Andriashev, 1987), and (ii) the communities are dominated by a single taxonomic group, the perciform suborder Notothenioidei, which accounts for about 35% of all Antarctic fish species (Eastman, 1993). In high Antarctic shelf areas, such as those of the eastern Weddell Sea, notothenioids form up to 98% of the total fish abundance and biomass (R. Knust and K. Mintenbeck, unpublished data). Groups typical of fish communities in temperate or boreal regions, such as clupeids, are absent. Non-notothenioid fish species inhabiting the South Polar Sea mostly belong to typical deep-sea groups such as zoarcids, liparids, macrourids and myctophids. The occurrence of these groups is largely restricted to the lower slope and the deep sea where notothenioid fishes, except for the two *Dissostichus* spp. (*Dissostichus eleginoides* and *Dissostichus mawsoni*), are absent (Boysen-Ennen and Piatkowski, 1988; Donnelly et al., 2004; Gon and Heemstra, 1990; Kock, 1992).

Close to 100 notothenioid fish species have been described from the South Polar Sea (Eastman and Eakin, 2000) but new species are still being discovered (see e.g. Eakin and Balushkin, 1998, 2000; Eakin and Eastman, 1998; Eakin et al., 2008). Most species belong to just five families: Nototheniidae (notothens), Channichthyidae (icefish), Artedidraconidae (plunderfish), Bathydraconidae (dragonfish) and Harpagiferidae (spiny plunderfish). Endemism within the suborder is extremely high, with 97% of notothenioid species being found only in the Antarctic (Andriashev, 1987).

The diversity of the demersal fish community differs regionally, with a latitudinal shift in species composition (Hureau, 1994; Kock, 1992; Mintenbeck et al., 2003, 2012; Permitin, 1977). In the ice-free zone, on the Sub-Antarctic island shelves, typical members of the demersal fish communities are the channichthyids *Chaenocephalus aceratus* and *Champsocephalus gunnari*, the nototheniids *Patagonotothen guntheri*, *Gobionotothen gibberifrons*, *Lepidonotothen* spp., *Notothenia* spp. and *D. eleginoides* (Patagonian toothfish).

At higher latitudes, in the seasonal sea ice zone, communities are dominated by *Lepidonotothen* spp., *Notothenia* spp., *Chionodraco rastrospinosus* (Channichthyidae) and some species of the genus *Trematomus* (see also Barrera-Oro, 2002). In inshore shallow waters, the harpagiferid *Harpagifer antarcticus* is also abundant (Barrera-Oro, 2002; Barrera-Oro and Casaux, 1998).

The demersal fish fauna in the high Antarctic zone is characterized by several *Trematomus* (Nototheniidae), artedidraconid and bathydraconid species, and the channichthyids *Chionodraco* spp. and *Cryodraco antarcticus* (Donnelly et al., 2004; Eastman and Hubold, 1999; Hubold, 1992; Schwarzbach, 1988). In high Antarctic shelf regions, such as the eastern Weddell Sea shelf, species diversity is much higher than on the Sub-Antarctic island shelves or west of the Antarctic Peninsula (Mintenbeck et al., 2012; Schröder et al., 2001). The major reasons for this high species diversity are the 3-dimensionality of the benthic habitat and the high between-habitat diversity shaped by grounding icebergs, both allowing for small-scale niche separation (horizontally and vertically) and thus for the coexistence of trophically similar species (Brenner et al., 2001; Gerdes et al., 2008; Knust et al., 2003).

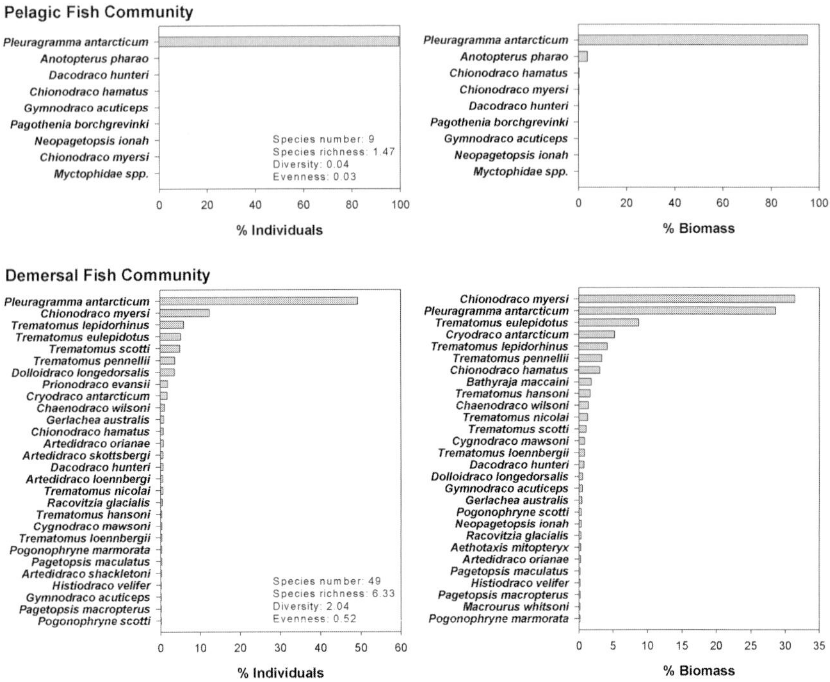

Figure 3 Composition of the pelagic and demersal fish communities on the eastern Weddell Sea shelf between 200 and 600 m water depth (samples from 26 otter trawl hauls and 10 hauls with a bentho-pelagic net taken between 1996 and 2004). Only the 28 out of 49 species contributing >0.15% to overall individuals and biomass are shown for the demersal community. Species number, species richness, diversity and evenness are given for the two communities using different scales.

The pelagic ichthyofauna of the South Polar Sea includes an oceanic and a neritic fish community. The oceanic pelagic communities off the shelves are mainly composed of several myctophid fish species (Barrera-Oro, 2002; Pusch et al., 2004). The neritic pelagic community differs significantly from the oceanic community and is extremely species poor compared with the demersal community on the shelf. In Fig. 3, this difference is exemplified by comparing the pelagic and the demersal fish communities on the eastern Weddell Sea shelf between 200 and 600 m water depth. The neritic pelagic fish community is composed of very few species, and most of them (e.g. the channichthyids *Chionodraco* spp., *Dacodraco hunteri*, *Neopagetopsis ionah* and the bathydraconid *Gymnodraco acuticeps*) are in fact demersal fishes that only occasionally move into the water column. The cryopelagic nototheniid *Pagothenia borchgrevinki* is closely associated with the underside of ice (e.g. Janssen et al., 1991) and is rarely found in open waters. Both the demersal and pelagic fish communities are distinctly dominated by a single nototheniid species, the Antarctic silverfish *P. antarcticum*. The only other species that attains higher biomass in the demersal community is the large icefish *Chionodraco myersi* (Fig. 3). *P. antarcticum* is an endemic species with circum-Antarctic distribution and is one of the few truly pelagic representatives of the entire suborder Notothenioidei. It is a shoaling species (Eastman, 1985a) and adults undertake diel vertical migrations from the sea floor towards the surface waters (Fuiman et al., 2002; K. Mintenbeck and R. Knust unpublished data; Plötz et al., 2001). This species dominates the pelagic fish biomass in coastal waters of the South Polar Sea by >90% (see also DeWitt, 1970; Donnelly et al., 2004; Hubold and Ekau, 1987). In the southern Weddell Sea, *P. antarcticum* accounts for most of the overall fish production, and stock density was estimated to amount at least 1 ton km^{-2} (Hubold, 1992). Though usually found in the free water column, its life cycle strategy, including its feeding dynamics, seems to be closely associated with the sea ice (Daniels, 1982; La Mesa and Eastman, 2012; Vacchi et al., 2004). Besides adult *P. antarcticum*, larvae and early juveniles of several notothenioid species dominate the neritic pelagic fish community numerically. The notothenioid ichthyoplankton community is also dominated by early life stages of *P. antarcticum*. Notothenioid larvae are mainly concentrated in the upper 50 m in well-stratified surface waters, while juveniles occur in slightly deeper waters (Granata et al., 2002; Hubold, 1984, 1985; Hubold and Ekau, 1987; Kellermann, 1986a, b; Morales-Nin et al., 1998). Due to their dominate role in Antarctic fish communities, this chapter largely focuses on notothenioids.

3.2. Evolution and adaptive radiation

The uniqueness of the Antarctic fish fauna with its high degree of endemism and a single dominant group is the result of a long evolutionary history of adaptive radiation in isolation at sub-zero temperatures. Though fossil records are scarce, there is some evidence that the fish fauna in the Antarctic during the Eocene differed substantially from the modern fauna and that the community was composed of species from many, and more cosmopolitan, families (Eastman, 1993, 2005; Eastman and Grande, 1989). Following the complete separation of Antarctica and the progressive cooling of the region's waters, most components of the Eocene fish fauna vanished from shelf areas. Local extinctions likely occurred due to habitat loss associated with the massive expansion of the ice sheet and changes in trophic structure (Eastman, 2005). After this period of extinctions, a multitude of niches were available for other species. These niches were filled by species of the suborder Notothenioidei, which have undergone a remarkable diversification by adaptive radiation on the isolated shelf of the Antarctic continent. The lack of competition from other fish groups allowed increased morphological and ecological diversification of notothenioid fish and expansion into various niches (e.g. Eastman and McCune, 2000; Ekau, 1988; Ptacnik et al., 2010). Accordingly, notothenioid fish species now occupy benthic, bentho-pelagic, pelagic as well as cryopelagic habitats. However, due to the lack of a swimbladder in their common ancestor, the majority of recent notothenioid species are demersal (Clarke and Johnston, 1996). Adaptive radiation of notothenioids also included trophic diversification (Ekau, 1988; Schwarzbach, 1988), and notothenioid fishes now occupy a multitude of trophic niches. Kock (1992) distinguished five main feeding types according to their principal prey: benthos feeders, fish and benthos feeders, plankton and fish feeders, plankton and benthos feeders, and plankton feeders. As some species, such as the channichthyid *D. hunteri*, rely almost exclusively on piscivory (Eastman, 1999; Schwarzbach, 1988), a sixth group of pure ‘fish feeders’ also exists.

The point at which the characteristic modern fauna became established exactly is unknown (Clarke and Johnston, 1996; Eastman, 2005). The few existing fossil records indicate first appearance of this group in the early Tertiary (38 Ma; Balushkin, 1994), and according to a recent phylogenetic study, radiation of notothenioids began near the Oligocene–Miocene transition (~24 Ma), coinciding with the enhanced formation of sea ice (Matschiner et al., 2011).

3.3. Adaptations and characteristics of notothenioid fishes

3.3.1 Physiological and morphological adaptations

Notothenioid fishes are characterized by a multitude of physiological adaptations to life within cold waters, but the key innovation enabling species to survive and diversify was most likely the evolution of antifreeze glycoproteins (AFGPs; Matschiner et al., 2011). AFGPs provide a highly efficient protection from freezing of hypoosmotic (compared to seawater) body fluids by adhering to and blocking the growth of ice crystals (DeVries, 1971; Fletcher et al., 2001). Concentrations of AFGP differ between species and depend on ambient water temperature, depth distribution, life cycle, activity and phylogeny (Wöhrmann, 1996, 1997). AFGP synthesis is most likely regulated by ambient water temperature (Wöhrmann, 1997).

Beside the risk of freezing, another problem of life at sub-zero temperatures is the temperature dependence of viscosity of body fluids in ectotherms. Viscosity and temperature are strongly, negatively correlated, that is, body liquids become more viscous when cold, which affects membrane fluidity, blood circulation, enzyme kinetics and gas diffusion (e.g. Hochachka and Somero, 2002). Cell membrane fluidity is maintained in the cold by homeoviscous adaptation (see Sinensky, 1974), which involves an increased content of unsaturated fatty acids and specific membrane phospholipids (e.g. Eastman, 1993; Hazel, 1995). An increased blood viscosity is offset by reduced haematocrit and haemoglobin concentrations in notothenioid fishes (Egginton, 1996, 1997a,b; Kunzmann, 1991). The only known exceptions of fish with a relatively high haematocrit are *Notothenia coriiceps* and *N. rossii* from the Antarctic Peninsula (Beers and Sidell, 2011; Mark et al., 2012; Ralph and Everson, 1968). Owing to the low metabolic demands of notothenioids (e.g. Clarke, 1983; Clarke and Johnston, 1996) and the increased physical oxygen solubility in seawater, blood and cytosol at cold temperatures, the reduction in haematocrit and respiratory pigment is not detrimental to aerobic performance. In species of the family Channichthyidae, the so-called icefishes or white-blood fishes, functional red blood cells are completely absent: these fishes do not possess any oxygen-binding pigment (haemoglobin) in their blood and some species also lack intracellular myoglobin (Montgomery and Clements, 2000; Sidell and O'Brien, 2006).

The limited oxygen-carrying capacity of the blood is compensated by a multitude of secondary adaptive body modifications in icefishes, for example, a larger ventricle, increased blood volume and cardiac output, and increased skin vascularity (Kock, 2005a; O'Brien and Sidell, 2000; O'Brien et al., 2003; Sidell, 1991). Molecular adaptations also include the absence of heat-shock protein expression in some notothenioids (Carpenter and Hofmann, 2000) and a rearrangement of the mitochondrial genome that may have supported cold adaptation of mitochondrial properties (Mark et al., 2012; Mueller et al., 2011; Papetti et al., 2007; Zhuang and Cheng, 2010).

The increased viscosity of body fluids, together with cold temperatures, affect enzyme kinetics and cytosolic diffusion processes (Sidell, 1991). Both gas diffusion and enzyme kinetics are temperature dependent and decelerate rapidly at cold temperatures. Mitochondrial oxidative capacity of notothenioid fishes is low compared with warm water species

(Johnston et al., 1994), and reduced diffusion of gas and metabolites to and from mitochondria entails an additional reduction of available energy and oxygen. The negative effects of temperature and viscosity on enzymes and diffusion are counterbalanced in Antarctic fishes by two metabolic adaptations. First, these fishes have increased quantities and capacities of intracellular enzymes (Crockett and Sidell, 1990), which reduce diffusion distance and increase efficiency (Pörtner et al., 2000). Second, they display mitochondrial proliferation, an increase in mitochondrial abundance and ultra-structural density (Guderley and Johnston, 1996). Up to 60% of muscle fibre volume of the slow-swimming, pelagic notothenioid *P. antarcticum* is occupied by mitochondria (Clarke and Johnston, 1996). Additionally, many species have relatively high intracellular concentrations of lipids which may be used as energy stores (Crockett and Sidell, 1990; Eastman and DeVries, 1981) and aid gas diffusion (Kamler et al., 2001). These intracellular lipids also play a role in buoyancy (see below).

Notothenioid fishes are thus well adapted to cold waters. Nevertheless, these adaptations apparently involve an extreme stenothermy of physiological functions and seem to result in narrow thermal tolerance windows of this group (Johnston, 2003; Mueller et al., 2011; Pörtner and Peck, 2010; Somero and DeVries, 1967).

However, the success of notothenioid species in the South Polar Sea ecosystem is not only based on physiological adaptations, but also on morphological modifications related to buoyancy (Eastman, 2005; Eastman et al., 2011). Notothenioids lack a swim bladder and, without this organ, the exploitation of bentho-pelagic or pelagic food sources is extremely energy consuming. To compensate for the lack of a swim bladder, some notothenioid species developed modifications in body structure, which allow them to inhabit and to exploit the pelagic realm without an energetic disadvantage. In these species, mineralization of skeleton and scales is reduced, and the skeleton contains a high proportion of cartilage, which is less dense than bone (DeVries and Eastman, 1978; Eastman and DeVries, 1981, 1982; Eastman et al., 2011). *P. antarcticum* has a persistent notochord and large amounts of lipids (accounting for ~39% of muscle dry mass) are stored in subcutaneous and intramuscular lipid sacs which provide static lift (DeVries and Eastman, 1978). *D. mawsoni* possess extensive lipid deposits in adipose cells, which account for ~23% white muscle dry weight (Eastman and DeVries, 1981). The lipid deposits in these species mainly consist of triglycerols (Eastman and DeVries, 1981, 1982; Hubold and Hagen, 1997). However, neutral buoyancy is rare in notothenioids and limited to very few species, for example, *P. antarcticum*, *D. mawsoni* and *Aethotaxis mitopteryx* (Eastman, 2005).

3.3.2 Growth, reproduction and development

Most notothenioids are characterized by a rather sluggish mode of life and high longevity. High Antarctic fish species typically reach ages of 15–21 years (see Kock, 1992; La Mesa and Vacchi, 2001 for review). The pelagic *P. antarcticum* was estimated to live more than 30 years (Radtke et al., 1993). Growth performance of most species is similar to species from boreal or temperate regions (Kock and Everson, 1998; La Mesa and Vacchi, 2001) but there seems to be a trend towards lower growth performance in the high Antarctic notothenioids, relative to their congeners from the seasonal sea ice zone (Kock, 1992). However, in adult fish, growth performance is apparently related to lifestyle and tends to increase from pelagic towards benthic lifestyles (La Mesa and Vacchi, 2001). Pelagic fishes such as *P. antarcticum* are thus characterized by slow growth (e.g. Hubold and Tomo, 1989). Many notothenioids show a distinct seasonal growth pattern with high growth rates in summer and low growth rates in winter (Hureau, 1970; North et al., 1980; White, 1991). The interspecific latitudinal and intraspecific seasonal differences in growth rates of notothenioids most likely (primarily) stem from variations in food supply and/or prey composition, and feeding intensity (Kock, 1992). For example, *N. coriiceps* undergoes winter metabolic suppression and enters a dormant stage with periodic arousals lasting only a few hours, resulting in a net loss of growth rate during the winter months (Campbell et al., 2008). However, most notothenioids seem to feed year round (e.g. Casaux et al., 1990; Hubold, 1992).

Sexual maturity is delayed in most Antarctic fishes. With a few exceptions, species reach maturity at 50–80% of their maximum age and size (Kock and Everson, 1998; La Mesa and Vacchi, 2001). The spawning season is species- and location specific: in the seasonal sea ice zone, most species spawn in autumn/winter and in the high Antarctic zone, most species are summer and autumn spawners (Kock and Kellermann, 1991). Some species have demersal eggs, which are often laid on rocks or in the cavity of sponges. Nest-guarding and other parental care behaviours have been reported increasingly in notothenioids (e.g. Barrera-Oro and Lagger, 2010; Detrich et al., 2005; Kock et al., 2006; Moreno, 1980). Others species, for example, *P. antarcticum*, have pelagic eggs (Faleyeva and Gerasimchuk, 1990; Vacchi et al., 2004). The eggs are usually large and yolky, so relative fecundity is low in most species, particularly in high Antarctic notothenioids (Hubold, 1992; Kock, 1992; Kock and Kellermann, 1991). The incubation period of eggs is long and usually takes several months (Hubold, 1992; Kock and Kellermann, 1991; North and White, 1987). Larvae of many species apparently hatch in spring and summer (Efremenko, 1983) when food conditions are best; however, some species also hatch in winter (Ekau, 1989; North and White, 1987). Larvae are large at hatching (Kellermann, 1990; North and White, 1987) and the mouth is well-developed, so that even early yolk-sac larvae are able to feed (Kellermann, 1986b). Most, if not all, notothenioid larvae are pelagic.

3.4. Threats to the fish community

Antarctic fish communities are threatened by climate change in multiple ways. On the one hand, fishes might be affected at the physiological level directly by increasing water temperatures and pCO_2, and reduced water salinity. Due to the numerous adaptations to life in the South Polar Sea, fishes are likely to be affected on different organizational levels, from the cellular level up to the population level and beyond. In particular, an increase of water temperatures might pose a major threat to stenothermal species (see e.g. Somero, 2010), whereas increasing concentrations of CO_2 might have more general detrimental effects across many fish species (e.g. Ishimatsu et al., 2005). Whether and to what extent fitness and survival are affected by such changes depends on individual or species-specific physiological plasticity.

Climate change can additionally affect fishes indirectly by secondary effects, such as those due to changes in the abiotic environment that will entail alterations in the food web, as also reported for instance in many freshwaters (Meerhoff et al., 2012). Unfavourable abiotic conditions as well as invasion of Antarctic waters by non-indigenous species may result in changes in trophic structure and dynamics by alterations in composition and population density of prey and predator communities (Woodward et al., 2010a).

Changes in prey species composition will involve alterations in the type of prey available to fishes, particularly in size structure and energy content: a shift from diatoms to cryptophytes is accompanied by a strong shift in size structure of primary producers, and ultimately with a potential size shift in secondary producers; a shift from a krill dominated zooplankton community towards a community dominated by salps involves a drastic decrease in nutritional value of potential prey for higher trophic level consumers (but see also Gili et al., 2006). In the marine Antarctic, where life cycles are closely coupled to seasonal sea ice dynamics, changes in water temperature itself but also reduced sea ice extent and duration might entail phenological shifts and a trophic mismatch between prey and consumer species (Hagen et al., 2012). Secondary effects of climate change may also involve a reduction of habitat, which could lead to fragmentation of the food web (Hagen et al., 2012): sea ice reduction means a loss of habitat for ice-associated pelagic species, but demersal fish are threatened by habitat reduction as well, as warmer temperatures will most likely result in enhanced disintegration of glaciers and ice shelves. An increased iceberg calving and breakup will lead to a higher frequency of iceberg scouring events and thus to increased seafloor and habitat destruction.

From other ecosystems, it is well-known that many fish species are sensitive to these types of threats, via mechanisms operating directly at the ecophysiological level (e.g. McFarlane et al., 2000; Pörtner and Peck, 2010; Pörtner et al., 2008) but also indirectly at the trophic level (Beaugrand et al., 2003; Benson and Trites, 2002; Drinkwater et al., 2010), as well as by alterations in habitat structure and heterogeneity (Hughes et al., 2002; Yeager et al., 2011).

4. PHYSIOLOGICAL VULNERABILITY OF ANTARCTIC FISHES

4.1. Sensitivity to changes in temperature and salinity

Antarctic fishes have very narrow thermal windows due to cold adaptation (Clarke, 1991; Somero et al., 1968; Wohlschlag, 1963), resulting in high stenothermy in this group (Gonzalez-Cabrera et al., 1995; Podrabsky and Somero, 2006; Robinson et al., 2011; Somero and DeVries, 1967). Most species, for example, the bottom-dwelling *Trematomus bernacchii*, *Trematomus hansoni* and *Trematomus pennellii*, have an upper lethal temperature between just 4 and 6 °C (Robinson, 2008; Somero and DeVries, 1967). Fish performance is already affected well below the lethal limit. However, the paradigm that all notothenioid species are extremely stenothermal without exceptions has recently been revised.

Some species such as *H. antarcticus* and young *N. coriiceps* are frequently found in tide pools in King George Island (South Shetland Islands), where during sunny days individuals are exposed to warm temperatures for many hours (E.R. Barrera-Oro and E. Moreira, personal observation). Thus, these species can at least cope with acute, relatively short-term temperature increases. A few species are apparently also able to compensate for chronic exposure to higher temperatures, for example, the cryopelagic *P. borchgrevinki* shows some metabolic plasticity: long-term warm acclimation of *P. borchgrevinki* to 4 °C results in a shift of the thermal tolerance window towards warmer temperatures (Bilyk and DeVries, 2011; Franklin et al., 2007; Robinson and Davison, 2008) owing to metabolic compensation (Seebacher et al., 2005) which leads to a reduced performance at low temperatures (Franklin et al., 2007). Recent measurements of routine metabolic rate of *Notothenia rossii* and *Lepidonotothen squamifrons* from the Scotia Arc shelf revealed a partial compensation after long-term acclimation to elevated temperatures (A. Strobel and F.C. Mark, unpublished data). Similarly, long-term warm acclimation of the Antarctic eelpout *Pachycara brachycephalum* involves metabolic rearrangements (Lannig et al., 2005) and indicates an improvement of hepatic metabolism accompanied by a shift of energy sources from lipids to carbohydrates (Brodte et al., 2006, Windisch et al., 2011).

However, these metabolic acclimations apparently do not result in a full compensation and cannot be generalized across all species (c.f. the deepwater zoarcid *Lycodichthys dearborni*; Podrabsky and Somero, 2006), but seem rather dependent on the physical capacities of the circulatory system: studies of energy allocation in isolated cells of Antarctic notothenioids and *P. brachycephalum* suggest that within a thermal range of about −1 to 12 °C, thermal tolerance limits are defined at the whole organism level (Fig. 4), for

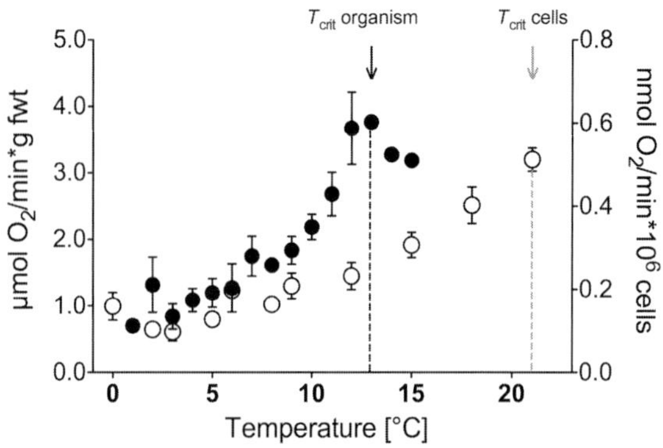

Figure 4 Temperature dependence of whole animal metabolic rate (filled symbols, left axis) and respiration rate of hepatocytes (open symbols, right axis) of the Antarctic eelpout *Pachycara brachycephalum*. Due to organismal complexity, acute whole animal critical temperatures (T_{crit}) that mark the onset of anaerobic metabolism are reached at lower temperatures (~13 °C, black arrow) than in isolated cells (>21 °C, grey arrow; *redrawn from Mark et al., 2002, 2005; with permission from Springer*).

example, by capacity limitations of the circulatory system rather than by a general failure of cellular energy metabolism (Mark et al., 2002, 2005). Due to their lower level of organizational complexity, thermal tolerance windows of organelles generally span a wider temperature range than those of the whole organism (Mark et al., 2005). Thus, acclimatory capacities are mainly defined by the degree of changes in cellular energy metabolism and depend strongly on the mitochondrial oxidative capacities. The acute thermal tolerance of oxidative capacity varies between species and tissues (Mark et al., 2012; Urschel and O'Brien, 2009). However, these acclimatory capacities are always constrained within the frame of the thermal window set by the whole organism's physiological plasticity.

Fishes possess the flexibility to respond to chronically elevated temperatures via mitochondrial proliferation—modifications of the amount and volume of mitochondria to adjust aerobic capacity (Tyler and Sidell, 1984; Urschel and O'Brien, 2008). Mitochondrial oxygen demand rises with increasing temperature and aerobic capacities need to be adjusted accordingly. Only a few studies have demonstrated a full compensation of the increased oxygen demand at the mitochondrial level after warm acclimation, and these studies were exclusively performed on non-Antarctic organisms (e.g. Dahlhoff and Somero, 1993; Sloman et al., 2008). This indicates that there are limitations to mitochondrial acclimation, ultimately co-defining the rather narrow bandwidth of thermal acclimation, especially in Antarctic fishes (Mark et al., 2006). Increased mitochondrial oxygen demand in warmer conditions may be met by altering haemoglobin affinities (Tetens et al., 1984) and raising the haematocrit to optimize the oxygen-carrying capacity of the blood. Yet, haematocrit levels are correlated with haemoglobin expression (Beers and Sidell, 2011). Thus, species with low or no haemoglobin levels have less capacity to adjust haematocrit. This is further exacerbated by the fact that rising temperatures result in lower levels of physically dissolved oxygen in the blood. Therefore, icefishes that lack haemoglobin will be more vulnerable to warming than red-blooded species, because they cannot increase the oxygen-carrying capacity of the blood. According to Beers and Sidell (2011), Antarctic fishes with higher haematocrit levels thus possess higher temperature acclimatory capacities than species with lower haematocrit.

Studies on the impact of salinity changes on notothenioid fishes are extremely scarce. O'Grady and DeVries (1982) investigated the capacity for osmoregulation of adult *P. borchgrevinki* and *Trematomus* spp. at a wide range of salinities (25–200% of the salinity in their natural habitat, 35 psu) and found these species to be rather tolerant towards even large fluctuations from 50% to 175% normal salinity. Blood serum osmolarities in Antarctic fishes are among the highest in marine teleosts, which has been interpreted as an additional antifreezing protection (O'Grady and DeVries, 1982). After warm acclimation, serum osmolarities reduced to the levels found in temperate teleosts were measured in *T. bernacchii*, *T. newnesi* and *P. borchgrevinki* (Gonzalez-Cabrera et al., 1995; Hudson et al., 2008; Lowe and Davison, 2005), and also in *N. rossii* and *L. squamifrons* (A. Strobel and F.C. Mark, unpublished data). Thus, adult notothenioids are apparently capable of efficient osmoregulation and seem able to adapt blood osmolarity to ambient environmental conditions.

Almost all studies on physiological sensitivity of notothenioid fishes to changing abiotic parameters have been conducted on adults. Data from non-Antarctic fish species suggest an ontogenetic shift in temperature tolerance, with narrow thermal tolerance windows in eggs and larvae (e.g. Pörtner and Farrell, 2008; Pörtner and Peck, 2010). Most teleost fishes are able to osmoregulate at hatch, but the efficiency seems to be higher in more advanced developmental stages (Varsamos et al., 2005). To our knowledge, there are no experimental data for early developmental stages of notothenioid fishes, but indirect evidence from abundance and distribution of *P. antarcticum* larvae and juveniles indicate that these early stages likely have limited ability to tolerate changes in temperature and salinity. Larvae and juveniles of this species are mostly found within water masses of particular temperature and salinity (e.g. Granata et al., 2002; Guglielmo et al., 1998; Hubold, 1984; Kellermann, 1986a). West of the Antarctic Peninsula, *P. antarcticum* larvae and juveniles were clearly confined to cold and high salinity water masses originating from the Weddell Sea (Slósarczyk, 1986). Based on combined datasets of fish abundances in waters of Weddell Sea and Bellinghausen Sea origin, Slósarczyk (1986) calculated the range of approximate optimum conditions for larvae and juveniles in the Bransfield Strait: Abundances suggested optimal ranges in temperature and salinity of -0.50 to $+0.45$ °C and $\sim$34.10–34.62 psu, respectively.

Though changes in salinity induced by climate change might be a locally restricted phenomenon, dense aggregations of larvae and juveniles that are concentrated close to the coast/shelf ice in the upper water layers might be significantly affected. Moline et al. (2004) found vast areas west of the Antarctic Peninsula covered by low salinity water (33.4–33.6 psu), and the meltwater plume extended to depths as great as 50 m (Dierssen et al., 2002). It still needs to be verified whether and to what degree temperatures and salinities outside the narrow ranges given by Slósarczyk (1986) limit physiological performance and survival of larvae and juveniles, but tolerance indeed seems to be low: *P. antarcticum* larvae acclimated to cold Weddell Sea water were observed to shrink and to die immediately at water temperatures >0 °C (Hubold, 1990). Whether or not this limited tolerance holds true for early stages of other species needs further investigation, but at least *P. antarcticum* larvae seem to be highly vulnerable to changes in the abiotic environment.

4.2. Sensitivity to increasing pCO_2

Ocean acidification, as an additional stressor in parallel to ongoing climate warming (Woodward et al., 2010a), may prove to be particularly threatening to polar ecosystems owing to enhanced CO_2 solubility in cold waters and body fluids. Thermal tolerance windows are narrow in most species and, thus, sensitivities to combined stressor effects are likely to be higher in cold-adapted polar compared to temperate species. Many notothenioids will eventually find themselves at the upper end of their thermal tolerance range, implying that they are energetically limited and their physiological performance is highly susceptible to further stressors, such as the increasing concentration of carbon dioxide (Pörtner and Peck, 2010).

Previous research on the effects of elevated CO_2 levels on marine fishes led to the general notion that fishes are not particularly vulnerable to the direct effects of ocean acidification alone, due to their powerful mechanisms of ion regulation (Fivelstad et al., 2003). Most adult fishes are able to compensate for acid–base disturbances (Larsen et al., 1997) and show only minimal effects of hypercapnia on physiological performance (Melzner et al., 2009) including an incomplete compensation of extracellular pH (Michaelidis et al., 2007).

However, several studies on different non-Antarctic fish species demonstrated chronic effects of environmental hypercapnia, with early developmental stages being particularly affected: exposure to elevated CO_2 concentrations impairs embryonic metabolism (Franke and Clemmesen, 2011), survival and growth of eggs and larvae (Baumann et al., 2012), and growth of juveniles (Moran and Støttrup, 2010), and causes severe to lethal tissue damage in many internal organs of larvae (Frommel et al., 2012). The sensitivity to ocean acidification may generally be enhanced by ocean warming, which has been confirmed in tropical fishes (Nilsson et al., 2009; Pörtner and Farrell, 2008). Still very little is known about how the physiology and distribution of Antarctic fishes and their various life stages may be altered by the additional effects of hypercapnia, but several current projects are dealing with this topic, and initial results indicate that chronic hypercapnia leads to significant reductions of mitochondrial capacities in *N. rossii* (A. Strobel and F.C. Mark, unpublished data), on top of the thermal sensitivity of its mitochondrial metabolism (see Section 4.1 above; Mark et al., 2012). Hypercapnia-induced regulatory shifts in intracellular metabolic pathways and capacities therefore may exacerbate the effects of increased temperature on cellular and whole animal metabolism.

We currently lack sufficient data on Antarctic fishes to be able to generalize as to whether all life stages respond similarly or whether early developmental stages represent potential bottlenecks for population survival. Another topic that needs to be addressed is how hypercapnia will modify interactions between species already affected by the warming trend. Recent findings in tropical coral reef fish demonstrated behavioural disturbances by moderate levels of ocean acidification (1050 ppm CO_2, pH 7.8, year 2100 scenario), presumably elicited by hypercapnia effects on the central nervous system (Munday et al., 2009). Hypercapnia therefore may also alter trophodynamic interactions in a particular ecosystem (cf. Ferrari et al., 2011), beyond those of the direct lethal effects on interacting organisms. These aspects have never been studied in Antarctic fishes and clearly need further research.

5. TROPHIC VULNERABILITY OF ANTARCTIC FISHES

5.1. Vulnerability to general changes in trophic structure and dynamics

The vulnerability of a particular species to changes in food web structure and dynamics depends on its ability to cope with both 'bottom-up' and 'top-down' effects (Jacob et al., 2011; Melian et al., 2011; O'Gorman and Emmerson, 2010). Trophic plasticity, that is, the capability to cope with fluctuations in resource availability, is positively related to prey diversity (specialist vs. generalist consumers; Johnson, 2000; Mihuc and Minshall, 1995). Predator-induced mortality is the principal 'top-down' effect, and suppression of a particular species strongly increases with increasing predator diversity (Snyder et al., 2006). Vulnerability to 'top-down' effects and resilience capability are thus related to the number of predator species. Accordingly, species vulnerability to food web-mediated alterations is expected to decrease with prey diversity and to increase with predator diversity. Whether and how the complete loss of one species will affect overall food web structure and ecosystem functioning depends on the community's capacity for functional compensation, that is, species trophic redundancy (Johnson, 2000; Naeem, 1998).

Here, the relative trophic vulnerability of the adult notothenioid fish community inhabiting the eastern Weddell Sea shelf (between 200 and 600 m water depth) to alterations in the food web was estimated. A simple, quantitative measure based on the number of feeding links to prey and predator species was used to assess vulnerability. Information on trophic linkages was extracted from the extensive trophic database published in Jacob et al. (2011) that includes information on feeding relations of 489 consumer and resource species from the Antarctic Weddell Sea (for detailed information and sources, see Jacob, 2005; Jacob et al., 2011). For seven more fish species, additional information on prey composition was collected from Foster et al. (1987), Gon and Heemstra (1990), La Mesa et al. (2004) and Schwarzbach (1988). All in all, information on prey composition and links to predators was available for 37 of the 42 notothenioid species inhabiting the shelf.

For each fish species i, the total number of prey species $\sum P_i$, the number of prey species belonging to the functional groups 'Benthos', 'Plankton' and 'Fish', $\sum P_{\mathrm{B},i}$, $\sum P_{\mathrm{P},i}$, $\sum P_{\mathrm{F},i}$, and the number of predators $\sum C_i$ were extracted from the database. Both $\sum P$ and $\sum C$ are common descriptors in theoretical food web ecology and usually referred to as 'generality' ($\sum P$) and 'vulnerability' ($\sum C$; see e.g. Memmot et al., 2000; Schoener, 1989). Here, both variables were combined to calculate consumers' relative trophic vulnerability, a comparative index with values located between 0 and 1. $\sum C$ was taken as a measure of vulnerability to top-down effects, and $\sum P$ as an (inverse) measure of vulnerability to bottom-up effects.

The relative trophic vulnerability VI_i of fish species i can thus be computed by

$$\mathrm{VI}_i = \frac{\sum_{i=1}^{m} C_i}{\sum_{i-1}^{n} P_i + \sum_{i=1}^{m} C_i} \qquad [1]$$

where m is the total number of consumer species and n is the total number of prey species of fish species i. $\sum P_i + \sum C_i \geq 1$ and $0 \leq \text{VI}_i \leq 1$; $\sum P_i = \sum P_{\text{B},i} + \sum P_{\text{P},i} + \sum P_{\text{F},i}$. In this basic equation (Eq. 1), each C_i and P_i count 1.

However, there is a difference in relative top-down and bottom-up effects depending on (i) whether a particular consumer is a generalist feeder or specialized on fish species i and (ii) whether a particular prey species is exclusively consumed by fish species i or exploited by a multitude of predators. To account for these differences, each consumer of fish species i, C_i, was weighted by the number of its own prey species (P_j) and each prey species of fish species i, P_i, was weighted by the number of its own consumer species (C_k). These weighted consumer and prey values of fish species i are referred to as WC_i and WP_i, respectively. Accordingly, the calculation of the relative trophic vulnerability index VI_i of fish species i was adapted by

$$\text{VI}_i = \frac{\sum_{i=1}^{m} \text{WC}_i}{\sum_{i=1}^{n} \text{WP}_i + \sum_{i=1}^{m} \text{WC}_i} \quad [2]$$

with

$$\text{WP}_{i,k} = \left(\frac{1}{\sum_{k=1}^{m} C_k} \right) \text{and}\, \text{WC}_{i,j} = \left(\frac{1}{\sum_{j=1}^{n} P_j} \right); \sum \text{WP}_i + \sum \text{WC}_i > 0$$

and $0 \leq \text{VI}_i \leq 1$; $\sum \text{WP}_i = \sum \text{WP}_{\text{B},i} + \sum \text{WP}_{\text{P},i} + \sum \text{WP}_{\text{F},i}$. Here, m is the total number of weighted consumer species and n is the total number of weighted prey species of fish species i. This index was used as an indicator of species' risk to be negatively affected by changes in the food web. Spearman's rank correlation was used to analyze relationships between all parameters with the aim to rank $\sum \text{WC}$, $\sum \text{WP}$, $\sum \text{WP}_\text{B}$, $\sum \text{WP}_\text{P}$ and $\sum \text{WP}_\text{F}$ according to their effect on VI.

The number of prey items $\sum P$ ranged from 5 in some planktivorous fishes to >100 in benthos feeders. The number of predators $\sum C$ ranged from 12 to 47 (Table A1). The majority of notothenioid fish species are benthos feeders and mixed feeders, consuming varying proportions of benthos and plankton (Fig. 5). The number of pure plankton feeders and mixed feeders of plankton and fish is comparatively low and pure piscivorous species are extremely scarce. Relative vulnerability VI is related to

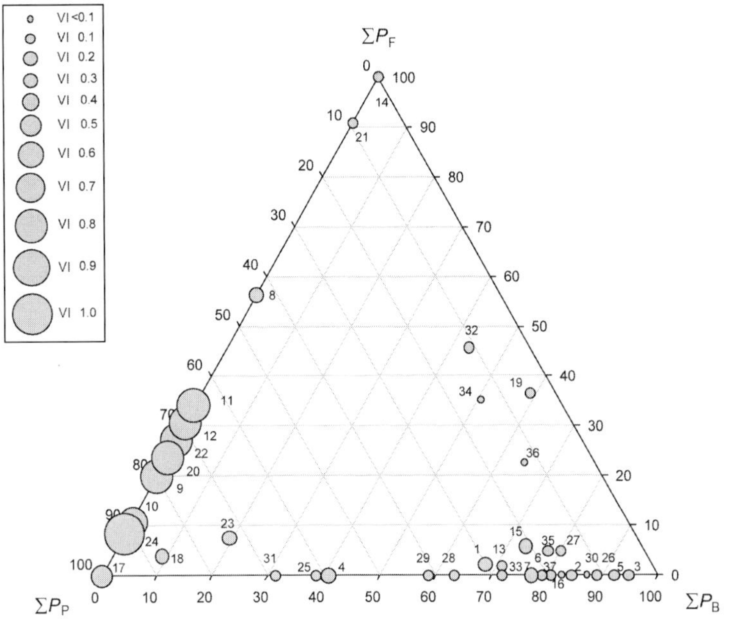

Figure 5 Relative proportions [%] of benthos ($\sum P_B$), plankton ($\sum P_P$) and fish ($\sum P_F$) in the diet of notothenioid fish species. Each circle represents one species; circle diameter indicates relative trophic vulnerability (VI). For species code numbers, see Table A1.

the distribution of prey species among the functional groups 'Benthos', 'Plankton' and 'Fish'. VI is lowest in benthos feeders (VI 0.02–0.11, mean ± SD 0.07 ± 0.03), fish feeders (VI 0.1, mean ± SD 0.1 ± 0) and benthos and fish feeders (VI 0.03–0.07, mean ± SD 0.05 ± 0.02), intermediate in mixed feeders of benthos and plankton (VI 0.02–0.28, mean ± SD 0.16 ± 0.07) and highest in species feeding almost exclusively on planktonic prey (VI 0.15–0.96, mean ± SD 0.56 ± 0.34) or on a mixture of plankton and fish (VI 0.28–0.77, mean ± SD 0.69 ± 0.20). The highest VI of 0.96 is found in the plankton-feeding *P. antarcticum,* followed by some channichthyid species such as *C. myersi* and *C. antarcticus* with VIs of 0.77 (see Table A1; Fig. 5). It appears that there is a certain accumulation of risk in the trophic group of plankton feeders. VI is correlated more strongly with $\sum$WP (Spearman's $r = -0.980$, $p < 0.0001$; Table 1) than to $\sum$WC ($r = 0.614$, $p < 0.0001$). In notothenioid fishes, differences in relative vulnerability VI between species are thus mainly determined by the number of prey items, that is, by the degree of generalism (see Table 1). The effect of predator diversity is of less significance, as most fish species share a similar number of potential predators that feed non-selectively.

Table 1 Spearman's rank correlations between relative trophic vulnerability index VI, weighted number of consumer species $\sum WC$ and weighted number of prey species $\sum WP$, with the functional prey groups benthos $\sum WP_B$, plankton $\sum WP_P$ and fish $\sum WP_F$

	VI	$\sum$WC	$\sum$WP	$\sum WP_B$	$\sum WP_P$
$\sum WC$	$r=0.614$ $p<0.0001$	–	–	–	–
$\sum WP$	$r=-0.980$ $p<0.0001$	$r=-0.505$ $p=0.0014$	–	–	–
$\sum WP_B$	$r=-0.861$ $p<0.0001$	$r=-0.648$ $p<0.0001$	$r=0.817$ $p<0.0001$	–	–
$\sum WP_P$	$r=-0.389$ $p=0.0174$	$r=-0.469$ $p=0.0034$	$r=0.347$ $p=0.0357$	$r=0.490$ $p=0.0021$	–
$\sum WP_F$	$r=0.186$ ns	$r=0.500$ $p=0.0016$	$r=-0.101$ ns	$r=-0.436$ $p=0.0070$	$r=-0.449$ $p=0.0053$

For each parameter combination, correlation coefficient r and level of significance (p value) are given (ns, not significant, that is, $p>0.05$). With respect to VI, r and p are interpreted as indicators of effect strength with signs indicating the direction of the effect.

Among functional prey groups, the number of benthic prey items $\sum WP_B$ exerts the strongest effect on VI ($r=-0.861$, $p<0.0001$), followed by planktonic prey $\sum WP_P$ ($r=-0.389$, $p=0.0174$). The number of fish species in the diet $\sum WP_F$ is not significantly related to VI ($p>0.05$). The pattern of high benthic biomass and diversity on the high Antarctic shelf (see Section 2.2) is obviously reflected in notothenioid prey diversity and thus in trophic vulnerability: the number of benthic prey species $\sum WP_B$ exerts by far the strongest effect on VI; the higher the share of benthic species in the diet, the lower is VI (Table 1; Fig. 5).

The resilience of the entire system, that is, to what extent the extinction of particular consumer species from the system impacts overall food web stability and ecosystem functioning, strongly depends on the systems' ability to compensate for the loss by co-occurring species (Johnson, 2000; Naeem, 1998). As the majority of species include a certain proportion of benthic prey in their diet, functional redundancy seems to be high among benthos feeders (see Fig. 5). Feeding on the benthos is associated with a high degree of trophic generalism and functional redundancy, and hence with a certain capability to adapt food choice to prey availability and to dampen bottom-up effects. Plankton consumers tend to have higher vulnerability: specializing on a comparatively narrow prey spectrum makes them more sensitive to changes in prey availability. As there are fewer plankton-feeding species in the system, the potential for functional compensability is lower, too, making this part of the food web particularly sensitive to change.

Larvae and juveniles were not considered in this analysis because information on diet composition of early stages is not as complete as that of the adults for most species (except for *P. antarcticum*; see e.g. Granata et al., 2009; Kellermann, 1987; Vallet et al., 2011). However, as early stages of most notothenioids are pelagic, it is most likely that the relative trophic vulnerability is high compared with adult benthic stages. The number of potential predators ($\sum C$) is presumably size- (see Section 5.2) and/or density-dependent (e.g. Woodward et al., 2010b). Early stages of *P. antarcticum*, for example, occur in dense aggregations and are heavily preyed upon by other notothenioids (Eastman, 1985b; Hubold and Ekau, 1990; La Mesa et al., 2011). The degree of trophic generalism ($\sum P$) seems to differ strongly among families. Nototheniid larvae (including *P. antarcticum*) feed mainly on early copepod stages and eggs, whereas early juveniles feed on small copepod species. Compared with nototheniids, larvae and juveniles of the family Channichthyidae (e.g. *C. myersi*, *C. antarcticus*) are trophic specialists, with a narrow food spectrum that is exclusively composed of early developmental stages of krill and fish fry (Hubold and Ekau, 1990; Kellermann, 1986b, 1987, 1989). Accordingly, relative trophic vulnerability of larval and juvenile channichthyids is expected to be very high.

5.2. Vulnerability to changes in size structure and prey quality

5.2.1 Prey size

Body size is one of the major factors determining who eats whom in aquatic food webs (e.g. Brose et al., 2006; Castle et al., 2011; Woodward et al., 2005, 2010a,b). In particular for early developmental stages of fishes, the size of their prey seems to play an important role as the upper limit of consumable prey size is strongly limited by mouth width (Kellermann, 1986b, 1987). Accordingly, early stages always feed on a relatively narrow prey size range (Hubold and Ekau, 1990; Kellermann, 1986b). However, size of ingested prey is not only determined by morphological constraints but may also be the result of selective feeding behaviour. In postlarval and juvenile *P. antarcticum*, prey selection was found to be a function of prey density: at low food density conditions, larger prey species were selectively chosen, but when food density was high, size-selective feeding behaviour was distinctly less pronounced (Kellermann, 1986b, 1987).

There is also some evidence of size-selective feeding in adult *P. antarcticum*, with a negative selection of highly abundant small prey (K. Mintenbeck, unpublished data). However, whether this is the result of density-dependent selection (as observed in early stages) or due to other restrictions still needs to be verified.

In adult fish, mouth gape is less restrictive for prey handling than in larvae and juveniles, but sensory capabilities might be a limiting factor for efficient detection of small-sized prey. Depending on species' sensory capabilities, detection, capture success and feeding efficiency are likely to vary with prey size. To test for the impact of prey size on fish detection capability and feeding efficiency, feeding experiments with two nototheniid species were carried out during the expedition ANT/XXVII-3 with RV *Polarstern* in 2011.

N. coriiceps and *N. rossii* were caught by means of baited traps in February 2011 in Potter Cove. Both species belong to the family Nototheniidae and are common components of the inshore demersal fish fauna in waters of the northwestern Antarctic Península region (Barrera-Oro, 2003). Fishes were held unfed in large tanks at a water temperature of 0 °C for about 2 months prior to the experiments. Ten individuals were selected of each species (size ranges: *N. coriiceps* 24.2–33.6 cm standard length (SL), *N. rossii* 24.0–35.1 cm SL) and transferred into individual 85-l aquaria (0 °C water temperature, dim light conditions) 24 h before the feeding experiment started. Five different prey size classes were offered to each fish (prey size categories 1 (small) to 5 (large); see Table 2), starting with the smallest size category. Prey density was constant with 30 prey individuals per fish in all feeding trials. Times of first reaction, and each

Table 2 Food used in the feeding experiments to test for the impact of prey size on fish detection capability and feeding efficiency included five different prey size categories

Size category	Type	Size range [cm] (min–max)
1	Cyclopoid copepods	0.8–1.3
2	Daphnia	2.2–3.0
3	Mysids	9.2–14.5
4	Juvenile euphausiids	12.2–18.9
5	Adult euphausiids	27.9–33.5

Categories 1–4 were commercial frozen fish food (Erdmann Frostfutter, Germany) and category 5 was adult ice krill (*Euphausia crystallorophias*) caught during the RV *Polarstern* expedition. The food was defrosted prior to experiments.

Table 3 Number (*N*) of feeding individuals (*N. coriiceps* and *N. rossii*), and mean times (min:s; ± standard deviation) until first reaction to prey addition (movement of head towards prey), first detection of a prey item that was followed by an attack and first consumption are given for each prey size class (categories 1–5)

Prey category	*N* of feeding fish	First reaction	First prey detection	First consumption
Notothenia coriiceps				
1	0	–	–	–
2	0	–	–	–
3	6	2:12 ± 2:53	3:34 ± 3:50	3:36 ± 3:50
4	10	1:27 ± 3:21	1:47 ± 3:18	1:48 ± 3:17
5	10	0:10 ± 0:16	0:10 ± 0:16	0:12 ± 0:15
Notothenia rossii				
1	0	–	–	–
2	1	0:37	0:37	0:40
3	5	4:12 ± 6:04	4:12 ± 6:04	4:13 ± 6:04
4	9	1:18 ± 3:19	1:31 ± 3:16	1:34 ± 3:15
5	10	0:10 ± 0:08	0:10 ± 0:08	0:12 ± 0:07

particular prey detection and consumption were registered; overall experimental duration was 15 min. Depending on the amount of food consumed, the time lag between particular experiments was up to 3 days to avoid an effect of satiation on feeding behaviour. The offered food was not alive, but prey items were in motion in the tanks all the time owing to aeration and steady inflow of fresh seawater.

None of the fish fed on the smallest prey (category 1) and only one small *N. rossii* consumed a single prey item of category 2 (Table 3). Except for this single individual, no reaction to prey of categories 1 or 2 was observed. Both *N. coriiceps* and *N. rossii* started to react when prey of size category 3 was offered: a total of six *N. coriiceps* and five *N. rossi* detected and consumed prey of this category. All but one fish fed on size categories 4 and 5. Time until first reaction to prey addition (movement of the head towards prey), time of first prey detection that was followed by an attack and time of first consumption were all inversely related to prey size in both species (Table 3).

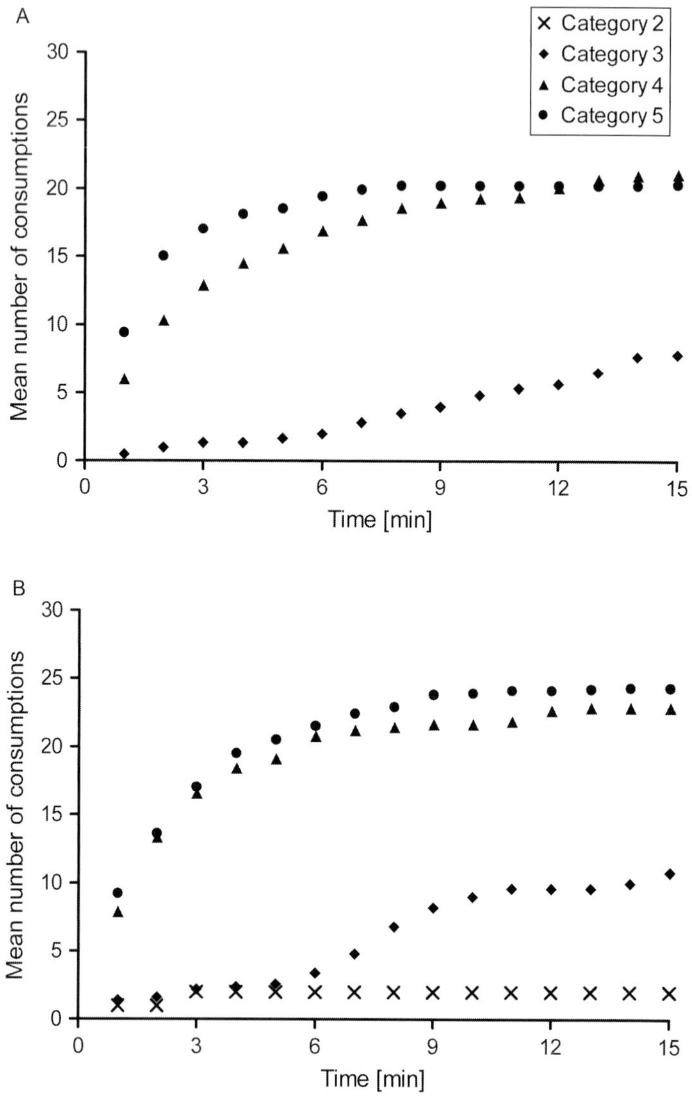

Figure 6 Mean consumption rate (number of consumptions $\times$ time^{-1}) in *N. coriiceps* (A) and *N. rossii* (B) depending on prey size category (2–5; 2: daphnia, 3: mysids, 4: juvenile euphausiids, 5: adult euphausiids). For the number (*N*) of feeding individuals, see Table 3.

The mean consumption rate (number of consumptions $\times$ time^{-1}) depended upon prey size category in both species (Fig. 6; Kruskal–Wallis ANOVA, *N. coriiceps*: $H = 29.45$, $p < 0.0001$; *N. rossii*: $H = 46.99$, $p < 0.0001$). The feeding rates were low when fish were offered size categories 2 and 3, and high for the two largest prey size categories. The feeding rates of fish offered the two small size categories (2 and 3) were significantly lower than the feeding rates of fish offered large prey (4 and 5; Dunn's post hoc test, $p < 0.01$), while no differences were found among the two small nor the two large categories (Dunn's post hoc test, $p > 0.05$). The mean time between detection and consumption was independent of prey size in both species (Kruskal–Wallis ANOVA, $p > 0.05$). Neither the total sum of consumed prey nor time until first reaction and mean time between detection and consumption were significantly correlated with fish size for any prey size category (Spearman's rank correlation, $p > 0.05$). However, the fish used in these experiments did not differ much in size, hence, further experiments using a broader range of fish sizes will be needed to verify the relationship between fish size and these parameters.

Nevertheless, these data clearly show that feeding rates depend strongly on prey size. Both species are obviously not capable (and/or willing) to attack small prey items and feeding efficiency is low below a certain prey size limit. This prey size-dependent detection and consumption rate are most likely not only found in these two species, but might be a limitation in many other Antarctic species (if not all).

5.2.2 Prey quality

The importance of prey quality for consumers is widely accepted, but only a few studies have dealt with this issue in fishes. Malzahn et al. (2007) found nutrient limitation of primary producers to propagate along the food chain, finally affecting condition of fish (larval herring *Clupea harengus*; condition assessed based on RNA/DNA ratios) feeding on herbivorous zooplankton. Based on the histology of the digestive organs, Koubbi et al. (2007) investigated the condition of larval *P. antarcticum* off Terre Adélie in relation to prey composition and found that larvae feeding on copepods were in better condition than those feeding on diatoms.

The energy contents of Antarctic and Sub-Antarctic species from various taxonomic groups are well studied (Ainley et al., 2003; Barrera-Oro, 2002; Clarke et al., 1992; Croxall and Prince, 1982; Donnelly et al., 1990, 1994; Eder and Lewis, 2005; Lea et al., 2002; Tierney et al., 2002; Torres et al., 1994) and are summarized in Fig. 7. Fishes and decapods have the highest energy contents; squid and crustaceans such as mysids, euphausiids and copepods have moderate energy contents. By far, the lowest energy content is found in gelatinous zooplankters such as salps, chaetognaths and cnidarians. Shifts in the zooplankton community with crustaceans being replaced by salps (see Section 1) thus involve a drastic decrease in energy density and nutritive value of prey for consumers such as fish.

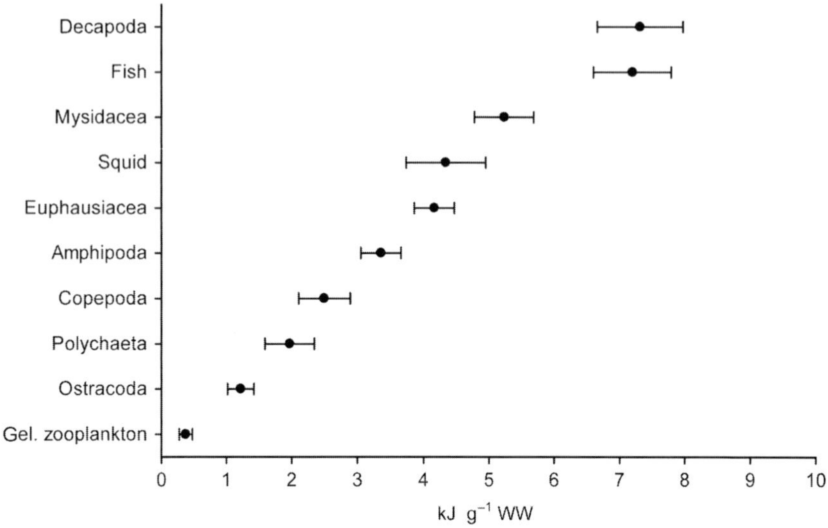

Figure 7 Energetic value (kJ g^{-1} wet weight; means ± SE) of Antarctic and Sub-Antarctic species belonging to several taxonomic groups (for details and data sources for each group, see Mintenbeck, 2008 and references therein).

Some species, such as *N. coriiceps*, feed on crustaceans (mainly krill and amphipods) and even on macroalgae (Iken, 1996; Iken et al., 1997). However, in feeding experiments, algae are only ingested in the absence of alternative animal prey (e.g. crustaceans; Fanta, 1999; K. Mintenbeck, unpublished data), providing evidence that macroalgae are not a favoured food source (see also Fanta et al., 2003). If krill is not available due to a shift in zooplankton composition, omnivorous fish species such as *N. coriiceps* might, however, be increasingly forced to feed on such low energy macroalgae.

Low energy food might affect survival, growth, body condition and reproductive output of consumers and ultimately might make fish species itself a low quality prey for its endothermic predators (see Österblom et al., 2006, 2008).

6. Vulnerability of Antarctic Fishes to Habitat Destruction

6.1. The impact of sea ice reduction

There is no doubt that a reduction in sea ice extent and duration of coverage due to climate-driven warming will affect the sympagic community living within the ice and invertebrates such as Antarctic krill, *E. superba*, whose life cycle is coupled to seasonal sea ice dynamics (see Moline et al., 2008 for review). However, there are also fish species whose life cycles and life styles are closely associated to the sea ice, namely *P. borchgrevinki* and *P. antarcticum*.

P. borchgrevinki is morphologically well adapted for a cryopelagic life (Eastman and DeVries, 1985) and is usually found closely associated with the underside of ice, where it frequently hides in crevices (e.g. Davis et al., 1999). Main prey items of this species include sympagic copepods (Hoshiai et al., 1989): sea ice thus provides the consumer with a habitat, refuge and feeding ground. Though *P. antarcticum* is usually found in the free water column, sea ice seems to be an important feeding ground for this species as well. Huge shoals of several thousand individuals have been observed feeding under the fast ice west off the Antarctic Peninsula (Daniels, 1982). The sea ice region is apparently also the spawning ground (La Mesa and Eastman, 2012) as the pelagic eggs were found floating under the sea ice (Vacchi et al., 2004). The reproductive cycle of *P. antarcticum* seems to be closely coupled to seasonal sea ice dynamics, and early stages depend on the temporal and spatial match with the seasonal zooplankton production (La Mesa and Eastman, 2012; La Mesa et al., 2010). The hatching period of *P. borchgrevinki* seems to be less strongly coupled to production peaks (Pankhurst, 1990), but both species are expected to be significantly affected by alterations in seasonal sea ice dynamics by loss of habitat/refuge and spawning ground and alterations at the base of the food web.

6.2. The impact of increased iceberg scouring

6.2.1 The role of habitat structure and disturbance events for species richness

For freshwater habitats (lakes and rivers) as well as for marine habitats, such as coral reefs and hard-substrate environments in the Mediterranean, a high diversity in habitat structures often promotes high species richness in the associated fish communities (Feld et al., 2011; Garcia and Ruzafa, 1998; Gratwicke and Speight, 2005; Guégan et al., 1998; Öhman and Rajasuriya, 1998). The benthic communities of the eastern Weddell Sea shelf in water depths between 200 and 450 m are characterized by a patchwork of structurally different successional stages; the two extremes are areas with a diverse epifauna forming the rich 3-dimensional habitat (Fig. 8A) on the one hand and desert-like areas with nearly no epibenthic fauna (Fig. 8B) on the other. This patchwork is the result of mechanical disturbance events by grounding icebergs, calving from the shelf ice and grounding at water depths between 200 and 500 m. This kind of

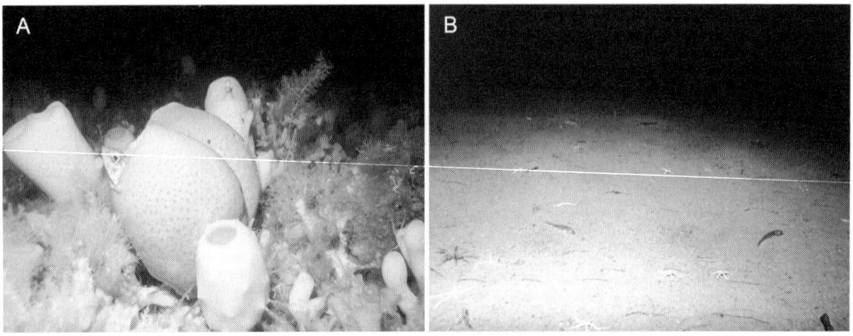

Figure 8 (A) Typical undisturbed site with a rich 3-dimensional habitat and *Trematomus* cf. *eulepidotus* hiding inside a large sponge; (B) Fresh iceberg scour habitat without any 3-dimensional megafauna species, only with a pycnogonid, some ophiuroids and *Prionodraco evansii*. Photos: ©Julian Gutt, AWI Bremerhaven.

disturbance by ice is a common phenomenon in polar regions and affects the sea floor up to a maximum water depth of 550 m (Conlan and Kvitek, 2005; Conlan et al., 1998; Gutt, 2001). Beside forest fires and hurricanes, iceberg scouring is one of the most significant natural physical perturbations imposed on ecosystems (Garwood et al., 1979; Gutt and Starmans, 2001; Peck et al., 1999) and is considered to be one of the strongest physical forces structuring the benthic environment in polar regions (Gutt and Piepenburg, 2003).

In the area of Austasen and Kapp Norvegia (eastern Weddell Sea), about 7% of the coastal zone (< 400 m) is covered by iceberg aggregations (Fig. 9; Knust et al., 2003). Gutt and Starmans (2001) analyzed video and photo material from this area and calculated that 42–70% of the sea bed is affected by iceberg scouring (Fig. 10). Depending on sea floor morphology, nearly 10–40% of the area showed young scour marks or early recolonization stages. On a smaller (local) scale, these iceberg disturbances completely destroy most of the in- and epifauna. However, on a regional scale, the sea floor destruction by icebergs creates new space for pioneer species that are specialized in recolonization (Gerdes et al., 2008; Teixidó et al., 2004), and the patchwork of structurally different successional stages increases the overall gamma diversity (Gray, 2000). Following the Intermediate Disturbance Hypothesis, which predicts a high diversity caused by intermediate levels of disturbance (Huston, 1979), Gutt and Piepenburg (2003) calculated an increased species diversity for the epibenthic megafauna on a regional scale (1–100 km) due to iceberg disturbances.

However, iceberg scouring not only affects benthic organisms, but also the demersal fish fauna, the composition of which differs significantly

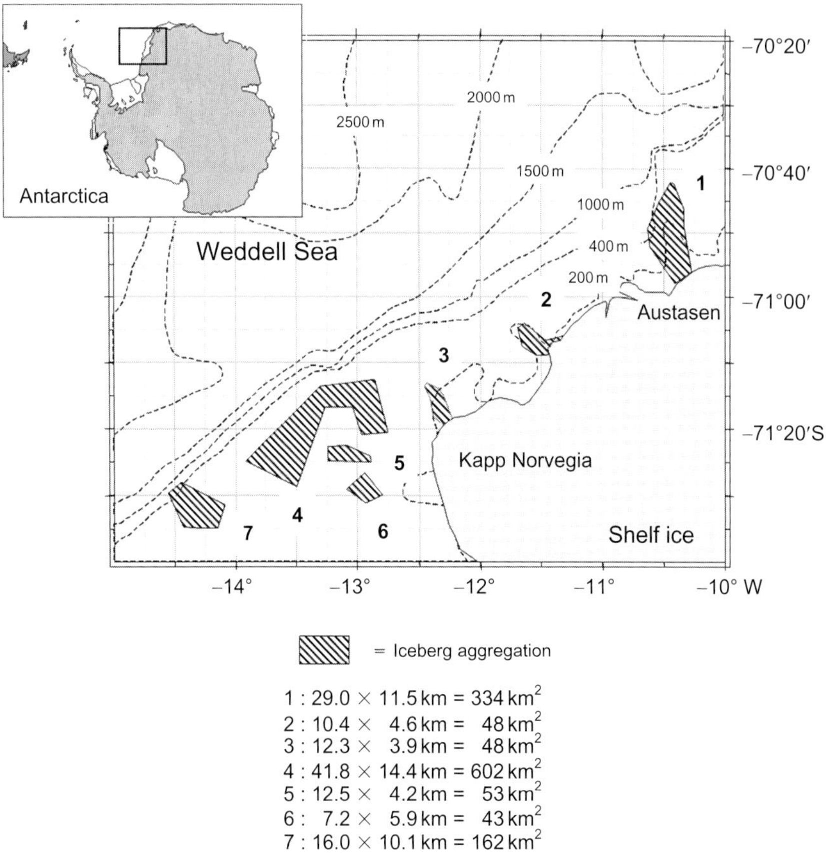

Figure 9 Iceberg aggregations on the high Antarctic shelf off Austasen and Kapp Norvegia (eastern Weddell Sea) in March 2000 *(redrawn from Knust et al., 2003; with permission from Backhuys Publishers).*

between undisturbed and recently disturbed sites, with higher species richness and a higher diversity (Shannon's diversity) in undisturbed areas, and in species identity (Fig. 11; Knust et al., 2003). Some species, such as *Trematomus scotti*, *C. antarcticus*, *Pagetopsis maculatus* and *Artedidraco loennbergi*, are typical members of the fish community in undisturbed areas, while *T. pennellii*, *Trematomus nicolai* and *Prionodraco evansii* are specialized to live at disturbed sites. Iceberg disturbance events play a key role in small-scale niche separation of fishes, as the structurally different habitats allow for the coexistence of trophically similar species (Brenner et al., 2001; Hagen et al., 2012; Ptacnik et al., 2010).

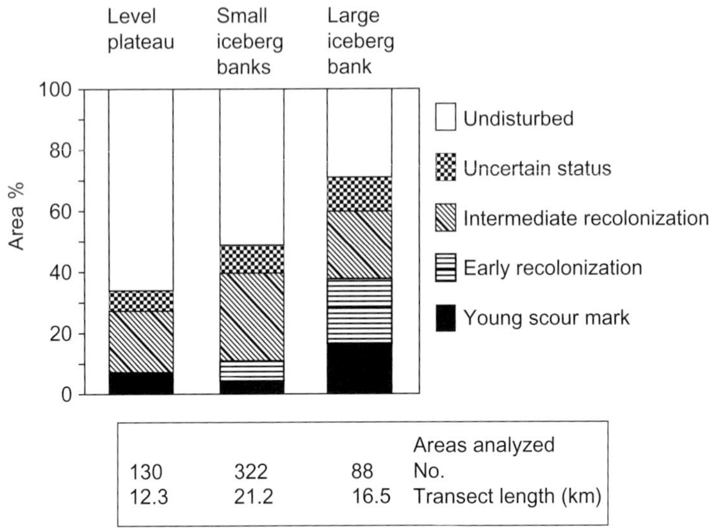

Figure 10 Proportions of different recolonization stages at different morphological condition in the shelf area off Austasen and Kapp Norvegia (eastern Weddell Sea) *(redrawn from Knust et al., 2003, based on data from Gutt and Starmans, 2001; with permission from Backhuys).*

In view of climate change, the major questions are (i) whether the pattern of increased diversity at intermediate disturbance levels found for benthic megafauna (e.g. Gutt and Piepenburg, 2003; see above) does also hold true for the demersal fish community and (ii) how diversity and community structure will respond to increased disturbance events. To answer these questions, we analyzed the relationship between the level of disturbance events and the gamma diversity of the demersal fish fauna on the eastern Weddell Sea shelf using a simple simulation model.

6.2.2 The disturbance simulation model

The calculations of the model are based on abundance data from several bottom trawls (standard otter trawl, 20-mm mesh in codend) taken in undisturbed and disturbed areas off Austasen and Kapp Norvegia (eastern Weddell Sea) during four expeditions between 1996 and 2004. The position of iceberg scours was identified by side scan sonar, ROV underwater video and underwater photography. The composition of invertebrate by-catches of the otter trawls was used as an indicator for the degree of disturbance. In particular, the presence/absence of large, old hexactinellid sponges indicated the disturbance level of the sea floor (Gerdes et al.,

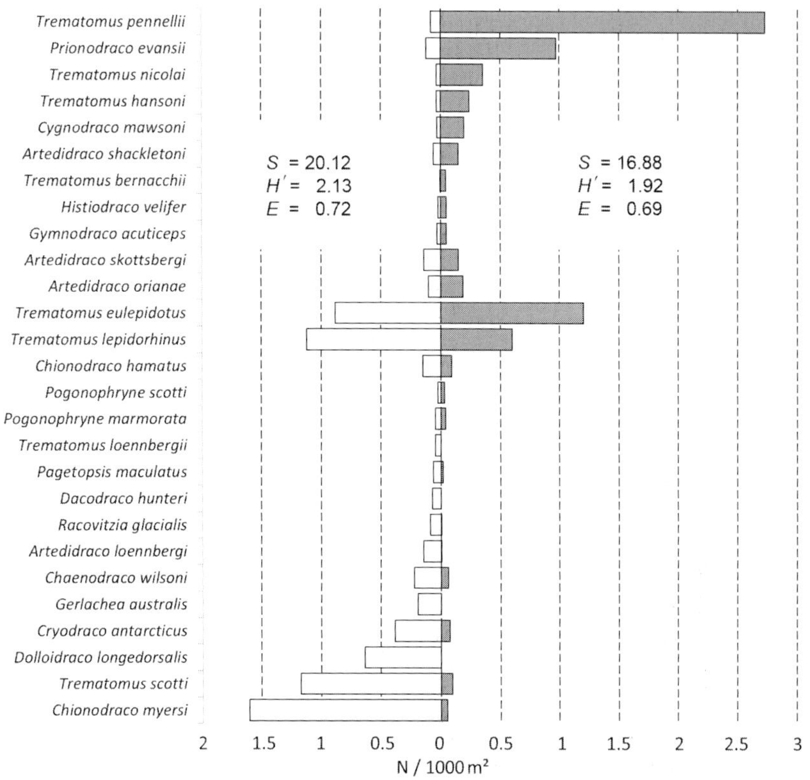

Figure 11 Mean abundance (N/1000 m^2) of the dominating fish species in undisturbed areas and on young iceberg scours (recalculated with data from Knust et al., 2003; R. Knust and K. Mintenbeck, unpublished data). For both communities, average number of species (*S*), average Shannon diversity (*H′*) and average evenness (*E*) are given.

2008; Knust et al., 2003). Overall 30 stations were sampled, 22 in undisturbed areas and 8 in disturbed areas. Fish abundance data were calculated based on 1000 m^2 swept area.

To simulate different disturbance levels, the abundance data of each catch from disturbed and undisturbed areas were randomly combined in a 200 station matrix by a Monte Carlo simulation. To simulate a completely undisturbed situation (disturbance level 0.0), data from exclusively undisturbed stations were randomly combined. The disturbance level 1.0 (the entire shelf is disturbed) was simulated by a random combination of catches from exclusively disturbed areas. For a disturbance level of 0.05 (5% of the shelf area is disturbed by iceberg scouring), 5% of the 200 stations in the matrix were filled with abundance data from disturbed stations and 95%

were filled with abundance data from undisturbed station. The stations were randomly selected. The disturbance level was increased in 0.05 steps. For each disturbance level, 100 iterations of random combination were computed.

For each disturbance level, the averages of total number of species (S), gamma diversity (Shannon's diversity, H') and evenness (Pielou, E) were calculated:

$$H' = \sum_{i=1}^{S} p_i \cdot \ln(p_i) \qquad [3]$$

where $p_i = n_i/N$, with n_i is the abundance of species i and N the sum of all individuals;

$$E = \frac{H'}{H'_{max}} \qquad [4]$$

where $H'_{max} = \log_2(S)$. To avoid an oversized effect of very rare species on the total number of species, only species with an abundance of >0.001 individuals per 1000 m^2 were taken into account for each combination.

The results of the Monte Carlo simulation are shown in Fig. 12. The results of the different catch combinations were fitted to a univariate, second-order polynomial (solid line), the dots represent the particular results of each combination. All three parameters increased with increasing disturbance level up to a maximum between disturbance values of 0.25–0.40, representing 25–40% of the shelf area disturbed by icebergs. Average species richness was highest (57.9 species) at a disturbance level of 0.25, average gamma diversity was highest ($H' = 2.80$) at a level of 0.34 and average evenness (0.69) was highest at a disturbance level of 0.40.

The comparison of these results with the estimated disturbance level on the shelf of the eastern Weddell Sea (Gutt and Starmans, 2001; see above) shows that the fish fauna on the eastern Weddell Sea shelf is obviously well adapted to this kind of mechanical disturbances and to the average level of disturbance occurrence there. The gamma diversity (and evenness) of the fish fauna is highest at the disturbance level we find nowadays, but rapidly decreases at higher levels according to the model predictions. Future climate scenarios suggest an increasing rate of iceberg calving in the shelf ice areas of Antarctica, with an enhanced risk of iceberg groundings and an increasing disturbance level in the benthic communities.

Such a reduction in habitat structure and heterogeneity means on the one hand a reduction of habitat and refuge for demersal fishes (e.g. Moreno et al.,

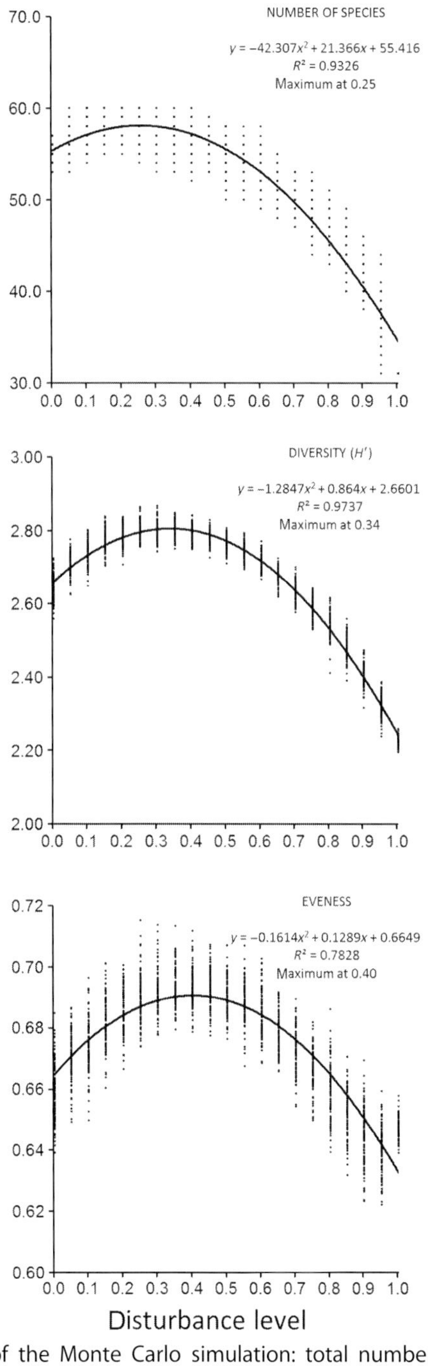

Figure 12 Results of the Monte Carlo simulation: total number of species, gamma diversity (H') and evenness (E) of the demersal fish community depending on disturbance rates. The calculations are based on fish abundance data from disturbed and undisturbed areas on the eastern Weddell Sea shelf (R. Knust and K. Mintenbeck unpublished data).

1982), with an increased predation risk for all developmental stages. On the other hand, a reduction in habitat diversity will result in an increased competition among species with overlapping trophic niches, which at the present disturbance level perfectly avoid competition by small-scale niche separation (Brenner et al., 2001). If the disturbance level increases in the future, a loss in species diversity seems to be inevitable.

7. DISCUSSION

7.1. The impact of climate change on Antarctic fish species

Notothenioid fishes are well adapted to their habitat, and alterations in the abiotic environment directly affect physiological functions. Increasing water temperatures, particularly in combination with ocean acidification, pose a major thread to the persistence of notothenioid fishes. Some species such as *P. borchgrevinki* and *Notothenia* spp. show some physiological plasticity and are able to compensate for increasing oxygen demand, for example, by mitochondrial proliferation and/or increased haematocrit. However, these compensatory mechanisms are limited and most notothenioid species are in fact stenothermal and are not capable to adjust metabolic functioning. Channichthyids are highly vulnerable to changes in the abiotic environment as they lack any capacity to adjust blood parameters to an increasing oxygen demand. Early developmental stages as well seem to be highly vulnerable to all kinds of abiotic alterations, including salinity.

General vulnerability to changes in food web structure and dynamics was analyzed using a conceptual approach, with a quantitative measure (VI) that served as an indicator of the risk of consumer species to be negatively affected by such changes. Relative trophic vulnerability was found to be low in all fish species that include a certain proportion of benthic organisms in their diet. Obviously, feeding on benthos goes along with a high degree of trophic generalism and, hence, with a certain capability to adapt food choice to prey availability and to dampen bottom-up effects. Plankton consumers displayed a distinctly higher vulnerability, as these species tend to specialize on a comparatively narrow prey spectrum, which makes them more sensitive to changes in prey availability. Thus, there exists an accumulation of risk in the trophic group of plankton feeders, making this part of the food web particularly sensitive to change. Highest trophic vulnerability was found in channichthyids, such as *C. myersi*, *C. antarcticus* and *P. maculatus*, which are all specialized on very few prey items, and in the nototheniid *P. antarcticum*, which not only has a narrow food spectrum but is additionally exploited by a multitude of predators. Relative trophic vulnerability of plankton-feeding larvae and early juveniles is expected to be very high, as well.

Trophic vulnerability to shifts in prey size was investigated in two fish species, *N. coriiceps* and *N. rossii*. In the two species studied, feeding efficiency and prey detection capability were found to strongly depend upon prey size, with a complete detection failure (or ignorance) of smallest prey. However, the general susceptibility to prey size shifts might differ among species and age/size classes. As large fish species/individuals usually have a larger size range of available prey than small species/specimens, large fish are expected to be less susceptible to shifts in prey size compared with small fish. Trophic vulnerability to shifts in energy content is likely similar for most fish species. Nutritive value varies strongly among taxonomic groups, with lowest energy contents found in gelatinous zooplankton. Salps with their guts filled with fresh phytoplankton are a valuable food source for some Antarctic benthic suspension feeders, which usually depend on more or less degraded POM (Gili et al., 2006). Nevertheless, in comparison with other zooplankton species such as decapods, euphausiaceans and copepods, their nutritive value for fishes is extremely low. Prey of inappropriate size and/or quality affects the nutritional status and condition of fishes (see also Beaugrand et al., 2003; Koubbi et al., 2007; Malzahn et al., 2007), and in the worst case even survival.

The loss of habitat poses a threat to the majority of Antarctic fish species. There is no doubt that a reduction in sea ice extent and duration due to climate-driven warming will particular affect the life stages of those fish species that are strongly associated with sea ice, namely *P. borchgrevinki* and *P. antarcticum*. Habitat structure and heterogeneity are of particular importance for the demersal fish community because their loss would imply a loss of refuge and shelter for juveniles, adults and eggs (see Barrera-Oro and Casaux, 1990; Moreno, 1980; Moreno et al., 1982), with the consequent increase in competition among trophically similar species (e.g. Brenner et al., 2001). Model simulations based on abundance data indicated that an increase in ice scouring will lead to a steep decrease in diversity and evenness, and to the loss of species.

7.2. Effects of climate change in other marine systems

In general, the changes detected so far in the South Polar Sea resemble many of those observed on a worldwide scale. The world's oceans are warming, atmospherical pCO_2 is rising, leading to potential ocean acidification

(IPCC, 2007), and seawater salinity is decreasing in the vicinity of melting ice and glaciers (e.g. Curry and Mauritzen, 2005). Fishes, for example, in temperate regions, have been shown to be significantly affected at the physiological level, in particular by increasing water temperatures, resulting in reduced growth performance, recruitment and abundance (e.g. Pörtner and Knust, 2007; Pörtner et al., 2008; Sirabella et al., 2001). In response to such environmental alterations, several fish species have already shifted their distributional ranges and have migrated into waters with more favourable conditions (Dulvy et al., 2008; Perry et al., 2005).

The risk of habitat loss and alteration of habitat structure and heterogeneity due to sea ice retreat (Stroeve et al., 2007) and increasing occurrence of iceberg scouring events (Conlan et al., 1998; Gutt et al., 1996) is comparable in the Arctic Ocean. Invasion by lithodids as observed west of the Antarctic Peninsula is also found in the northern hemisphere. The most popular example is the red king crab (*Paralithodes camtschaticus*), which was introduced into the Barents Sea and subsequently invaded Norwegian waters. In the presence of these invaders, benthic biomass and diversity is drastically reduced; these crabs thus remove potential prey for benthos feeding fish species and even prey upon fish eggs (reviewed in Falk-Petersen et al., 2011).

Alterations in plankton community composition due to climate forcing are common in many ecosystems and often include a shift from larger to smaller phytoplankton (Yvon-Durocher et al., 2011) and zooplankton species, in particular in copepods (e.g. in the North Sea, Beaugrand et al., 2003; Helaouët and Beaugrand, 2007; and in the Humboldt Current ecosystem reviewed in Alheit and Niquen, 2004). Substantial increases in gelatinous zooplankton have been observed in different marine systems in recent years (e.g. Attril et al., 2007; Brodeur et al., 1999; Purcell, 2005). High abundances of gelatinous zooplankton are often related to water temperature and salinity, suggesting that population density of gelatinous zooplankton will further increase under future climate change scenarios (reviewed in Purcell, 2005). In the central North Sea, the occurrence of jellyfish was also negatively correlated with seawater pH, thus, future levels of CO_2 may synergistically promote the presence of gelatinous zooplankton (Attril et al., 2007). Gelatinous zooplankton prey upon fish eggs and larvae (Doyle et al., 2008; Purcell, 1985) and also represent strong competitors for plankton-feeding fish by efficiently removing potential prey, such as copepods (Purcell and Decker, 2005) or euphausiids (Suchman et al., 2008). Some gelatinous zooplankton such as salps or ctenophores are occasionally consumed by fishes (including notothenioids), but appear to be a form of 'survival food' when preferred zooplankton prey are not abundant (Kashkina, 1986; Mianzan et al., 2001).

Taking a closer look at the effects of climatic shifts in the world's oceans, an intriguing pattern seems to emerge: the most severe (or most rapid) effects of climate forcing on marine biota appear to be found within pelagic communities (Alheit, 2009; Alheit and Niquen, 2004; Alheit et al., 2005; Arntz, 1986; Beaugrand et al., 2003; Benson and Trites, 2002). These observed changes in the pelagic realm are not restricted to one or two trophic levels, or to specific species, but usually involve strong alterations in food web structure and ecosystem functioning and dynamics. In the central Baltic and the North Sea (Alheit et al., 2005; Beaugrand et al., 2003), in the Bering Sea and the North Pacific (Benson and Trites, 2002) as well as in the Humboldt Current ecosystem (Alheit and Niquen, 2004; Arntz, 1986; Arntz and Fahrbach, 1991), all trophic levels in the pelagial, from primary producers to apex predators, were affected by direct and/or indirect climate forcing. In particular, pelagic fishes with short plankton-based food chains, such as clupeids, may undergo strong fluctuations in stock density (Alheit and Niquen, 2004), with severe consequences for their endothermic consumers (e.g. Alheit, 2009; Arntz, 1986; Cury et al., 2000).

To our knowledge, there are no reports on such extensive severe effects of climatic shifts on benthic biota, suggesting that they are less affected, but whether this is really a common pattern or just due to a greater focus on the pelagic realm still needs to be verified.

Overall, the general effects of climate forcing and the potential direct and indirect impact on marine living communities thus appear to be similar in the South Polar Sea compared with marine systems worldwide. Nevertheless, there are some significant differences: (i) fishes inhabiting temperate and tropical regions often have the opportunity to emigrate into waters with more favourable abiotic and biotic (prey) conditions (see Arntz, 1986; Dulvy et al., 2008; Perry et al., 2005). For notothenioid fishes, particularly for high Antarctic species, emigration is strongly limited by stenothermy and the lack of alternative habitats, as they are already living at the highest latitudes. (ii) Fish species such as clupeids in upwelling systems are evolutionarily adapted to strong environmental fluctuations by possessing traits associated with fast growth (Cubillos et al., 2002) and high fecundity (Alheit and Alegre, 1986), both facilitating population recovery after stock decline. In tropical reef fish, a rapid transgenerational acclimation to increasing water temperatures has also been observed (Donelsen et al., 2012). In contrast, notothenioid fish species are characterized by slow development rates and low fecundity, and their recovery potential is thus strongly limited. Due to the low recovery capacity, even a modest increase in the amplitude of interannual climate fluctuations could affect long-term population dynamics of notothenioid fish, with ramifications that would ripple through the wider food web. Given the rate of alterations due to climate change observed off the Antarctic Peninsula and fish species' life history characteristics, an evolutionary adaptation of notothenioid fishes that keeps pace with the rate of change in conditions is unlikely, if not impossible (see Somero, 2010).

7.3. Antarctic fish community persistence—Winners and losers

Given the current state of knowledge, it is unlikely that there will be any true 'winners' of climate change among notothenioid fish species. There will be only 'survivors' and many 'losers'. *P. borchgrevinki*, for example, might be among the survivors in a changing South Polar Sea as this species is characterized by relatively high metabolic plasticity and a wide thermal tolerance window (Robinson and Davison, 2008), and low trophic vulnerability. But still, *P. borchgrevinki* is threatened by habitat reduction owing to retreat of sea ice and so far it is unknown whether this species can cope with loss of its ice habitat. Demersal fish species (except for plankton-feeding channichthyids) show low relative trophic vulnerability, but will be significantly affected by the loss of habitat structure and diversity. Which species will ultimately survive and which will not in the long run depends on a combination of species-specific physiological and trophical plasticity, and population dynamics, in addition to higher-level food web effects. Trophic plasticity is apparently high in benthos feeders, but acclimation capacity seems to differ strongly among species: some, such as *N. rossii*, possess partially compensatory mechanisms (A. Strobel and F.C. Mark, unpublished data), whereas the potential for acclimation is apparently low for many others (e.g. high Antarctic Trematomus spp.; Robinson, 2008). However, which demersal species exactly will be lost due to habitat loss on the high Antarctic shelf remains unknown for now. Population dynamic parameters such as relative fecundity and growth rate are rather similar among most demersal high Antarctic notothenioids (e.g. Kock and Kellermann, 1991; La Mesa and Vacchi, 2001), but much more work is needed on the effects that changes in the abiotic environment exert on population dynamics.

Icefishes, however, are one group that will almost certainly be on the losing side in a warming South Polar Sea with increasing levels of CO_2, as the oxygen-carrying capacity of their blood is limited, and there is no potential for physiological acclimation to satisfy the increasing tissue oxygen demand. Moreover, many channichtyids, such as *C. myersi*, are specialist consumers with a high relative trophic vulnerability, making this group additionally susceptible to changes in food web structure and dynamics. Another potential 'loser' which is affected by direct and indirect effects of climate change is the currently dominant pelagic species, *P. antarcticum*, which will be most likely affected by sea ice reduction. It also seems to be highly vulnerable to alterations in the food web, and indirect evidence suggests that at least larvae and juveniles are highly vulnerable to abiotic changes. Theoretically, all fish species are threatened by a shift in prey size structure and a decrease in prey nutritive value. However, the pelagic realm in the South Polar Sea (and elsewhere) will likely be among the first to react to climate fluctuations, and the shifts in question have thus so far only been observed in the plankton community. Plankton consumers are, therefore, especially vulnerable. Shifts in size distribution from large to small phytoplankton organisms in the marine Antarctic (as observed west off the Antarctic Peninsula; Moline et al., 2004) will thus most likely favour the prevalence of small zooplankton species, such as cyclopoid copepods (as observed in other marine systems). Given the size dependency of prey detection and feeding efficiency in notothenioid fish, it is questionable whether plankton consumers such as *P. antarcticum* but also icefish can cope with such a prey size shift. Feeding on low quality 'survival food' such as salps is not a suitable alternative to energy-rich crustaceous zooplankton in the long run. In particular, at the edge of their thermal tolerance window, fish species will be highly sensitive to such additional stressors.

However, the major bottleneck for the persistence of most (if not all) species' populations is most likely the survival of early developmental stages. Eggs and larvae appear particularly sensitive to alterations in the abiotic environment, some might be affected by sea ice reduction (e.g. early stages of *P. antarcticum*), and larvae and juveniles are apparently vulnerable to indirect food-web-mediated effects of climate change. For non-Antarctic pelagic larval fish, three key parameters were identified: prey abundance, prey type and seasonal timing (Beaugrand et al., 2003). Shifts in any of these three parameters might significantly compromise larval condition and survival. Most notothenioid larvae depend on seasonal timing as well (Efremenko, 1983; La Mesa and Eastman, 2012; La Mesa et al., 2010), and the capacity to avoid a mismatch depends on species' plasticity in their reproductive cycle. In some species, a certain plasticity to adapt the reproductive cycle might exist, as indicated by differences in spawning time among populations in different locations (see Kock and Kellermann, 1991), but it is unknown whether all species are able to adapt their reproductive cycle and by which factors the timing is triggered. Without offspring sustaining the stocks, populations will progressively age and density will inexorably decline.

So far, our knowledge on species-specific vulnerability, potential plasticity, and acclimation and/or adaptation potential is still limited. However, based on what we already know from the Antarctic and from what we can observe in other systems, we can at least identify some potential bottlenecks. Nevertheless, there is an urgent need for more experimental studies on a broad range of species to gain a cause-and-effect understanding of the consequences of the potentially complex interactions between abiotic and biotic mechanisms (Woodward et al., 2010a).

7.4. Consequences of fish species loss for the marine Antarctic ecosystem

Fishes of the South Polar Sea will be affected by climate change in multiple ways, with the potential vulnerability and mechanisms differing among species, developmental stages and habitats. What we know about the effects of climate change and the vulnerability of Antarctic fish species leaves little doubt that the population density of many species will decline and some species will go extinct in the long run. The resilience of the entire system, that is, to what extent the decline or extinction of particular notothenioid fish species from the system impacts overall food web stability and ecosystem functioning, depends strongly on the species' functional role and the systems' ability to absorb for the loss by compensatory mechanisms among co-occurring species (Johnson, 2000; Naeem, 1998).

In the South Polar Sea, demersal fish communities are characterized by relatively high species richness in comparison with the very limited diversity of the pelagic ichthyofauna. The majority of demersal fish species are opportunistic generalist consumers with high trophic niche overlap, indicating a high functional redundancy. It is therefore likely that within the demersal community, the decline or loss of some species can be compensated for by others. However, the model on the impact of disturbance events on demersal fish communities indicated not only a future decrease in species number but also in evenness. Thus, one or a few species might become extremely dominant, and the impact of such an alteration in community composition on overall ecosystem functioning ultimately depends on a species' identity, its specific traits and its potential to serve as valuable prey for top predators. As the role of demersal fish as a food source for endothermic predators is more important in inshore compared to offshore areas (Barrera-Oro, 2002), the effects of alterations in the demersal community on fish consumers might be stronger in shallow coastal zones. In inshore waters of the South Shetland Islands, for example, *N. coriiceps* proliferated and became the dominant demersal fish species after the stocks of the demersal *N. rossii* and *G. gibberifrons* were drastically diminished in the early 1980s due to anthropogenic actions (Barrera-Oro et al., 2000; see Section 7.5 below). Simultaneously, the breeding populations of one of their most important consumers in this area, the Antarctic shag *Phalacrocorax bransfieldensis*, now mainly preying upon *N. coriiceps*, steadily declined. Thus, though *N. coriiceps* is similar to *N. rossii* and *G. gibberifrons* in terms of ecology and body size, it could apparently not fully compensate for the reduction of the two fish species within the food web (Casaux and Barrera-Oro, 2006).

The extent to which the loss of channichthyids such as *C. myersi* or *C. antarcticus* can be compensated by other fish species is not clear. Channichthyids feed almost exclusively on krill and fish (reviewed in Hureau, 1994; Kock, 2005a; La Mesa et al., 2004). While a multitude of other fish species also feed on krill, true piscivores are rare among other notothenioids (e.g. Hureau, 1994). Some species occasionally feed on fish (see Fig. 5; La Mesa et al., 2004) but to a much lesser extent compared to channichthyids. Hence, the loss of channichthyids from the system might release some fish populations from top-down control. Though icefish are an abundant component in Antarctic fish communities, they seem to be of minor importance in the diet of endothermic predators, but their importance increases regionally when other prey, such as krill, is scarce (Kock, 2005a,b and references therein). Compared with most other notothenioids, adult channichthyid fishes are large and show only weak escape responses (authors' personal observation from ROV videos), making them an easy-to-catch prey for large consumers. At our current state of knowledge, we can only speculate, but it is likely that the loss of channichthyids will have detrimental effects on many components of the Antarctic marine food web.

The pelagic fish community is composed of very few species only, and the whole community is dominated by a single fish species, *P. antarcticum* (see above), which on the high Antarctic shelf seems to occupy a similar ecological key role in the food web as Antarctic krill in the seasonal sea ice zone

(Hubold, 1992; Hureau, 1994; La Mesa et al., 2004; Takahashi and Nemoto, 1984). It is one of the principal consumers of zooplankton, and all developmental stages are among the most important food sources for a multitude of predators, in particular for endotherms inhabiting Antarctic shelf areas (e.g. Daneri and Carlini, 2002; Hureau, 1994; La Mesa et al., 2004; Plötz, 1986). This pelagic fish species occurs in loose shoals (Eastman, 1985a; Fuiman et al., 2002) and undertakes nocturnal migrations into upper water layers (K. Mintenbeck and R. Knust, unpublished data; Plötz et al., 2001), where it provides a rich and easily accessible food source. It is thus of critical importance in the Antarctic marine food web. No other species, neither other pelagic notothenioids nor invertebrates (e.g. squid or krill), may be able to provide full functional compensation in the case of its extinction or reduction of the stock, in particular because none combines a pelagic shoaling life style and vertical migration with a comparable size spectrum and energy content (see e.g. Ainley et al., 2003). In its appearance (Fig. 13) and life style, as well as in its central role in a relatively simply structured and highly productive pelagic system, *P. antarcticum* strongly resembles shoaling clupeid fishes in upwelling systems (see Section 7.2 above).

Figure 13 Catch of *Pleuragramma antarcticum* in the eastern Weddell Sea (Photo by J. Plötz, AWI Bremerhaven).

In the eastern South Pacific, for example, El Niño events involve strong reductions in stocks of anchovy and sardine (owing to direct and indirect climate forcing in combination with fisheries effects), causing starvation and mortality in the very top predators, birds and seals (e.g. Arntz, 1986). But life history traits of clupeids and the nototheniid *P. antarcticum* (see Table 4) and recovery potential differ significantly: population doubling time was estimated to be often less than 15 months in clupeid fish species and 5–14 years in *P. antarcticum* (Froese et al., 2002), making populations of the latter extremely vulnerable to any kind of disturbance or systemic shifts.

Nevertheless, seals and penguins do not depend exclusively on pelagic prey but also prey upon demersal fishes (e.g. Casaux et al., 2006; Coria et al., 2000; Plötz et al., 1991). In many high Antarctic shelf areas, exploitation of this resource requires deep diving. Though Weddell seal and Emperor penguin are both excellent divers (Burns and Kooyman, 2001; Wienecke et al., 2007), exploitation of fishes at great depth is energetically disadvantageous for these air-breathing endothermic predators as it involves an increased swimming effort, shorter times at feeding depth, and/or longer dives followed by longer recovery phases (Kooyman, 1989; Kooyman and Kooyman, 1995; Wilson and Quintana, 2004). Moreover, foraging efficiency is higher in shallow dives (Croxall et al., 1985), while encounter rates are probably lower in light-depleted deep waters, as indicated by a lower number of feeding events at depth (see Liebsch et al., 2007; Plötz et al., 2005). Hence, declining stocks or complete loss of *P. antarcticum* will in either case severely affect the top predators in the Antarctic marine ecosystem.

Table 4 Life history traits of clupeid fishes (sardines and anchovies) and the nototheniid *P. antarcticum*

	Clupeids	*Pleuragramma antarcticum*
Von Bertalanffy growth constant K	0.5–0.8[3]	0.05–0.07[5]
Age at first spawning (years)	1–1.5[2,8]	7–9[7]
Relative fecundity (eggs g^{-1} wet weight)	550–600[1]	70–160[4]
Duration of larval phase (days)	~37–74[6]	180–365[5]

Data sources are indicated by superscripts: [1]Alheit and Alegre (1986), [2]Cubillos and Claramunt (2009), [3]Cubillos et al. (2002), [4]Gerasimchuk (1988), [5]Hubold and Tomo (1989), [6]Houde and Zastrow (1993), [7]Kock and Kellermann (1991) and [8]Whitehead (1985).

What are the future perspectives for Antarctic fish communities? There is no doubt that fishes still will be an important and abundant component of the Antarctic marine ecosystem in the future, but the composition of communities will change significantly in the long run. It is likely that, with an ongoing warming trend, Sub-Antarctic demersal fish species such as *Notothenia* spp. (but also non-notothenioids) will move southwards into high Antarctic shelf areas, taking over the role of extinct or declining species in the present-day food web. Possible future scenarios for the pelagic community are the occupation of the 'small pelagic zooplankton feeder' niche by myctophid fishes or by clupeids such as the Falkland sprat, *Sprattus fuegensis*. Whether myctophids or clupeids can effectively replace *P. antarcticum* in its functional role in the food web, however, remains to be seen.

7.5. Final thoughts—Is climate change exclusively to blame?

Though our knowledge is steadily improving, we are in fact just starting to comprehend the structure, dynamics and functioning of the Antarctic marine ecosystem, while the system apparently has already started to respond to climate change. This, however, is not the only threat to marine living communities in the South Polar Sea, and human activities have already caused significant alterations in the past and still affect Antarctic communities today. Commercial fisheries in the Antarctic started in the late 1960s/early 1970s (see Kock et al., 2007 for review), and the destructive impact of bottom trawling on benthic communities is comparable to the impact of iceberg scouring (discussed in Barnes and Conlan, 2007). Commercial sealing and whaling activities in the South Polar Sea in the nineteenth and twentieth centuries (see Kock, 2007; Laws, 1977) resulted in large-scale and long-term alterations of food web structure and population dynamics of prey and competitors (Ainley et al., 2007, 2010; Laws, 1985). Industrial exploitation of Antarctic fish species (and krill) between the 1970s and 1990s resulted in dramatic stock decreases and rapid overexploitation of some species (reviewed in Ainley and Blight, 2008; Kock, 1992, 2007). Since 1982, the fisheries are regulated by CCAMLR (Commission for the Conservation of Antarctic Marine Living Resources), and many were closed between 1985 and 1990 due to overexploitation (Ainley and Blight, 2008; Kock et al., 2007). Nevertheless (to provide some examples only), regular monitoring of the stocks of many commercially exploited fish species, such as *N. rossii* and *G. gibberifrons*, around the South Shetland Islands, indicates a lack of recovery more than three decades after the end of fisheries (Barrera-Oro and Marschoff, 2007; Barrera-Oro et al., 2000). Similarly, stocks of the channichthyid *C. gunnari* in the Indian Ocean did not recover to pre-exploitation levels after the fishery had ceased for many years (Kock, 2005b). One reason for the slow stock recovery may be the low fecundity and slow development (see Section 3.3.2 above) of many species. However, two additional factors that may adversely affect stock recovery of *C. gunnari* were proposed (Kock, 2005b and references therein): (i) an increase in top-down pressure, that is, increased predation by seals owing to fluctuations in alternative prey (krill) and (ii) possible direct effects of climate change, in particular increasing water temperature. Owing to strict regulation, the numbers of some whale species seem to be increasing again in the South Polar Sea (Branch, 2006, 2007), which is on the one hand desirable, but on the other hand might entail an additional increased top-down pressure on zooplankton and fish communities.

These examples emphasize the complexity of relationships among human activities (historic and current), abiotic climate forcing and altered trophic structure, and how these factors can interact to control fish populations in the South Polar Sea. Thus, multiple drivers act synergistically to affect a particularly sensitive ecosystem, and projecting the future trajectories of fish stocks is particularly challenging, but we are better placed than ever before to start to anticipate and respond to likely scenarios of future change.

ACKNOWLEDGEMENTS

We thank the crew and officers of RV *Polarstern* for professional support in fisheries during several expeditions, and Tomas Lundälv and Julian Gutt for providing the underwater photographs. We are deeply grateful to Carlos Bellisio, Lena Rath, Luis Vila, Nils Koschnick, Oscar González, Timo Hirse and Tina Sandersfeld for their help in field activities, in the lab and in experimental work. We appreciate the reviewers' efforts and comments that helped to improve the manuscript. The work of K. M. and A. S. was funded by the German Research Foundation (DFG, SSP 1158; projects MI 1391/1-1 and PO 273/13-1).

APPENDIX

Table A1 Species names, family and code no. (see Fig. 5) of the species used for the analysis and comparison of the trophic vulnerability to general changes in food web structure and dynamics

No.	Species name	Family	ΣP	ΣC	VI	TG
1	*Aethotaxis mitopteryx*	Nototheniidae	53	14	0.20	BP
2	*Akarotaxis nudiceps*	Bathydraconidae	79	13	0.09	B
3	*Artedidraco loennbergi*	Artedidraconidae	108	14	0.06	B
4	*Artedidraco orianae*	Artedidraconidae	27	14	0.21	BP
5	*Artedidraco shackletoni*	Artedidraconidae	110	14	0.07	B
6	*Artedidraco skottsbergi*	Artedidraconidae	86	13	0.09	BP
7	*Bathydraco marri*	Bathydraconidae	47	13	0.17	BP
8	*Chaenodraco wilsoni*	Channichthyidae	16	15	0.28	PF
9	*Chionobathyscus dewitti*	Channichthyidae	5	14	0.77	PF
10	*Chionodraco hamatus*	Channichthyidae	10	15	0.67	P
11	*Chionodraco myersi*	Channichthyidae	5	15	0.77	PF
12	*Cryodraco antarcticus*	Channichthyidae	5	15	0.77	PF
13	*Cygnodraco mawsoni*	Bathydraconidae	55	14	0.14	BP
14	*Dacodraco hunteri*	Channichthyidae	37	15	0.10	F
15	*Dissostichus mawsoni*	Nototheniidae	52	21	0.28	BP
16	*Dolloidraco longedorsalis*	Artedidraconidae	142	14	0.04	B
17	*Gerlachea australis*	Bathydraconidae	14	14	0.46	P
18	*Gymnodraco acuticeps*	Bathydraconidae	33	14	0.15	P
19	*Histiodraco velifer*	Artedidraconidae	85	13	0.07	BF
20	*Neopagetopsis ionah*	Channichthyidae	5	14	0.77	PF
21	*Pagetopsis macropterus*	Channichthyidae	43	15	0.10	F
22	*Pagetopsis maculatus*	Channichthyidae	5	15	0.77	PF
23	*Pagothenia borchgrevinki*	Nototheniidae	17	12	0.27	BP
24	*Pleuragramma antarcticum*	Nototheniidae	12	47	0.96	P
25	*Pogonophryne marmorata*	Artedidraconidae	45	14	0.13	BP

Table A1 Species names, family and code no. (see Fig. 5) of the species used for the analysis and comparison of the trophic vulnerability to general changes in food web structure and dynamics—cont'd

No.	Species name	Family	ΣP	ΣC	VI	TG
26	*Pogonophryne permitini*	Artedidraconidae	79	14	0.10	B
27	*Pogonophryne scotti*	Artedidraconidae	78	14	0.11	B
28	*Prionodraco evansii*	Bathydraconidae	88	14	0.08	BP
29	*Racovitzia glacialis*	Bathydraconidae	89	14	0.08	BP
30	*Trematomus bernacchii*	Nototheniidae	93	14	0.02	B
31	*Trematomus eulepidotus*	Nototheniidae	45	14	0.12	BP
32	*Trematomus hansoni*	Nototheniidae	81	14	0.06	BF
33	*Trematomus lepidorhinus*	Nototheniidae	71	14	0.10	BP
34	*Trematomus loennbergii*	Nototheniidae	105	14	0.05	BF
35	*Trematomus nicolai*	Nototheniidae	88	14	0.09	B
36	*Trematomus pennellii*	Nototheniidae	164	14	0.03	BF
37	*Trematomus scotti*	Nototheniidae	121	14	0.06	B

All species are members of the fish community on the eastern Weddell Sea shelf. Species are listed in alphabetical order; for authorities, please consult Gon and Heemstra (1990). For each notothenioid species, the number of prey ($\sum P$) and consumer species ($\sum C$), the relative trophic vulnerability (VI) and trophic group (TG) are given. The index of relative vulnerability VI was calculated from the weighted number of prey species ($\sum$WP) and weighted number of consumer species ($\sum$WC) (see Eq. 2). Data on trophic links are part of the database published in Jacob et al. (2011). Trophic groups were assigned according to main food components as shown in Fig. 5, with B, benthos; P, plankton; F, fish.

REFERENCES

Ainley, D.G., Blight, L.K., 2008. Ecological repercussions of historical fish extraction from the Southern Ocean. Fish Fish. 9, 1–26.

Ainley, D.G., Ballard, G., Barton, K.J., Karl, B.J., Rau, G., Ribic, C.A., Wilson, P.R., 2003. Spatial and temporal variation of diet within a presumed metapopulation of Adélie penguins. Condor 105, 95–106.

Ainley, D., Ballard, G., Ackley, S., Blight, L.K., Eastman, J.T., Emslie, S.D., Lescroël, A., Olmastroni, S., Townsend, S.E., Tynan, C.T., Wilson, P., Woehler, E., 2007. Paradigm lost, or is top-down forcing no longer significant in the Antarctic marine ecosystem? Antarct. Sci. 19, 283–290.

Ainley, D., Ballard, G., Blight, L.K., Ackley, S., Emslie, S.D., Lescroël, A., Olmastroni, S., Townsend, S.E., Tynan, C.T., Wilson, P., Woehler, E., 2010. Impacts of cetaceans on the structure of Southern Ocean food webs. Mar. Mamm. Sci. 26, 482–498.

Alheit, J., 2009. Consequences of regime shifts for marine food webs. Int. J. Earth Sci. 98, 261–268.

Alheit, J., Alegre, B., 1986. Fecundity of Peruvian anchovy, *Engraulis ringens*. ICES CM 1986/H, 60.

Alheit, J., Niquen, M., 2004. Regime shifts in the Humbold Current ecosystem. Prog. Oceanogr. 60, 201–222.
Alheit, J., Möllmann, C., Dutz, J., Kornilovs, G., Loewe, P., Mohrholz, V., Wasmund, N., 2005. Synchronous ecological regime shifts in the central Baltic and the North Sea in the late 1980s. ICES J. Mar. Sci. 62, 1205–1215.
Andriashev, A.P., 1987. A general review of the Antarctic bottom fish fauna. In: Kullander, S.O., Fernholm, B. (Eds.), Proc. V Congr. Europ. Ichthyol. Stockholm 1985, 357–372.
Arntz, W.E., 1986. The two faces of El Niño 1982-83. Meeresforschung 31, 1–46.
Arntz, W.E., Fahrbach, E., 1991. El Niño—Klimaexperiment der Natur. Physikalische Ursachen und biologische Folgen, Birkhäuser Verlag, Basel.
Arntz, W.E., Brey, T., Gallardo, V., 1994. Antarctic zoobenthos. Oceanogr. Mar. Biol. 32, 241–304.
Arntz, W.E., Lovrich, G.A., Thatje, S. (Eds.), 2005. The Magellan-Antarctic connection: links and frontiers at high southern latitudes. Sci. Mar. 69, 1–373.
Aronson, R.B., Thatje, S., Clarke, A., Peck, L.S., Blake, D.B., Wilga, C.D., Seibel, B.A., 2007. Climate change and invasibility of the Antarctic benthos. Annu. Rev. Ecol. Evol. Syst. 38, 129–154.
Arrigo, K.R., Worthen, D.L., Lizotte, M.P., Dixon, P., Dieckmann, G., 1997. Primary production in Antarctic sea ice. Science 276, 394–397.
Arrigo, K.R., Weiss, A.M., Smith Jr., W.O., 1998. Physical forcing of phytoplankton dynamics in the southwestern Ross Sea. J. Geophys. Res. 103, 1007–1021.
Atkinson, A., Ward, P., Williams, R., Poulet, S.A., 1992. Diel vertical migration and feeding of copepods at an oceanic site near South Georgia. Mar. Biol. 113, 583–593.
Atkinson, A., Siegel, V., Pakhomov, E., Rothery, P., 2004. Long-term decline in krill stock and increase in salps within the Southern Ocean. Nature 432, 100–103.
Attril, M.J., Wright, J., Edwards, M., 2007. Climate-related increases in jellyfish frequency suggest a more gelatinous future for the North Sea. Limnol. Oceanogr. 52, 480–485.
Balushkin, A.V., 1994. Fossil notothenioid, and not gadiform, fish *Proeleginops grandeastmanorum* gen. nov. sp. nov. (Perciformes, Notothenioidei, Eleginopidae) from the late Eocene found in Seymour Island (Antarctica). J. Ichthyol. 34, 298–307.
Barnes, D.K.A., 2002. Invasions by marine life on plastic debris. Nature 416, 808–809.
Barnes, D.K.A., Conlan, K.E., 2007. Disturbance, colonization and development of Antarctic benthic communities. Philos. Trans. R. Soc. B 362, 11–38.
Barrera-Oro, E.R., 2002. The role of fish in the Antarctic marine food web: differences between inshore and offshore waters in the southern Scotia Arc and west Antarctic Peninsula. Antarct. Sci. 14, 293–309.
Barrera-Oro, E.R., 2003. Analysis of dietary overlap in Antarctic fish (Notothenioidei) from the South Shetland Islands: no evidence of food competition. Polar Biol. 26, 631–637.
Barrera-Oro, E.R., Casaux, R.J., 1990. Feeding selectivity in *Notothenia neglecta*, Nybelin, from Potter Cove, South Shetland Islands, Antarctica. Antarct. Sci. 2, 207–213.
Barrera-Oro, E.R., Casaux, R.J., 1998. Ecology of demersal fish species from Potter Cove. In: Wiencke, C., Ferreyra, G., Arntz, W.E., Rinaldi, C. (Eds.), The Potter Cove Coastal Ecosystem, Antarctica, Ber. Polarforsch.299, 156–167.
Barrera-Oro, E.R., Lagger, C., 2010. Egg-guarding behaviour in the Antarctic bathydraconid dragonfish *Parachaenichthys charcoti*. Polar Biol. 33, 1585–1587.
Barrera-Oro, E.R., Marschoff, E.R., 2007. Information on the status of fjord *Notothenia rossii, Gobionotothen gibberifrons* and *Notothenia coriiceps* in the lower South Shetland Islands, derived from the 2000-2006 monitoring program at Potter Cove. CCAMLR Sci. 14, 83–87.
Barrera-Oro, E.R., Marschoff, E.R., Casaux, R.J., 2000. Trends in relative abundance of fjord *Notothenia rossii, Gobionotothen gibberifrons* and *Notothenia coriiceps* at Potter Cove, South Shetland Islands, after commercial fishing in the area. CCAMLR Sci. 7, 43–52.

Bathmann, U.V., Fischer, G., Müller, P.J., Gerdes, D., 1991. Short-term variations in particulate matter sedimentation off Kapp Norvegia, Weddell Sea, Antarctica: relation to water mass advection, ice cover, plankton biomass and feeding activity. Polar Biol. 11, 185–195.
Baumann, H., Talmage, S.C., Gobler, C.J., 2012. Reduced early life growth and survival in a fish in direct response to increased carbon dioxide. Nat. Clim. Chang. 2, 38–41.
Beaugrand, G., Brander, K.M., Lindley, J.A., Souissi, S., Reid, P.C., 2003. Plankton effect on cod recruitment in the North Sea. Nature 426, 661–664.
Beers, J.M., Sidell, B.D., 2011. Thermal tolerance of Antarctic notothenioid fishes correlates with level of circulating hemoglobin. Physiol. Biochem. Zool. 84, 353–362.
Benson, A.J., Trites, A.W., 2002. Ecological effects of regime shifts in the Bering Sea and eastern North Pacific Ocean. Fish Fish. 3, 95–113.
Bilyk, K.T., DeVries, A.L., 2011. Heat tolerance and its plasticity in Antarctic fishes. Comp. Biochem. Phys. A 158, 382–390.
Bodungen, B.v, Nöthig, E.M., Sui, Q., 1988. New production of phytoplankton and sedimentation during summer 1985 in the southeastern Weddell Sea. Comp. Biochem. Physiol. B 90, 475–487.
Boyd, I.L., 2002. Antarctic marine mammals. In: Perrin, W., Würstig, B., Thewissen, J.G.M. (Eds.), Encyclopedia of Marine Mammals. Academic Press, San Diego, pp. 30–36.
Boysen-Ennen, E., Piatkowski, U., 1988. Meso- and macrozooplankton communities in the Weddell Sea, Antarctica. Polar Biol. 9, 17–35.
Branch, T.A., 2006. Humpback abundance south of 60°S from three completed sets of IDCR/SOWER circumpolar surveys. IWC Paper SC/A06/HW6, 1–14.
Branch, T.A., 2007. Abundance of Antarctic blue whales south of 60°S from three complete circumpolar sets of surveys. J. Cetacean Res. Manag. 9, 253–262.
Brenner, M., Buck, B.H., Cordes, S., Dietrich, L., Jacob, U., Mintenbeck, K., Schröder, A., Brey, T., Knust, R., Arntz, W., 2001. The role of iceberg scours in niche separation within the Antarctic fish genus *Trematomus*. Polar Biol. 24, 502–507.
Brey, T., Gerdes, D., 1997. Is Antarctic benthic biomass really higher than elsewhere? Antarct. Sci. 9, 266–267.
Brodeur, R.D., Mills, C.E., Overland, J.E., Walters, G.E., Schumacher, J.D., 1999. Evidence for a substantial increase in gelatinous zooplankton in the Bering Sea, with possible links to climate change. Fish. Oceanogr. 8, 296–306.
Brodte, E., Knust, R., Pörtner, H.O., 2006. Temperature-dependent energy allocation to growth in Antarctic and boreal eelpout (Zoarcidae). Polar Biol. 30, 95–107.
Brose, U., Jonsson, T., Berlow, E.L., Warren, P., Banasek-Richter, C., Bersier, L.F., Blanchard, J.L., Brey, T., Carpenter, S.R., Cattin Blandenier, M.F., Cushing, L., Dawah, H.A., Dell, T., Edwards, F., Harper-Smith, S., Jacob, U., Ledger, M.E., Martinez, N.D., Memmot, J., Mintenbeck, K., Pinnegar, J.K., Rall, B.C., Rayner, T.S., Reuman, D.C., Ruess, L., Ulrich, W., Williams, R.J., Woodward, G., Cohen, J.E., 2006. Consumer-resource body-size relationships in natural food webs. Ecology 87, 2411–2417.
Burns, J.M., Kooyman, G.L., 2001. Habitat use by Weddell seals and Emperor penguins foraging in the Ross Sea, Antarctica. Am. Zool. 41, 99–112.
Caldeira, K., Wickett, M.E., 2003. Anthropogenic carbon and ocean pH. Nature 425, 365.
Caldeira, K., Wickett, M.E., 2005. Ocean model predictions of chemistry changes from carbon dioxide emissions to the atmosphere and ocean. J. Geophys. Res. Oceans 110, 1–12.
Campbell, H.A., Fraser, K.P.P., Bishop, C.M., Peck, L.S., Egginton, S., 2008. Hybernation in an Antarctic fish: on ice for winter. PLoS One 3, e1743.
Carpenter, C.M., Hofmann, G.E., 2000. Expression of 70 kda heat shock proteins in Antarctic and New Zealand notothenioid fish. Comp. Biochem. Phys. A 125, 229–238.

Casaux, R.J., Barrera-Oro, E.R., 2006. Shags in Antarctica: their feeding behavior and ecological role in the marine food web. Antarct. Sci. 18, 3–14.
Casaux, R.J., Baroni, A., Ramón, A., 2006. The diet of the Weddell Seal *Leptonychotes weddellii* at the Danco Coast, Antarctic Peninsula. Polar. Biol. 29, 257–262.
Casaux, R.J., Mazzotta, A.S., Barrera-Oro, E.R., 1990. Seasonal aspects of the biology and diet of nearshore nototheniid fish at Potter Cove, South Shetland Islands, Antarctica. Polar Biol. 11, 63–72.
Castle, M.D., Blanchard, J.L., Jennings, S., 2011. Predicted effects of behavioural movement and passive transport on individual growth and community size structure in marine ecosystems. Adv. Ecol. Res. 45, 41–66.
Ciattaglia, L., Rafanelli, C., Araujo, J., Rodriguez, H., 2008. Long-term measurements of the atmospheric carbon dioxide concentration measured at Jubany Station indicate a relationship with "El Niño" In: Wiencke, C., Ferreyra, G., Abele, D., Marenssi, S. (Eds.), The Antarctic Ecosystem of Potter Cove, King-George Island (Isla 25 de Mayo), Ber. Polarforsch. Meeresforsch.571, 390–396.
Clarke, A., 1983. Life in cold water: the physiological ecology of polar marine ectotherms. Oceanogr. Mar. Biol. Ann. Rev. 21, 341–453.
Clarke, A., 1985. Energy flow in the Southern Ocean food web. In: Siegfried, W.R., Condy, P.R., Laws, R.M. (Eds.), Antarctic Nutrient Cycles and Food Webs. Springer Verlag, Berlin, pp. 573–580.
Clarke, A., 1991. What is cold adaptation and how should we measure it? Am. Zool. 31, 81–92.
Clarke, A., Johnston, I.A., 1996. Evolution and adaptive radiation of Antarctic fishes. Trends Ecol. Evol. 11, 212–218.
Clarke, A., Holmes, L.J., Gore, D.J., 1992. Proximate and elemental composition of gelatinous zooplankton from the Southern Ocean. J. Exp. Mar. Biol. Ecol. 155, 55–68.
Conlan, K.E., Kvitek, R.G., 2005. Recolonization of soft-sediment ice scours on an exposed Arctic coast. Mar. Ecol. Prog. Ser. 286, 21–42.
Conlan, K.E., Lenihan, H.S., Kvitek, R.G., Oliver, J.S., 1998. Ice scour disturbance to benthic communities in the Canadian high Arctic. Mar. Ecol. Prog. Ser. 166, 1–16.
Coria, N., Libertelli, M., Casaux, R., Darrieu, C., 2000. Inter-annual variation in the autumn diet of the gentoo penguin at Laurie Island, Antartica. Waterbirds 23, 511–517.
Crockett, E.L., Sidell, B.D., 1990. Some pathways of energy-metabolism are cold adapted in Antarctic fishes. Physiol. Zool. 63, 472–488.
Croxall, J.P., Prince, P.A., 1982. Caloric content of squid (Mollusca: Cephalopoda). Br. Antarct. Surv. Bull. 55, 27–31.
Croxall, J.P., Everson, I., Kooyman, G.L., Ricketts, C., Davis, R.W., 1985. Fur seal diving behaviour in relation to vertical distribution of krill. J. Anim. Ecol. 54, 1–8.
Cubillos, L.A., Claramunt, G., 2009. Length-structured analysis of the reproductive season of anchovy and common sardine off central southern Chile. Mar. Biol. 156, 1673–1680.
Cubillos, L.A., Bucarey, D.A., Canales, M., 2002. Monthly abundance estimation for common sardine *Strangomera bentincki* and anchovy *Engraulis ringens* in the central-southern area off Chile (34-40°S). Fish. Res. 57, 117–130.
Curran, M.A.J., van Ommen, T.D., Morgan, V.I., Phillips, K.L., Palmer, A.S., 2003. Ice core evidence for Antarctic sea ice decline since the 1950s. Science 302, 1203–1206.
Curry, R., Mauritzen, C., 2005. Dilution of the northern North Atlantic Ocean in recent decades. Science 308, 1772–1774.
Cury, P., Bakun, A., Crawford, R.J.M., Jarre, A., Quiñones, R.A., Shannon, L.J., Verheye, H.M., 2000. Small pelagics in upwelling systems: patterns of interaction and structural changes in "wasp-waist" ecosystems. ICES J. Mar. Sci. 57, 603–618.

Dahlhoff, E., Somero, G.N., 1993. Effects of temperature on mitochondria from abalone (genus *Haliotis*): adaptive plasticity and its limits. J. Exp. Biol. 185, 151–168.
Daneri, G.A., Carlini, A.R., 2002. Fish prey of southern elephant seals, *Mirounga leonina*, at King George Island. Polar Biol. 25, 739–743.
Daniels, R.A., 1982. Feeding ecology of some fishes of the Antarctic Peninsula. Fish. B.-NOAA 80, 575–589.
Davis, R.W., Fuiman, L.A., Williams, T.M., Collier, S.O., Hagey, W.P., Kanatous, S.B., Kohin, S., Horning, M., 1999. Hunting behaviour of a marine mammal beneath the Antarctic fast ice. Science 283, 993–996.
Dayton, P.K., 1990. Polar benthos. In: Smith, W.O. (Ed.), Polar Oceanography, Part B: Chemistry, Biology, and Geology. Academic Press, Boston, pp. 631–685.
Dayton, P.K., Robilliard, G.A., Paine, R.T., Dayton, L.B., 1974. Biological accommodation in the benthic community at McMurdo Sound, Antarctica. Ecol. Monogr. 44, 105–128.
Dayton, P.K., Mordida, B.J., Bacon, F., 1994. Polar marine communities. Am. Zool. 34, 90–99.
Deacon, G., 1984. The Antarctic Circumpolar Ocean. Cambridge University Press, Cambridge.
Detmer, A.E., Bathmann, U.V., 1997. Distribution patterns of autotrophic pico- and nanoplankton and their relative contribution to algal biomass during spring in the Atlantic sector of the Southern Ocean. Deep Sea Res. II 44, 299–320.
Detrich III, H.W., Jones, C.D., Kim, S., North, A.W., Thurber, A., Vacchi, M., 2005. Nesting behaviour of the icefish *Chaenocephalus aceratus* at Bouvetøya, Southern Ocean (CCAMLR Subarea 48.6). Polar Biol. 28, 828–832.
DeVries, A.L., 1971. Glycoproteins as biological antifreeze agents in Antarctic fishes. Science 172, 1152–1155.
DeVries, A.L., Eastman, J.T., 1978. Lipid sacs as a buoyancy adaptation in an Antarctic fish. Nature 271, 352–353.
DeWitt, H.H., 1970. The character of the midwater fish fauna of the Ross Sea, Antarctica. In: Holdgate, M.W. (Ed.), Antarctic Ecology, vol. 1. Academic Press, London, pp. 305–314.
Dierssen, H.M., Smith, R.C., Vernet, M., 2002. Glacial meltwater dynamics in coastal waters west of the Antarctic Peninsula. Proc. Natl. Acad. Sci. U.S.A. 99, 1790–1795.
DiTullio, G.R., Grebmeier, J.M., Arrigo, K.R., Lizotte, M.P., Robinson, D.H., Leventer, A., Barry, J.P., VanWoert, M.L., Dunbar, R.B., 2000. Rapid and early export of *Phaeocystis antarctica* blooms in the Ross Sea, Antarctica. Nature 404, 595–598.
Domack, E., Leventer, A., Burnett, A., Bindschadler, R., Convey, P., Kirby, M. (Eds.), 2003. Antarctic Peninsula climate variability: historical and paleoenvironmental perspectives. Antarct. Res. Ser. 79, 260.
Domack, E., Duran, D., Leventer, A., Ishman, S., Doane, S., McCallum, S., Amblas, D., Ring, J., Gilbert, R., Prentice, M., 2005. Stability of the Larsen B ice shelf on the Antarctic Peninsula during the Holocene epoch. Nature 436, 681–685.
Donelsen, J.M., Munday, P.L., McCormick, M.I., Pitcher, C.R., 2012. Rapid transgenerational acclimation of a tropical reef fish to climate change. Nat. Clim. Chang. 2, 30–32.
Donnelly, J., Torres, J.J., Hopkins, T.L., Lancraft, T.M., 1990. Proximate composition of Antarctic mesopelagic fishes. Mar. Biol. 106, 13–23.
Donnelly, J., Torres, J.J., Hopkins, T.L., Lancraft, T.M., 1994. Chemical composition of Antarctic zooplankton during austral fall and winter. Polar Biol. 14, 171–183.
Donnelly, J., Torres, J.J., Sutton, T.T., Simoniello, C., 2004. Fishes of the eastern Ross Sea, Antarctica. Polar Biol. 27, 637–650.
Doyle, T.K., De Haas, H., Cotton, D., Dorschel, B., Cummins, V., Houghton, J.D.R., Davenport, J., Hays, G.C., 2008. Widespread occurrence of the jellyfish *Pelagia noctiluca* in Irish coastal and shelf waters. J. Plankton Res. 30, 963–968.

Drinkwater, K.F., Beaugrand, G., Kaeriyma, M., Kim, S., Ottersen, G., Perry, R.I., Pörtner, H.O., Polovina, J.J., Takasuka, A., 2010. On the processes linking climate to ecosystem changes. J. Mar. Syst. 79, 374–388.
Dulvy, N.K., Rogers, S.I., Jennings, S., Stelzenmüller, V., Dye, S.R., Skjoldal, H.R., 2008. Climate change and deepening of the North Sea fish assemblage: a biotic indicator of warming seas. J. Appl. Ecol. 45, 1029–1039.
Eakin, R.R., Balushkin, A.V., 1998. A new species of toadlike Plunderfish *Pogonophryne orangiensis* sp. nova (Artedidraconidae, Notothenioidei) from the Weddell Sea, Antarctica. J. Ichthyol. 38, 800–803.
Eakin, R.R., Balushkin, A.V., 2000. A new species of *Pogonophryne* (Pisces: Perciformes: Artedidraconidae) from East Antarctica. Proc. Biol. Soc. Wash. 113, 264–268.
Eakin, R.R., Eastman, J.T., 1998. New species of *Pogonophryne* (Pisces, Artedidraconidae) from the Ross Sea, Antarctica. Copeia 4, 1005–1009.
Eakin, R.R., Eastman, J.T., Matallanas, J., 2008. New species of Pogonophryne (Pisces, Artedidraconidae) from the Bellinghausen Sea, Antarctica. Polar Biol. 31, 1175–1179.
Eastman, J.T., 1985a. The evolution of neutrally buoyant notothenioid fishes: their specialization and potential interactions in the Antarctic marine food web. In: Siegfried, W.R., Condy, P.R., Laws, R.M. (Eds.), Antarctic Nutrient Cycles and Food Webs. Springer Verlag, Berlin, pp. 430–436.
Eastman, J.T., 1985b. *Pleuragramma antarcticum* (Pisces, Nototheniidae) as food for other fishes in McMurdo Sound, Antarctica. Polar Biol. 4, 155–160.
Eastman, J.T., 1991. The fossil and modern fish faunas of Antarctica: evolution and diversity. In: Di Prisco, G., Maresca, B., Tota, B. (Eds.), Biology of Antarctic Fish. Springer Verlag, Berlin, pp. 116–130.
Eastman, J.T., 1993. Antarctic Fish Biology—Evolution in a Unique Environment. Academic Press, San Diego.
Eastman, J.T., 1999. Aspects of the biology of the icefish *Dacodraco hunteri* (Notothenioidei, Channichthyidae) in the Ross Sea, Antarctica. Polar Biol. 21, 194–196.
Eastman, J.T., 2005. The nature of the diversity of Antarctic fishes. Polar Biol. 28, 93–107.
Eastman, J.T., DeVries, A.L., 1981. Buoyancy adaptations in a swim-bladderless Antarctic fish. J. Morphol. 167, 91–102.
Eastman, J.T., DeVries, A.L., 1982. Buoyancy studies of notothenioid fishes in McMurdo Sound, Antarctica. Copeia 2, 385–393.
Eastman, J.T., DeVries, A.L., 1985. Adaptations for cryopelagic life in the Antarctic notothenioid fish *Pagothenia borchgrevinki*. Polar Biol. 4, 45–52.
Eastman, J.T., Eakin, R.R., 2000. An updated species list for notothenioid fish (Perciformes; Notothenioidei), with comments on Antarctic species. Arch. Fish. Mar. Res. 48, 11–20.
Eastman, J.T., Grande, L., 1989. Evolution of the Antarctic fish fauna with emphasis on the recent notothenioids. In: Crame, J.A. (Ed.), Origins and Evolution of the Antarctic Biota. Geological Society, London, pp. 241–252 Special Publication No. 47.
Eastman, J.T., Hubold, G., 1999. The fish fauna of the Ross Sea, Antarctica. Antarct. Sci. 11, 293–304.
Eastman, J.T., McCune, A.R., 2000. Fishes on the Antarctic continental shelf: evolution of a marine species flock? J. Fish Biol. 57, 84–102.
Eastman, J.T., Barrera-Oro, E.R., Moreira, E., 2011. Adaptive radiation at a low taxonomic level: divergence in buoyancy of the ecologically similar Antarctic fish *Notothenia coriiceps* and *N. rossii*. Mar. Ecol. Prog. Ser. 438, 195–206.
Eder, E.B., Lewis, M.N., 2005. Proximate composition and energetic value of demersal and pelagic prey species from the SW Atlantic Ocean. Mar. Ecol. Prog. Ser. 291, 43–52.
Efremenko, V.N., 1983. Atlas of fish larvae of the Southern Ocean. Cybium 7, 1–74.

Egginton, S., 1996. Blood rheology of Antarctic fishes: viscosity adaptations at very low temperatures. J. Fish Biol. 48, 513–521.
Egginton, S., 1997a. A comparison of the response to induced exercise in red- and white-blooded Antarctic fishes. J. Comp. Physiol. B 167, 129–134.
Egginton, S., 1997b. Control of tissue blood flow at very low temperatures. J. Therm. Biol. 22, 403–407.
Ehrmann, W.U., Mackensen, A., 1992. Sedimentological evidence for the formation of an East Antarctic ice sheet in the Eocene/Oligocene time. Palaeogeogr. Palaeoclimatol. Palaeoecol. 93, 85–112.
Eicken, H., 1992. The role of sea ice in structuring Antarctic ecosystems. Polar Biol. 12, 3–13.
Eicken, H., 1995. Wie polar wird das Polarmeer durch das Meereis? In: Hempel, I., Hempel, G. (Eds.), Biologie der Polarmeere. Gustav Fischer Verlag, Jena, pp. 58–76.
Ekau, W., 1988. Ökomorphologie nototheniider Fische aus dem Weddellmeer, Antarktis. Ber. Polarforsch. 51, 1–140.
Ekau, W., 1989. Egg development of *Trematomus eulepidotus* Regan, 1914 (Nototheniidae, Pisces) from the Weddell Sea, Antarctica. Cybium 13, 213–219.
Estrada, M., Delgado, M., 1990. Summer phytoplankton distributions in the Weddell Sea. Polar Biol. 10, 441–449.
Faleyeva, T.I., Gerasimchuk, V.V., 1990. Features of reproduction in the Antarctic sidestripe, *Pleuragramma antarcticum* (Nototheniidae). J. Ichthyol. 30, 67–79.
Falk-Petersen, J., Renaud, P., Anisimova, N., 2011. Establishment and ecosystem effects of the alien invasive red king crab (*Paralithodes camtschaticus*) in the Barents Sea—a review. ICES J. Mar. Sci. 68, 479–488.
Fanta, E., 1999. Laboratory test on feeding interactions of some Antarctic fish from Admiralty Bay (King George Island, South Shetlands). Pol. Polar Res. 20, 17–28.
Fanta, E., Rios, F.S., Donatti, L., Cardoso, W.E., 2003. Spatial and temporal variation in krill consumption by the Antarctic fish *Notothenia coriiceps*, in Admirality Bay, King George Island. Antarct. Sci. 15, 458–462.
Feely, R.A., Sabine, C.L., Lee, K., Berelson, W., Kleypas, J., Fabry, V.J., Millero, F.J., 2004. Impact of anthropogenic CO_2 on the $CaCO_3$ system in the oceans. Science 305, 362–366.
Feld, C.K., Birk, S., Bradley, D.C., Hering, D., Kail, J., Marzin, A., Melcher, A., Nemitz, D., Pedersen, M.L., Pletterbauer, F., Pont, D., Verdonschot, P.F.M., Friberg, N., 2011. From natural to degraded rivers and back again: a test of restoration ecology theory and practice. Adv. Ecol. Res. 44, 119–210.
Ferrari, M.C.O., McCormick, M.I., Munday, P.L., Meekan, M.G., Dixson, D.L., Lonnstedt, Ö., Chivers, D.P., 2011. Putting prey and predator into the CO_2 equation—qualitative and quantitative effects of ocean acidification on predator-prey interactions. Ecol. Lett. 14, 1143–1148.
Fischer, G., 1989. Stabile Kohlenstoff-Isotope in partikulärer organischer Substanz aus dem Südpolarmeer (Atlantischer Sektor). Berichte aus dem Fachbereich Geowissenschaften der Universität Bremen No. 5, 161pp. Dissertation, Universität Bremen.
Fivelstad, S., Olsen, A.B., Asgard, T., Baeverfjord, G., Rasmussen, T., Vindheim, T., Stefansson, S., 2003. Long-term sublethal effects of carbon dioxide on Atlantic salmon smolts (*Salmo salar* L.): ion regulation, haematology, element composition, nephrocalcinosis and growth parameters. Aquaculture 215, 301–319.
Fletcher, G.L., Hew, C.L., Davies, P.L., 2001. Antifreeze proteins of teleost fishes. Annu. Rev. Physiol. 63, 359–390.
Foster, B.A., Cargill, J.M., Montgomery, J.C., 1987. Planktivory in *Pagothenia borchgrevinki* (Pisces: Nototheniidae) in McMurdo Sound, Antarctica. Polar Biol. 8, 49–54.

Franke, A., Clemmesen, C., 2011. Effect of ocean acidification on early life stages of Atlantic herring (*Clupea harengus* l.). Biogeosci. Discuss. 8, 7097–7126.

Franklin, C.E., Davison, W., Seebacher, F., 2007. Antarctic fish can compensate for rising temperatures: thermal acclimation of cardiac performance in *Pagothenia borchgrevinki*. J. Exp. Biol. 210, 3068–3074.

Froese, R., Palomares, M.L.D., Pauly, D., 2002. Estimation of Life-History Key Facts. Version of 17 July 2002, http://www.fishbase.org.

Frommel, A., Maneja, R., Lowe, D., Malzahn, A., Geffen, A., Folkvord, A., Piatkowski, U., Reusch, T., Clemmesen, C., 2012. Severe tissue damage in Atlantic cod larvae under increasing ocean acidification. Nat. Clim. Chang. 2, 43–46.

Fuiman, L., Davis, R., Williams, T., 2002. Behaviour of midwater fishes under the Antarctic ice: observations by a predator. Mar. Biol. 140, 815–822.

Gambi, M.C., Lorenti, M., Russo, G.F., Scipione, M.B., 1994. Benthic associations of the shallow hard bottoms off Terra Nova Bay, Ross Sea: zonation, biomass and population structure. Antarct. Sci. 6, 449–462.

García Raso, J.E., Manjón-Cabeza, M.E., Ramos, A., Olaso, I., 2005. New record of Lithodidae (Crustacea: Decapoda: Anomura) from the Antarctic (Bellingshausen Sea). Polar Biol. 28, 642–646.

Garcia, J.A., Ruzafa, P., 1998. Correlation between habitat structure and a rocky reef assemblage in the Southwest Mediterranean. Mar. Ecol. 19, 111–128.

Garwood, N.C., Janos, D.P., Brokaw, N., 1979. Earthquake-caused landslides: a major disturbance to tropical forests. Science 205, 997–999.

Gerasimchuk, V.V., 1988. On the fecundity of Antarctic sidestripe, *Pleuragramma antarcticum*. J. Ichthyol. 28, 98–100.

Gerdes, D., Isla, E., Knust, R., Mintenbeck, K., Rossi, S., 2008. Response of Antarctic benthic communities to disturbance: first results from the artificial Benthic Disturbance Experiment on the eastern Weddell Sea shelf, Antarctica. Polar Biol. 31, 1469–1480.

Gili, J.-M., Rossi, S., Pages, F., Orejas, C., Teixidó, N., López-González, P.J., Arntz, W.E., 2006. A new trophic link between the pelagic and benthic system on the Antarctic shelf. Mar. Ecol. Prog. Ser. 322, 43–49.

Gille, S.T., 2002. Warming of the Southern Ocean since the 1950s. Science 295, 1275–1277.

Gille, S.T., 2008. Decadal-scale temperature trends in the southern hemisphere ocean. J. Clim. 21, 4749–4765.

Gon, O., Heemstra, P.C., 1990. Fishes of the Southern Ocean. J.L.B Smith Institute of Ichthyology, Grahamstown 462pp.

Gonzalez-Cabrera, P.J., Dowd, F., Pedibhotla, V.K., Rosario, R., Stanley-Samuelson, D., Petzel, D., 1995. Enhanced hypo-osmoregulation induced by warm-acclimation in Antarctic fish is mediated by increased gill and kidney Na+/K (+)-ATPase activities. J. Exp. Biol. 198, 2279–2291.

Gordon, A.L., Goldberg, R.D., 1970. Circumpolar characteristics of Antarctic waters. In: Bushnell, V.C. (Ed.), Antarctic Map Folio Series, Folio 13. American Geographical Society, New York, pp. 1–5.

Gordon, A.L., Chen, C.T.A., Metcalf, W.G., 1984. Winter mixed layer entrainment of Weddell deep water. J. Geophys. Res. 89, 637–640.

Granata, A., Cubeta, A., Guglielmo, L., Sidoti, O., Greco, S., Vacchi, M., La Mesa, M., 2002. Ichthyoplankton abundance and distribution in the Ross Sea during 1987-1996. Polar Biol. 25, 187–202.

Granata, A., Zagami, G., Vacchi, M., Guglielmo, L., 2009. Summer and spring trophic niche of larval and juvenile *Pleuragramma antarcticum* in the Western Ross Sea, Antarctica. Polar Biol. 32, 369–382.

Gratwicke, B., Speight, M.R., 2005. The relationship between fish species richness, abundance and habitat complexity in a range of shallow tropical marine habitats. J. Fish Biol. 66, 650–667.

Gray, J.S., 2000. The measurement of marine species diversity, with an application to the benthic fauna of the Norwegian continental shelf. J. Exp. Mar. Biol. Ecol. 250, 23–49.

Guderley, H., Johnston, I., 1996. Plasticity of fish muscle mitochondria with thermal acclimation. J. Exp. Biol. 199, 1311–1317.

Guégan, J.F., Lek, S., Oberdorff, T., 1998. Energy availability and habitat heterogeneity predict global riverine fish diversity. Nature 391, 382–384.

Guglielmo, L., Granata, A., Greco, S., 1998. Distribution and abundance of postlarval and juvenile *Pleuragramma antarcticum* (Pisces, Nototheniidae) off Terra Nova Bay (Ross Sea, Antarctica). Polar Biol. 19, 37–51.

Gutt, J., 2000. Some "driving forces" structuring communities of the sublittoral Antarctic macrobenthos. Antarct. Sci. 12, 297–313.

Gutt, J., 2001. On the direct impact of ice on marine benthic communities, a review. Polar Biol. 24, 553–564.

Gutt, J., Piepenburg, D., 2003. Scale-dependent impact on diversity of Antarctic benthos caused by grounding of icebergs. Mar. Ecol. Prog. Ser. 253, 77–83.

Gutt, J., Starmans, A., 1998. Structure and biodiversity of megabenthos in the Weddell and Lazarev Seas (Antarctica): ecological role of physical parameters and biological interactions. Polar Biol. 20, 229–247.

Gutt, J., Starmans, A., 2001. Quantification of iceberg impact and benthic recolonisation patterns in the Weddell Sea (Antarctica). Polar Biol. 24, 615–619.

Gutt, J., Starmans, A., Diekmann, G., 1996. Impact of iceberg scouring on polar benthic habitats. Mar. Ecol. Prog. Ser. 137, 311–316.

Gutt, J., Sirenko, B.I., Smirnov, I.S., Arntz, W.E., 2004. How many macrozoobenthic species might inhabit the Antarctic shelf? Antarct. Sci. 16, 11–16.

Hagen, M., Kissling, W.D., Rasmussen, C., De Aguiar, M.A.M., Brown, L., Carstensen, D.W., Alves-Dos-Santos, I., Dupont, Y.L., Edwards, F.K., Genini, J., Guimarães Jr., P.R., Jenkins, G.B., Jordano, P., Kaiser-Bunbury, C.N., Ledger, M., Maia, K.P., Marquitti, F.M.D., Mclaughlin, Ó., Morellato, L.P.C., O'Gorman, E.J., Trøjelsgaard, K., Tylianakis, J.M., Vidal, M.M., Woodward, G., Olesen, J.M., 2012. Biodiversity, species interactions and ecological networks in a fragmented world. Adv. Ecol. Res. 46, 89–210.

Hazel, J.R., 1995. Thermal adaptation in biological membranes: is homeoviscous adaptation the explanation? Annu. Rev. Physiol. 57, 19–42.

Helaouët, P., Beaugrand, G., 2007. Statistical study of the ecological niche of *Calanus finmarchicus* and *C. helgolandicus* in the North Atlantic Ocean and adjacent seas. Mar. Ecol. Prog. Ser. 345, 147–165.

Hempel, G., 1985. Antarctic marine food webs. In: Siegfried, W.R., Condy, P.R., Laws, R.M. (Eds.), Antarctic Nutrient Cycles and Food Webs. Springer Verlag, Berlin, pp. 266–270.

Hochachka, P.W., Somero, G.N., 2002. Biochemical Adaptation. Oxford University Press, New York.

Hopkins, T.L., 1985. Food web of an Antarctic midwater ecosystem. Mar. Biol. 89, 197–212.

Hoshiai, T., Tanimura, A., Fukuchi, M., Watanabe, K., 1989. Feeding by the nototheniid fish, *Pagothenia borchgrevinki* on the ice-associated copepod, *Paralabidocera antarctica*. Polar Biol. 2, 61–64.

Houde, E.D., Zastrow, C.E., 1993. Ecosystem- and taxon-specific dynamic and energetics properties of fish larvae assemblages. Bull. Mar. Sci. 53, 290–335.

Hubold, G., 1984. Spatial distribution of *Pleuragramma antarcticum* (Pisces: Nototheniidae) near the Filchner- and Larsen ice shelves (Weddell Sea/Antarctica). Polar Biol. 3, 231–236.
Hubold, G., 1985. The early life-history of the high-Antarctic Silverfish, *Pleuragramma antarcticum*. In: Siegfried, W.R., Condy, P.R., Laws, R.M. (Eds.), Antarctic Nutrient Cycles and Food Webs. Springer Verlag, Berlin, pp. 446–451.
Hubold, G., 1990. Seasonal patterns of ichthyoplankton distribution and abundance in the southern Weddell Sea. In: Kerry, K.R., Hempel, G. (Eds.), Antarctic Ecosystems. Ecological Change and Conservation. Springer Verlag, Berlin, pp. 149–158.
Hubold, G., 1992. Zur Ökologie der Fische im Weddellmeer. Ber. Polarforsch. 103, 1–157.
Hubold, G., Ekau, W., 1987. Midwater fish fauna of the Weddell Sea, Antarctica. In: Kullander, S.O., Fernholm, B. (Eds.), Proc. V Congr. Europ. Ichthyol. Stockholm 1985, 391–396.
Hubold, G., Ekau, W., 1990. Feeding patterns of post-larval and juvenile notothenioids in the Southern Weddell Sea (Antarctica). Polar Biol. 10, 255–260.
Hubold, G., Hagen, W., 1997. Seasonality of feeding and lipid content of *Pleuragramma antarcticum* (Nototheniidae) in the southern Weddell Sea. In: Battaglia, B., Valencia, J., Walton, D.W.H. (Eds.), Antarctic Communities: Species, Structure and Survival. Cambridge University Press, Cambridge, pp. 277–283.
Hubold, G., Tomo, A.P., 1989. Age and growth of Antarctic silverfish *Pleuragramma antarcticum* Boulenger, 1902, from the southern Weddell Sea and Antarctic Peninsula. Polar Biol. 9, 205–212.
Hudson, H.A., Brauer, P.R., Scofield, M.A., Petzel, D.H., 2008. Effects of warm acclimation on serum osmolality, cortisol and hematocrit levels in the Antarctic fish, *Trematomus bernacchii*. Polar Biol. 31, 991–997.
Hughes, J.E., Deegan, L.A., Wyda, J.C., Weaver, M.J., Wright, A., 2002. The effects of eelgrass habitat loss on estuarine fish communities of southern New England. Estuaries 25, 235–249.
Huntley, M.E., Sykes, P.F., Marin, V., 1989. Biometry and trophodynamics of *Salpa thompsoni* foxton (Tunicata: Thaliacea) near the Antarctic Peninsula in austral summer, 1983-1984. Polar Biol. 10, 59–70.
Hureau, J.C., 1970. Biologie compareé de quelques poissons antarctiques (Nototheniidae). Bull. Inst. Oceanogr. Monaco 68, 1–244.
Hureau, J.C., 1994. The significance of fish in the marine Antarctic ecosystems. Polar Biol. 14, 307–313.
Huston, M.A., 1979. A general hypothesis of species diversity. Am. Nat. 113, 81–101.
Iken, K., 1996. Trophische Beziehungen zwischen Makroalgen und Herbivoren in der Potter Cove (King George-Insel, Antarktis). Ber. Polarforsch. 201, 1–214.
Iken, K., Barrera-Oro, E.R., Quartino, M.L., Casaux, R.J., Brey, T., 1997. Grazing in the Antarctic fish *Notothenia coriiceps*: evidence for selective feeding on macroalgae. Antarct. Sci. 9, 386–391.
IPCC, 2007. Fourth Assessment Report of the Intergovernmental Panel on Climate Change. Cambridge University Press, Cambridge.
Ishimatsu, A., Hayashi, M., Lee, K.S., Kikkawa, T., Kita, J., 2005. Physiological effects on fishes in a high-CO_2 world. J. Geophys. Res. 110, C09S09.
Ivany, L.C., Lohmann, K.C., Hasiuk, F., Blake, D.B., Glass, A., Aronson, R.B., Moody, R.N., 2008. Eocene climate record of a high southern latitude continental shelf: Seymour Island, Antarctica. Geol. Soc. Am. Bull. 120, 659–678.
Jacob, U., 2005. Trophic dynamics of Antarctic shelf ecosystems—food webs and energy flow budgets. PhD Thesis, University of Bremen, 141pp.
Jacob, U., Thierry, A., Brose, U., Arntz, W.E., Berg, S., Brey, T., Fetzer, I., Jonsson, T., Mintenbeck, K., Möllmann, C., Petchey, O., Riede, S., Dunne, J.A., 2011. The role of body size in complex food webs: a cold case. Adv. Ecol. Res. 45, 181–223.

Jacobs, S.S., Comiso, J.C., 1997. Climate variability in the Amundsen and Bellinghausen Seas. J. Clim. 10, 697–709.

Jacobs, S.S., Giulivi, C.F., Mele, P.A., 2002. Freshening of the Ross Sea during the late 20th century. Science 297, 386–389.

Janssen, J., Sideleva, V., Montgomery, J., 1991. Under-ice observation of fish behaviour at McMurdo Sound. Antarct. J. U.S. 26, 174–175.

Jarre-Teichmann, A., Brey, T., Bathmann, V.V., Dahm, C., Dieckmann, G.S., Gorny, M., Klages, M., Pagés, F., Plötz, J., Schnack-Schiel, S.B., Stiller, M., Arntz, W.E., 1995. Trophic flows in the benthic shelf community of the eastern Weddell Sea, Antarctica. In: Battaglia, B., Valencia, J., Walton, D. (Eds.), Antarctic Communities: Species, Structure and Survival. Cambridge University Press, Cambridge, pp. 118–134.

Johnson, K.H., 2000. Trophic-dynamic considerations in relating species diversity to ecosystem resilience. Biol. Rev. Camb. Philos. Soc. 75, 347–376.

Johnston, I., 2003. Muscle metabolism and growth in Antarctic fishes (suborder Notothenioidei): evolution in a cold environment. Comp. Biochem. Physiol. B 136, 701–713.

Johnston, I., Guderley, H., Franklin, C., Crockford, T., Kamunde, C., 1994. Are mitochondria subject to evolutionary temperature adaptation? J. Exp. Biol. 195, 293–306.

Kamler, E., Krasicka, B., Rakusa-Suszczewski, S., 2001. Comparison of lipid content and fatty acid composition in muscle and liver of two notothenioid fishes from Admiralty Bay (Antarctica): an eco-physiological perspective. Polar Biol. 24, 735–743.

Kashkina, A.A., 1986. Feeding of fishes on salps (Tunicata, Thaliacea). J. Ichthyol. 26, 57–64.

Kawaguchi, S., Kuruhara, H., King, R., Hale, L., Berli, T., Robinson, J.P., Ishida, A., Wakita, M., Virtue, P., Nicol, S., Ishimatsu, A., 2011. Will krill fare well under Southern Ocean acidification? Biol. Lett. 7, 288–291.

Kellermann, A., 1986a. Geographical distribution and abundance of postlarval and juvenile *Pleuragramma antarcticum* (Pisces, Notothenioidei) off the Antarctic Peninsula. Polar Biol. 6, 111–119.

Kellermann, A., 1986b. Zur Biologie der Jungstadien der Notothenioidei (Pisces) an der Antarktischen Halbinsel. Ber. Polarforsch. 31, 1–155.

Kellermann, A., 1987. Food and feeding ecology of postlarval and juvenile *Pleuragramma antarcticum* (Pisces, Notothenioidei) in the seasonal pack ice zone off the Antarctic Peninsula. Polar Biol. 7, 307–315.

Kellermann, A., 1989. Food and feeding of early stage *Chionodraco rastrospinosus* DeWitt & Hureau 1979 (Pisces; Notothenioidei) off the Antarctic Peninsula. Pesq. Antárt. Bras. 1, 25–30.

Kellermann, A., 1990. Identification key and catalogue of larval Antarctic fishes. Ber. Polarforsch. 67, 1–136.

Knox, G.A., 1970. Antarctic marine ecosystems. In: Holdgate, M.W. (Ed.), Antarctic Ecology, vol. 1. Academic Press, London, pp. 69–96.

Knust, R., Arntz, W.E., Boche, M., Brey, T., Gerdes, D., Gutt, J., Mintenbeck, K., Schröder, A., Starmans, A., Teixido, N., 2003. Iceberg scouring on the eastern Weddell Sea shelf (Antarctica). A benthic system shaped by physical disturbance. In: Huiskes, A.H.L., Gieskes, W.W.C., Rozema, J., Schorno, R.M.L., van der Vies, S.M., Wolff, W.J. (Eds.), Antarctic Biology in a Global Context. Backhuys Publishers, Leiden, pp. 96–101.

Kock, K.-H., 1992. Antarctic Fish and Fisheries. Cambridge University Press, Cambridge 359pp.

Kock, K.-H., 2005a. Antarctic icefishes (Channichthyidae): a unique family of fishes. A review, Part I. Polar Biol. 28, 862–895.

Kock, K.-H., 2005b. Antarctic icefishes (Channichthyidae): a unique family of fishes. A review, Part II. Polar Biol. 28, 897–909.

Kock, K.-H., 2007. Antarctic marine living resources—exploitation and its management in the Southern Ocean. Antarct. Sci. 19, 231–238.

Kock, K.-H., Everson, I., 1998. Age, growth and maximum size of Antarctic notothenioid fish—revisited. In: Di Prisco, G., Pisano, E., Clarke, A. (Eds.), Fishes of Antarctica—A Biological Overview. Springer Verlag Italia, Milano, pp. 29–40.

Kock, K.-H., Kellermann, A., 1991. Reproduction in Antarctic notothenioid fish. Antarct. Sci. 3, 125–150.

Kock, K.-H., Pshenichnov, L.K., DeVries, A.L., 2006. Evidence for egg brooding and parental care in icefish and other notothenioids in the Southern Ocean. Antarct. Sci. 18, 223–227.

Kock, K.-H., Reid, K., Croxall, J., Nicol, S., 2007. Fisheries in the Southern Ocean: an ecosystem approach. Philos. Trans. R. Soc. B 362, 2333–2349.

Kooyman, G.L., 1989. Diverse Divers: Physiology and Behaviour. Springer Verlag, Berlin.

Kooyman, G.L., Kooyman, T.G., 1995. Diving behaviour of Emperor penguins nurturing chicks at Coulman Island, Antarctica. Condor 97, 536–549.

Koubbi, P., Vallet, C., Razouls, S., Grioche, A., Hilde, D., Courcot, L., Janquin, M.A., Vacchi, M., Hureau, J.C., 2007. Condition and diet of larval *Pleuragramma antarcticum* (Nototheniidae) from Terre Adélie (Antarctica) during summer. Cybium 31, 67–76.

Kremer, P., Madin, L.P., 1992. Particle retention efficiency of salps. J. Plankton Res. 14, 1009–1015.

Kunzmann, A., 1991. Blood physiology and ecological consequences in Weddell Sea fishes (Antarctica). Ber. Polarforsch. Meeresforsch. 91, 1–79.

Kurihara, H., 2008. Effect of CO_2-driven ocean acidification on the early developmental stages of invertebrates. Mar. Ecol. Prog. Ser. 373, 275–284.

La Mesa, M., Eastman, J.T., 2012. Antarctic silverfish: life strategies of a key species in the high-Antarctic ecosystem. Fish Fish. 13, 241–266.

La Mesa, M., Vacchi, M., 2001. Age and growth of high Antarctic notothenioid fish—review. Antarct. Sci. 13, 227–235.

La Mesa, M., Eastman, J.T., Vacchi, M., 2004. The role of notothenioid fish in the food web of the Ross Sea shelf waters: a review. Polar Biol. 27, 321–338.

La Mesa, M., Catalano, B., Russo, A., Greco, S., Vacchi, M., Azzali, M., 2010. Influence of environmental conditions on spatial distribution and abundance of early life stages of Antarctic silverfish, *Pleuragramma antarcticum* (Nototheniidae), in the Ross Sea. Antarct. Sci. 22, 243–254.

La Mesa, M., Catalano, B., Greco, S., 2011. Larval feeding of *Chionodraco hamatus* (Pisces, Channichthyidae) in the Ross Sea and its relation to environmental conditions. Polar Biol. 34, 127–137.

Lannig, G., Storch, D., Pörtner, H.O., 2005. Aerobic mitochondrial capacities in Antarctic and temperate eelpout (Zoarcidae) subjected to warm versus cold acclimation. Polar Biol. 28, 575–584.

Larsen, B.K., Pörtner, H.O., Jensen, F.B., 1997. Extra- and intracellular acid-base balance and ionic regulation in cod (*Gadus morhua*) during combined and isolated exposures to hypercapnia and copper. Mar. Biol. 128, 337–346.

Laws, R.M., 1977. Seals and whales of the Southern Ocean. Philos. Trans. R. Soc. B 279, 81–96.

Laws, R.M., 1985. The ecology of the Southern Ocean. Am. Sci. 73, 26–40.

Lea, M.-A., Nichols, P.D., Wilson, G., 2002. Fatty acid composition of lipid-rich myctophids and mackerel icefish (*Champsocephalus gunnari*)—Southern Ocean food web implications. Polar Biol. 25, 843–854.

Lee, J.E., Chown, S.L., 2007. *Mytilus* on the move: transport of an invasive bivalve to the Antarctic. Mar. Ecol. Prog. Ser. 339, 307–310.

Lewis, P.N., Riddle, M.J., Smith, S.D.A., 2005. Assisted passage or passive drift: a comparison of alternative transport mechanisms for non-indigenous coastal species into the Southern Ocean. Antarct. Sci. 17, 183–191.
Lewis, P.N., Bergstrom, D.M., Whinam, J., 2006. Barging in: a temperate marine community travels to the Subantarctic. Biol. Invasions 8, 787–795.
Liebsch, N., Wilson, R.P., Bornemann, H., Adelung, D., Plötz, J., 2007. Mouthing off about fish capture: jaw movement in pinnipeds reveals the real secrets of ingestion. Deep Sea Res. II 54, 256–269.
Lizotte, M.P., 2001. The contributions of sea ice algae to Antarctic marine primary production. Am. Zool. 41, 57–73.
Loeb, V., Siegel, V., Holm-Hansen, O., Hewitt, R., Fraser, W., Trivelpiece, W., Trivelpiece, S., 1997. Effects of sea-ice extent and krill or salp dominance on the Antarctic food web. Nature 387, 897–900.
Lowe, C., Davison, W., 2005. Plasma osmolarity, glucose concentration and erythrocyte responses of two Antarctic nototheniid fishes to acute and chronic thermal change. J. Fish Biol. 67, 752–766.
Madin, L.P., 1974. Field observations on the feeding behaviour of salps (Tunicata: Thaliacea). Mar. Biol. 25, 143–147.
Malzahn, A.M., Aberle, N., Clemmesen, C., Boersma, M., 2007. Nutrient limitation of primary producers affects planktivorous fish condition. Limnol. Oceanogr. 52, 2062–2071.
Mark, F.C., Bock, C., Pörtner, H.O., 2002. Oxygen-limited thermal tolerance in Antarctic fish investigated by MRI and 31P-MRS. Am. J. Physiol. Regul. Integr. Comp. Physiol. 283, R1254–R1262.
Mark, F.C., Hirse, T., Pörtner, H.O., 2005. Thermal sensitivity of cellular energy budgets in some Antarctic fish hepatocytes. Polar Biol. 28, 805–814.
Mark, F., Lucassen, M., Pörtner, H., 2006. Thermal sensitivity of uncoupling protein expression in polar and temperate fish. Comp. Biochem. Phys. D 1, 365–374.
Mark, F.C., Lucassen, M., Strobel, A., Barrera-Oro, E., Koschnick, N., Zane, L., Patarnello, P., Pörtner, H.O., Papetti, C., 2012. Mitochondrial function in Antarctic notothenioids with nd6 translocation. PLoS One 7, e31860.
Marshall, H.P., 1988. The overwintering strategy of Antarctic krill under the pack-ice of the Weddell Sea. Polar Biol. 9, 129–135.
Marshall, G.J., Orr, A., van Lipzig, N.P.M., King, J.C., 2006. The impact of a changing Southern Hemisphere Annular Mode on Antarctic Peninsula summer temperatures. J. Clim. 19, 5388–5404.
Matschiner, M., Hanel, R., Salzburger, W., 2011. On the origin and trigger of the notothenioid adaptive radiation. PLoS One 6, e18911.
McFarlane, G.A., King, J.R., Beamish, R.J., 2000. Have there been recent changes in climate? Ask the fish. Prog. Oceanogr. 47, 147–169.
Meerhoff, M., Teixeira-de Mello, F., Kruk, C., Alonso, C., González-Bergonzoni, I., Pacheco, J.P., Lacerot, G., Arim, M., Beklioğlu, M., Brucet, S., Goyenola, G., Iglesias, C., Mazzeo, N., Kosten, S., Jeppesen, E., 2012. Environmental warming in shallow lakes: a review of effects on community structure as evidenced from space-for-time substitution approach. Adv. Ecol. Res. 46, 259–350.
Melian, C.J., Vilas, C., Baldo, F., Gonzalez-Ortegon, E., Drake, P., Williams, R.J., 2011. Eco-evolutionary dynamics of individual-based food webs. Adv. Ecol. Res. 45, 225–268.
Melnikov, I.A., 1998. Winter production of sea ice algae in the western Weddell Sea. J. Mar. Syst. 17, 195–205.
Melzner, F., Göbel, S., Langenbuch, M., Gutowska, M.A., Pörtner, H.O., Lucassen, M., 2009. Swimming performance in Atlantic cod (*Gadus morhua*) following long-term (4-12 months) acclimation to elevated seawater pCO_2. Aquat. Toxicol. 92, 30–37.

Memmot, J., Martinez, N.D., Cohen, J.E., 2000. Predators, parasitoids and pathogens: species richness, trophic generality and body sizes in a natural food web. J. Anim. Ecol. 69, 1–15.

Meredith, M.P., King, J.C., 2005. Rapid climate change in the ocean west of the Antarctic Peninsula during the second half of the 20th century. Geophys. Res. Lett. 32, L19604.

Mianzan, H., Pajaro, M., Alvarez Colombo, G., Madirolas, A., 2001. Feeding on survival-food: gelatinous plankton as a source of food for anchovies. Hydrobiologia 451, 45–53.

Michaelidis, B., Spring, A., Pörtner, H.O., 2007. Effects of long-term acclimation to environmental hypercapnia on extracellular acid-base status and metabolic capacity in Mediterranean fish *Sparus aurata*. Mar. Biol. 150, 1417–1429.

Mihuc, T.B., Minshall, G.W., 1995. Trophic generalists vs. trophic specialists: implications for food web dynamics in post-fire streams. Ecology 76, 2361–2372.

Mincks, S.L., Smith, C.R., Jeffreys, R.M., Sumida, P.Y., 2008. Trophic structure on the West Antarctic Peninsula shelf: detritivory and benthic inertia revealed by $\delta^{13}C$ and $\delta^{15}N$ analysis. Deep Sea Res. II 55, 2502–2514.

Mintenbeck, K., 2008. Trophic interactions within high Antarctic shelf communities—food web structure and the significance of fish. PhD Thesis, University of Bremen, 137pp. http://nbn-resolving.de/urn:nbn:de:gbv:46-diss000109584.

Mintenbeck, K., Alarcón, R., Brodte, E., Vanella, F., Knust, R., 2003. Ecology, biodiversity, biogeography and evolution—demersal fish. In: Arntz, W.E., Brey, T. (Eds.), The Expedition ANTARKTIS XIX/5 (LAMPOS) of RV "Polarstern" in 2002, Ber. Polarforsch. Meeresforsch.462, 55–58.

Mintenbeck, K., Damerau, M., Hirse, T., Knust, R., Koschnick, N., Matschiner, M., Rath, L., 2012. Biodiversity and zoogeography of demersal fish. In: Knust, R., Mintenbeck, K. (Eds.), The Expedition of the Research Vessel "Polarstern" to the Antarctic in 2011 (ANT XXVII/3) (CAMBIO), Ber. Polarforsch. Meeresforsch.644, 43–54.

Moline, M.A., Claustre, H., Frazer, T.K., Schofield, O., Vernet, M., 2004. Alteration of the food web along the Antarctic Peninsula in response to a regional warming trend. Glob. Change Biol. 10, 1973–1980.

Moline, M.A., Karnovsky, N.J., Brown, Z., Divoky, G.J., Frazer, T.K., Jacoby, C.A., Torres, J.T., Fraser, W.R., 2008. High latitude changes in ice dynamics and their impact on polar marine ecosystems. Ann. N.Y. Acad. Sci. 1134, 267–319.

Montgomery, J., Clements, K., 2000. Disaptation and recovery in the evolution of Antarctic fishes. Trends Ecol. Evol. 15, 267–270.

Morales-Nin, B., Garcia, M.A., Lopez, O., 1998. Distribution of larval and juvenile *Nototheniops larseni* and *Pleuragramma antarcticum* off the Antarctic Peninsula in relation to oceanographic conditions. Cybium 22, 69–81.

Moran, D., Støttrup, J., 2010. The effect of carbon dioxide on growth of juvenile Atlantic cod *Gadus morhua* l. Aquat. Toxicol. 102, 24–30.

Moreno, C.A., 1980. Observations on food and reproduction in *Trematomus bernacchii* (Pisces: Nototheniidae) from the Palmer Archipelago, Antarctica. Copeia 1, 171–173.

Moreno, C.A., Zamorano, J.H., Duarte, W.E., Jara, H.F., 1982. Abundance of Antarctic juvenile fishes on soft-bottom substrates: the importance of the refuge. Cybium 6, 37–41.

Mueller, I.A., Grim, J.M., Beers, J.M., Crockett, E.L., O'Brien, K.M., 2011. Interrelationship between mitochondrial function and susceptibility to oxidative stress in red- and white-blooded Antarctic notothenioid fishes. J. Exp. Biol. 214, 3732–3741.

Munday, P.L., Dixson, D.L., Donelson, J.M., Jones, G.P., Pratchett, M.S., Devitsina, G.V., Doving, K.B., 2009. Ocean acidification impairs olfactory discrimination and homing ability of a marine fish. Proc. Natl. Acad. Sci. U.S.A. 106, 1848–1852.

Murase, H., Matsuoka, K., Ichii, T., Nishiwaki, S., 2002. Relationship between the distribution of euphausiids and baleen whales in the Antarctic (35°E–145°W). Polar Biol. 25, 135–145.
Murphy, R.C., 1962. The oceanic life of the Antarctic. Sci. Am. 207, 186–210.
Murphy, J.M., Mitchel, J.F.B., 1995. Transient response of the Hadley Centre coupled ocean-atmosphere model to increasing carbon dioxide. Part II: spatial and temporal structure of response. J. Clim. 8, 57–80.
Murphy, E.J., Trathan, P.N., Watkins, J.J., Reid, K., Meredith, M.P., Forcada, J., Thorpe, S.E., Johnston, N.M., Rothery, P., 2007. Climatically driven fluctuations in Southern Ocean ecosystems. Proc. R. Soc. B 274, 3057–3067.
Naeem, S., 1998. Species redundancy and ecosystem reliability. Conserv. Biol. 12, 39–45.
Nicol, S., Allison, I., 1997. The frozen skin of the Southern Ocean. Am. Sci. 85, 426–439.
Nilsson, G.E., Crawley, N., Lunde, I.G., Munday, P.L., 2009. Elevated temperature reduces the respiratory scope of coral reef fishes. Glob. Change Biol. 15, 1405–1412.
North, A.W., White, M.G., 1987. Reproductive strategies of Antarctic fish. In: Kullander, S.O., Fernholm, B. (Eds.), Proc. V Congr. Europ. Ichthyol. Stockholm 1985, 381–390.
North, A.W., White, M.G., Burchett, M.S., 1980. Age determination of Antarctic fish. Cybium 8, 7–15.
Nöthig, E.-M., Bodungen, B.v., 1989. Occurrence and vertical flux of faecal pellets of probably protozoan origin in the southeastern Weddell Sea (Antarctica). Mar. Ecol. Prog. Ser. 56, 281–289.
Nöthig, E.-M., Bodungen, B.v., Siu, Q., 1991. Phyto- and protozooplankton biomass during austral summer in surface waters of the Weddell Sea and vicinity. Polar Biol. 11, 293–304.
O'Gorman, E., Emmerson, M., 2010. Manipulating interaction strengths and the consequences for trivariate patterns in a marine food web. Adv. Ecol. Res. 42, 301–419.
O'Brien, K., Sidell, B., 2000. The interplay among cardiac ultrastructure, metabolism and the expression of oxygen-binding proteins in Antarctic fishes. J. Exp. Biol. 203, 1287–1297.
O'Brien, K.M., Skilbeck, C., Sidell, B.D., Egginton, S., 2003. Muscle fine structure may maintain the function of oxidative fibres in haemoglobinless Antarctic fishes. J. Exp. Biol. 206, 411–421.
O'Grady, S.M., DeVries, A.L., 1982. Osmotic and ionic regulation in polar fishes. J. Exp. Mar. Biol. Ecol. 57, 219–228.
Öhman, M.C., Rajasuriya, A., 1998. Relationship between habitat structure and fish communities on coral and sand reefs. Environ. Biol. Fish. 53, 19–31.
Olbers, D., Gouretski, V., Seiss, G., Schroeter, J., 1992. Hydrographic Atlas of the Southern Ocean. Alfred Wegener Institute, Bremerhaven, Germany, hdl:10013/epic.12913.
Øresland, V., 1995. Winter population structure and feeding of the chaetognath *Eukronia hamata* and the copepod *Euchaeta antarctica* in Gerlache Strait, Antarctic Peninsula. Mar. Ecol. Prog. Ser. 119, 77–86.
Orsi, A.H., Whitworth III, T., Nowling Jr., W.D., 1995. On the meridional extent and fronts of the Antarctic Circumpolar Current. Deep Sea Res. I 42, 641–673.
Österblom, H., Casini, M., Olsson, O., Bignert, A., 2006. Fish, seabirds and trophic cascades in the Baltic Sea. Mar. Ecol. Prog. Ser. 323, 233–238.
Österblom, H., Olsson, O., Blenckner, T., Furness, R.W., 2008. Junk-food in marine ecosystems. Oikos 117, 967–977.
Ozaki, H., Obata, H., Naganobu, M., Gamo, T., 2009. Long-term bottom water warming in the north Ross Sea. J. Oceanogr. 65, 235–244.
Pakhomov, E.A., Froneman, P.W., Perissinotto, R., 2002. Salp/krill interactions in the Southern Ocean: spatial segregation and implications for the carbon flux. Deep Sea Res. II 49, 1881–1907.
Pankhurst, N.W., 1990. Growth and reproduction of the Antarctic nototheniid fish *Pagothenia borchgrevinki*. Polar Biol. 10, 387–391.

Papetti, C., Lio, P., Ruber, L., Patarnello, T., Zardoya, R., 2007. Antarctic fish mitochondrial genomes lack nd6 gene. J. Mol. Evol. 65, 519–528.

Pearson, P.N., Foster, G.L., Wade, B.S., 2009. Atmospheric carbon dioxide through the Eocene–Oligocene climate transition. Nature 461, 1110–1113.

Peck, L.S., Brockington, S., Vanhove, S., Beghyn, M., 1999. Community recovery following catastrophic iceberg impacts in a soft-sediment shallow-water site at Signy Island, Antarctica. Mar. Ecol. Prog. Ser. 186, 1–8.

Perissinotto, R., Pakhomov, E.A., 1998a. Contribution of salps to carbon flux of marginal ice zone of the Lazarev Sea, southern ocean. Mar. Biol. 131, 25–32.

Perissinotto, R., Pakhomov, E.A., 1998b. The trophic role of the tunicate *Salpa thompsoni* in the Antarctic marine ecosystem. J. Mar. Syst. 17, 361–374.

Permitin, Y.Y., 1977. Species composition and zoogeographical analysis of the bottom fish fauna of the Scotia Sea. J. Ichthyol. 17, 710–726.

Perry, A.L., Low, P.J., Ellis, J.R., Reynolds, J.D., 2005. Climate change and distribution shifts in marine fishes. Science 308, 1912–1915.

Plötz, J., 1986. Summer diet of Weddell seals (*Leptonychotes weddelli*) in the eastern and southern Weddell Sea, Antarctica. Polar Biol. 6, 97–102.

Plötz, J., Gerdes, D., Gräfe, M., Klages, N., Reijnders, P., Steinmetz, R., Zegers, K., 1987. Weddell seals and emperor penguins in the Drescher Inlet. In: Schnack-Schiel, S. (Ed.), The Winter Expedition of RV "Polarstern" to the Antarctic (ANT V/1-3), Ber. Polarforsch.39, 222–227.

Plötz, J., Ekau, W., Reijnders, P.J.H., 1991. Diet of Weddell seals *Leptonychotes weddellii* at Vestkapp, eastern Weddell Sea (Antarctica), in relation to local food supply. Mar. Mamm. Sci. 7, 136–144.

Plötz, J., Bornemann, H., Knust, R., Schröder, A., Bester, M., 2001. Foraging behaviour of Weddell seals, and its ecological implications. Polar Biol. 24, 901–909.

Plötz, J., Bornemann, H., Liebsch, N., Watanabe, Y., 2005. Foraging ecology of Weddell seals. In: Arntz, W.E., Brey, T. (Eds.), The expedition ANTARKTIS XXI/2 (BENDEX) of RV "Polarstern" in 2003/2004, Ber. Polarforsch.503, 63–67.

Podrabsky, J.E., Somero, G.N., 2006. Inducible heat tolerance in Antarctic notothenioid fishes. Polar Biol. 30, 39–43.

Pörtner, H.O., Farrell, A.P., 2008. Physiology and climate change. Science 322, 690–692.

Pörtner, H.O., Knust, R., 2007. Climate change affects marine fishes through the oxygen limitation of thermal tolerance. Science 315, 95–97.

Pörtner, H.O., Peck, M.A., 2010. Climate change effects on fishes and fisheries: towards a cause-and-effect understanding. J. Fish Biol. 77, 1745–1779.

Pörtner, H.O., van Dijk, P.L.M., Hardewig, I., Sommer, A., 2000. Levels of metabolic cold adaptation: tradeoffs in eurythermal and stenothermal ectotherms. In: Davison, W., Williams, C.W. (Eds.), Antarctic Ecosystems: Models for a Wider Understanding. Caxton Press, Christchurch, pp. 109–122.

Pörtner, H.O., Bock, C., Knust, R., Lannig, G., Lucassen, M., Mark, F.C., Sartoris, F.J., 2008. Cod and climate in a latitudinal cline: physiological analyses of climate effects in marine fishes. Clim. Res. 37, 253–270.

Ptacnik, R., Moorthi, S.D., Hillebrand, H., 2010. Hutchinson reversed, or why there need to be so many species. Adv. Ecol. Res. 43, 1–44.

Purcell, J.E., 1985. Predation on fish eggs and larvae by pelagic cnidarians and ctenophores. Bull. Mar. Sci. 37, 739–755.

Purcell, J.E., 2005. Climate effects on formation of jellyfish and ctenophore blooms: a review. J. Mar. Biol. Assoc. U.K. 85, 461–476.

Purcell, J.E., Decker, M.B., 2005. Effects of climate on relative predation by scyphomedusae and ctenophores on copepods in Chesapeake Bay during 1987-2000. Limnol. Oceanogr. 50, 376–387.

Pusch, C., Hulley, P.A., Kock, K.H., 2004. Community structure and feeding ecology of mesopelagic fishes in the slope waters of King George Island (South Shetland Islands, Antarctica). Deep Sea Res. I 51, 1685–1708.
Quartino, M.L., Boraso de Zaixso, A.L., 2008. Summer macroalgal biomass in Potter Cove, South Shetland Islands, Antarctica: its production and flux to the ecosystem. Polar Biol. 31, 281–294.
Radtke, R.L., Hubold, G., Folsom, S.D., Lenz, P.H., 1993. Otolith structural and chemical analysis: the key to resolving age and growth of the Antarctic silverfish, *Pleuragramma antarcticum*. Antarct. Sci. 5, 51–62.
Ralph, R., Everson, I., 1968. The respiratory metabolism of some Antarctic fish. Comp. Biochem. Physiol. 27, 299–307.
Riebesell, U., Schloss, I., Smetacek, V., 1991. Aggregation of algae released from melting sea ice: implications for seeding and sedimentation. Polar Biol. 11, 239–248.
Rignot, E., Bamber, J.L., van den Broeke, M.R., Davis, C., Li, Y., van de Berg, W.J., van Meijgaard, E., 2008. Recent Antarctic ice mass loss from radar interferometry and regional climate modelling. Nat. Geosci. 1, 106–110.
Robertson, R., Vosbeck, M., Gordon, A.L., Fahrbach, E., 2002. Long-term temperature trends in the deep waters of the Weddell Sea. Deep Sea Res. II 49, 4791–4806.
Robinson, E.E., 2008. Antarctic fish: thermal specialists or adaptable generalists? PhD Thesis, University of Canterbury, Christchurch, 229pp.
Robinson, E., Davison, W., 2008. The Antarctic notothenioid fish *Pagothenia borchgrevinki* is thermally flexible: acclimation changes oxygen consumption. Polar Biol. 31, 317–326.
Robinson, E., Egginton, S., Davison, W., 2011. Warm-induced bradycardia and cold-induced tachycardia: mechanisms of cardiac and ventilatory control in a warm-acclimated Antarctic fish. Polar Biol. 34, 371–379.
Rodhouse, P.G., White, M.G., 1995. Cephalopods occupy the ecological niche of epipelagic fish in the Antarctic Polar Frontal Zone. Biol. Bull. 189, 77–80.
Sabine, C.L., Feely, R.A., Gruber, N., Key, R.M., Lee, K., Bullister, J.L., Wanninkhof, R., Wong, C.S., Wallace, D.W.R., Tilbrook, B., Millero, F.J., Peng, T.H., Kozyr, A., Ono, T., Rios, A.F., 2004. The oceanic sink for anthropogenic CO_2. Science 305, 367–371.
Sahade, R., Tatián, M., Kowalke, J., Kühne, S., Esnal, G.B., 1998. Benthic faunal associations on soft substrates at Potter Cove, King George Island, Antarctica. Polar Biol. 19, 85–91.
Sahade, R., Tarantelli, S., Tatían, M., Mercuri, G., 2008. Benthic community shifts: a possible linkage to climate change? In: Wiencke, C., Ferreyra, G., Abele, D., Marenssi, S. (Eds.), The Antarctic Ecosystem of Potter Cove, King-George Island (Isla 25 de Mayo), Ber. Polarforsch. Meeresforsch. 571, 331–337.
Scharek, R., Nöthig, E.M., 1995. Das einzellige Plankton im Ozean der Arktis und der Antarktis. In: Hempel, I., Hempel, G. (Eds.), Biologie der Polarmeere. Gustav Fischer Verlag, Jena, pp. 116–127.
Scharek, R., Tupas, L.M., Karl, D.M., 1999. Diatom fluxes to the deep sea in the oligotrophic North Pacific gyre at Station ALOHA. Mar. Ecol. Prog. Ser. 182, 55–67.
Schloss, I.R., Ferreyra, G.A., González, O., Atencio, A., Fuentes, V., Tosonotto, G., Mercuri, G., Sahade, R., Tatián, M., Abele, D., 2008. Long-term hydrographic conditions and climate trends in Potter Cove. In: Wiencke, C., Ferreyra, G., Abele, D., Marenssi, S. (Eds.), The Antarctic Ecosystem of Potter Cove, King-George Island (Isla 25 de Mayo), Ber. Polarforsch. Meeresforsch. 571, 382–389.
Schoener, T.W., 1989. Food webs from the small to the large. Ecology 70, 1559–1589.
Schröder, A., Artigues, B., Gonzales, J., Mintenbeck, K., Knust, R., 2001. Biodiversity and biogeography. Demersal fish fauna. In: Arntz, W., Brey, T. (Eds.), The Expedition ANTARKTIS XVII/3 (EASIZ III) of RV "Polarstern" in 2000, Ber. Polarforsch. Meeresforsch.402, 70–74.

Schwarzbach, W., 1988. The demersal fish fauna of the eastern and southern Weddell Sea: geographical distribution, feeding of fishes and their trophic position in the food web. Ber. Polarforsch. 54, 1–93.

Seebacher, F., Davison, W., Lowe, C.J., Franklin, C.E., 2005. A falsification of the thermal specialization paradigm: compensation for elevated temperatures in Antarctic fishes. Biol. Lett. 1, 151–154.

Sidell, B.D., 1991. Physiological roles of high lipid content in tissues of Antarctic fish species. In: Di Prisco, G., Maresca, B., Tota, B. (Eds.), Biology of Antarctic Fish. Springer Verlag, Berlin, pp. 221–231.

Sidell, B.D., O'Brien, K.M., 2006. When bad things happen to good fish: the loss of hemoglobin and myoglobin expression in Antarctic icefishes. J. Exp. Biol. 209, 1791–1802.

Siegel, V., Skibowski, A., Harm, U., 1992. Community structure of the epipelagic zooplankton community under the sea-ice of the northern Weddell Sea. Polar Biol. 12, 15–24.

Sinensky, M., 1974. Homeoviscous adaptation—a homeostatic process that regulates the viscosity of membrane lipids in *Escherichia coli*. Proc. Natl. Acad. Sci. U.S.A. 71, 522–525.

Sirabella, P., Giuliani, A., Colosimo, A., Dippner, J.W., 2001. Breaking down the climate effects on cod recruitment by principal component analysis and canonical correlation. Mar. Ecol. Prog. Ser. 216, 213–222.

Sloman, K.A., Mandic, M., Todgham, A.E., Fangue, N.A., Subrt, P., Richards, J.G., 2008. The response of the tidepool sculpin, *Oligocottus maculosus*, to hypoxia in laboratory, mesocosm and field environments. Comp. Biochem. Physiol. A 149, 284–292.

Słósarczyk, W., 1986. Attempts at a quantitative estimate by trawl sampling of distribution of postlarval and juvenile notothenioids (Pisces, Perciformes) in relation to environmental conditions in the Antarctic Peninsula region during SIBEX 1983-84. Mem. Natl. Inst. Polar Res. 40, 299–315.

Smale, D.A., Brown, K.M., Barnes, D.K.A., Fraser, K.P.P., Clarke, A., 2008. Ice scour disturbance in Antarctic waters. Science 321, 371.

Smetacek, V., Scharek, R., Nöthig, E.M., 1990. Seasonal and regional variation in the pelagial and its relationship to the life history cycle of krill. In: Kerry, K.R., Hempel, G. (Eds.), Antarctic Ecosystems. Ecological Change and Conservation. Springer Verlag, Berlin, pp. 103–114.

Smith Jr., W.O., Nelson, D.M., 1986. Importance of ice edge phytoplankton production in the Southern Ocean. Bioscience 36, 251–257.

Smith Jr., W.O., Sakshaug, E., 1990. Polar phytoplankton. In: Smith Jr., W.O. (Ed.), Polar Oceanography, Part B: Chemistry, Biology, and Geology. Academic Press, San Diego, pp. 477–525.

Smith, C.R., Mincks, S., DeMaster, D.J., 2006. A synthesis of bentho-pelagic coupling on the Antarctic shelf: food banks, ecosystem inertia and global climate change. Deep Sea Res. II 53, 875–894.

Smith, C.R., Grange, L.J., Honig, D.L., Naudts, L., Huber, B., Guidi, L., Domack, E.A., 2012. A large population of king crabs in Palmer Deep on the west Antarctic Peninsula shelf and potential invasive impacts. Proc. R. Soc. B 279, 1017–1026.

Snyder, W.E., Snyder, G.B., Finke, D.L., Straub, C.S., 2006. Predator biodiversity strengthens herbivore suppression. Ecol. Lett. 9, 789–796.

Somero, G.N., 2010. The physiology of climate change: how potentials for acclimatization and genetic adaptation will determine 'winners' and 'losers'. J. Exp. Biol. 213, 912–920.

Somero, G.N., DeVries, A.L., 1967. Temperature tolerance of some Antarctic fishes. Science 156, 257–258.

Somero, G.N., Giese, A.C., Wohlschlag, D.E., 1968. Cold adaptation of the Antarctic fish *Trematomus bernacchii*. Comp. Biochem. Physiol. 26, 223–233.

Stammerjohn, S.E., Martinson, D.G., Smith, R.C., Iannuzzi, R.A., 2008a. Sea ice in the western Antarctic Peninsula region: spatio-temporal variability from ecological and climate change perspectives. Deep Sea Res. II 55, 2041–2058.

Stammerjohn, S.E., Martinson, D.G., Smith, R.C., Yuan, X., Rind, D., 2008b. Trends in Antarctic annual sea ice retreat and advance and their relation to El Niño-southern oscillation and southern annular mode variability. J. Geophys. Res. 113, C03S90.

Stroeve, J., Holland, M.M., Meier, W., Scambos, T., Serreze, M., 2007. Arctic sea ice decline: faster than forecast. Geophys. Res. Lett. 34, L09501.

Suchman, C.L., Daly, E.A., Keister, J.E., Peterson, W., Brodeur, R.D., 2008. Feeding patterns and predation potential of scyphomedusae in a highly productive upwelling region. Mar. Ecol. Prog. Ser. 358, 161–172.

Tada, S., Sato, T., Sakurai, H., Arai, H., Kimpara, I., Kodama, M., 1996. Benthos and fish community associated with clumps of submerged drifting algae in Fildes Bay, King George Island, Antarctica. Proc. NIPR Symp. Polar Biol. 9, 243–251.

Takahashi, M., Nemoto, T., 1984. The food of some Antarctic fish in the western Ross Sea in summer 1979. Polar Biol. 3, 237–239.

Takeuchi, I., Watanabe, K., 2002. Mobile epiphytic invertebrates inhabiting the brown macroalga, *Desmarestia chordalis*, under the coastal fast ice of Lützow-Holm Bay, East Antarctica. Polar Biol. 25, 624–628.

Tatián, M., Sahade, R., Esnal, G.B., 2004. Diet composition in the food of Antarctic ascidians living at low levels of primary production. Antarct. Sci. 16, 123–128.

Tavares, M., De Melo, G.A.S., 2004. Discovery of the first known benthic invasive species in the Southern Ocean: the North Atlantic spider crab *Hyas araneus* found in the Antarctic Peninsula. Antarct. Sci. 16, 129–131.

Teixidó, N., Garrabou, J., Gutt, J., Arntz, W.E., 2004. Recovery in Antarctic benthos after iceberg disturbance: trends in benthic composition, abundance and growth forms. Mar. Ecol. Prog. Ser. 278, 1–16.

Tetens, V., Wells, R.M.G., Devries, A.L., 1984. Antarctic fish blood: respiratory properties and the effects of thermal acclimation. J. Exp. Biol. 109, 265–279.

Thatje, S., Anger, K., Calcagno, J.A., Lovrich, G.A., Pörtner, H.O., Arntz, W.E., 2005. Challenging the cold: crabs reconquer the Antarctic. Ecology 86, 612–625.

Thatje, S., Hall, S., Hauton, C., Held, C., Tyler, P., 2008. Encounter of lithodid crab *Paralomis birsteini* on the continental slope off Antarctica, sampled by ROV. Polar Biol. 31, 1143–1148.

Tierney, N., Hindell, M.A., Goldsworthy, S., 2002. Energy content of mesopelagic fish from Macquarie Island. Antarct. Sci. 14, 225–230.

Tomczak, M., Godfrey, J.S., 1994. Regional Oceanography: An Introduction. Pergamon, Oxford.

Torres, J.J., Donnelly, J., Hopkins, T.L., Lancraft, T.M., Aarset, A.V., Ainley, D.G., 1994. Proximate composition and overwintering strategies of Antarctic micronektonic Crustacea. Mar. Ecol. Prog. Ser. 113, 221–232.

Tranter, D.J., 1982. Interlinking of physical and biological processes in the Antarctic Ocean. Oceanogr. Mar. Biol. Annu. Rev. 20, 11–35.

Turner, J., Colwell, S.R., Marshall, G.J., Lachlan-Cope, T.A., Carleton, A.M., Jones, P.D., Lagun, V., Reid, P.A., Iagovkina, S., 2005. Antarctic climate change during the last 50 years. Int. J. Climatol. 25, 279–294.

Tyler, S., Sidell, B., 1984. Changes in mitochondrial distribution and diffusion distances in muscle of goldfish upon acclimation to warm and cold temperatures. J. Exp. Zool. 232, 1–9.

Urschel, M.R., O'Brien, K.M., 2009. Mitochondrial function in Antarctic notothenioid fishes that differ in the expression of oxygen-binding proteins. Polar Biol. 32, 1323–1330.

Urschel, M.R., O'Brien, K.M., 2008. High mitochondrial densities in the hearts of Antarctic icefishes are maintained by an increase in mitochondrial size rather than mitochondrial biogenesis. J. Exp. Biol. 211, 2638–2646.

Vacchi, M., La Mesa, M., Dalu, M., MacDonald, J., 2004. Early life stages in the life cycle of Antarctic silverfish, *Pleuragramma antarcticum* in Terra Nova Bay, Ross Sea. Antarct. Sci. 16, 299–305.

Vallet, C., Beans, C., Koubbi, P., Courcot, L., Hecq, J.H., Goffart, A., 2011. Food preferences of larvae of Antarctic silverfish *Pleuragramma antarcticum* Boulenger, 1902 from Terre Adélie coastal waters during summer 2004. Polar Sci. 5, 239–251.

Van Franeker, J.A., Bathmann, U.V., Mathot, S., 1997. Carbon fluxes to Antarctic top predators. Deep Sea Res. II 44, 435–455.

Varsamos, S., Nebel, C., Charmantier, G., 2005. Ontogeny of osmoregulation in postembryonic fish: a review. Comp. Biochem. Physiol. A 141, 401–429.

Vaughan, D.G., Marshall, G.J., Connolley, W.M., Parkinson, C., Mulvaney, R., Hodgson, D.A., King, J.C., Pudsey, C.J., Turner, J., 2003. Recent rapid regional climate warming on the Antarctic Peninsula. Clim. Chang. 60, 243–274.

Voss, J., 1988. Zoogeography and community analysis of macrozoobenthos of the Weddell Sea (Antarctica). Ber. Polarforsch. 45, 1–145.

White, M.G., 1991. Age determination in Antarctic fish. In: Di Prisco, G., Maresca, B., Tota, B. (Eds.), Biology of Antarctic Fish. Springer Verlag, Berlin, pp. 87–100.

Whitehead, P.J.P., 1985. FAO species catalogue. Vol. 7. Clupeoid fishes of the world (suborder Clupeioidei). An annotated and illustrated catalogue of the herrings, sardines, pilchards, sprats, shads, anchovies and wolf-herrings. Part 1-Chirocentridae, Clupeidae and Pristigasteridae. FAO Fish. Synop. 125, 1–303.

Wienecke, B., Robertson, G., Kirkwood, R., Lawton, K., 2007. Extreme dives by free-ranging emperor penguins. Polar Biol. 30, 133–142.

Wilson, R.P., Quintana, F., 2004. Surface pauses in relation to dive duration in emperial cormorants; how much time for a breather? J. Exp. Biol. 207, 1789–1796.

Windisch, H.S., Kathöver, R., Pörtner, H.O., Frickenhaus, S., Lucassen, M., 2011. Thermal acclimation in Antarctic fish: transcriptomic profiling of metabolic pathways. Am. J. Physiol. Regul. Integr. Comp. Physiol. 301, R1453–R1466.

Wohlschlag, D.E., 1963. An Antarctic fish with unusually low metabolism. Ecology 44, 557–564.

Wöhrmann, A., 1996. Antifreeze glycopeptides and peptides in Antarctic fish species from the Weddell Sea and the Lazarev Sea. Mar. Ecol. Prog. Ser. 130, 47–59.

Wöhrmann, A., 1997. Freezing resistance in Antarctic and Arctic fishes: its relation to mode of life, ecology and evolution. Cybium 21, 423–442.

Woodward, G., Ebenman, B., Emmerson, M., Montoya, J.M., Olesen, J.M., Valido, A., Warren, P.H., 2005. Body-size in ecological networks. Trends Ecol. Evol. 20, 402–409.

Woodward, G., Benstead, J.P., Beveridge, O.S., Blanchard, J., Brey, T., Brown, L., Cross, W.F., Friberg, N., Ings, T.C., Jacob, U., Jennings, S., Ledger, M.E., Milner, A.M., Montoya, J.M., O'Gorman, E.O., Olesen, J.M., Petchey, O.L., Pichler, D.E., Reuman, D.C., Thompson, M.S., Van Veen, F.J.F., Yvon-Durocher, G., 2010a. Ecological networks in a changing climate. Adv. Ecol. Res. 42, 72–138.

Woodward, G., Blanchard, J., Lauridsen, R.B., Edwards, F.K., Jones, J.I., Figueroa, D., Warren, P.H., Petchey, O.L., 2010b. Individual-based food webs: species identity, body size and sampling effects. Adv. Ecol. Res. 43, 211–266.

Yeager, L.A., Layman, C.A., Allgeier, J.E., 2011. Effects of habitat heterogeneity at multiple spatial scales on fish community assembly. Oecologia 167, 157–168.

Yvon-Durocher, G., Montoya, J.M., Trimmer, M., Woodward, G., 2011. Warming alters the size spectrum and shifts the distribution of biomass in freshwater ecosystems. Glob. Change Biol. 17, 1681–1694.

Zhou, M., Dorland, R.D., 2004. Aggregation and vertical migration behaviour of *Euphausia superba*. Deep Sea Res. II 51, 2119–2137.

Zhuang, X., Cheng, C.H., 2010. Nd6 gene "lost" and found: evolution of mitochondrial gene rearrangement in Antarctic notothenioids. Mol. Biol. Evol. 27, 1391–1403.

Zwally, H.J., Comiso, J.C., Parkinson, C.L., Campbell, W.J., Carsaey, F.D., Gloersen, P., 1983. Antarctic sea ice 1973-1976: satellite passive-microwave observations. NASA SP 459, Washington, 206pp.

A Complete Analytic Theory for Structure and Dynamics of Populations and Communities Spanning Wide Ranges in Body Size

Axel G. Rossberg[*,†,1]
*Centre for Environment, Fisheries and Aquaculture Science (Cefas), Lowestoft Laboratory, Suffolk, United Kingdom
†Medical Biology Centre, School of Biological Sciences, Queen's University Belfast, Belfast, United Kingdom
[1]Corresponding author: e-mail address: axel.rossberg@cefas.co.uk

Contents

Advances in Ecological Research, Volume 46
ISSN 0065-2504
http://dx.doi.org/10.1016/B978-0-12-396992-7.00008-3

Abstract

The prediction and management of ecosystem responses to global environmental change would profit from a clearer understanding of the mechanisms determining the structure and dynamics of ecological communities. The analytic theory presented here develops a causally closed picture for the mechanisms controlling community and population size structure, in particular community size spectra, and their dynamic responses to perturbations, with emphasis on marine ecosystems. Important implications are summarised in non-technical form. These include the identification of three different responses of community size spectra to size-specific pressures (of which one is the classical trophic cascade), an explanation for the observed slow recovery of fish communities from exploitation, and clarification of the mechanism controlling predation mortality rates. The theory builds on a community model that describes trophic interactions among size-structured populations and explicitly represents the full life cycles of species. An approximate time-dependent analytic solution of the model is obtained by coarse graining over maturation body sizes to obtain a simple description of the model steady state, linearising near the steady state, and then eliminating intraspecific size structure by means of the quasi-neutral approximation. The result is a convolution equation for trophic interactions among species of different maturation body sizes, which is solved analytically using a novel technique based on a multiscale expansion.

1. INTRODUCTION

1.1. Orientation on a changing planet

We have entered a period of rapid, global environmental change, in which anthropogenic climate change is one important factor (Solomon et al., 2007), which is complicated by additional drivers such as elevated nutrient discharge and the direct impacts of overexploitation on natural resources (Millennium Ecosystem Assessment, 2005). The planet's ecosystems may never before have experienced this type of environmental change. Knowing how they will respond could help us prepare for and, if possible, manage the consequences of these developments (Woodward et al., 2010a).

Experimental and comparative methods, the dominant modes of scientific enquiries in ecological research, have obvious limitations when addressing scenarios of unprecedented environmental change at global or even regional scales. The relevant temporal and spacial scales are too large to conduct experiments (O'Gorman et al., 2012), and the data required for direct comparative analyses are often simply not yet available (Twomey et al., 2012).

The natural response, therefore, is to resort to models. These models can, for example, be (i) experimental systems such as microcosms (Reiss et al., 2010) or mesocosms (Yvon-Durocher et al., 2011), (ii) statistical models to extrapolate ecological data to the environmental conditions expected in future, (iii) numerical models aimed at simulating the relevant aspects of future scenarios directly, or (iv) analytic models that seek to isolate the implications of general ecological and biological principles that are sufficiently fundamental not to be overruled by global change. Common to any such modelling is the expectation that some aspect of the scenarios one seeks to predict are sufficiently insensitive to ecological detail that such detail need not be reproduced in the models (Rossberg, 2007). These robust aspects are those one can be predicted, for which one can prepare, and which one might attempt to manage. Those aspects of future system states that do depend on many details (Friberg et al., 2011), however, might simply be beyond control. Common to all modelling is the problem of knowing which aspects of future system states can be predicted and which system details are relevant for this.

A common approach to identifying aspects of complex systems amenable to modelling is to search for coherent patterns in data. Obviously, if clear patterns arise similarly in data from different systems, the mechanisms generating these patterns are unlikely to depend on many details (Riede et al., 2010). It is, therefore, natural for research addressing the large-scale ecological impacts of global change to concentrate on the major known macroecological patterns. The patterns at the focus of the present work are those found in the distributions of the biomasses of individuals or species over wide ranges in body size. The first step, before addressing specific problems in the face of global change, is to ask which ecological details are important for generating these patterns in reality and what the underlying mechanisms are; these will be the details that any type of model, for example (i)–(iv), would need to capture to be reliable. Answering these questions is the central aim of the present work.

Technically, this is done by deriving an approximate analytic solution of a detailed community model. In doing so, determining which approximations are feasible without much affecting the predicted body-size distributions means understanding which ecological details are unimportant for the outcome. The resulting, approximate descriptions of community structure and dynamics encapsulate the mechanisms by which those details that do matter bring about the dominating patterns and their responses to pressures. Therefore, judiciously applied approximations, rather than meaning loss of information, actually reveal what is important. This is why the route along which the approximate solution of the community model is derived is as important as the outcome. The step-by-step derivation, therefore, forms the main body of this chapter.

1.2. Size spectra

Much of the early empirical and theoretical work on size distributions related to plant communities (see Koyama and Kira, 1956). Regularities in size spectra have now been investigated, for example, for benthic communities (Duplisea, 2000; Gerlach et al., 1985), terrestrial soil (Mulder and Elser, 2009; Mulder et al., 2008; Reuman et al., 2008), and birds (Thibault et al., 2011). The most striking form of the phenomenon, however, is found in pelagic communities. Sheldon et al. (1972) suggested, based on measurements and literature data, that pelagic biomass is approximately evenly distributed over the logarithmic body-size axis "from bacteria to whales", spanning up to 20 orders of magnitude in body mass. The core of this conclusion has remained unchallenged, although intensive human harvesting makes its validation for the largest size classes difficult (Gaedke, 1992b; Jennings and Blanchard, 2004; Sheldon et al., 1977). Over shorter size ranges, similar phenomena are observed for benthic communities (Duplisea, 2000; Gerlach et al., 1985) or in terrestrial soil ecosystems (Mulder and Elser, 2009; Mulder et al., 2008, 2011; Reuman et al., 2008).

The work of Sheldon et al. (1972) drew attention to the usefulness of characterising community structure through *size spectra*. In the form they used, size structure is represented irrespective of species identity, trophic level, or life-history stage. In a common, though by no means exclusive, protocol to determine size spectra in aquatic communities, individuals are assigned to logarithmic size classes by carbon content, for example, ..., 1–2 pgC, 2–4 pgC, ... using various optical and mechanical techniques. The density of biomass in each size class is then determined, for example, as a value averaged over a full year and a specific range of the water column.

Subsequent investigations revealed deviations from the hypothesis formulated by Sheldon et al. (1972) in the detail (Kerr and Dickie, 2001). As illustrated in Fig. 1, one often finds approximate power-law relationships between abundance and body mass as the basic pattern, that is, straight lines on double-logarithmic plots, with "slopes" (actually exponents) close to but different from the one corresponding to the Sheldon hypothesis (which is -2, -1, or 0, depending on the specific choices of dependent and independent variables; see Blanco et al., 1994). On top of this, size spectra often deviate from power laws by exhibiting several maxima (or "domes") spaced more or less evenly along the logarithmic body-size axis (see, e.g. Fig. 1, or Kerr and Dickie, 2001).

The idea that the size distribution of individuals can indicate the "health" of an ecological community (e.g. Castle et al., 2011; Jennings and Blanchard, 2004; Kerr and Dickie, 2001; Rice et al., 2011; Sheldon et al., 1977) has recently been inscribed in legal documents. Arguably, the simplest way of characterising a size distribution is to compare the relative proportions of biomass in two wide body-size classes. In a series of reports, reviewed by Greenstreet et al. (2011), the International Council for the Exploration of the Sea developed, as an indicator for community state, the Large Fish Indicator (LFI), defined as the proportion among all fish in a community of

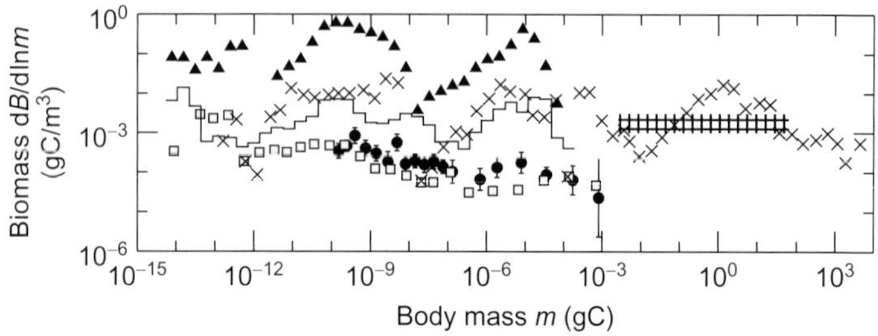

Figure 1 Five typical planktonic size spectra. Triangles represent data for the highly eutrophic lake Müggelsee averaged over 3 years, after Gaedke et al. (2004); stairs represent a seasonal average for Überlingersee, a division of the meso-eutrophic Lake Constance (Bodensee), after Gaedke (1992b), including an estimate for the contribution by fish (shaded area); crosses represent yearly averages for Lake Ontario, after Sprules and Goyke (1994), using a nominal depth of 50 m to convert to volume density; circles with 1 SD error bars represent the oligotrophic waters of the North Pacific Central Gyre averaged over 3 years of sampling by Rodriguez and Mullin (1986a), where measurements are compatible with a perfect power law; squares represent the oligotrophic open waters at a station near the Yakutat Seamount in the Northwest Atlantic, after Quiñones et al. (2003). For further examples, see Boudreau and Dickie (1992), among many others. The density of biomass along the logarithmic size axis d*B*/d ln *m* was estimated from the published data as $B_i m_i/(\Delta m_i)$, where B_i is the biomass of individuals in body-mass interval *i* (actually, their biomass per unit volume), m_i the midpoint of this body-mass interval on a logarithmic axis, and Δm_i is the linear width of the interval. This representation combines the advantage of Sheldon size spectra (Sheldon et al., 1972) of visualising structural details with the advantage of high intercomparability between empirical datasets (often highlighted for normalised spectra $B_i/\Delta m_i$), because $B_i m_i/(\Delta m_i)$ approximates the protocol-independent quantity d*B*/d ln *m* up to a relative error declining as fast as $(\Delta m_i/m_i)^2$ for small $\Delta m_i/m_i$. Axes are isometric to ease visual appreciation of the varying degrees of uniformity of the spectra.

individuals longer than a given length threshold. The Convention for the Protection of the Marine Environment of the North-East Atlantic proposed the LFI as a means to set an Ecological Quality Objective (OSPAR, 2006). EU legislation (European Commission, 2010) includes the LFI in a list of "Criteria to be used by the Member States to assess the extent to which good environmental status is being achieved". Among the important open issues in this context is, for example, the question of how fast the LFI will return to natural levels when fishing pressure is relaxed (OSPAR, 2006). Predictions appear to depend sensitively on the way communities are modelled (ICES, 2010; Shephard et al., 2012), and observed recovery is slower than was previously thought (Fung et al., 2012a). A better understanding of the underlying fundamental processes is thus required (Arim et al., 2011; Castle et al., 2011; Gilljam et al., 2011; Jacob et al., 2011; Melian et al., 2011).

In view of such current, practical questions regarding marine size spectra and their dynamics in response to anthropogenic pressures, the theory presented here is phrased with reference to marine communities and their perturbations by fishing. However, it should apply similarly to other cases of size-structured communities, for example, soil ecosystems, and partially also to communities of competing plants.

1.3. Mathematical size-spectrum models

As its starting point, any mathematical modelling requires a basic formal description of the system. This should be sufficiently detailed to contain or imply descriptions of the main phenomena of interest. Deciding how much detail one needs to start with is often a matter of trial and error. Typically, one will start with very simple basic descriptions and add model ingredients only upon the suspicion that this might lead to different conclusions. However, the implications of these more complete models can easily become difficult to understand. In particular, the application of formal analytic tools may become impossible. The history of the theory of community size spectra is no exception.

The earliest models developed to understand size spectra distinguished species by their trophic levels and considered the balance of energy flow through food chains (Kerr, 1974; Sheldon et al., 1977). Later theories, however, tended to follow the empirically successful paradigm of simplifying the description of community structure by disregarding species identity. This led to models in which individuals enter the community at a fixed lower cut-off size at constant rate, grow, consume other individuals, and eventually die through predation, fishing, or other causes (e.g. Benoît and Rochet, 2004; Blanchard et al., 2009; Camacho and Solé, 2001; Datta et al., 2010; Law et al., 2009; Platt and Denman, 1978; Silvert and Platt, 1978, 1980; Zhou and Huntley, 1997). In reality, the growth of individuals slows when they reach maturation, at a size dependent on their species identity, after which a large proportion of food intake is invested into reproductive efforts. To take interspecific variability in life histories into account, recent size-spectrum models often go back to distinguishing individuals by species (Hall et al., 2006) or maturation size class (Andersen and Beyer, 2006). To overcome another limitation of earlier theories, Arino et al. (2004) and Maury et al. (2007a,b) made the production of offspring dependent on available resources. Models such as those of Andersen and Ursin (1977), Shin and Cury (2004), Pope et al. (2006), and Hartvig et al. (2011) combine these lines of thought by resolving individuals by both body size (or age) and species, and accounting for the full life cycle of each species separately. The complexity of these models, however, renders their mathematical analysis challenging (Hartvig et al., 2011).

The model analysed here was developed by Hartvig et al. (2011), building on the ideas of Andersen and Beyer (2006). The structure of the model is illustrated in Fig. 2: populations of different species (not necessarily fish) are coupled through feeding interactions. Some of the energy intake by feeding is used to cover maintenance costs, the rest invested into growth. After body mass reaches a maturation threshold, which depends on the species, increasingly larger proportions of surplus energy are invested into reproduction so that growth ultimately comes to a halt. Fecundity, therefore, depends on food availability, but not on the density of reproducing adults, as some models in fisheries science assume. Feeding interactions are particularly strong for predator individuals that are by a certain factor larger than their prey, but interaction strengths may in addition depend on species identities. Predation is the dominating cause of mortality. Each of these model components is represented by fairly simple submodels designed to keep the number of model parameters low. Where applicable, the size-dependence of biological and ecological rates is represented through allometric scaling laws. For a mathematical description of the model, see Section 4.

The model of Hartvig et al. (2011) represents, in simplified form, processes operating on a number of different levels of organisation: the short-term energy balance of individuals, the full life cycle including growth, maturation, and reproduction, the dynamics of intraspecific population structures, population dynamics driven by trophic interactions, and the resulting structure and dynamics of the community as a whole. The model is complete in that it contains, except for the energy input at the bottom of

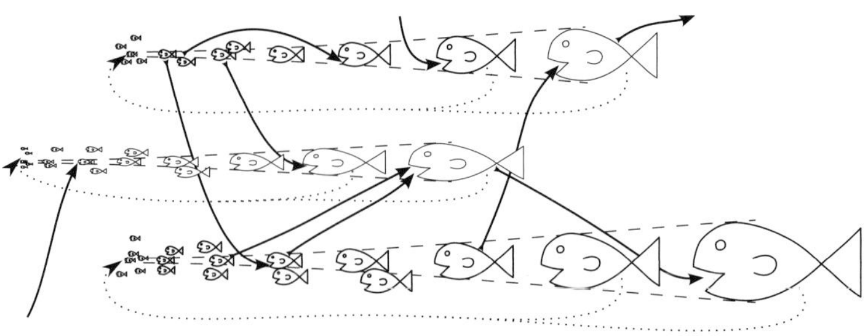

Figure 2 Schematic illustration of the model structure. Shown are three populations of species with different maturation body sizes that feed on and are fed on by other species and themselves. Dashed lines indicate growth of individuals, dotted lines reproduction, and thick arrows feeding interactions.

the food chain, no "loose ends;" the balance of biologically available energy within the system and the balance of individuals within each population are fully accounted for.

1.4. Approximations

Legitimate approximations of community models, here those that do not much affect predicted structure and dynamics of the size spectrum, inform one of those details of community dynamics of which community structure is largely independent. The approximation techniques used here will be discussed in detail when they are applied in the formal analysis. The following provides a non-technical overview.

The basic structure of the mathematical analysis follows from a standard technique used throughout the scientific literature: system dynamics is considered in a *linearised* form valid for small deviations from an equilibrium state. As a result, the analysis technically separates into three steps: (i) the derivation of the equilibrium state, (ii) linearization of system dynamics near this equilibrium state, and (iii) evaluation of this linearised description. Of course, approximations employed in one of the earlier steps remain in place at the later steps. The premise implied when applying this technique is that, although there will be corrections to system dynamics through non-linear effects, its semi-quantitative nature is captured already at the linear level. Whether this is indeed the case is not analysed here, although preliminary comparisons of results with empirical data such as in Fig. 1 are encouraging. What will be discussed in Section 9.6 is the question of whether the linearization is self-consistent in the sense that small deviations from equilibrium remain small in the future (and eventually die out).

An approximate analytic description of the model's equilibrium state becomes possible mainly by combining two standard techniques: *coarse graining* (Perry and Enright, 2006) along the maturation body-size axis and a *mean-field approximation* (Keitt, 1997; Wilson et al., 2003) of trophic interaction strengths. As explained in Section 5.1, these two approximations combined lead to a picture in which species are distinguished only by their characteristic size, for example, maturation body size, and individuals by body size and maturation size, where the community is described by a continuum of species of different maturation sizes, and where trophic interaction strengths depend only on the body sizes of predator and prey individuals. The validity of these approximations supports studies of model systems that focus on the body sizes of species and individuals (Belgrano and Reiss, 2011). Although formal derivation of these approximations for the present model is conceivable, their justification here comes only from their frequent successful application in other contexts. One must therefore be mindful of possible artefacts. Indeed, as discussed in Sections 2.8 and 6.4, while the intricacies of real community food webs appear negligible for size-spectrum dynamics, the stabilising effect resulting from their structure with distinct species linked through a sparse network of feeding interactions cannot be ignored entirely.

Even with these two approximations, the direct mathematical analysis of linearised system dynamics would still be challenging. To simplify the problem further, the *Quasi-Neutral Approximation* (QNA; Rossberg and Farnsworth, 2011) of population dynamics is employed. The QNA assumes that variations through time in density-dependent life-history parameters, and the resulting modifications of intraspecific population structures, are rather small. Common mathematical techniques exploiting small parameters can then be applied. This allows direct calculation of the effective growth (or decay) rates of entire populations, which result from interactions between individuals of different populations at different life stages, without having to take the changes in population structures involved in these processes explicitly into account. When the QNA is valid, it justifies the use of community models that do not resolve different life stages within populations. The common concern that broad intraspecific size structure complicates community dynamics (Gilljam et al., 2011) should therefore not lead to outright dismissal of such models.

A last step required to bring linearised community dynamics into a form simple enough to derive analytic time-dependent solutions is to recognise that, when the offspring of a species are much smaller than the adults, its actual body size has little influence on dynamics under the approximations used here. Offspring may just as well be considered "infinitely small". The resulting, highly simplified description of community dynamics is called the Species Size-Spectrum Model. The approximation is verified by performing QNA and model linearization initially in a formulation where offspring size does depend on maturation body size and carefully analysing the implications of letting the ratio of offspring to adult size go to zero. Incompatibility of unicellular organisms with this condition is one of the reasons why microcosm experiments using such taxa (Reiss et al., 2010) are unsuitable for modelling marine size spectra.

An approximate analytic description of the response of the Species Size-Spectrum Model to external pressures (e.g. fishing) is then derived using a novel technique, related to a *singular perturbation expansion* (Kevorkian and Cole, 1996) of special solutions of the model. These special solutions correspond to the trophic cascades and the bending of size spectra described in Section 2. The validity of the approximation (verified numerically) highlights the centrality of these phenomena for community dynamics.

This analytic theory for size spectrum structure and dynamics is complete to the degree that the underlying model of Hartvig et al. (2011) is complete. It explicitly specifies conditions for jointly attaining and maintaining equilibrium of all essential feedback loops that link individual physiology with community structure. These are the metabolic equilibrium of individuals, the demographic equilibrium of population structures, population-dynamical equilibrium, and the ecological balance of community size spectra.

1.5. Structure of the paper

After briefly reviewing relevant mathematical methods (Section 3) and providing a formal description of the model (Section 4), the approximate steady-state solution of the model is derived in Section 5. Using this solution, the QNA is applied to the model and linearised near the equilibrium in Section 6, leading to the Species Size-Spectrum Model. Novel mathematical techniques are then used to find approximate analytic solutions of that model (Section 7) and verified by comparison with simulations of the same model (Section 8). Section 9 provides technical discussion of further implications of the results and points to questions and research problems arising. Non-technical summaries of a number of noteworthy aspects of the theory are provided in Section 2.

2. SOME ASPECTS OF THE ANALYTIC THEORY EXPLAINED IN NON-MATHEMATICAL LANGUAGE

The formal mathematical analysis of the model by Hartvig et al. (2011) leads to several novel insights into structure and dynamics of size-structured communities and the underlying mechanisms. This section summarises and discusses some of these findings in a non-mathematical language. To avert the risk of misunderstanding inherent in such undertakings, consultation of the referenced parts of the theory for details and specifics is encouraged. Table 1 summarises the crucial mechanisms succinctly.

2.1. Two different scenarios generate power-law community size spectra

The analysis reveals two different scenarios that lead to formation of power-law size spectra (Section 5.2). In the first scenario, primary producers are sufficiently abundant for all grazers and all species higher up in the food chain to feed *ad libitum*. The second scenario corresponds to communities in which

Table 1 Summary of major regulating constraints controlling the model community's steady state

	Regulating constraint	
Phenomenon	**Eutrophic regime**	**Oligotrophic regime**
Size-spectrum exponent (slope)	Physiological mortality rate	Constant satiation
Mortality	Growth and physiological mortality rates	
Individual growth rate	Satiation	
Satiation	Maximal	Size-spectrum coefficient and search/attack rate
Size-spectrum coefficient (intercept)	Producer abundance	Physiological mortality rate
Physiological mortality rate	Intraspecific size structure	
Intraspecific size structure	Population-dynamic equilibrium	
Search/attack rate	Unknown (see Section 9.6)	

typical individuals at all sizes find enough food so as not to die of starvation but if more food was available they could grow faster and generate more offspring. (The crossover between the two scenarios is not considered.)

Figure 1 illustrates that biomass densities of individuals in all size classes can be much higher in eutrophic systems than they are in oligotrophic systems, so situations in which consumers effectively feed *ad libitum* are more likely. The first scenario is therefore here called the *eutrophic regime* here and the second the *oligotrophic regime*. Of course, this nomenclature is not meant to imply that trophic status is defined by consumer satiation. The labelling just hints at the fact that, the more eutrophic a system is, the more likely the regime with *ad libitum* feeding will be encountered.[1]

Perfect oligotrophic power-law size spectra require the "right" abundance of primary producers or a corresponding tuning of other biological or ecological parameters. One conceivable mechanism for this is the evolutionary adaptation of attack (or search) rates to avoid overexploitation of resources (Rossberg et al., 2008). The size-spectrum is determined solely

[1] Bolt labelling of mathematically distinct regimes is common scientific practice. Solid-state physicists, for example, refer to scenarios unfolding at the temperature of liquid nitrogen (-196°C) as the "high-temperature regime" Pratap et al., 1999).

by the allometric exponents for metabolic and attack rates (Andersen and Beyer, 2006 and Section 5.2), typically leading to slopes close to the value suggested by Sheldon et al. (1972). With somewhat over-abundant producers, size spectra will bend upward on double-logarithmic scales, and *vice versa* (Section 9.6).

Eutrophic power-law size spectra appear to impose fewer constraints on parameters. In the eutrophic regime, the size-spectrum slope depends on the efficiency of energy transfer from smaller to larger species. Depending on ecological parameters, the slope can become larger or smaller than in the oligotrophic case, and for real systems, it can vary along the size axis. Absolute abundances scale with the abundances of primary producers.

As the oligotrophic regime is characterised by *ad libitum*, and therefore density-independent, feeding, model populations in this regime are regulated entirely through density-dependent mortality, so it might be a surprise that, despite this pure "top-down control", the abundance of primary producers can determine population abundances at higher trophic levels. This becomes possible because overabundance of species in any given size class releases the early life stages of larger species from predation mortality. As a result, the populations of larger species grow and deplete smaller species through increased mortality—until a new equilibrium is reached (Section 8.4.5).

This new, differentiated picture of the controlling mechanisms might help to understand the role of trophic status in shaping observed size spectra. A general empirical pattern seems to be (Marquet et al., 2005) that in oligotrophic systems, size spectra closely follow power laws with slopes (i.e. exponents) in line with Sheldon's hypothesis Sprules and Munawar (1986), Rodriguez and Mullin (1986a), Gaedke (1992b), Quiñones et al. (2003), and that with increasing nutrient load, size spectra become more irregular, exhibiting weak trends in the best-fitting slopes, which vary between studies (Ahrens and Peters, 1991; Bays and Crisman, 1983; Bourassa and Morin, 1995; Dortch and Packard, 1989; Gaedke et al., 2004; Jeppesen et al., 2000; Pace, 1986; Sprules and Munawar, 1986), with some tendency to favouring higher relative abundance of larger individuals at high nutrient loads (Ahrens and Peters, 1991; Sprules and Munawar, 1986, see also Fig. 1) as expected from the present theory. Mulder and Elser (2009) and Mulder et al. (2009) found the same trend in soil communities. Indeed, such patterns are suggested already by comparisons of the size spectra recorded by Sheldon et al. (1972) around the American continent with corresponding maps of nutrient loading (Garcia et al., 2006). Related to this may be the observation by Yvon-Durocher et al. (2011) that warming increases size-spectrum slopes in aquatic mesocosms.

2.2. Feeding alone couples size classes

An intermediate result of the analytic calculations is the Species Size-Spectrum Model, Eq. (78), a balance equation for the biomasses pertaining to species in different size classes, which are coupled through feeding interactions. The Species Size-Spectrum Model only accounts implicitly for intraspecific size structure. By showing that such a model can describe community dynamics and, by extension, the dynamics of the classical individual size spectrum, the present analysis differs from some earlier approaches. It implies that somatic growth does not essentially contribute to the ecological coupling along the size axis, notwithstanding its important contributions in controlling biomasses and population structures of species. This result provides firm foundations to related criticism by Kerr and Dickie (2001), who argue that the "*characteristics [of size spectra] form an inexorable consequence of mathematical relations between size-dependent allometries of the acquisition and transmission of energy through predation processes and the attendant metabolic dissipation.*" (p. 138). They consider that "mechanisms operating at the physiological level" enter just by determining the "scaling of the efficiency coefficients" (p. 139) in this process. Indeed, the observed deep gaps between domes that Kerr and Dickie (2001) highlight would become tight bottlenecks were size classes primarily coupled through somatic growth. This insight parallels a corresponding observation that, despite the wide body-size ranges species span during growth, the individual size spectrum is highly correlated with the species size spectrum (Shephard et al., 2012).

2.3. Population-level predator–prey size-ratio windows are wide

Most verbal models and qualitative models of trophic interactions in aquatic communities[2] operate at the population or species level, as do many quantitative models. Understanding the general size-dependence of trophic interactions at the population level is therefore important not only for size-spectrum theory.

[2] Such models abound in the literature. Two randomly chosen examples are the statement "*Norway lobster, while not quite as an important part of the cod diet as whiting, is a slow-growing species and consequently more sensitive to the predation release produced by the removal of cod*" made by Speirs et al. (2010) or the graphs of food-web topologies by Yodzis (1998).

Under the simplifying assumption that the size-dependence of trophic interactions is controlled, apart from allometric scaling laws, only by relative sizes, it can be characterised by a predator–prey size-ratio window. This gives, for each ratio of predator to prey size, the probability or intensity of trophic interactions. The window can be defined either for interactions between individuals, with the relative sizes given, for example, in terms of the individuals' body masses, or for interactions between populations of species, with size then measured, for example, in terms of maturation body mass.

It is shown in Section 9.3 that the structure of (average) population-level predator–prey size-ratio windows is mostly determined by the intraspecific size structures of prey and predator. The individual-level predator–prey size-ratio window plays only a minor role. The predicted population-level predator–prey size-ratio window (Fig. 9) is wide, in line with empirical analyses by Neubert et al. (2000) and Woodward and Warren (2007). Under the idealisations invoked in the analytic theory, it decays with an exponent near zero (≈ -0.05) with increasing maturation body mass ratio (Fig. 9). In reality, the windows will, of course, be truncated above maturation size ratios where the newborns or hatchlings of one species are so large that they feed on the adults of the other.

2.4. Simple mass-balance models explain only equilibrium

It turns out (Section 9.2) that predicted equilibrium size spectra of communities perturbed by size-selective fishing can consistently be interpreted by verbal arguments or simplified, static mass-balance models (Christensen and Pauly, 1992) where "large species eat small species" without regard to the broad population-level predator–prey size-ratio windows resulting from somatic growth. However, when such simplified models would be used to predict dynamics (akin to the Ecopath with Ecosim approach; Pauly et al., 2000), this would lead to very different results from when size structure was, at least implicitly, included (Section 9.2, Fig. 10).

2.5. Both upward and downward cascades form in response to size-specific perturbations

The dynamics of complex systems can often efficiently be analysed and described in terms of their responses to controlled perturbations of system states. Adapting this idea to the ecological context, Bender et al. (1984) distinguish between short *pulse perturbations* and long-lasting *press perturbations* of constant intensity. Here, press perturbations will be considered, if not stated otherwise. An example of such a perturbation is sustained fishing with a constant effort.

Trophic cascades, in which a perturbation of abundance at a higher trophic level leads through trophic top-down effects to alternating increases and decreases of abundances towards lower trophic levels, are a well understood phenomenon. They naturally arise in size-spectrum models (Benoît and Rochet, 2004). Andersen and Pedersen (2010) demonstrated in simulations of a size spectrum perturbed by fishing that, together with these "downward" cascades, another type of cascade can emerge, with alternating abundance increase and decrease towards higher trophic levels. The cascades found by Andersen and Pedersen (2010) were always damped, that is, the relative change in abundances became smaller with increasing distance from the externally perturbed size class. Periodic maxima ("domes") in size spectra, as observed, for example, by Schwinghammer (1981), Sprules et al. (1983), and Sprules and Goyke (1994), and documented in Fig. 1, are likely to correspond either to upward or to downward trophic cascades.

The Species Size-Spectrum Model accounts for both types of cascades (Section 8.4.1). Its analysis reveals that upward and downward cascades are mathematically independent phenomena (Section 7.4), and that both can be damped or not (i.e. amplifying), sensitively depending on model parameters (Sections 8.4.4 and 8.4.5). The intuitive mechanism underlying upward cascades is the more-or-less pronounced formation of a trophic ladder, with the steps corresponding to the cascade's local maxima and the space between steps to the minima. In simple variants of the Species Size-Spectrum Model with narrow individual-level predator–prey size-ratio windows (not analysed here), therefore one finds that in upward cascades subsequent abundance maxima are separated by the predator–prey size ratio, whereas for downward cascades subsequent maxima are separated by the square of the predator–prey size ratio (two steps down ladder). In the more realistic model variants considered in Section 8.4, this situation is realised for some parameter choices (e.g. Fig. 7A and B). However, for the standard parameter set used here (Section 8.1), which builds on that proposed by Hartvig et al. (2011), the distinction of upward and downward cascades by characteristic size ratios is blurred (Figs. 4 and 6).

The separation of subsequent local maxima for standard parameters corresponds in both directions to a factor $\approx 10^5$ in body mass (Fig. 4). Perhaps fortuitously, this agrees well with the separation of subsequent "domes" or maxima in empirical size spectra (Fig. 1; Sprules and Goyke, 1994; Sprules et al., 1983). To classify cascades empirically, it might therefore be necessary to determine the direction of motion of perturbation responses in temporally resolved size spectra (Rodriguez and Mullin, 1986b; Rodriguez et al., 1987; Vasseur and Gaedke, 2007).

In the light of the observations in Section 2.4, it is conceivable, though by no means established, that the theoretical interpretation of "domes" in size spectra by Thiebaux and Dickie (1993) through equilibrium mass–balance equations is compatible with formation of these structures by one of the two mechanisms leading to periodic modulations in the present theory. The difference would be that, whereas Thiebaux and Dickie (1993) address the fully developed non-linear form of the domes, the present theory describes their dynamic emergence in the linear response to perturbations. The relationship between the two theories would then be similar to that encountered for water waves (Johnson, 1997), where linear and weakly non-linear theories describe low-amplitude waves, which are necessarily sinusoidal, whereas high-amplitude waves, which, as we know from experience, can assume entirely different shapes, are described by a different, non-linear theory.

The observation that different phyla are often associated with different domes (Gaedke, 1992a) does not necessarily stand in opposition to explanations of these structures in terms of general principles. In fact, evolution might naturally lead to this kind of specialisation within uneven size spectra.

2.6. Trophic cascades form slowly

Upward and downward trophic cascades in communities with broad intrapopulation size distributions will emerge slower than would be expected from models that do not account for intrapopulation size structure (Section 9.4). The reason is that pairs of species for which the adults are in a clear predator–prey relation to each other might interact as prey–predator or as competitors at other life-history stages. Figure 4, for example, documents formation of a downward cascade in the model with a delay corresponding to the threefold maturation age of the main size class targeted by fishing. Lags of 6–12 years between time-series of size-spectrum characteristics and of exploitation rates, which Daan et al. (2005) observed in the North Sea but found "*hard to account for*", might be explained by this observation. The slow recovery of the LFI shown by Fung et al. (2012a,b) in data and simulations is likely to reflect this phenomenon too.

2.7. Depletion of species higher up in the food chain is fast

For the scenario labelled as the oligotrophic regime, the Species Size-Spectrum Model predicts that upward trophic cascades are superimposed with an overall decline of species larger than the size class targeted by size-selective harvesting (Sections 8.4.1 and 8.4.4) and that this effect is the stronger the larger species are relative to the targeted size class (Section 9.4). The underlying mechanism is the same as that operating in simple food-chain models: insufficient supply of energy. However, with broad intrapopulation size distributions, the effect can propagate faster to larger species than it would in simpler models, because the indirect responses of larger species propagating through the food chain are combined with direct responses of similar magnitude attributable to harvesting or starvation of their juveniles. The timescale for the population response of large species to size-selective harvesting of smaller species or individuals is therefore of the order of magnitude of their maturation times.

2.8. Food-web structure essentially affects size-spectrum structure and dynamics

The derivation of the Species Size-Spectrum Model depicts communities as if there was a continuum of interacting species of all sizes and assumes that life-history traits and feeding interactions are determined by maturation sizes and body sizes alone. An immediate consequence is that species of comparable size are also similar in all their ecological traits and therefore compete with each other. Without appropriate regularisation of the model, the resulting competitive exclusion dynamics lead to instabilities of size spectra in which increases in the abundances of some species are compensated by decreases in the abundances of other species of very similar size (when coarsening the resolution of size classes, the spectra are generally stable). In reality, stable size-structured communities can form because ecological traits differ even between species of equal maturation size (Jacob et al., 2011), and, perhaps most importantly, in general, feeding interactions do not depend only on the sizes of prey and predator (Naisbit et al., 2012). Species coexist by forming rather sparse food webs: for example, Rossberg et al. (2011) show that only about 10 species contribute more than 1% to the diet of the average fish species, independent of local species richness. This, it is argued in Section 6.4, modifies continuum size-spectrum dynamics at size resolutions corresponding to the maturation body-size ratios among species in the diets of typical predators.

Because, as a result of strongly overlapping intraspecific size distributions, size-spectrum dynamics rather slow, these food-web effects are likely to be sufficiently strong to modify dynamics substantially, even on coarser scales. In the present analysis, food-web effects are derived, parametrised, and included in the model based on heuristic arguments rather than on first principles. The theory, while allowing many other insights, is therefore unsuitable for reliable quantitative predictions of community dynamics. This would require models that quantitatively account for food-web topology.

2.9. Body mass approximates reproductive value

Fisher (1930) defined the reproductive value of an individual as its expected contribution "*to the ancestry of future generations*". Recently, Rossberg and Farnsworth (2011) developed a method to extend this empirically useful concept from the density-independent (linear) dynamics of a single population to complex communities of interacting populations. They argued, with fisheries management in mind, that a quantification of the size of a population by the summed reproductive values of its members should allow better predictions of future population (or "stock") sizes than other measures of population size, for example, the commonly used spawning-stock biomass.

Remarkably, under a certain set of approximations, the present theory evaluates the reproductive value of individuals to be exactly equal to their body mass (Section 6.2). This result holds for both mature and immature individuals. The summed reproductive value of a population would then simply be its total biomass. Tests of this prediction by computing reproductive values from empirically determined life-history parameters of a population might lead to more reliable characterisations of stock sizes and help to resolve the question whether the smaller or the larger individuals of a stock should be harvested preferentially to maximise economic output under given constraints on ecological impact.

2.10. Physiological mortality is constrained by population dynamic equilibrium

Beyer (1989) recognised the importance of the specific physiological mortality rate, a dimensionless parameter defined as a cohort's mortality rate divided by its specific growth rate (i.e. the relative body mass increase per unit time). Physiological mortality as a function of body size determines a population's internal size structure. Here, it is shown that, under a certain set of approximations, specific physiological mortality must be exactly 1 for immature cohorts in population-dynamic equilibrium (Section 5.4). Greater mortality leads to the decline, less mortality to the growth of populations. Even under relaxed assumptions, a value close to one can still be expected (Section 5.7). Indeed, the density dependence of physiological mortality alone essentially controls population dynamics (Section 6.3, Eqs. (61) and (63)). Real-world complications will modify these results, especially for species with rather small ratios of adult to newborn (or hatchling) body sizes, but overall a close relation between population-dynamic equilibrium and specific physiological mortality is expected. This insight might help fisheries managers interpreting the outcomes of virtual population analyses (Lassen and Medley, 2001) and relating them to stock dynamics.

To the extent that the specific physiological mortality of immature individuals attains a universal value, the results of Beyer (1989) imply that the size structure of the immature component of populations is universal, except for a truncation below the size of newborns or hatchlings (Section 6.2). An empirical test of this approximate universality of size distributions and a better understanding of the magnitude and the causes for deviations from it might lead to other generalisations regarding structure and dynamics of interacting size-structured populations (Houde, 2009).

2.11. Solutions of "size-spectrum" equations follow general characteristics

In Section 7, this chapter develops a general method to approximate the solutions of a wide class of equations that describe the dynamics of size-structure communities, subject to two general principles that are often invoked in ecological theory. The first principle is allometric scaling of ecological rates with a fixed power n of body size ($n=3/4$ is often used), and the second principle posits that, apart from the allometric scaling of rates, interaction strengths between individuals depend only on their relative, but not their absolute, sizes.

The approximate general solution of this class of equations implies that, in response to press perturbations targeting a narrow range of size classes, fronts are emitted from the perturbed size range that travel at constant speed on a scale given by body mass raised to the power $(1-n)$, either towards larger or towards smaller size classes. The fronts leave behind themselves either a static, modulated structure with a wavelength that is constant on the logarithm body mass axis (e.g. trophic cascades) or regions of either consistently raised or reduced abundances. The amplitude of the modulation or of the rise/decline can either increase away from the perturbed size class, or decrease. The effects of a pulse perturbation can be described by linear combinations of press perturbations (Section 9.6). A graphic method to decide for a given system which of the various conceivable scenarios will be realised is introduced in Section 8.4.3.

3. METHODS

This section briefly recalls some standard concepts of functional analysis used in developments of the theory below. The text by Boccara (1990) covers most of these topics in mathematically rigorous yet accessible form.

For a real- or complex-valued function $f(x)$, define its Fourier transform $\hat{f}(\xi)$ such that

$$\hat{f}(\xi) = \int_{-\infty}^{\infty} e^{-i\xi x} f(x)\mathrm{d}x, \quad f(x) = \int_{-\infty}^{\infty} \frac{e^{i\xi x}}{2\pi}\hat{f}(\xi)\mathrm{d}\xi, \qquad [1]$$

where i is the imaginary unit ($i^2 = -1$). The convolution $f*g$ of two functions is defined as the function h given by

$$h(x) = [f*g](x) = \int_{-\infty}^{\infty} f(y-x)g(y)\mathrm{d}y. \qquad [2]$$

With normalisations as in Eq. (1), Fourier transformations directly carry convolutions over into multiplication, that is, for any two functions f, g, one has $\widehat{f*g} = \hat{f}\hat{g}$. This relation implies that $f*g = g*f$, $(f*g)*h = f*(g*h)$, and $f*(g+h) = f*g + f*h$.

Depending on the convergence of the defining integral, the value of a Fourier transform $\hat{f}(\xi)$ can be computed via Eq. (1) not only for any real ξ but also for some complex-valued ξ. Consider, for example, the Gaussian window

$$W_\sigma(y) = \exp\left(-\frac{y^2}{2\sigma^2}\right). \qquad [3]$$

Its Fourier integral evaluates for arbitrary complex ξ to

$$\hat{W}_\sigma(\xi) = \sqrt{2\pi}\sigma \exp(-\xi^2\sigma^2/2). \qquad [4]$$

As another example, the function defined by $f(x) = 0$ for $x < 0$ and $f(x) = e^{-2x}$ for $x \geq 0$ has, by Eq. (1), the Fourier transform $\hat{f}(\xi) = (2 + i\xi)^{-1}$ for any complex ξ such that $\mathrm{Im}\{\xi\} < 2$. (Real and imaginary parts of complex numbers ξ are denoted by $\mathrm{Re}\{\xi\}$ and $\mathrm{Im}\{\xi\}$, respectively.) When $\mathrm{Im}\{\xi\} \geq 2$, the integral in Eq. (1) does not converge. However, the definition of $\hat{f}(\xi)$ is naturally extended to any complex $\xi \neq 2i$ by analytic continuation. Then, the formula $\hat{f}(\xi) = (2 + i\xi)^{-1}$ is valid for all complex ξ except $\xi = 2i$. Such analytic continuations of Fourier transforms will play an important role below.

Dirac's delta function $\delta(x)$ symbolises the idealised case of a function that is zero for all x except $x = 0$, and so (infinitely) large at $x = 0$ that $\int_{-a}^{a} \delta(x)\mathrm{d}x = 1$ for any $a > 0$. It follows, for example, that $\int_{-a}^{a} f(x)\delta(x)\mathrm{d}x = f(0)$ for any $a > 0$.

A functional derivative $\delta F[g]/\delta g(y)$ is the generalisation of the notion of a gradient to vectors with continuous-valued indices y, that is, functions $g(y)$. Standard rules of differential calculus translate straightforwardly to functional derivatives, except that, because the dependence of an expression $F[g]$ in $g(x)$ on the single point of $g(x)$ where $x = y$ is generally tiny, functional derivatives contain an additional factor $\delta(y)$. For example, $\delta g(x)/\delta g(y) = \delta(x - y)$, and $\delta \int_{-\infty}^{\infty} f(g(z))\mathrm{d}z/\delta g(y) = \mathrm{d}f(u)/\mathrm{d}u$ evaluated at $u = g(y)$.

Define a scalar product for real-valued functions as

$$\langle f|g\rangle = \int_0^\infty f(x)g(x)\mathrm{d}x. \qquad [5]$$

The adjoint of a linear operator L (a linear mapping from functions $g(x)$ onto functions $[Lg](x)$) can then be defined as the linear operator L^+ that satisfies for all functions f and g (with some mathematical constraints):

$$\langle L^+ f|g\rangle = \langle f|Lg\rangle. \qquad [6]$$

It can be computed as

$$L^+ f(x) = \frac{\delta\langle f|Lg\rangle}{\delta g(x)}. \qquad [7]$$

4. MODEL

The model underlying this theory has been motivated and derived by Hartvig et al. (2011). Its major strengths are a consistent synthesis of dynamic descriptions of size-structured communities, populations, individual growth, and bioenergetics based on general biological principles; the model's small set of parameters, all of which are estimated from empirical data; and its demonstrated structural stability. This section provides only a technical summary and highlights minor deviations from the formulation of Hartvig et al. (2011). Table 2 lists symbols used throughout this work and their interpretation.

Table 2 List of important symbols

Symbol	Defined near equation	Interpretation
α	(11)	Assimilation efficiency
β		Preferred predator–prey size ratio
$\tilde{\beta}(w)$	(71)	Equilibrium population structure on log-scale
$\tilde{\beta}^{-}(w)$	(73)	$\tilde{\beta}(-w)$
$\tilde{\beta}_{v}^{-}(w)$	(126)	$e^{vw}\tilde{\beta}^{-}(w)$
γ	(10)	Coefficient of search/attack rate
ϵ	(12)	Efficiency of reproduction
η	(11)	Maturation- over asymptotic mass
$\kappa_0(w)$	(70)	Individual-level interaction kernel on log-scale
λ	(16)	Exponent of community size spectrum
$\mu_{\mathrm{p}}(m)$	(13)	Predation mortality
μ_0	(14)	Coefficient of natural mortality
$\tilde{\mu}$	(26)	Coefficient of predation mortality
v	(73)	$3-\lambda-n$
σ_F	(77)	Width of harvested size range
$\phi(m)$	(19)	Density of prey available
$\tilde{\phi}$	(19)	Coefficient of $\phi(m)$
$\Phi(u)$	(79)	Fishing pressure on log-scale
$\psi(x)$	(11)	Reproduction selection function
$a_j(u, t)$	(93)	Delocalised response to fishing pressure
$a(m, m_*)$	(60)	Specific physiological mortality
$\tilde{a}$	(35)	Scale-free juvenile specific physiological mortality
$b(u)$	(75)	Perturbation of equilibrium biomass distribution over log-scale
$b_0(u)$	(89)	Fourier solution of Species Size-Spectrum model
$b_c(u)$	(93)	Core response to fishing pressure

Continued

Table 2 List of important symbols—cont'd

Symbol	Defined near equation	Interpretation
$B(m_*)$	(56)	Reduced dynamic variables of QNA
$B_{\text{tot}}(m_*)$	(40)	Scale-invariant equilibrium biomass per maturation size class
$\tilde{B}_{\text{tot}}$	(40)	Coefficient of $B_{\text{tot}}(m_*)$
B_{m_*}	(49)	Equilibrium biomass per maturation size class
C_{m_*}	(53)	Normalisation constant of reproductive value
F	(77)	Maximal fishing mortality
$f(m)$	(10)	Degree of satiation
$\tilde{g}(x)$	(24)	Scale-free growth rate
$\tilde{g}_r(x)$	(24)	Scale-free fecundity
$g(m, m_*)$	(11)	Growth rate
$g_r(m, m_*)$	(11)	Fecundity
$\tilde{g}_0$	(24)	Coefficient of growth rate/fecundity
h	(11)	Coefficient of maximal food intake
$J_{-m_*}(m)$	(52)	Equilibrium flow of reproductive value
k	(11)	Coefficient of metabolic loss rate
$K_{\text{tot}}(m_*)$	(42)	Metabolic losses per maturation size class
$\tilde{K}_{\text{tot}}$	(42)	Coefficient of $K_{\text{tot}}(m_*)$
$K(m_*, m'_*)$	(62)	Population-level interaction kernel on linear scales
$\tilde{K}(w)$	(74)	Population-level interaction kernel on log-scales
L_{m_*}	(47)	McKendrick–von Foerster operator
$L^+_{m_*}$	(51)	Adjoint McKendrick–von Foerster operator
m		Body size
M	(75)	Arbitrary unit mass
m_*		Body size at maturation
m_0, $m_0(m_*)$	(16)	Size of offspring

Table 2 List of important symbols—cont'd

Symbol	Defined near equation	Interpretation
m_{F}	(79)	Size class targeted by fishing
n	(10)	Exponent for physiological rates
$N(m, m_*)$	(16)	Distribution of individuals by body and maturation size
$N_i(m)$	(9)	Intraspecific size distribution
$\tilde{N}(x)$	(16)	Scale-free intraspecific size distribution
$\mathcal{N}(m)$	(17)	Community size spectrum
$\tilde{\mathcal{N}}$	(17)	Coefficient of community size spectrum
q	(10)	Exponent for search/attack rate
$s(x)$	(8)	Predator–prey size-ratio window
u_{F}	(108)	Logarithmic size class targeted by fishing
$V_{m_*}(m)$	(51)	Reproductive value
$W_{\sigma}(y)$	(3)	Gaussian window
$W_{m_*}(m)$	(48)	Equilibrium population structure
x_0	(29)	Relative size of offspring

The model describes trophic interactions between reproductively isolated populations of heterotrophic species in large, size-structured aquatic communities. It distinguishes between individuals of different body mass m within each population j, and assumes all individuals of the same size and the same population to be equivalent. Each population is therefore characterised by a size distribution $N_j(m)$ of individuals. Specifically, following Hartvig et al. (2011), the number of individuals of population j within the infinitesimally small body mass interval $[m; m+dm]$ is given by $\mathcal{V}N_j(m)dm$, where $\mathcal{V}$ is the volume of the system considered (a formulation in terms of areal density would equally be possible). In general, $N_j(m)$ will be time-dependent.

Energy and biomass are thought to flow overwhelmingly from smaller to larger individuals (Gilljam et al., 2011; Woodward et al., 2010b). The demographics of the smallest species in the community are not modelled explicitly. Rather, the idealisation is invoked that all species are much larger than the smallest individuals in the community, so that the relevant (log m)-axis stretches infinitely into both directions. This admits a description in which the prey species of every consumer are again consumers, and all species are consumed by some other, larger species so that, mathematically, there is an infinite hierarchy of species sizes in the community. The original model of Hartvig et al. (2011) instead summarises the density of individuals of small species by a *resource spectrum* $N_R(m)$.

The biomass density of food available to an individual of species j with body mass m is given by

$$\phi_j(m) = \int_0^\infty m_p \mathcal{N}_j(m_p) s\left(\ln \frac{m}{m_p}\right) dm_p, \qquad [8]$$

where $\mathcal{N}_j(m_p)$ is the density of prey of size m_p available to j, and the function $s(\cdot)$ characterises the individual-level predator–prey size-ratio window on a logarithmic scale.

The theory admits essentially arbitrary predator–prey size-ratio windows, but the following plausible constraints shall be imposed: $s(u)$ is non-negative and decays faster than exponentially as $u \to \pm\infty$. The Fourier transform $\hat{s}(\xi)$ of $s(u)$ is then an entire function (i.e. it is finite for any complex ξ). An example satisfying this condition is given by Eqs. (109) and (110).

The density of prey entering Eq. (8) is given by

$$\mathcal{N}_j(m) = \sum_k \theta_{j,k} N_k(m), \qquad [9]$$

with the weights $\theta_{j,k} \in [0;\ 1]$ characterising dietary preferences, that is, the food web. Dependence of $\theta_{j,k}$ on life stages could be incorporated into the formalism but is not considered here. On the contrary, the analysis will largely be based on a mean-field approximation, that is, the ecosystem mean of $\theta_{j,k}$ is absorbed into the search rate (see below), and all $\theta_{j,k}$ are set to 1. Then $\mathcal{N}_j(m) = \mathcal{N}(m)$ is the same for all predators j, and so the density of available food $\phi_j(m) = \phi(m)$.

Food intake is determined by the density of prey $\phi(m)$, the effective search rate, assumed to be of the form γm^q (with constants γ, $q > 0$), and the physiological maximum food-intake rate of individuals. The latter is assumed to scale as hm^n, with constants $h > 0$ and $0 < n < q$. For simplicity, n is identified with the allometric exponent for metabolic rates. Typically, $n \approx 0.75$ (Peters, 1983), but here only $n < 1$ is required. Specifically, individuals of size m feed at a rate $f(m)hm^n$, where the degree of satiation $f(m) \in [0;\ 1]$ is given by

$$f(m) = \frac{\gamma m^q \phi(m)}{\gamma m^q \phi(m) + h m^n}. \qquad [10]$$

This corresponds to an individual-level Type II functional response with handling time $(hm^n)^{-1}$ (dimensions: Time/Mass). In the limit $h \to \infty$, the special case of a linear functional response $f(m)hm^n \to \gamma m^q \phi(m)$ is recovered.

Food intake is discounted by the assimilation efficiency $0 < \alpha < 1$ and metabolic losses at a rate km^n (with constant $k > 0$), leaving a net uptake of energy $\alpha f(m)hm^n - km^n$ available for somatic growth and reproduction. It is apportioned between these two uses according to a reproduction-selection function $\psi(m/m_{*j}) \in [0, 1]$, with m_{*j} characterising the size of species j at maturation. That is, individuals grow at a rate $g(m, m_{*j})$ (dimension Mass/Time) and invest into offspring at a rate $g_r(m, m_{*j})$, with

$$\begin{aligned} g(m, m_{*j}) &= [1 - \psi(m/m_{*j})][\alpha f(m)h - k]m^n, \\ g_r(m, m_{*j}) &= \psi(m/m_{*j})[\alpha f(m)h - k]m^n. \end{aligned} \qquad [11]$$

Individuals produce offspring, assumed to be of size m_{0j}, at a rate $(\epsilon/2m_{0j})g_r\,(m, m_{*j})$ (dimension 1/Time), with the factor 1/2 representing the assumed proportion of females in a population and the reproduction efficiency ϵ discounting for all additional losses from production of eggs to birth or hatching. With the case of fish in mind, Hartvig et al. (2011) assumed the size of hatchlings $m_{0j} = m_0$ to be effectively the same for all species, independent of maturation size m_{0j}. Here, m_{0j} values that vary between species will be admitted. Integrating over spawners of all sizes, species j produces offspring at a rate

$$R_j = \frac{\epsilon}{2m_{0j}} \int_0^\infty N_j(m) g_r(m, m_{*j}) \mathrm{d}m. \qquad [12]$$

Many analytic results derived here are independent of the specific form of the reproduction-selection function $\psi(x)$, but the following general characteristics will be used below: whereas maturation is a gradual process, it is sufficiently well defined that, as $x \to 0$, $\psi(x)$ goes to zero, with most of the transition taking place near $x \approx 1$. In order for individuals to have well-defined growth trajectories, $\psi(x)$ should be Lipschitz continuous.[3] Further, it is assumed that $\psi(\eta^{-1}) = 1$ for some constant $0 < \eta < 1$, and that $0 \leq \psi(x) < 1$ for $0 < x < \eta^{-1}$.

[3] A function $f(x)$ is called Lipschitz continuous if there is a positive constant K such that $|f(x) - f(y)| < K\,|x - y|$ for all x, y in the domain of f.

These conditions ensure that individuals cannot reach the size m_*/η, which therefore becomes a sharp upper bound on body size.

The predation mortality of individuals of size m_p follows from the model of food intake above. It is given by

$$\mu_\mathrm{p}(m_\mathrm{p}) = \int_0^\infty s\left(\ln\frac{m}{m_\mathrm{p}}\right)[1-f(m)]\gamma m^q \mathcal{N}(m)\mathrm{d}m. \qquad [13]$$

Mortality from all other causes is summarised in a species-dependent background mortality

$$\mu_j = \mu_0 m_{*j}^{n-1}, \qquad [14]$$

chosen such as to be proportional to inverse generation time. It is assumed here that, as Hartvig et al. (2011) suggested, the parameter $\mu_0 \geq 0$ is sufficiently small that μ_j becomes relevant only for adults. Hartvig et al. (2011) also included starvation mortality in their model, but the situations where this becomes relevant lie beyond the scope of the current theory.

Combining the effects of growth, mortality, and reproduction, one obtains a balance equation for the distribution $N_j(m)$ of the individuals forming a population over body sizes (McKendrick, 1926; von Foerster, 1959):

$$\frac{\partial N_j(m)}{\partial t} + \frac{\partial}{\partial m}\left[g(m, m_{*j})N_j(m)\right] = -\left[\mu_\mathrm{p}(m) + \mu_0 m_{*j}^{n-1}\right]N_j(m). \qquad [15]$$

The boundary condition $R_j = g(m_{0j}, m_{*j})N_j(m_{0j})$, equating offspring production with outgrowth, closes the model.

5. PROPERTIES OF THE SCALE-INVARIANT COMMUNITY STEADY STATE

5.1. Discussion of the underlying approximations

To obtain an analytic characterisation of the solution of this model, two additional approximations are made. Both are motivated by the empirical fact that the community size spectrum $\mathcal{N}(m)$ often follows a power law $\mathcal{N}(m) \propto m^{-\lambda}$ over a wide range of body sizes m. (Sheldon's hypothesis corresponds to $\lambda \approx 2$.) The first approximation is to coarse-grain over species with similar maturation size m_{*j}, giving a smooth distribution $N(m, m_*)$ of individuals over body sizes m and maturation sizes m_* (Andersen and Beyer, 2006). In this coarse-grained description, the number of individuals in the infinitesimal body size interval $[m; m+\mathrm{d}m]$ belonging to species with maturation body sizes in the infinitesimal interval $[m_*; m_* + dm_*]$ is given by $\mathcal{V}N(m,m_*)\,dm\,dm_*$ with $\mathcal{V}$ again denoting system volume. It is assumed that the size of hatchlings or newborns m_0 is determined largely by maturation size m_*, so that the former becomes a function of the latter. This can be a linear relationship, a functional relationship where m_0 is constant over some range in m_*, or any other type of non-linear functional relationship. With this and the foregoing mean-field approximation in place, species with similar m_* will typically be similar in all their ecological characteristics. The competitive exclusion dynamics resulting from this approximation is discussed in Section 6.4.

The second approximation is motivated by the scale-invariance of the community size spectrum, highlighted by Camacho and Solé (2001) and Capitán and Delius (2010).[4] It suggests that the underlying intraspecific size distributions $N_j(m)$ are, to some approximation, also scale-invariant when factoring out the dependence on maturation body size m_{*j}. Specifically, conditions will be derived for which $N_j(xm_{*j}) \propto N_k(xm_{*k})$ to a good approximation over a wide range in dimensionless body masses x for most pairs of species (j, k). The obvious truncation of intraspecific size structure below the size of hatchlings m_{0j}, that is, for[5] $x < x_{0j} \stackrel{\text{def}}{=} m_{0j}/m_{*j}$, violates this approximation and will therefore receive special attention below. It turns out that the limit $x_{0j} \to 0$ is often non-singular and taking it yields ecologically consistent results.

The combination of these two approximations leads to an ansatz

$$N(m, m_*) = \begin{cases} m_*^{-\lambda-1}\tilde{N}(m/m_*) & \text{for } m \geq m_0, \\ 0 & \text{for } m < m_0 \end{cases} \qquad [16]$$

for the coarse-grained, scale-invariant distribution of individuals, where m_0 is a function of m_* and the scale-free size distribution $\tilde{N}(x)$ remains to be determined.[6]

In the small-hatchling limit $m_0/m_* \to 0$, the coarse-grained community size spectrum Eq. (9) can be evaluated, by using first Eq. (16) and then substituting $m_* = m/x$, as

4 Recall that scale invariance in the simple form $\mathcal{N}(cm) = c^{-\lambda}\mathcal{N}(m)$ is equivalent to the power-law condition $\mathcal{N}(m) \propto m^{-\lambda}$: The power law obviously implies scale invariance. The converse can be seen by differentiating the scaling laws with respect to the scale factor c at $c=1$ and solving the resulting differential equation. As both conditions imply each other, they are equivalent.

5 The notation $A \stackrel{\text{def}}{=} B$ indicates the definition of A as B.

6 To streamline notation, a tilde (~) is used throughout this work to indicate the "scale-invariant part" of some mass-dependent function. This leads to the following modification of the notation used by Andersen and Beyer (2006) and Hartvig et al. (2011): $a \to \tilde{a}, N_c(m) \to \mathcal{N}(m), \kappa_c \to \tilde{\mathcal{N}}, \hbar \to \tilde{g}_0, \alpha_p \to \tilde{\mu}$.

$$\begin{aligned}\mathcal{N}(m) &= \int_0^\infty N(m, m_*)\mathrm{d}m_* \\ &= \int_0^\infty m_*{}^{-\lambda-1}\tilde{N}(m/m_*)\mathrm{d}m_* \\ &= \int_0^\infty x^{\lambda-1}m^{-\lambda}\tilde{N}(x)\mathrm{d}x \\ &= m^{-\lambda}\tilde{\mathcal{N}},\end{aligned} \qquad [17]$$

with a constant

$$\tilde{\mathcal{N}} = \int_0^\infty x^{\lambda-1}\tilde{N}(x)\mathrm{d}x. \qquad [18]$$

Equation (17) confirms the supposed relation between ansatz (16) and power-law size spectra, Eq. (17). Consistency of this result requires that the integral in Eq. (18) converges, which will be verified after computing $\tilde{N}(x)$ and λ in Section 5.4.

5.2. Scale-invariant demographics

The steady-state demographic rates are now evaluated assuming scale-invariance (Eq. (17)) and making use of the mean-field approximation $\mathcal{N}_j(m) = \mathcal{N}(m)$. Prey availability, Eq. (8), then becomes

$$\begin{aligned}\phi(m) &= \int_0^\infty m_\mathrm{p}\mathcal{N}(m_\mathrm{p})s\left(\ln\frac{m}{m_\mathrm{p}}\right)\mathrm{d}m_\mathrm{p} \\ &= m^{2-\lambda}\tilde{\phi},\end{aligned} \qquad [19]$$

where the constant

$$\tilde{\phi} = \tilde{\mathcal{N}}\int_0^\infty y^{\lambda-3}s(\ln y)\mathrm{d}y \qquad [20]$$

is obtained from Eq. (19) using the substitution $m_\mathrm{p} = m/y$. Doing another substitution, $y = e^x$, then noting the formal similarity with a Fourier integral, the integral in the last equation above can be expressed in terms of the Fourier transform $\tilde{s}(\xi)$ of $s(x)$:

$$\begin{aligned}\int_0^\infty y^{\lambda-3}s(\ln y)\mathrm{d}y &= \int_{-\infty}^\infty e^{(\lambda-2)x}s(x)\mathrm{d}x \\ &= \tilde{s}(i(\lambda-2)).\end{aligned} \qquad [21]$$

Below, similar transformations will be applied to evaluate other integrals.

The consumer satiation resulting from the prey availability $\phi(m)$ computed above is

$$f(m) = \frac{\gamma m^{2+q-\lambda}\tilde{\phi}}{\gamma m^{2+q-\lambda}\tilde{\phi} + hm^n}. \quad [22]$$

According to Eq. (11), consumers neither grow nor reproduce when satiation $f(m) < k/\alpha h$. Exclusion of this situation leaves two scenarios for scale-invariance of $f(m)$. In the first case, $\gamma m^{2+q-\lambda}\tilde{\phi} \gg hm^n$ over the body-size (m) range of interest, so $f(m) \approx 1$. This regime breaks down at body sizes $m \approx [h/(\tilde{\phi}\gamma)]^{1/(2+q-\lambda-n)}$, where a crossover to a non-scale-invariant regimes occurs. However, this point can be moved mathematically to arbitrarily large or small m, respectively, by an appropriate choice of γ. In the second case, $2+q-\lambda=n$, so that $f(m)=\tilde{f}$ with constant $\tilde{f} \stackrel{\text{def}}{=} \gamma\tilde{\phi}/(\gamma\tilde{\phi}+h)$. This fixes the size-spectrum exponent as $\lambda = -2+q-n$ (Andersen and Beyer, 2006). Mindful of the caveats discussed in Section 2.1, the first case is called the eutrophic regime and the second the oligotrophic regime.[7]

Both cases lead to scale-invariant somatic and reproductive growth rates:

$$g(m, m_*) = \tilde{g}(m/m_*)m^n, \quad g_r(m/m_*) = \tilde{g}_r(m/m_{r*})m^n, \quad [23]$$

where

$$\tilde{g}(x) = [1 - \psi(x)]\tilde{g}_0, \quad \tilde{g}_r(x) = \psi(x)\tilde{g}_0 \quad [24]$$

with

$$\tilde{g}_0 = \alpha f_0 h - k, \quad [25]$$

and $f_0 = 1$ or $f_0 = \tilde{f}$, depending on the case considered.

To evaluate predation mortality, Eq. (13), in the eutrophic regime, the lowest-order correction to satiation, $f(m) = 1 - hm^{n+\lambda-2-q}\gamma^{-1}\tilde{\phi}^{-1} + \cdots$, needs to be taken into account, giving

$$\begin{aligned} \mu_p(m_p) &= \int_0^\infty s\left(\ln\frac{m}{m_p}\right) hm^{n-2}\tilde{\phi}^{-1}\tilde{\mathcal{N}}\, dm \\ &= \tilde{\mu} m_p^{n-1}, \end{aligned} \quad [26]$$

with a constant

[7] Maury et al. (2007b) investigated a size-spectrum model in which the allometric exponent for maximal ingestion $\propto m^{2/3}$ is smaller than the exponent for metabolic losses $\propto m$, so that ingestion limits are relevant only for the largest organisms. Then other scaling regimes become possible, e.g. such that ingestion balances maintenance cost by having feeding levels $\propto m^{1-2/3} = m^{1/3}$. Their Fig. 5e might represent this scenario for body lengths <0.1 m. It is not considered here.

$$\tilde{\mu} = \frac{h\tilde{\mathcal{N}}}{\tilde{\phi}} \hat{s}(i(n-1)), \qquad [27]$$

expressed in terms of the Fourier transform of $s(x)$.

In the oligotrophic regime, predation mortality evaluates to a power law of the same form as Eq. (26), but now with coefficient

$$\tilde{\mu} = (1-\tilde{f})\gamma\tilde{\mathcal{N}}\hat{s}(i(n-1)) = \frac{\gamma h \tilde{\mathcal{N}}}{\gamma\tilde{\phi}+h} \hat{s}(i(n-1)). \qquad [28]$$

It is worth noting that, although the demographic parameters derived in this section depend on the exponent λ and coefficient $\tilde{\mathcal{N}}$ characterising the community size spectrum, Eq. (17), the scale-free size distribution $N(x)$ itself does not enter the results. Hence, the results hold also when the community size spectrum follows a power law but is not apportioned to scale-free intrapopulation size distributions. With such a situation in mind, it is instructive to consider the boundary condition $g_j(m_{0j})N_j(m_{0j}) = R_j$ of the McKendrick–von Foerster Equation, while making use of the scale-invariant expressions for the demographic parameters derived above but allowing arbitrary population structures $N_j(m_{0j})$. With $m_{0j} = x_0 m_{*j}$, this yields by Eqs. (12), (23), and (24):

$$[1-\psi(x_0)]\tilde{g}_0 x_0^n m_{*j}^n N_j(x_0 m_{*j}) = \frac{\epsilon \tilde{g}_0 m_{*j}^n}{2x_0} \int_0^\infty N_j(x m_{*j})\psi(x)x^n \mathrm{d}x. \qquad [29]$$

Observing that the maturation selection function for hatchlings $\psi(x_0)$ is generally zero, and cancelling common factors on both sides of the equation, this reduces to

$$N_j(x_0 m_{*j}) = \frac{\epsilon x_0^{-1-n}}{2} \int_0^\infty N_j(x m_{*j})\psi(x)x^n \mathrm{d}x. \qquad [30]$$

The exponent $-1-n$ of x_0 decomposes into a contribution -1, arising from simple book-keeping, and a contribution $-n$ for the ratio of the rates of metabolic activity of adults and hatchlings. Little more ecology enters this power law, Eq. (30), for the scaling of the ratio of hatchling and adult abundance with x_0. It should therefore be robust with respect to variations in the structure of the model and its solutions.

5.3. Scale-free size distribution

The scale-free size distribution $\tilde{N}(x)$ can now be obtained as the equilibrium solution of the McKendrick–von Foerster Equation (15). Taking above results for growth and mortality into account,

$$\frac{\partial}{\partial m}\left[\tilde{g}(m/m_*)m^n\tilde{N}(m/m_*)\right] = -(\tilde{\mu}m^{n-1} + \mu_0 m_*^{n-1})\tilde{N}(m/m_*), \qquad [31]$$

or, substituting $m = m_* x$ and cancelling a factor m_*^{n-1},

$$\frac{\partial}{\partial x}\left[\tilde{g}(x)x^n\tilde{N}(x)\right] = -(\tilde{\mu}x^{n-1} + \mu_0)\tilde{N}(x). \qquad [32]$$

The equation is solved by

$$\tilde{N}(x) = \frac{\tilde{N}_0}{\tilde{g}(x)x^n}\exp\left[-\int_1^x \frac{\tilde{\mu}}{\tilde{g}(x')x'} + \frac{\mu_0}{\tilde{g}(x')x'^n}\mathrm{d}x'\right] \qquad [33]$$

for $x < \eta^{-1}$ and $\tilde{N}(x) = 0$ for $x \geq \eta^{-1}$. The normalisation factor $\tilde{N}_0$ is related to $\tilde{\mathcal{N}}$ via Eq. (18). Differentiability of $\tilde{N}(x)$ requires the integral in Eq. (33) to diverge at $x = \eta^{-1}$, which is the case when the function $\psi(x)$ is Lipschitz continuous at the point $\psi(\eta^{-1}) = 1$, as demanded above.

To understand the structure of solution (33), recall that for body sizes much smaller than maturation size ($x \ll 1$), all available energy is used for growth ($\psi(x) = 0$), so $\tilde{g}(x)$ reduces to $\tilde{g}_0$ by Eq. (24). The integral in Eq. (33) can then be evaluated, giving

$$\tilde{N}(x) \propto \frac{1}{x^n}x^{-\frac{\tilde{\mu}}{\tilde{g}_0}}\exp\left[-\frac{\mu_0(x^{1-n} - 1)}{\tilde{g}_0(1-n)}\right] \quad (\text{for}\, x \ll 1).$$

With $n < 1$, and therefore $x^{1-n} \to 0$ as $x \to 0$

$$\tilde{N}(x) \propto x^{-n-\tilde{a}} \quad (\text{for}\, x \ll 1), \qquad [34]$$

where the constant

$$\tilde{a} = \frac{\tilde{\mu}}{\tilde{g}_0}, \qquad [35]$$

is the specific physiological mortality (Beyer, 1989) of immature individuals.

5.4. Implications of the boundary condition (part I)

To assure demographic equilibrium, the boundary condition, Eq. (30), still needs to be evaluated. Putting $N_j(m) \propto \tilde{N}(m/m_{*j})$ into this equation and substituting Eq. (34) for the left-hand side, one obtains

$$x_0^{-n-\tilde{a}} \propto x_0^{-1-n}. \qquad [36]$$

As x_0 takes different values for different maturation size classes, existence of a scale-free intrapopulation size distribution $\tilde{N}(x)$ requires that

$$\tilde{a} = 1. \qquad [37]$$

The mechanism by which this condition can be satisfied depends on the size-spectrum regime. In the oligotrophic case, the results of Section 5.2 combine with $\tilde{a} = 1$ to the condition

$$\tilde{\mathcal{N}} = \frac{k}{\gamma[(\alpha - kh^{-1})I_1 - I_2]}, \qquad [38]$$

where $I_1 \stackrel{\text{def}}{=} \hat{s}(i(\lambda - 2))$ and $I_2 \stackrel{\text{def}}{=} \hat{s}(i(n-1))$. Therefore, $\tilde{a} = 1$ implies a constraint on physiological parameters or on the coefficient of the community size spectrum $\tilde{\mathcal{N}}$. For the eutrophic regime, Section 5.2 with $\tilde{a} = 1$ implies

$$(\alpha - kh^{-1})I_1 - I_2 = 0. \qquad [39]$$

By varying the size-spectrum slope λ, the value of I_1 can be adjusted to satisfy this condition. Hence, power-law size spectra impose a condition on the size-spectrum slope in the eutrophic regime, and on absolute abundances (with a slope fixed at $\lambda = 2 + q - n$) in the oligotrophic regime.

To interpret Eqs. (38) and (39) ecologically, observe first that $\alpha - kh^{-1}$ is the conversion efficiency at *ad libitum* feeding (assimilation minus metabolic losses). Now, consider the special case of a sharply defined predator–prey size ratio β, that is, $s(x) = \delta(x - \ln \beta)$, $\hat{s}(\xi) = \beta^{-i\xi}$. Equation (39) is then equivalent to $\alpha - kh^{-1} = \beta^{n-1}\beta^{2-\lambda}$, which has the form of an energy-balance equation as encountered in early size-spectrum theory (e.g. Platt and Denman, 1978): losses through conversion $\alpha - kh^{-1}$ are matched by a corresponding reduction of metabolic activity at the next trophic level β^{n-1}, corrected by the ratio of predator to prey biomass abundance $\beta^{2-\lambda}$. In the oligotrophic regime, where hypothetical *ad libitum* feeding leads to an energy surplus (a positive denominator in Eq. (38)), a power-law size spectrum is attained with less-efficient feeding at lower abundances. Remarkably, despite the fact that reproduction played a crucial role in deriving Eqs. (38) and (39), the result could have been obtained similarly when disregarding species identity and reproduction.

It is also noteworthy that, to the degree that (i) natural death is negligible so that conversion efficiency equals the Lindeman's trophic transfer efficiency (commonly found to be ≈ 0.1) and (ii) Sheldon's hypothesis $\lambda = 2$ holds, the energy balance discussed above necessarily leads to a predator–prey size ratio $\beta = 0.1^{1/(n-1)} \approx 1000$, which is similar to observed values, without any reference to mechanical constraints on feeding interactions.

One might therefore wonder whether the size of a fish's gape is determined by community-level constraints, rather than *vice versa*.

We can now ask under which conditions the integral in Eq. (18) converges for small x, in other words, under which conditions the contribution of large species to the abundance of small individuals in a community is small. By Eqs. (34) and (37), this requires $\lambda > 1 + n$. Therefore, there is with $n \approx 3/4$ little scope for size-spectrum slopes much smaller than the Sheldon slope $\lambda = 2$.

5.5. Total biomass per size class

From Eq. (16), the total biomass of species belonging to a small maturation size interval $[m_*; m_* + \Delta m_*]$ can be computed for the scale-invariant solution as $\mathcal{V} B_{\text{tot}}(m_*) \Delta m_*$, with

$$\begin{aligned} B_{\text{tot}}(m_*) &= \int_{m_0}^{\infty} mN(m, m_*)\mathrm{d}m \\ &= \int_{m_0}^{\infty} m\, m_*^{-\lambda-1} \tilde{N}(m/m_*)\mathrm{d}m \\ &= m_*^{-\lambda+1} \tilde{B}_{\text{tot}}(m_0/m_*) \end{aligned} \qquad [40]$$

and

$$\tilde{B}_{\text{tot}}(x_0) = \int_{x_0}^{\infty} x\tilde{N}(x)\mathrm{d}x. \qquad [41]$$

From the asymptotic form of $\tilde{N}(x)$, Eq. (34), it follows that the integrand above scales as x^{-n} for small x, implying that the contribution from small individuals to the total biomass is small as long as $n < 1$. The biomass of a maturation size class or species is dominated by its adults.

5.6. Total metabolic loss rate per size class

Contrasting with population biomass, the rate of metabolic losses of a maturation size class or species contains a large contribution from small individuals and diverges in the limit $x_0 \to 0$. Specifically, the total losses km^n of individuals belonging to a small maturation size interval $[m_*; m_* + \Delta m_*]$ evaluates to $\mathcal{V} K_{\text{tot}}(m_*) \Delta m_*$, with

$$\begin{aligned} K_{\text{tot}}(m_*) &= \int_{m_0}^{\infty} km^n N(m, m_*)\mathrm{d}m \\ &= \int_{m_0}^{\infty} km^n m_*^{-\lambda-1} \tilde{N}(m/m_*)\mathrm{d}m \\ &= m_*^{-\lambda+n} \tilde{K}_{\text{tot}}(m_0/m_*) \end{aligned} \qquad [42]$$

and

$$\tilde{K}_{\mathrm{tot}}(x_0) = k\int_{x_0}^{\infty} x^n \tilde{N}(x)\mathrm{d}x. \qquad [43]$$

By Eqs. (34) and (37), the integrand above is proportional to x^{-1} for small x, implying a logarithmic divergence of the integral as $x_0 \to 0$. It is easily verified that a similar logarithmic divergence arises for biomass loss by predation mortality and biomass gain by somatic growth. Consistency of model solutions in the limit $x_0 \to 0$ requires that these contributions cancel each other out.

5.7. Implications of the boundary condition (part II)

The conditions derived in the foregoing section leave no free parameter that could be adjusted to ensure that boundary condition Eq. (29) is satisfied not only by order of magnitude but also exactly. Below, it is shown that exact satisfaction is possible only under special, artificial assumptions, so deviations of the population structure $N(m, m_*)$ from the ideal scale-invariant form $m_*^{-\lambda-1}N(m/m_*)$ must be expected. Then, the qualitative nature of the expected deviations is discussed.

Consider first the case of vanishing background mortality $\mu_0 = 0$ and assume $\tilde{a} = 1$ (i.e. $\tilde{\mu} = \tilde{g}_0$) and scale-invariance: $N_0(m) \propto \tilde{N}(m/m_{*j})$ for $m_0 \le m \le \eta^{-1} m_*$ and $N_j(m) = 0$ otherwise. Define the function

$$E(x) = \exp\left[-\int_1^x \frac{1}{[1-\psi(x')]x'}\mathrm{d}x'\right], \qquad [44]$$

which satisfies the identify

$$\begin{aligned}\int_{x_0}^{\eta^{-1}} \frac{\psi(x)}{1-\psi(x)}E(x)\mathrm{d}x &= \int_{x_0}^{\eta^{-1}}\left[\frac{1}{1-\psi(x)}-1\right]E(x)\mathrm{d}x \\ &= \int_{x_0}^{\eta^{-1}} -x\frac{\mathrm{d}E(x)}{\mathrm{d}x} - E(x)\mathrm{d}x \\ &= -\int_{x_0}^{\eta^{-1}} \frac{\mathrm{d}}{\mathrm{d}x}[xE(x)]\mathrm{d}x = x_0E(x_0).\end{aligned} \qquad [45]$$

Using Eqs. (24), (33), (44), and (45), one can re-evaluate the general boundary condition (29) as

$$
\begin{aligned}
[1-\psi(x_0)]\tilde{N}(x_0)=\frac{\tilde{N}_0 x_0^{-n}}{\tilde{g}_0}E(x_0) &= \frac{\epsilon x_0^{-1-n}}{2}\int_{x_0}^{\eta^{-1}} \tilde{N}(x)\psi(x)x^n\,\mathrm{d}x \\
&= \frac{\epsilon \tilde{N}_0 x_0^{-1-n}}{2\tilde{g}_0}\int_{x_0}^{\eta^{-1}} \frac{\psi(x)}{1-\psi(x)}E(x)\,\mathrm{d}x \\
&= \frac{\epsilon \tilde{N}_0 x_0^{-n}}{2\tilde{g}_0}E(x_0).
\end{aligned}
\qquad [46]
$$

Hence, boundary condition (46) will be satisfied for scale-invariant solutions only when $\epsilon/2=1$. This requires an energetically impossible reproduction efficiency $\epsilon>1$. When natural mortality (μ_0) is taken into account, the value of the left-hand side in Eq. (46) would reduce further, requiring even larger values of ϵ. Perfect scale-invariant solutions are ecologically infeasible, so what happens instead?

Losses caused by reproduction efficiencies $\epsilon<2$ need to be compensated by accelerated growth or reduced mortality, that is, reduced physiological mortality over some phase of life history, typically in the immature phase. In the current model, specific physiological mortality before maturation depends only on body size, not on species identity. Therefore, if specific physiological mortality is reduced for one species over a given body-size range, so as to increase abundances by $2/\epsilon$ compared with the unmodified case, it will lead to an increase of abundances by $2/\epsilon$ for all other species covering this range too, effectively compensating their inefficiency in reproduction. Species covering a larger body-size range can therefore exploit conditions generated by other species contained within this range (this may be one of the reasons why large body-size ranges are encountered in nature).

The resulting picture for the case of fish is illustrated in Fig. 3: as hatchling size, m_0, is approximately the same for all fish (Cury and Pauly, 2000), independent of maturation size, specific physiological mortality will be reduced only within a size range covered by the smallest species. Larger species are affected only in early life history. For most fish species, setting $\epsilon=2$ and assuming the universal population structure given by Eq. (33) will therefore yield legitimate approximations of the true population structure, provided it can be shown that the effects of deviations from this structure at small body sizes are negligible.

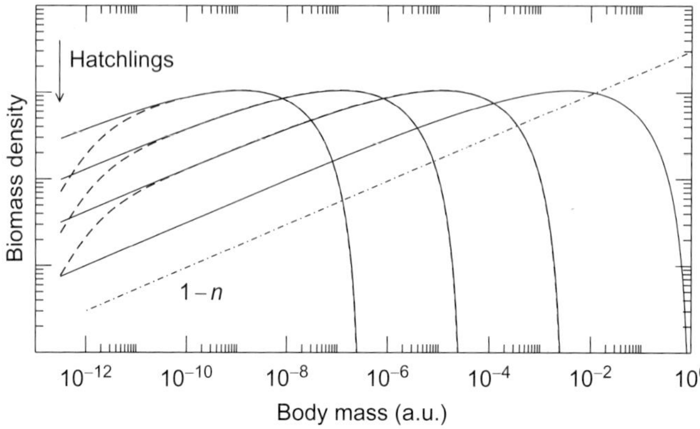

Figure 3 Schematic representation of intrapopulation size distributions. Body-mass ranges are exaggerated. Curves show the density of a population's biomass along the logarithmic size axis, given by $m^2 N_j(m)$ up to a constant factor. Solid lines correspond to scale-invariant distributions, which have a small-size tail scaling as m^{1-n} (dash-dotted line) by Eqs. (16), (34), and (37). Dashed lines represent conceivable corrections due to reproductive losses. Because juvenile specific physiological mortality, which controls the local slope of the curves, does not depend on species identity, corrections for all species occur in the same size range.

This picture explains the observed discrepancies between the simulation results of Hartvig et al. (2011) and their analytic equilibrium theory, which predicts another value for $\tilde{a}$. Specifically, the scenario sketched in Fig. 3 agrees with simulations by Hartvig et al. (2011) (i) in the predicted deformation of population structures for the smallest individuals, (ii) in its prediction for the scaling of offspring abundance with maturation size (corresponding to $\tilde{a}=1$), (iii) in its prediction of power-law scaling of survival to a given size corresponding to $\tilde{a}=1$, after a short phase of lower mortality. See Hartvig et al. (2011) for details.

Outside the range covered by fish, one can expect other regions on the logarithmic body-size axis where $\tilde{a}$ is depleted relative to 1, so as to compensate for inefficient reproduction of the species overlapping these regions, and values of $\tilde{a}$ near 1 in between. The locations and relative proportions of these regions on the body-size axis are currently unclear. On the condition that most species in a community cover wide ranges in body size, few regions of depleted $\tilde{a}$ will be necessary to compensate for $\epsilon < 2$, and these will, for most maturation size classes, fall outside their adult size ranges.

This completes the characterisation of the model's steady state. In what follows, the response of this steady state to perturbations will be evaluated in a linear approximation. To understand the impact of short regions of depleted $\tilde{a}$ on the logarithmic body-size axis, some critical calculations below will be carried out along two tracks: first for the *general* case of arbitrary $\epsilon \leq 2$, $m_0(m_*)$ and resulting forms for $g(m, m_*)$, $\mu_p(m)$, and $N(m, m_*)$, taking the considerations above into account, and then for the *scale-invariant* case of $\epsilon = 2$, $m_0(m_*) \ll m_*$, with scale-invariant $N(m, m_*)$ and life-history parameters. It will be argued that, as far as regions of depleted specific physiological mortality fall outside the adult body-size ranges of species, the *scale-invariant* case is a reasonable approximation of the *general* case. To aid orientation of the reader, the terms *general* and *scale-invariant* are italicised in the subsequent Section 6 when used with these particular meanings. Later, an additional simplification $\mu_0 = 0$ is introduced for the *scale-invariant* case, which is then explicitly stated.

6. DERIVATION OF THE SPECIES SIZE-SPECTRUM MODEL

6.1. General framework

As an effective and accurate method to reduce models for the dynamics of interacting structured populations to models in which each population is described by a single variable only (e.g. population biomass), Rossberg and Farnsworth (2011) introduce the QNA. The QNA for a system of S structured populations, described by vectors $\boldsymbol{n}_j = \mathbf{n}_j(t)$ of abundances by stage $(1 \leq j \leq S)$, satisfying $d\mathbf{n}_j/dt = \mathbf{A}_j(\mathbf{n}_1, \ldots, \mathbf{n}_S)\mathbf{n}_j$ with a density-dependent population matrix $\mathbf{A}_j(\mathbf{n}_1, \ldots, \mathbf{n}_S)$, can be carried out by following a simple recipe. (1) Obtain an approximate description of the community steady state and use it to construct for each species j an approximation of its steady-state population matrix $\underline{\mathbf{A}}_j$, that is, a constant matrix of transition rates between different population stages in the community steady state. Ensure that, because the system is in steady state and populations neither grow nor decay on average, all matrices $\underline{\mathbf{A}}_j$ have an eigenvalue zero. (2) Compute the eigenvector $\mathbf{w}_j$ and adjoint eigenvector $\mathbf{v}_j$ corresponding to eigenvalue zero for each $\underline{\mathbf{A}}_j$, choosing normalisations such that the equilibrium population structure $\mathbf{w}_j$ corresponds to a population of unit size (e.g. 1 kg of biomass), and that $\mathbf{v}_j^{\mathrm{T}}\mathbf{w}_j = 1$. (3) Approximate the dynamics of population sizes, defined by $B_j(t) = \mathbf{v}_j^{\mathrm{T}}\mathbf{n}_j(t)$, as $\mathrm{d}B_j/\mathrm{d}t = \mathbf{v}_j^{\mathrm{T}}\mathbf{A}(\mathbf{w}_1 B_1, \ldots, \mathbf{w}_S B_S)\mathbf{w}_j B_j$, which is an ordinary differential equation in the variables B_j. Using this result, approximate the full community dynamics as $\mathbf{n}_j(t) \approx \mathbf{w}_j B_j(t)$. The components of the vectors $\mathbf{v}_j$ are interpreted as the reproductive values of the corresponding population stage.

6.2. Operators and eigenfunctions

Applying the QNA to obtain a simplified description of the dynamics of the density of individuals, $N(m, m_*)$, requires adapting it to a continuum of life-history stages indexed by body mass m and a continuum of populations characterised by maturation size m_*. Functional analysis (Section 3) provides the necessary formal tools. The first step of the recipe (approximation of the steady state) has been completed in the Section 5. To obtain a description of dynamics in a form appropriate for the second step, the McKendrick–von Foerster Equation (15) is written as $\mathrm{d}N(m, m_*)/\mathrm{d}t = L_{m_*}N(m, m_*)$, where the linear operator L_{m_*}, parametrised by m_*, is modified such as to include the production of hatchlings explicitly:

$$L_{m_*}N(m,m_*) = -\frac{\partial}{\partial m}\left[g(m,m_*)N(m,m_*)\right] - \left[\mu_p(m) + \mu_0 m_*^{n-1}\right]N(m,m_*) + \delta(m-m_0)\frac{\epsilon}{2m_0}\int_0^\infty N(m',m_*)g_r(m',m_*)\mathrm{d}m'. \quad [47]$$

Here, m_0 is again understood to be a function of m_*. The boundary condition for $N(m, m_*)$ then becomes $N(m, m_*) = 0$ for any $m < m_0$. The facts that the last term correctly enforces the original boundary condition $g(m, m_*)$ $N(m, m_*) = R_{m_*}$ (with R_{m_*} given after obvious adjustments by Eq. (12)) is readily verified by integrating Eq. (47) over a small interval $[m_0 - \Delta m/2; m_0 + \Delta m/2]$.

The *general* steady-state McKendrick-von Foerster operator $\underline{L}_{m_*}$ is obtained from L_{m_*} by replacing the functions $g(m, m_*)$, $g_r(m, m_*)$, and $\mu_p(m)$ with the results $\underline{g}(m,m_*)$, $\underline{g}_r(m,m_*)$, and $\underline{\mu}(m)$ for a particular *general* steady-state community size spectrum $\mathcal{N}(m) = \underline{\mathcal{N}}(m)$. The null-eigenvectors required by the QNA, that is, the equilibrium population structures $W_{m_*}(m)$ satisfying $\underline{L}_{m_*} W_{m_*}(m) = 0$, can be obtained as

$$W_{m_*}(m) = B_{m_*}^{-1}\underline{N}(m,m_*), \quad [48]$$

where $\underline{N}(m,m_*)$ is $N(m, m_*)$ evaluated at a *general* steady state (which may or may not be scale-invariant), and a normalisation to unit biomass (actually biomass density) is obtained by setting

$$B_{m_*} = \int_0^\infty m\underline{N}(m,m_*)dm. \quad [49]$$

As for the *scale-invariant* case in Section 5.5, this integral is dominated by adult individuals in the *general* case as well. Therefore, for maturation size classes m_* where the *general* population structures $\underline{N}(m, m_*)$ do not differ much from the *scale-invariant* case in the adult range of m, the *scale-invariant* form of W_{m_*} is a good approximation of *general* W_{m_*} in the adult range. In the *scale-invariant* case, W_{m_*} is obtained from Eqs. (48), (16), (49), and (40) as

$$W_{m_*}(m) = \begin{cases} m_*^{-2}\tilde{B}_{\text{tot}}^{-1}\tilde{N}(m/m_*) & \text{for } m \geq m_0 \\ 0 & \text{for } m < m_0, \end{cases} \quad [50]$$

with $\tilde{N}(m/m_*)$ given by Eq. (33) and $\tilde{B}_{\text{tot}}$ standing as shorthand for $\tilde{B}_{\text{tot}}(m_0/m_*)$.

For populations described by a finite number of stages (Section 6.1), the steady-state population matrices $\underline{\mathbf{A}}_j$ only have a finite number of eigenvalues. The eigenvalue 0, assumed to be the one with the largest real part and to have multiplicity one, is therefore always separated by a gap from the real parts of the other eigenvalues. This leads to a separation between the timescales of intra- and interspecific population dynamics that is exploited in the QNA. However, when populations are described by a continuum of life-history stages, as is the case here, the corresponding linear operator $\underline{L}_{m_*}$ has an infinite number of eigenvalues. Thus, the existence of a spectral gap between 0 and the other eigenvalues is not guaranteed. In Appendix, the spectrum of operator $\underline{L}_{m_*}$ is studied for a numerical example of the *scale-invariant* case. It is concluded that indeed 0 is the unique eigenvalue of $\underline{L}_{m_*}$ with largest real part, and that all other eigenvalues have real parts smaller than $-1 \times$ (mortality of the largest individuals of the population). These observations can plausibly be expected to generalise to the *general* class of operators $\underline{L}_{m_*}$ considered here, so the eigenvalue zero will be separated from the subdominant eigenvalues by a gap of size $\underline{\mu}(m_*/\eta) + \mu_0 m_*^{n-1}$. The resulting separation of timescales is estimated in Section 9.5.

To obtain the adjoint eigenvectors (here eigenfunctions), the adjoint linear operator $\underline{L}_{m_*}^+$ of $\underline{L}_{m_*}$ needs to be computed. Application of the method described in Section 3 yields

$$\begin{aligned} \underline{L}_{m_*}^+ V(m) &= \underline{g}(m, m_*)\frac{\partial V(m)}{\partial m} - \left[\underline{\mu}(m) + \mu_0 m_*^{n-1}\right] V(m) \\ &\quad + \underline{g}_r(m, m_*)\frac{\epsilon V(m_0)}{2m_0}. \end{aligned} \quad [51]$$

Following the general prescription (Section 6.1), the reproductive value of an individual of size m and maturation size m_* is given by the solution $V_{m_*}(m)$ of $\underline{L}^+_{m_*} V(m) = 0$, subject to the normalisation condition $\langle V_{m_*} | W_{m_*} \rangle = 1$. As, for any species of maturation size class m_*, the equilibrium density of individuals along the m axis for a population of unit biomass is given by $W_{m_*}(m)$, the equilibrium density of reproductive value for a unit population along the m axis is given by the product $V_{m_*}(m) W_{m_*}(m)$. Multiplying this with the equilibrium growth rate, $\underline{g}(m, m_*)$ gives the equilibrium flow of reproductive value along the size axis for a unit population, $\underset{-m_*}{J}(m) \stackrel{\text{def}}{=} \underline{g}(m, m_*) V_{m_*}(m) W_{m_*}(m)$. It is readily verified using $L_{m_*} W_{m_*} = L^+_{m_*} V_{m_*} = 0$ and Eqs. (47) and (51) that for $m > m_0$

$$\frac{d\underset{-m_*}{J}}{dm} = \frac{d}{dm}\left[\underline{g}(m, m_*) V_{m_*}(m) W_{m_*}(m)\right] = -\frac{\epsilon V_{m_*}(m_0)}{2m_0} \underset{-r}{g}(m, m_*) W_{m_*}(m). \quad [52]$$

That is, until maturation $\underset{-m_*}{J}(m)$ is constant and then it declines. Obviously, $\underset{-m_*}{J}(m) = 0$ for $m < m_0$ and $m \geq m_*/\eta$. Therefore, one can compute *general* reproductive values from a given *general* population structure $W_{m_*}(m)$ by integrating Eq. (52) as

$$V_{m_*}(m) = \frac{1}{C_{m_*} \underline{g}(m, m_*) W_{m_*}(m)} \int_m^{m_*/\eta} \underset{-r}{g}(m', m_*) W_{m_*}(m') dm', \quad [53]$$

where the correct normalisation $\langle V_{m_*} | W_{m_*} \rangle = 1$ is assured by setting

$$\frac{2m_0}{\epsilon V_{m_*}(m_0)} = C_{m_*} = \int_{m_0}^{m_*/\eta} \frac{1}{\underline{g}(m, m_*)} \int_m^{m_*/\eta} \underset{-r}{g}(m', m_*) W_{m_*}(m') dm' \, dm. \quad [54]$$

An important observation now is that, because $\underline{g}(m, m_*)$ scales as m^n with $n < 1$ for $m \ll m_*$, and $\underset{-r}{g}(m, m_*)$ is localised near m_*, the integral over m in Eq. (54) is dominated by contributions from large m. The value of C_{m_*} therefore depends only on $W_{m_*}(m)$ and the size spectrum near m_*, and these will often be well approximated by the *scale-invariant* forms.

If natural mortality (μ_0) is negligible in the *scale-invariant* case, Eq. (53) evaluates by manipulations similar to Eq. (45) to the simple form

$$V_{m_*}(m) = m. \quad [55]$$

In this approximation, reproductive value therefore exactly equals body mass. This result is most easily verified by directly confirming $\underline{L}_{m_*}^{+} V(m) = 0$ and $\langle V_{m_*} \mid W_{m_*} \rangle = 1$. The simplicity of this result allows carrying out the QNA analytically, which is done in the next section.

A generalisation of Eq. (55) to the case with natural mortality gives an enhancement of reproductive value by an amount of the order of magnitude $m\mu_0 / \tilde{g}_0$ for old individuals, and a corresponding proportional reduction for all others. As this does not much affect the overall structure of $V_{m_*}(m)$, natural mortality is not considered further in detail below.

6.3. Reduced dynamics

Following the recipe of the QNA, define a reduced description $B(m_*)$ of the community state as

$$B(m_*) = \left\langle V_{m_*} \middle| N(\cdot, m_*) \right\rangle = \int_0^\infty V_{m_*}(m) N(m, m_*) \mathrm{d}m \qquad [56]$$

for an arbitrary, time-dependent distribution of individuals $N(m, m_*)$. In the approximation that reproductive value equals body mass, the function $B(m_*)$ approximates the time-dependent distribution of biomass over maturation body sizes. To the accuracy of the QNA, the dynamics of $B(m_*)$ follow

$$\begin{aligned} \frac{1}{B(m_*)} \frac{\partial B(m_*)}{\partial t} &= \left\langle V_{m_*} \middle| L_{m_*} W_{m_*} \right\rangle \\ &= \int_0^\infty V_{m_*}(m) \left\{ -\frac{\partial}{\partial m} \left[g(m, m_*) W_{m_*}(m) \right] - \left[\mu_\mathrm{p}(m) + \mu_0 m_*^{n-1} \right] W_{m_*}(m) \right\} \mathrm{d}m, \\ &\quad + \frac{\epsilon V_{m_*}(m_0)}{2m_0} \int_0^\infty W_{m_*}(m) g_r(m, m_*) \mathrm{d}m \end{aligned} \qquad [57]$$

where $g(m, m_*)$, $g_r(m, m_*)$, and $\mu_\mathrm{p}(m)$ are evaluated for a variable size spectrum given by

$$\mathcal{N}(m) = \int_0^\infty B(m_*) W_{m_*}(m) \mathrm{d}m_*. \qquad [58]$$

By Eq. (57), one can read $\langle V_{m_*} \mid L_{m_*} W_{m_*} \rangle$ as the momentary growth rate of biomass in maturation size class m_*.

In the *scale-invariant* case without natural mortality ($\mu_0 = 0$), the simple results (50) and (55) for the null-eigenfunctions hold and the right-hand side of Eq. (57) simplifies through integration by parts to

$$\left\langle V_{m_*} | L_{m_*} W_{m_*} \right\rangle = \int_{m_0}^{\infty} \left[g(m, m_*) + g_r(m, m_*) - m\mu_p(m) \right] \frac{\tilde{N}(m/m_*)}{m_*^2 \tilde{B}_{\text{tot}}} \mathrm{d}m$$
$$= \int_{m_0}^{\infty} \left[(\alpha f(m) h - k) m^n - m\mu_p(m) \right] \frac{\tilde{N}(m/m_*)}{m_*^2 \tilde{B}_{\text{tot}}} \mathrm{d}m. \quad [59]$$

The second step implies that the density-dependencies of investments into somatic and reproductive growth are here equivalent in their population-dynamic effects. This is remarkable when recalling that the causal chains through which these effects are achieved are fundamentally different.

Careful inspection reveals that Eq. (59) holds to a good degree also in the *general* case. To see this, observe first that, using $\underline{L}_{m_*}^{+} V_{m_*}(m) = 0$, Eqs. (51), and (54),

$$\frac{\partial V_{m_*}(m)}{\partial m} = V_{m_*}(m) \left[\frac{\underline{a}(m, m_*)}{m} - \frac{\underline{g}_r(m, m_*)}{C_{m_*} \underline{g}(m, m_*) V_{m_*}(m)} \right], \quad [60]$$

where $\underline{a}(m, m_*) \stackrel{\text{def}}{=} m \left[\underline{\mu}_p(m) + \mu_0 m_*^{n-1} \right] / \underline{g}(m, m_*)$ is the steady-state specific physiological mortality. Then integrate the somatic growth term in Eq. (57) by parts, eliminate $\partial V_{m_*}/\partial m$ through Eq. (60), and, finally, eliminate ϵ using Eq. (54), to obtain

$$\left\langle V_{m_*} | L_{m_*} W_{m_*} \right\rangle = \int_0^{\infty} g(m, m_*) V_{m_*}(m) W_{m_*}(m) \frac{\underline{a}(m, m_*) - a(m, m_*)}{m} \mathrm{d}m$$
$$+ \int_0^{\infty} \frac{W_{m_*}(m)}{C_{m_*}} \left[\frac{g_r(m, m_*) \underline{g}(m, m_*) - \underline{g}_r(m, m_*) g(m, m_*)}{\underline{g}(m, m_*)} \right] \mathrm{d}m$$
$$= \int_0^{\infty} g(m, m_*) V_{m_*}(m) W_{m_*}(m) \frac{\underline{a}(m, m_*) - a(m, m_*)}{m} \mathrm{d}m$$
$$= \int_0^{\infty} \underline{J}_{m_*}(m) \frac{\underline{a}(m, m_*) - a(m, m_*)}{m} \mathrm{d}m + \text{h.o.t.} \quad [61]$$

with $a(m, m_*) \stackrel{\text{def}}{=} m \left[\mu_p(m) + \mu_0 m_*^{n-1} \right] / g(m, m_*)$. In the second step the second integral vanishes because the term in brackets becomes zero when inserting the definitions of $g(m, m_*)$ and $g_r(m, m_*)$ from Eqs. (23) and (24). Below, only effects linear in deviations of growth, reproduction and mortality from the steady-state rates will be investigated, and to linear order one can substitute $g(m, m_*)$ by $\underline{g}(m, m_*)$ in the remaining integral, so $\underline{g}(m, m_*)V_{m_*}(m)W_{m_*}(m)$ reduces to $\underline{J}_{-m_*}(m)$ in the last line, up to higher-order contributions indicated by h.o.t.

Denote by $\underline{B}(m_*)$ the equilibrium values of the reduced dynamic variables $B(m_*)$ defined in Eq. (56). To investigate the community response to perturbations of this equilibrium state, the linearization of $\langle V_{m_*} | L_{m_*} W_{m_*} \rangle$ for small deviations $\Delta B(m_*) = B(m_*) - \underline{B}(m_*)$ from this state is now constructed. Specifically, an integral kernel $K(m_*, m'_*)$ is sought such that

$$\left\langle V_{m_*} | L_{m_*} W_{m_*} \right\rangle = \int_0^\infty K(m_*, m'_*) \Delta B(m'_*) \mathrm{d}m'_* + \text{h.o.t.}, \qquad [62]$$

with h.o.t. denoting higher-order terms in ΔB. As $\langle V_{m_*} | L_{m_*} W_{m_*} \rangle$ depends on $B(m_*)$ only through the size spectrum $\mathcal{N}(m)$, the chain rule for functional derivatives can be applied before using Eqs. (58) and (61) to obtain

$$\begin{aligned} K\left(m_*, m'_*\right) &= \frac{\delta\left\langle V_{m_*} | L_{m_*} W_{m_*} \right\rangle}{\delta B(m'_*)} \\ &= \int_0^\infty \frac{\delta\left\langle V_{m_*} | L_{m_*} W_{m_*} \right\rangle}{\delta \mathcal{N}(m')} \frac{\delta \mathcal{N}(m')}{\delta B(m'_*)} \mathrm{d}m' \\ &= \int_0^\infty \int_0^\infty \frac{\underline{J}_{-m_*}(m)}{m} \frac{\delta a(m, m_*)}{\delta \mathcal{N}(m')} W_{m'_*}(m') \mathrm{d}m \mathrm{d}m' \\ &= \int_0^\infty \int_0^\infty \frac{\underline{J}_{-m_*}(m)}{m\left[1 - \psi(m/m_*)\right]} \frac{\delta\left[1 - \psi\left(m/m_*\right)\right] a\left(m, m_*\right)}{\delta \mathcal{N}(m')} W_{m'_*}(m') \mathrm{d}m' \mathrm{d}m. \end{aligned} \qquad [63]$$

This expression needs to be evaluated at the equilibrium state. By multiplying $a(m, m_*)$ with $[1 - \psi(m/m_*)]$ in the last step, a divergence of $a(m, m_*)$ at the size of the largest adults caused by cessation of growth, Eq. (24), is suppressed. This effectively removes the distinction between somatic and reproductive growth. The division by $[1 - \psi(m/m_*)]$ in the first factor of the integrand cancels with a corresponding factor entering $\underline{J}_{-m_*}(m) = \underline{g}(m, m_*)V_{m_*}(m)W_{m_*}(m)$ through $\underline{g}(m, m_*)$ and, therefore, does not cause a divergence for large adults. Equations (57), (62), and (63) together specify the QNA for the *general* case.

To understand to what extent deviations from scale-invariance affect Eq. (63), it is useful to investigate how the value of the integrand scales for small m and m' while $m > m_0(m_*)$, $m' > m_0(m'_*)$. As $\underline{J}_{-m_*}(m)$ is constant for small m, the first factor scales as m^{-1}. The second factor describes the dependence of the specific physiological mortality of individuals of size m on the density of individuals of size m'. The body-size ratios of interacting individuals are constrained by the predator–prey size-ratio window $s(x)$. The second factor will therefore be significantly different from zero only when m and m' are of roughly similar magnitude. For such a situation, the functional derivative operator $\delta/\delta\mathcal{N}(m')$ shows the same scaling behaviour as $\delta(m-m')/\mathcal{N}(m')$, with $\delta(m-m')$ denoting Dirac's delta functional (as can be verified by evaluating the derivative). The operand $[1-\psi(m/m_*)]a(m,m_*)$ will be of the order of magnitude of one, even in the *general* case. In the *scale-invariant* case, the last factor grows for small m' as $(m')^{-n-1}$ by Eqs. (50), (34), and (37). In the *general* case, it may grow slower to accommodate inefficient reproduction (Section 5.7), but it will never grow faster. After performing the integration over m', the integrand does therefore not increase faster than $m^{-1}\times m^{\lambda}\times m^{-n-1}=m^{\lambda-n-2}$ for small m. Small m dominates the integral if the integrand increases faster than m^{-1}, implying that $\lambda<1+n$. This is unlikely to be the case, because it violates a condition for the existence of a *scale-invariant* steady state (Section 5.4). Therefore, the integrand is generally dominated by contributions where m and m' are of similar order of magnitude as m'_*.

These considerations have two important implications. As explained in Section 6.2, *general* $\underline{J}_{m_*}(m)$ and $W_{m_*}(m)$ can be approximated by their *scale-invariant* forms in the adult range, at least as long as m_* is not in a size range of depleted specific physiological mortality. Further, $\underline{J}_{m_*}(m)$ attains in the juvenile range a constant value that depends only on the life-history parameter in the adult range. The first implication is therefore that, except for cases where m_* or m'_* are in or near one of the regions of depleted specific physiological mortality described in Section 5.7, the right-hand side of (63) can be approximated by the corresponding *scale-invariant* form. Because contributions from small m and m' do not dominate the integral, a second consequence is that the mathematical limit $m_0(m_*), m_0(m'_*)\to 0$ of infinitely small offspring is regular and approximates the case of general but small $m_0(m_*)/m_*$. The calculations are therefore continued from here on using the simplifying approximations of *scale-invariant* population structures with $m_0(m_*)\to 0$ for any m_*. To simplify matters further, natural mortality is not taken into consideration ($\mu_0=0$), as it has only minor effects on dynamics (see also Hartvig et al., 2011).

Therefore, starting from Eq. (59) and making use of Eq. (50) for *scale-invariant* population structures,

$$
\begin{aligned}
K\left(m_*, m'_*\right) &= \frac{\delta\left\langle V_{m_*} | L_{m_*} W_{m_*} \right\rangle}{\delta B\left(m'_*\right)} \\
&= \int_0^\infty \frac{\delta\left\langle V_{m_*} | L_{m_*} W_{m_*} \right\rangle}{\delta \mathcal{N}(m')} \frac{\delta \mathcal{N}(m')}{\delta B\left(m'_*\right)} \mathrm{d}m' \\
&= \int_0^\infty \left\{ \int_0^\infty \left[\alpha h m^n \frac{\delta f(m)}{\delta \mathcal{N}(m')} - m \frac{\delta \mu_\mathrm{p}(m)}{\delta \mathcal{N}(m')} \right] \frac{\tilde{N}\left(m/m_*\right)}{m_*^2 \tilde{B}_\mathrm{tot}} \mathrm{d}m \right\} \frac{\tilde{N}\left(m'/m'_*\right)}{m'^2_* \tilde{B}_\mathrm{tot}} \mathrm{d}m'.
\end{aligned}
\tag{64}
$$

The functional derivatives of $f(m)$ and $\mu_\mathrm{p}(m)$ can be evaluated based on results of Section 5.2. For the eutrophic regime, one obtains

$$
\frac{\delta f(m)}{\delta \mathcal{N}(m')} = \frac{h}{\gamma \tilde{\phi}^2} m' m^{2\lambda+n-q-4} s\left(\ln \frac{m}{m'}\right) \tag{65}
$$

and

$$
\frac{\delta \mu_\mathrm{p}(m)}{\delta \mathcal{N}(m')} = \frac{h}{\tilde{\phi}} \left[s\left(\ln \frac{m'}{m}\right) - \frac{1}{I_1} \Sigma\left(\ln \frac{m'}{m}, \lambda - 4 + n\right) \right] m'^{\lambda+n-2}, \tag{66}
$$

using $I_1 = \hat{s}(i(\lambda - 2))$ as above and the abbreviation

$$
\begin{aligned}
\Sigma(y, c) &= \int_0^\infty x^c s(\ln x) s(y + \ln x) \mathrm{d}x \\
&= \int_{-\infty}^\infty \mathrm{e}^{(c+1)u} s(u) s(y + u) \mathrm{d}u.
\end{aligned}
\tag{67}
$$

In Eq. (66), the first term in brackets describes direct predation mortality, the second term a release from predation pressure in the presence of other prey of similar size. It follows from the condition $\gamma m^{2+q-\lambda} \tilde{\phi} \gg h m^n$, which defines the eutrophic regime, that the functional derivative of $f(m')$ in Eq. (64) will generally be negligible compared with that of $\mu_\mathrm{p}(m')$. It is discarded hereafter.

In the oligotrophic regime,

$$
\frac{\delta f(m)}{\delta \mathcal{N}(m')} = \frac{\gamma h}{\left(\gamma \tilde{\phi} + h\right)^2} m' m^{q-n} s\left(\ln \frac{m}{m'}\right) \tag{68}
$$

and, using Eq. (38),

$$\frac{\delta\mu_{\rm p}(m)}{\delta\mathcal{N}(m')}=\frac{\gamma h}{\gamma\tilde{\phi}+h}\left[s\left(\ln\frac{m'}{m}\right)-\frac{k}{h(\alpha I_1-I_2)}\Sigma\left(\ln\frac{m'}{m},q-2\right)\right]m'^{q}. \qquad [69]$$

To simplify Eq. (64), define first the individual-level interaction kernel

$$\kappa_0\left(\ln\frac{m}{m'}\right)=\frac{m'^{-\lambda-n+2}}{m}\left[\alpha h m^{n}\frac{\delta f(m)}{\delta\mathcal{N}(m')}-m\frac{\delta\mu_{\rm p}(m)}{\delta\mathcal{N}(m')}\right]. \qquad [70]$$

It follows from Eqs. (66) to (69) that the right-hand side of Eq. (70) does indeed depend only on the ratio m/m'. Further, let

$$\tilde{\beta}(w)=\tilde{B}_{\rm tot}^{-1}{\rm e}^{2w}\tilde{N}({\rm e}^{w}) \qquad [71]$$

describe the scale-free distribution of a species' biomass over the logarithmic body mass axis for $m_0, x_0\to 0$, normalised to $\int_{-\infty}^{\infty}\tilde{\beta}(w){\rm d}w=1$. Equation (64) can then be rewritten in terms of convolution integrals by performing the substitutions $m={\rm e}^{-w}m_*$ and $m'={\rm e}^{-v}m_*$, which yield, after some calculation,

$$\begin{aligned}K\left(m_*,m_*'\right)&=m_*^{-\nu}\int_{-\infty}^{\infty}\int_{-\infty}^{\infty}{\rm e}^{\nu v}\kappa_0(v-w)\tilde{\beta}(-w)\tilde{\beta}\left(-v+\ln\frac{m_*}{m_*'}\right){\rm d}w\,{\rm d}v\\&=m_*^{-\nu}\left[\kappa*\tilde{\beta}\right]\left(\ln\frac{m_*}{m_*'}\right),\end{aligned} \qquad [72]$$

with the abbreviations $\nu=3-\lambda-n$ and

$$\kappa(v)={\rm e}^{\nu v}\left[\kappa_0*\tilde{\beta}^{-}\right](v), \qquad [73]$$

where $\tilde{\beta}^{-}(v)\overset{\rm def}{=}\tilde{\beta}(-v)$.

In order to fully go over to logarithmic scales, introduce the scale-invariant interaction kernel

$$\tilde{K}(w)=\tilde{B}_{\rm tot}\left[\kappa*\tilde{\beta}\right](w), \qquad [74]$$

and a logarithmic mass axis $u=\ln(m_*/M)$, where M denotes a unit mass (e.g. $M=1$ g). Deviations $b(u)$ of the distribution of biomass on the u-axis from the power-law steady state are given by

$$b(u)=M{\rm e}^{u}\Delta B(M{\rm e}^{u}). \qquad [75]$$

The density-dependent growth rate, given by Eq. (62), then becomes, up to higher-order terms,

$$\begin{aligned}\left\langle V_{m_*} \middle| L_{m_*} W_{m_*} \right\rangle &= \tilde{B}_{\mathrm{tot}}^{-1} m_*^{-\nu} \left[\tilde{K} * b\right]\left(\ln \frac{m_*}{M}\right) \\ &= \frac{m_*^{n-2}}{B_{\mathrm{tot}}(m_*)} \left[\tilde{K} * b\right]\left(\ln \frac{m_*}{M}\right),\end{aligned} \quad [76]$$

where Eq. (40) was applied in the second step.

The response of size spectra to external perturbations is of great practical interest: here, we concentrate on the effects of fishing. Assume a size-selective fishing regime that enhances mortality of individuals by up to F over a range centred at m_{F} with logarithmic width σ_{F}; that is, next to predation and natural mortality, there is additional mortality $\mu_{\mathrm{F}}(m) = F W_{\sigma_{\mathrm{F}}}(\ln m/m_{\mathrm{F}})$ with the Gaussian window $W_{\sigma_{\mathrm{F}}}(x)$ defined in Section 3. Putting this into Eq. (57) and using Eqs. (50) and (71) leads, in the limit $m_0 \to 0$, in addition to natural community dynamics, to a pressure on populations with maturation size m_*, corresponding to a rate of decline (dimension 1/Time):

$$\begin{aligned}\int_0^\infty m\mu_{\mathrm{F}}(m) W_{m_*}(m) \mathrm{d}m &= F \int_{-\infty}^{\infty} W_{\sigma_{\mathrm{F}}}\left(u + \ln \frac{m_*}{m_{\mathrm{F}}}\right) \tilde{\beta}(u) \mathrm{d}u \\ &= F \int_{-\infty}^{\infty} W_{\sigma_{\mathrm{F}}}\left(-u + \ln \frac{m_*}{m_{\mathrm{F}}}\right) \tilde{\beta}^{-}(u) \mathrm{d}u \\ &= F \left[W_{\sigma_{\mathrm{F}}} * \tilde{\beta}^{-}\right]\left(\ln \frac{m_*}{m_{\mathrm{F}}}\right).\end{aligned} \quad [77]$$

One can then, by applying Eqs. (76), (75), and $B(m_*) = \Delta B(m_*) + B_{\mathrm{tot}}(m_*)$ to Eq. (57) and including of the effect of fishing, Eq. (77), express system dynamics in response to fishing pressure that sets in at $t = 0$ by the integro-differential equation

$$\frac{\partial b(u)}{\partial t} = (M\mathrm{e}^u)^{n-1} \left[\tilde{K} * b\right](u) - (M\mathrm{e}^u)^{2-\lambda} \Phi(u) \Theta(t), \quad [78]$$

where

$$\Phi(u) = \tilde{B}_{\mathrm{tot}} F \left[W_{\sigma_{\mathrm{F}}} * \tilde{\beta}^{-}\right]\left(u - \ln \frac{m_{\mathrm{F}}}{M}\right) \quad [79]$$

describes the fishing pressure, and $\Theta(t)$ denotes the unit step function: $\Theta(t)=0$ for $t<0$ and $\Theta(t)=1$ for $t\geq 1$. If fishing was specific to the maturation size of species rather than to the size of individuals, Eqs. (79) would be modified to

$$\Phi(u)=\tilde{B}_{\text{tot}}FW_{\sigma_{\text{F}}}\left(u-\ln\frac{m_{\text{F}}}{M}\right). \qquad [80]$$

Equation (78) represents the Species Size-Spectrum Model. It describes the dynamics of deviations of the size spectrum from the steady state in the linearised QNA. Linearised versions of other size-spectrum models proposed in the literature would assume similar forms. The main difference between these models lies in the interaction kernels they contain. In practice, the interaction kernel derived here, $\tilde{K}(w)=\tilde{B}_{\text{tot}}\left[\kappa*\tilde{\beta}\right](w)$, is best evaluated via its Fourier transform

$$\hat{K}(\xi)=\tilde{B}_{\text{tot}}\hat{\kappa}(\xi)\hat{\beta}(\xi)=\tilde{B}_{\text{tot}}\hat{\kappa}_0(\xi+\nu i)\hat{\beta}(-\xi-\nu i)\hat{\beta}(\xi), \qquad [81]$$

which follows from Eqs. (73) and (74) and some manipulations of convolution integrals.

Making use of Eq. (70), the preceding explicit formulae (65) to (69) for the functional derivatives, and the observation that the Fourier transform of $\Sigma(y,c)$ with respect to y is

$$\hat{\Sigma}(\xi,c)=\hat{s}(-\xi+(c+1)i)\hat{s}(\xi) \qquad [82]$$

give the Fourier transform of the individual-level interaction kernel

$$\hat{\kappa}_0(\xi)=\frac{h}{\tilde{\phi}}\left[-[\hat{s}(\xi^*)]^*+\frac{\hat{s}(\xi-\nu i)[\hat{s}(\xi^*)]^*}{\hat{s}((\lambda-2)i)}\right] \qquad [83]$$

in the eutrophic regime and

$$\begin{aligned}\hat{\kappa}_0(\xi)=&\frac{\alpha h^2\gamma}{(\gamma\tilde{\phi}+h)^2}\hat{s}(\xi-\nu i)-\frac{\gamma h}{\gamma\tilde{\phi}+h}[\hat{s}(\xi^*)]^*\\&+\frac{\gamma^2\tilde{N}h}{(\gamma\tilde{\phi}+h)^2}\hat{s}(\xi-\nu i)[\hat{s}(\xi^*)]^*\end{aligned} \qquad [84]$$

in the oligotrophic regime. The last expression can be brought into alternative forms by using the identity

$$\frac{\gamma\tilde{N}}{\gamma\tilde{\phi}+h}=\frac{k}{h(\alpha I_1-I_2)}, \qquad [85]$$

which follows from results in Section 5.

6.4. Food-web effects

Simulations show that Eq. (78) with the kernel given by (74) is structurally unstable, as indicated in Section 5.1. As a result of combining coarse graining with the mean-field approximation, the equation effectively describes competition among a continuum of species along the species-size axis. This destabilises the system by mechanisms similar to those involved in the scenarios described by Pigolotti et al. (2007), leading ultimately to a breakup of the continuum into a set of distinct species (isolated spikes on the size axis). In real communities, this distinct set of species is already established and, as a result, dynamics will be modified. Deriving these modifications of dynamics from first principles would require a better understanding of food-web structure and dynamics than we currently have. But the following heuristic arguments suggest the general pattern that can be expected.

Consider perturbations of $b(u)$ of the form $\iota \sin(\xi u + \psi)$ with small amplitude ι and arbitrary phase ψ. For species *randomly* distributed along the u-axis and sufficiently small wavelengths $2\pi/\xi$, the perturbations received by each species become uncorrelated. Such short-wavelength perturbations of the size spectrum are therefore effectively equivalent to random perturbations with a distribution independent of the wave number ξ and, relative to species biomass, independent and identical among species. The relaxation of the system from such perturbations to the equilibrium state (or the attractor) can be expected to be rather complicated in detail and involves a range of timescales (Rossberg and Farnsworth, 2011), but, as a simple model, linear relaxation at a constant rate proportional with m_*^n and independent of ξ (if large enough) should provide an appropriate picture. Unfortunately, these heuristic considerations do not provide a value for the relaxation rate, except for the general principle that rates become larger for larger $\tilde{B}_{\text{tot}}$. Because the community effect of perturbations of food-web nodes is typically diffuse (Yodzis, 1998) and decays swiftly with the degree of separation from the perturbed node (Berlowa et al., 2009), a periodic perturbation has effectively a random effect when the wavelength $2\pi/\xi$ is small compared to the typical distance in u among the main prey and predators of a species. As wavelengths become larger—and wave numbers ξ smaller—these food-web effects will gradually subside. Food-web effects might therefore be modelled by adding a correction of the form $-r\tilde{B}_{\text{tot}}\left[1 - \exp(-\sigma_r^2\xi^2/2)\right]$ to the Fourier transform of the interaction kernel $\hat{K}(\xi)$. The constant σ_r controls the cut-off-wavelength for food-web effects. The relaxation-rate constant r has the same dimensions as the search-rate coefficient γ, and it may therefore be assumed to be of comparable magnitude, presumably somewhat smaller (Rossberg and Farnsworth, 2011).

Back-transforming, this leads to a corrected interaction kernel

$$\tilde{K}(w) = \tilde{B}_{\mathrm{tot}}\left\{\left[\kappa * \tilde{\beta}\right](w) + X(w)\right\} \quad [86]$$

with food-web effects being given by

$$X(w) = -r\left[\frac{1}{\sqrt{2\pi}\sigma_r}\exp\left(-\frac{w^2}{2\sigma_r^2}\right) - \delta(w)\right]. \quad [87]$$

7. SOLUTION OF THE SPECIES SIZE-SPECTRUM MODEL

The factor $(Me^u)^{n-1}$ in Eq. (78) expresses the slowing-down of effective ecological activity of larger species attributable to allometric scaling of metabolic rates. This factor makes the solution of Eq. (78), either numerically or analytically, difficult. Datta et al. (2010) and Capitán and Delius (2010) avoided this problem by imposing conditions equivalent to $n=1$, i.e. absence of allometric scaling of metabolism. Three analytical approaches that do not require this constraint will be explored here.

7.1. Approximation for times shortly after onset of fishing

For the time before the onset of fishing, assume an unperturbed power-law size spectrum, $b(u)=0$. Shortly after the onset at $t=0$, when $b(u)$ is still close to zero, the first term on the right-hand side of Eq. (78) will be small compared with the second term. Disregarding the first term, the solution of Eq. (78) can be approximated as $b(u) \approx -(Me^u)^{2-\lambda}\,\Phi(u)t$. This corresponds to a depletion of populations in response to fishing without consideration of density-dependent effects. Early density-dependent responses by prey and predators can be obtained by putting this first-order approximation into Eq. (78) and integrating in time, which leads to the second-order approximation

$$b(u) = -(Me^u)^{2-\lambda}\Phi(u)t - \frac{1}{2}M^{n+1-\lambda}e^{(n-1)u}\left\{\tilde{K}(u) * \left[e^{(2-\lambda)u}\Phi(u)\right]\right\}t^2 + \text{h.o.t.} \quad [88]$$

Here, h.o.t. stands for higher-order terms in t, which could be computed by iterating this procedure. However, it is quite likely that a series expansion in t would not converge for any value of t. Inclusion of higher-order terms would then not generally improve the approximation. It is valid only asymptotically, as $t \to 0$. To describe the system's long-term response to fishing, other approaches are required.

7.2. Tentative steady-state solution

An idea that comes to mind for understanding the long-term dynamics of Eq. (78) is to consider its steady-state ($\partial b/\partial t=0$). A formal steady-state solution $b(u)=b_0(u)$ is easily derived. Dividing the right-hand side of Eq. (78) by $(Me^u)^{n-1}$ and Fourier transforming, it becomes

$$0=\hat{K}(\xi)\hat{b}_0(\xi)-M^{\nu}\hat{\Phi}(\xi+\nu i), \qquad [89]$$

with $\hat{K}(\xi)$ denoting the Fourier transform of $\tilde{K}(u)$ and $\hat{\Phi}(\xi)=\tilde{B}_{\text{tot}}F\hat{W}_{\sigma_{\text{F}}}(\xi)\hat{\beta}(-\xi)\exp\left(-i\xi\ln\frac{m_{\text{F}}}{M}\right)$ the Fourier transform of $\Phi(u)$. As $\hat{K}(\xi)$ generally has no zeros on the real ξ axis, we can solve for

$$\hat{b}_0(\xi)=\frac{M^{\nu}\hat{\Phi}(\xi+\nu i)}{\hat{K}(\xi)}, \qquad [90]$$

from which $b_0(u)$ is obtained by back-transforming. However, care must be taken, because not all steady-state solutions $b(u)$ necessarily have a Fourier transform. To understand why, the analytic properties of $\hat{K}(\xi)$ need to be discussed.

7.3. The roles of complex poles and zeros of $\hat{K}(\xi)$

Consider, for some complex $\xi\in\mathbb{C}$, the convolution integral

$$\tilde{K}(u)*e^{i\xi u}=\int_{-\infty}^{\infty}\tilde{K}(v)e^{i(u-v)\xi}dv=e^{iu\xi}\int_{-\infty}^{\infty}\tilde{K}(v)e^{-i\xi v}dv. \qquad [91]$$

The integral exists if and essentially only if $\left|\tilde{K}(u)\right|$ decays faster than $|e^{i\xi u}|=e^{-u\,\text{Im}\{\xi\}}$ as $u\to\pm\infty$, with $\pm$ given by the sign of $\text{Im}\{\xi\}$. If the integral exists, it equals $e^{iu\xi}$ times the analytic continuation of the Fourier transform $\hat{K}(\xi)$ into the complex plane. The range of Im $\{\xi\}$ values over which Eq. (91) converges determines a stripe in the complex plane, parallel to and covering the real axis. Typically, as is the case here, the stripe is limited by poles of $\hat{K}(\xi)$, which correspond to exponentially decaying "tails" of $\tilde{K}(u)$ as $u\to\pm\infty$. To find these poles, recall that

$$\hat{K}(\xi)=\tilde{B}_{\text{tot}}\hat{\kappa}_0(\xi+\nu i)\hat{\beta}(-\xi-\nu i)\hat{\beta}(\xi)-r\tilde{B}_{\text{tot}}\left[1-\exp\left(-\sigma_r^2\xi^2/2\right)\right], \qquad [92]$$

and that $\hat{\beta}(\xi)$ has a pole at $\xi = -(1-n)i$ because of the hatchling tail of $\tilde{N}(x)$, Eq. (34), and possibly more poles farther from the real axis. This is the only source of poles of $\hat{K}(\xi)$. The Fourier transform $\hat{k}_0(\xi)$ of $k_0(w)$ is an entire function by Eqs. (83) and (84) and the assumption that $\hat{s}(\xi)$ is an entire function. In the oligotrophic regime $\hat{K}(\xi)$ therefore has poles at $\xi = -(1-n)i$ and $\xi = (q-n)i$, corresponding to decays of $\tilde{K}(u)$ as $e^{(1-n)u}$ for $u \to -\infty$ and as $e^{-(q-n)u}$ for $u \to +\infty$, and these two poles generally limit the range over which the integral (91) exists. In the eutrophic regime the pole $\xi = (1-n-\nu)i$ of $\hat{\beta}(-\xi - \nu i)$ coincides with a zero of $\hat{k}_0(\xi + \nu i)$ by Eq. (83), eliminating the corresponding pole of $\hat{K}(\xi)$, but other poles of $\hat{\beta}(-\xi - \nu i)$ may persist. Denote by Ξ the maximal (open) stripe in the complex plane parallel to the real axis for which $\hat{K}(\xi)$ is analytic ($\Xi = \{\xi : -(1-n) < \mathrm{Im}\{\xi\} < q-n\}$ in the oligotrophic regime).

Now, if $\xi_0 \in \Xi$ is a complex zero of $\hat{K}(\xi)$, and A_0 is any complex number, then with every solution $b_0(u)$ of Eq. (78), $b(u) = b_0(u) + A_0 \exp(i\xi_0 u)$ is, by Eq. (91), a solution of Eq. (78), too. Such additional exponential terms do not have finite Fourier integrals for any ξ and therefore cannot be found using Eq. (90). The correct choice of these corrections to the Fourier transform solution $b_0(u)$ of (78) follows, as for corresponding problems in physics (Jackson, 1962), from causality considerations: the corrections should be added if localised perturbations trigger waves or fronts that increase in magnitude as they propagate towards $|u| \to \infty$. However, because of the factor $(Me^u)^{n-1}$ in Eq. (78), standard methods for deciding when to make these corrections cannot be applied directly. Instead, this problem will here be (partially) solved by first investigating explicitly time-dependent solutions of Eq. (78).

7.4. Analytic approximation of time-dependent size-spectrum dynamics

Above, it was explained how complex zeros of $\hat{K}(\xi)$ lead to complications when solving Eq. (78). The idea of the approach described in this section is that, under certain conditions, the dynamics of $b(u)$ can indeed be dominated by the effects of these zeros. For example, it is evident from Eq. (90) that the Fourier transform $\hat{b}_0(\xi)$ of the steady-state solution $b_0(u)$ derived there generally has poles wherever $\hat{K}(\xi)$ has zeros. These poles are likely to have strong effects when the inhomogeneity (the fishing term) in Eq. (78) is not too broadly spread out on the logarithmic species size axis, that is, if its Fourier transform is not too localised along the real ξ axis. Specifically, the spread σ_F should be smaller than the typical logarithmic predator–prey size ratio, that is, fishing should not target more than about two trophic levels. This is often the case in practice.

Firstly, only zeros of $\hat{K}(\xi)$ within the stripe Ξ are considered. Denote by $\xi_1, \xi_2, \ldots$ an enumeration of these zeros (which may be finite or infinite in number). To isolate the effect of these zeros on the time-dependent solution of Eq. (78), we seek a representation of the solution in the form

$$b(u,t) = b_c(u,t) + \sum_j a_j(u,t)\exp(i\xi_j u), \qquad [93]$$

with the functions $b_c(u, t)$ and $a_j(u, t)$ to be specified below. The condition that $b(u, t)$ is real-valued requires that $a_j(u, t) = (a_k(u, t))^*$ when $\xi_j = -\xi_k^*$ (the asterisk * denotes complex conjugation).

Eliminating b in Eq. (78) by Eq. (93), one obtains an equation for b_c,

$$\frac{\partial b_c(u,t)}{\partial t} = (Me^u)^{n-1}\left[\tilde{K} * b_c\right](u,t) + (Me^u)^{n-1}\Psi(u,t), \qquad [94]$$

with

$$\begin{aligned}\Psi(u,t) = &-(Me^u)^{\nu}\Phi(u)\Theta(t) + \sum_j \int_{-\infty}^{\infty} \tilde{K}(u-v)a_j(v,t)e^{i\xi_j v}\,dv \\ &-\sum_j (Me^u)^{1-n}\frac{\partial a_j(u,t)}{\partial t}e^{i\xi_j u}.\end{aligned} \qquad [95]$$

The condition for choosing the functions $a_j(u, t)$ is now that they should vary slowly in u and t in a sense explained below, and that, even in the limit $t \to \infty$, the Fourier transform $\hat{b}_c(\xi, t)$ of $b_c(u, t)$ exists and can be continued analytically into Ξ. This implies that, even in the limit $t \to \infty$, $b_c(u, t)$ decays at least as fast as $\tilde{K}(u)$ for $u \to +\infty$ and $-\infty$, that is, $b_c(u, t)$ is at least as localised as $\tilde{K}(u)$. The impacts of fishing on species much smaller or larger than the target species should then be completely captured by the functions $a_j(u, t)$.

The strategy to achieve this goal is to assure that the Fourier transform $\hat{\Psi}(\xi, t)$ of $\Psi(u, t)$ has zeros at the points $\xi_k \in \Xi$ at all times, so as not to excite these modes in $b_c(u, t)$. This gives the condition

$$0 = \hat{\Psi}(\xi_k, t), \qquad [96]$$

and one can require $\hat{\Psi}(\xi, t)$ to be small also in the vicinity of these points. An interesting implication of this condition stems from the Fourier transform of the convolution integral in Eq. (95). Let ζ be an arbitrary complex number in Ξ.

Using information about the exponential decay of $\tilde{K}(u)$ as $u \to \pm\infty$ derived above and the fact that $\int_{-\infty}^{\infty} \tilde{K}(u)\exp(-i\xi_j u)\mathrm{d}u = \hat{K}(\xi_j) = 0$, one finds that

$$\int_{-\infty}^{\infty} \mathrm{e}^{-i\zeta u} \int_{-\infty}^{\infty} \tilde{K}(u-\upsilon)a_j(\upsilon,t)\mathrm{e}^{i\xi_j\upsilon}\mathrm{d}\upsilon\,\mathrm{d}u$$

becomes, after multiplying and dividing by $\mathrm{e}^{i\xi_j u}$

$$= \int_{-\infty}^{\infty} \mathrm{e}^{i(\xi_j-\zeta)u} \int_{-\infty}^{\infty} \tilde{K}(u-\upsilon)a_j(\upsilon,t)\mathrm{e}^{i\xi_j(u-\upsilon)}\mathrm{d}\upsilon\,\mathrm{d}u,$$

after integrating by parts in υ

$$= \int_{-\infty}^{\infty} \mathrm{e}^{i(\xi_j-\zeta)u} \int_{-\infty}^{\infty}\int_{-\infty}^{u-\upsilon} \tilde{K}(w)\mathrm{e}^{-i\xi_j u}\mathrm{d}w \frac{\partial a_j(\upsilon,t)}{\partial \upsilon}\mathrm{d}\upsilon\,\mathrm{d}u,$$

after changing order of integration and substituting $u \to u+\upsilon$

$$\int_{-\infty}^{\infty}\int_{-\infty}^{\infty} \mathrm{e}^{i(\xi_j-\zeta)(u+\upsilon)} \int_{-\infty}^{u} \tilde{K}(w)\mathrm{e}^{-i\xi_j w}\mathrm{d}w \frac{\partial a_j(\upsilon,t)}{\partial \upsilon}\mathrm{d}u\,\mathrm{d}\upsilon,$$

and after factoring out a constant

$$= \left(\int_{-\infty}^{\infty} \mathrm{e}^{i(\xi_j-\zeta)u} \int_{-\infty}^{u} \tilde{K}(w)\mathrm{e}^{-i\xi_j w}\mathrm{d}w\,\mathrm{d}u\right)\left(\int_{-\infty}^{\infty} \mathrm{e}^{i(\xi_j-\zeta)u}\frac{\partial a_j(\upsilon,t)}{\partial \upsilon}\mathrm{d}\upsilon\right).$$

If $\xi_j \neq \zeta$, the first factor evaluates, integrating by parts in u, to

$$\begin{aligned}\frac{1}{i(\zeta-\xi_j)}\int_{-\infty}^{\infty} \mathrm{e}^{i(\xi_j-\zeta)u}\tilde{K}(u)\mathrm{e}^{-i\xi_j u}\mathrm{d}u &= \frac{1}{i(\zeta-\xi_j)}\int_{-\infty}^{\infty}\mathrm{e}^{-i\zeta u}\tilde{K}(u)\mathrm{d}u \\ &= \frac{\hat{K}(\zeta)}{i(\zeta-\xi_j)},\end{aligned} \qquad [97]$$

whereas for $\xi_j = \zeta$, it becomes, again integrating by parts,

$$-\int_{-\infty}^{\infty} u\mathrm{e}^{-i\xi_j u}\tilde{K}(u)\mathrm{d}u = -i\frac{\mathrm{d}}{\mathrm{d}\xi}\int_{-\infty}^{\infty}\mathrm{e}^{-i\xi u}\tilde{K}(u)\mathrm{d}u|_{\xi=\xi_j} = -i\hat{K}'(\xi_j), \qquad [98]$$

with the prime denoting the derivative with respect to the argument. Therefore, in the vicinity of the points $\zeta = \xi_k$, the Fourier transform of the convolution integrals is either zero (or small) by Eq. (97) or approximated by Eq. (98) when $\xi_k = \xi_j$.

For reasons to become clear later, define

$$v_k = i(1-n)M^{n-1}\hat{K}'(\xi_k). \quad [99]$$

With respect to a Fourier transform near ξ_k, the sum over convolution integrals in Eq. (95) is then equivalent to a term $-(1-n)^{-1}v_k e^{i\xi_k u}M^{1-n}\partial a_k(u,t)/\partial u$. One can now attempt to choose the functions $a_k(u,t)$ such that, with this substitution, the right-hand side of Eq. (95) for $\Psi(u)$ vanishes identically for all u. This leads, for each k, to the condition

$$\frac{\partial a_k(u,t)}{\partial t} = -\frac{v_k e^{u(n-1)}}{1-n}\frac{\partial a_k(u,t)}{\partial u} - (Me^u)^{2-\lambda}e^{-i\xi_k u}\Phi(u)\Theta(t) - \sum_{j\neq k} e^{i(\xi_j-\xi_k)u}\frac{\partial a_j(u,t)}{\partial t}. \quad [100]$$

An alternative route to Eq. (100), currently under development, employs a multiple-scale singular perturbation formalism (Kevorkian and Cole, 1996). In this approach, the functions $a_k(u,t)$ vary formally on slow spatial and temporal scales, admitting a gradient expansion of the convolution, of which the derivative in u above is the lowest-order contribution.

Now, a new independent variable $z = e^{u(1-n)}$ is introduced, so, because $dz/du = (1-n)e^{u(1-n)} = (1-n)z$, one gets $\partial f/\partial u = (\partial f/\partial z)(dz/du) = (1-n)z\,\partial f/\partial z$ for any function $f(u)$. Writing $A_k(z,t) = a_k(u,t)$, Eq. (100) then becomes

$$\frac{\partial A_k(z,t)}{\partial t} = -v_k\frac{\partial A_k(z,t)}{\partial z} - M^{2-\lambda}z^{\frac{2-\lambda-i\xi_k}{1-n}}\Phi\left(\frac{\ln z}{1-n}\right)\Theta(t) - \sum_{j\neq k} z^{\frac{i(\xi_j-\xi_k)}{1-n}}\frac{\partial A_j(z,t)}{\partial t}. \quad [101]$$

When reading this equation as

$$\frac{\partial A_k(z,t)}{\partial t} = -v_k\frac{\partial A_k(z,t)}{\partial z} + H_k(z,t), \quad [102]$$

it formally describes fronts that are generated by the inhomogeneity $H_k(z,t)$ and move away from it at a velocity v_k. However, complicating matters, the velocity v_k can here be a complex numbers. Yet, for $H_k(z,t)$ analytic in z, one can write a formal solution of Eq. (102),

$$A_k(z,t) = \int_0^\infty H_k(z - v_k\tau, t-\tau)\,d\tau, \quad [103]$$

which is easily verified. The solution is causal: $A(z,t)$ depends on $H_k(z',t')$ only for times $t' \leq t$. To understand this result better, consider first the hypothetical special case that $H_k(z,t) = \exp(-(z-z_0)^2/2\sigma^2)\Theta(t)/\sqrt{2\pi\sigma^2}$, with arbitrary $\sigma, z_0 > 0$. Then

$$A_k(z,t) = \frac{\Theta(t)}{2v_k}\left[\operatorname{erf}\left(\frac{z-z_0}{\sqrt{2}\sigma}\right) - \operatorname{erf}\left(\frac{z-z_0-v_k t}{\sqrt{2}\sigma}\right)\right]. \qquad [104]$$

The error function erf(x) for complex arguments x approaches $\pm 1 = \operatorname{sign}\operatorname{Re}\{x\}$ for large $|x|$ when $|\operatorname{Re}\{x\}| > |\operatorname{Im}\{x\}|$, but for $|\operatorname{Re}\{x\}| < |\operatorname{Im}\{x\}|$, it oscillates heavily. It follows that, as $H_k(z,t)$ converges to $\delta(z-z_0)\Theta(t)$ for small σ, one can approximate

$$\begin{aligned} A_k(z,t) &\approx v_k^{-1}\Theta(t)[\Theta(z-z_0) - \Theta(z-z_0-\operatorname{Re}\{v_k\}t)] \\ &= \begin{cases} +v_k^{-1} & \text{for} \operatorname{Re}\{v_k\} > 0, t > 0 \text{ and } z_0 < z < z_0 + \operatorname{Re}\{v_k\}t, \\ -v_k^{-1} & \text{for} \operatorname{Re}\{v_k\} < 0, t > 0 \text{ and } z_0 + \operatorname{Re}\{v_k\}t < z < z_0, \\ 0 & \text{otherwise} \end{cases} \end{aligned} \qquad [105]$$

when $|\operatorname{Re}\{v_k\}| > |\operatorname{Im}\{v_k\}| \geq 0$. This describes a sharp front moving at velocity $\operatorname{Re}\{v_k\}$ away from $z = z_0$. The heavy oscillations in z and t arising when $|\operatorname{Re}\{v_k\}| < |\operatorname{Im}\{v_k\}|$ violate the assumptions that $A_k(z,t)$ varies slowly along z and bring $A_k(z,t)$ outside the range of validity of Eq. (102). Generally, one would assume such oscillations to be suppressed by higher-order derivative terms not included in Eq. (102), resulting, again, in a front moving at velocity $\approx \operatorname{Re}\{v_k\}$. Below, Eq. (105) will be used to approximate the response of $A_k(z,t)$ to any sharply localised perturbation δ-$(z-z_0)\Theta(t)$ "switched on" at $t = 0$, as long as $\operatorname{Re}\{v_k\} \neq 0$. The case $\operatorname{Re}\{v_k\} = 0$ does not generally arise and is not considered here. The approximate response of $A_k(z,t)$ to general $H_k(z,t)$ can be obtained as a linear combination of solutions of the form (105).

According to Eq. (105), $\partial A_k/\partial t$ is localised at the tip of the travelling fronts generated by inhomogeneities elsewhere on the z-axis. The last term in Eq. (101) describes secondary excitations of fronts/waves by fronts pertaining to other zeros ξ_j. In the singular perturbation formalism, it arises as a higher-order correction. Its effect can be observed in simulations of the system (see Section 8.4.1) but are disregarded in the following analytic calculations, for simplicity. The term containing the fishing pressure $\Phi(\cdot)$ in Eq. (100) is then the only contribution to $H_k(z,t)$ and, using approximation (101), one obtains

$$A_k(z,t) \approx -\frac{M^{2-\lambda}\Theta(t)}{\nu_k}\int_{z_1}^{z} y^{\frac{2-\lambda-i\xi_k}{1-n}}\Phi\left(\frac{\ln y}{1-n}\right)\mathrm{d}y \qquad [106]$$

where $z_1 = \max(0, z - \mathrm{Re}\{\nu_k\}t)$.

On the logarithmic maturation size scale, this becomes

$$a_k(u,t) \approx -\frac{M^{2-\lambda}\Theta(t)(1-n)}{\nu_k}\int_{u_1}^{u} \mathrm{e}^{(\nu-i\xi_k)w}\Phi(w)\mathrm{d}w, \qquad [107]$$

with $u_1 = u_1(u,t) = (1-n)^{-1}\ln[\mathrm{e}^{(1-n)u} - \mathrm{Re}\{\nu_k\}t]$ if the expression in brackets is positive, and $u_1 = -\infty$ otherwise. For fishing pressure specific to maturation size rather than individual size (i.e. using Eq. (80)), the integral can be evaluated analytically, leading to

$$a_k(u,t) \approx -\frac{\sqrt{\pi}\tilde{B}_{\mathrm{tot}}M^{2-\lambda}F\sigma_{\mathrm{F}}\Theta(t)(1-n)}{\sqrt{2}\nu_k}\exp\left[(\nu-i\xi_k)u_{\mathrm{F}} + \frac{(\nu-i\xi_k)^2\sigma_{\mathrm{F}}^2}{2}\right]$$
$$\times\left\{\mathrm{erf}\left[\frac{(\nu-i\xi_k)\sigma_{\mathrm{F}}^2 + u_{\mathrm{F}} - u_1}{\sqrt{2}\sigma_{\mathrm{F}}}\right] - \mathrm{erf}\left[\frac{(\nu-i\xi_k)\sigma_{\mathrm{F}}^2 + u_{\mathrm{F}} - u}{\sqrt{2}\sigma_{\mathrm{F}}}\right]\right\}, \qquad [108]$$

where $u_{\mathrm{F}} = \ln(m_{\mathrm{F}}/M)$ and u_1 is as defined above. Once the front generated by the onset of fishing has run over a given point u on the logarithmic species-size axis, which happens at time $t \approx (\mathrm{e}^{(1-n)u} - \mathrm{e}^{(1-n)u_{\mathrm{F}}})/\mathrm{Re}\{\nu_k\}$, the expression in braces evaluates to $2\,\mathrm{sign}\,\mathrm{Re}\{\nu_k\}$, and the amplitude of the community response to fishing is controlled by the constant of dimension Mass/Volume given by the first two factors in Eq. (108). The first factor accounts for the available biomass ($B_{\mathrm{tot}}M^{2-\lambda}$ of dimensions Mass/Volume), the fishing pressure applied to it ($F\sigma_{\mathrm{F}}$ dimension 1/Time), and the rate at which this pressure is diverted to other size classes (ν_k, dimension 1/Time). The exponential function in the second factor combines the allometric scaling laws of response rates $(1-n)$ and available biomass ($2-\lambda$, recall $\nu = 3-\lambda-n$), a weight and phase factor $\exp(-i\xi_k u_{\mathrm{F}})$ attributable the modulation of the kth mode, and a correction factor relevant when fishing is spread out over a broad body mass range.

The time-dependent community response to size-selective fishing can be approximated analytically by inserting Eq. (108) into Eq. (93) and setting the localised core contribution $b_{\mathrm{c}}(u,t)$ to zero. According to this approximation, fronts travelling towards smaller body sizes ($\mathrm{Re}\{\nu_k\} < 0$) accelerate and reach $u = -\infty$, that is, $z, m_* = 0$, after the finite time $t = e^{(1-n)u_F} / |\text{Re}\{\nu_k\}| = (m_F/M)^{(1-n)} / |\text{Re}\{\nu_k\}|$. Fronts travelling towards larger body sizes ($\text{Re}\{\nu_k\} > 0$) decelerate logarithmically on the u-axis. All fronts travel at constant speed on the $z = (m_*/M)^{(1-n)}$ axis. If $\text{Re}\{\nu_k\}$ and $\text{Im}\{\xi_k\}$ have the same sign, the front described by $a_k(u,t)\exp(-i\xi_k u)$ travels into the direction in which $|\exp(-i\xi_k u)|$ decreases. The contribution from $a_k(u,t)\exp(-i\xi_k u)$ is then bounded and its Fourier transform exists. In the opposite case (sign $\text{Re}\{\nu_k\} \neq$ sign $\text{Im}\{\xi_k\}$), the contribution by $a_k(u,t)\exp(-i\xi_k u)$ grows beyond all bounds as time proceeds. These contributions are of the amplifying type not captured by the Fourier transform steady-state solution Eq. (90). This phenomenon is closely related to what the physics and engineering literature calls a *convective instability*. It is characterised by the fact that, although the system response to a pulse perturbation remains finite at any fixed point on the u-axis, it grows beyond all bounds applicable uniformly for the entire infinite u-axis.[8] In reality, logarithmic size u is limited from both above and below, and the community response to perturbations remains finite. Convective instabilities are distinguished from *absolute instabilities*, where perturbation responses grow beyond all bounds for some given value of u. Such instabilities are conceivable also for size spectra and can be observed in simulations for specific parameter values, but the analysis described here is unable to capture them. An adaptation of standard methods for identifying absolute instabilities (e.g. Akhiezer and Polovin, 1971) to the particular form of Eq. (78) is therefore highly desirable.

7.5. Extension of the method to all zeros of $\hat{K}(\xi)$

It turns out that the approximation constructed above can be improved further by including also the zeros of $\hat{K}(\xi)$ outside the stripe Ξ, although some care needs to be taken. For zeros ξ_k outside the strip, the exponentially decaying tails of $\hat{K}(u)$ are "flat" compared with the rate of increase or decrease of $e^{-i\xi_k}$. The tails can therefore mediate long-range interactions and delocalise the dynamics of the modes $a_k(u,t)$. An analysis of the special case where $\hat{K}(\xi)$ is the ratio of two linear polynomials (not discussed here) reveals that this can lead to the suppression of convective instabilities even when sign $\text{Re}\{\nu_k\} \neq$ sign $\text{Im}\{\xi_k\}$: the response of $a_k(u,t)$ is then in the direction on the u-axis opposite to $\text{Re}\{\nu_k\}$. To take this into account, the following heuristics suppressing convective instabilities are employed. In the rare cases

[8] As here press perturbations are considered here rather than pulse perturbation, not all unbounded responses necessarily correspond to convective instabilities.

where $\text{Re}\{\nu_k\}$ and $\text{Im}\{\xi_k\}$ have opposite signs for zeros $\xi_k \notin \Xi$, the value of u_1 in Eq. (107) or (108) is, for all times $t>0$, set to $+\infty$ for $\text{Re}\{\nu_k\}>0$ and to $-\infty$ for $\text{Re}\{\nu_k\}<0$. Because of the rapid decay of these modes along u, the discontinuity of $b(u)$ in t resulting from this approximation is hardly noticeable. If $\text{Re}\{\nu_k\}$ and $\text{Im}\{\xi_k\}$ have the same signs, $a_k(u,t)$ is computed according to Eq. (107) or (108) without modifications. Although convective instabilities for zeros outside Ξ have not been observed in simulations, there is some risk that exceptions to these heuristics exist.

Provided the extension of this method to all zeros ξ_k of $\tilde{K}(\xi)$ is successful so that $\lim_{t\to\infty}\hat{\Psi}(\xi_k,t)=0$, one can expect that, by Eq. (94), the Fourier transform of the residual $b_c(u,t)$ converges to an entire function as $t\to\infty$. That is, $|\lim_{t\to\infty} b_c(u,t)|$ decays faster than exponentially for $u\to\pm\infty$. The steady-state residual $\lim_{t\to\infty} b_c(u,t)$ unaccounted for by the approximation developed above is then, in this sense, strongly localised near the size class targeted by fishing. When using the approximation method below, the contribution by $b_c(u,t)$ is not included.

8. COMPARISON OF ANALYTIC THEORY AND SIMULATIONS

8.1. A specific parametrisation

To visualise typical responses of size spectra to fishing and to test the analytic predictions of Section 7.4 by comparison with simulations, specific choices for the free model parameters need to be made. This includes, in particular, choices for the predator–prey size-ratio window $s(x)$ and the maturation selection function $\psi(x)$. Standard values used for the scalar parameters are listed in Table 3. The choices follow Hartvig et al. (2011), and the reader is referred to their work for detailed motivations. All body masses are measured in units of the body mass class targeted by fishing ($M=m_F$). To report simulation results independent of the specific choice of m_F, time is measured in units of the approximate age of the targeted size class at maturation (i.e. the age when $m=m_*=m_F$), computed for $\psi(x)=0$ and $m_0\to 0$ for simplicity. The growth trajectories of individuals then follow $\mathrm{d}m/\mathrm{d}t=\tilde{g}_0 m^n$ and time to maturation is $T_{\text{mat}}=m_F^{1-n}\tilde{g}_0^{-1}(1-n)^{-1}$. For $m_F=1$ kg, this equals 3.0 years with standard parameters. Fishing pressure F was fixed so that, in response to the onset of fishing, species of size m_F initially decline at a rate $F=0.1\,T_{\text{mat}}^{-1}$.

Table 3 Standard parameters used in simulations

Symbol	Value	Unit	Description
n	3/4		Exponent of physiological activity
h	85	$g^{1-n}yr^{-1}$	Coefficient of maximal food intake
k	10	$g^{1-n}yr^{-1}$	Coefficient of metabolic loss rate
q	0.8		Exponent of search/attack rate
γ	Cancels out	$g^{1-q}m^{-3}\,yr^{-1}$	Coefficient search/attack rate
α	0.6		Conversion efficiency
β	100		Preferred predator–prey mass ratio
σ_s	1		Width of predator–prey size-ratio window
F	See text	yr^{-1}	Fishing mortality
σ_F	ln 10		Width of harvested size range
η	0.25		Maturation over asymptotic mass
σ_r	0.5		Cut-off length for food-web effects
r	0.5γ	$g^{1-q}m^{-3}\,yr^{-1}$	Strength of food-web effects

The parameters values $r=0.5\gamma$ and $\sigma_r=0.5$ of the heuristic submodel for food-web effects where chosen, following the reasoning of Section 6.4, according to the observation that γ is the only other model parameter of the same dimension as r, so that both might be of similar magnitude, and that $\exp(\sigma_r)$ corresponds to the typical mass ratio between prey in the diet of a consumer.

8.1.1 Choice of the predator–prey size-ratio window

With β standing for the preferred predator–prey mass ratio (not to be confused with the size distribution $\tilde{\beta}(w)$) and x for the actual logarithmic predator–prey mass ratio, the predator–prey size-ratio window is chosen following Hartvig et al. (2011) as a Gaussian window:

$$s(x) = \exp\left[-\frac{(x-\ln\beta)^2}{2\sigma_s^2}\right]. \qquad [109]$$

Its Fourier transform is

$$\hat{s}(\xi) = (2\pi)^{1/2}\sigma_s \exp\left[-\frac{\sigma_s^2\xi^2}{2}\right]\beta^{-i\xi}. \qquad [110]$$

8.1.2 Choice of the reproduction-selection function and its implications

Hartvig et al. (2011) showed that with a reproduction-selection function of the form

$$\psi(x) = (\eta x)^{1-n}, \tag{111}$$

growth of individuals follows a von Bertalanffy trajectory. The coefficient η stands for the ratio of maturation size to asymptotic size. Hartvig et al. (2011) multiplied this form with a smoothed step function to describe the onset of maturation at $x=1$, but here I stick with Eq. (111) for the sake of analytic tractability. In fact, population structures resulting from $\psi(x)$ with or without this additional factor do not differ much.

For the special case that the allometric exponent for the metabolic loss rate is exactly $n=3/4$, and that background mortality is negligible ($\mu_0=0$), the scale-free population size structure $\tilde{N}(x)$ (Eqs. 33 and 24) with $\psi(x)$ given by Eq. (111), and $\tilde{\mu}=\tilde{g}_0$ (i.e. $\tilde{a}=1$) evaluates to

$$\tilde{N}(x) = \frac{\tilde{N}_0\left[1-(\eta x)^{1/4}\right]^3}{\tilde{g}_0\left[1-\eta^{1/4}\right]^4 x^{7/4}} \tag{112}$$

for $x \leq \eta^{-1}$ and $\tilde{N}(x)=0$ otherwise. This result follows after verifying that $E(x)$, defined by Eq. (44), is given by $E(x)=[1-(\eta x)^{1/4}]^4[1-\eta^{1/4}]^{-4}x$ (compute $E'(x)/E(x)$ to see this).

From Eq. (112) one gets, according to Eq. (41),

$$\begin{aligned}\tilde{B}_{\text{tot}} = \tilde{B}_{\text{tot}}(0) &= \int_0^\infty x\tilde{N}(x)\mathrm{d}x \\ &= \frac{\tilde{N}_0}{\tilde{g}_0\left[1-\eta^{1/4}\right]^4\eta^{1/4}}.\end{aligned} \tag{113}$$

Consequently, the distribution of a species' biomass over the logarithmic size axis is given, following Eq. (71), by

$$\tilde{\beta}(w) = (\eta e^w)^{1/4}\left[1-(\eta e^w)^{1/4}\right]^3 \tag{114}$$

for $w \leq -\ln\eta$ and $\tilde{\beta}(w)=0$ otherwise. This function has the Fourier transform

$$\hat{\beta}(\xi) = \frac{\eta^{i\xi}}{(i+\xi)(i+4\xi/3)(i+2\xi)(i+4\xi)}. \tag{115}$$

Next to the pole at $\xi = -i/4$, expected from the general asymptotic form of $\tilde{N}(x)$ for small x, Eq. (34), the analytic continuation of $\hat{\beta}(\xi)$ has poles also at $\xi = -i/2$, $\xi = -3i/4$, and $\xi = -i$. Interestingly, all poles are located at multiples of $-i(1-n)$.

As Eq. (113) contains the unknown $\tilde{N}_0$, it does not give a numerical value for $\tilde{B}_{\text{tot}}$. To compute $\tilde{B}_{\text{tot}}$, use Eq. (71) to express $\tilde{N}(x)$ by $\tilde{\beta}(w)$ in Eq. (18). The resulting equation can be solved for $\tilde{B}_{\text{tot}}$, to obtain

$$\tilde{B}_{\text{tot}} = \frac{\tilde{\mathcal{N}}}{\hat{\beta}(i(\lambda - 2))}. \quad [116]$$

As $\tilde{\beta}(0) = 1$ by normalisation, the two coefficients $\tilde{B}_{\text{tot}}$ and $\tilde{\mathcal{N}}$ are therefore nearly identical when $\lambda \approx 2$.

8.2. Simulation technique

Simulation of the Species Size-Spectrum Model, Eq. (78), over a large range in u is difficult because, by the factor $(Me^u)^{n-1}$, a broad range of timescales has to be covered. The u-axis was discretized to 256 points at distance 0.5, ranging from $u = u_{\text{min}} = -128/3$ to $u = 254.5/3$. This corresponds to a range from about 10^{-19}M to 10^{37}M in maturation size, which is much larger than the size range covered in real size spectra. The reason for choosing this broad range is to isolate the effects of scale-invariant dynamics, which can be approximated analytically, from conceivable complications due to boundary effects. To gain some qualitative understanding of the modifications of dynamics that could result from boundary effects, a variant of the model (constrained-domain variant) was simulated where Eq. (78) was applied only over the interval $e^u M = 10^{-15}$M to 10^3M, and $b(u) = 0$ held fixed otherwise.

The right-hand-side of Eq. (78) was evaluated using a pseudospectral method: the convolution was evaluated in Fourier space by multiplication of $\hat{b}(\xi)$ with the result for $\hat{K}(\xi)$, Eq. (92), with either Eq. (83) or (84). To reduce aliasing, the u-axis was extended by another 256 points for this operation. To suppress numerical instabilities developing at the lower edge of the u-axis, a stabilising extra term $-8(Me^u)^{n-1}b(u)/\{1 + \exp[0.6(u - u_{\text{min}} - \ln\beta)]\}$ was added to Eq. (78). Over most of the u-axis, this term has no effect. The discretized system was then solved using the implicit ODE solver CVODE included in the SUNDIALS package (Hindmarsh et al., 2005). This solver automatically adjusts the approximation order and the step size to achieve a prescribed accuracy (here 10^{-4} per step). In addition, upper limits on step size, scaling as the square-root of time since the start of simulations, were imposed to suppress numerical instabilities.

8.3. Numerical evaluation of the analytic approximation

The analytic formula for the Fourier transform of the interaction kernel, $\tilde{K}(\xi)$, was converted into a subroutine which takes arbitrary complex ξ as arguments and outputs $\tilde{K}(\xi)$. The zeros of $\tilde{K}(\xi)$ in the vicinity of $\xi = 0$ were then determined using a simple secant search algorithm, initiated at all points on a grid with resolution $\pi/(8 \ln \beta)$ spanning the range $0 \leq \mathrm{Re}\xi \leq 8\pi/\ln \beta$, $|\mathrm{Im}\,\xi| \leq 4\pi/\beta$. It was verified that enlarging this range did not change the results markedly. Zeros found repeatedly were discarded. The derivatives of $\tilde{K}(\xi)$ at the zeros, required to evaluate the analytic approximation of dynamics, are most easily computed numerically.

The programming language used (C++) did not include an implementation of the error function for complex arguments in its mathematical library. This function is required when computing the analytic approximation of size-spectrum dynamics using Eq. (108). The function was therefore implemented using the method proposed by Hui et al. (1978). Alternatively, the integrals in Eq. (107) could have been evaluated numerically. This second approach has the advantage of being applicable for arbitrary external perturbations $\Phi(w)$ but is slightly more computation-intensive.

8.4. Case studies

8.4.1 Standard parameters

Equation (78) for the oligotrophic case was solved with standard parameters both numerically and in the analytic approximation described in Section 8.1. The evolution of the system state is shown in Fig. 4. One observes the formation of two trophic cascades, one downwards from m_F to smaller species, the other one upwards, towards larger species. In a later phase, an upward-moving front emerges that bends the size spectrum downward and carries an additional downward cascade in its wake.

The initial dynamics are not reproduced well by the analytic result in Eqs. (93) and (108). The theory overestimates the initial decay of the targeted size class (which is well described by the small-t approximation, Eq. (88)) and is late in predicting the formation of a weak downward trophic cascade. The analytic solution predicts the timing of the saturation of the downward cascade around 3.3–10 T_{mat} reasonably well. It makes an excellent prediction of its final, saturated state.

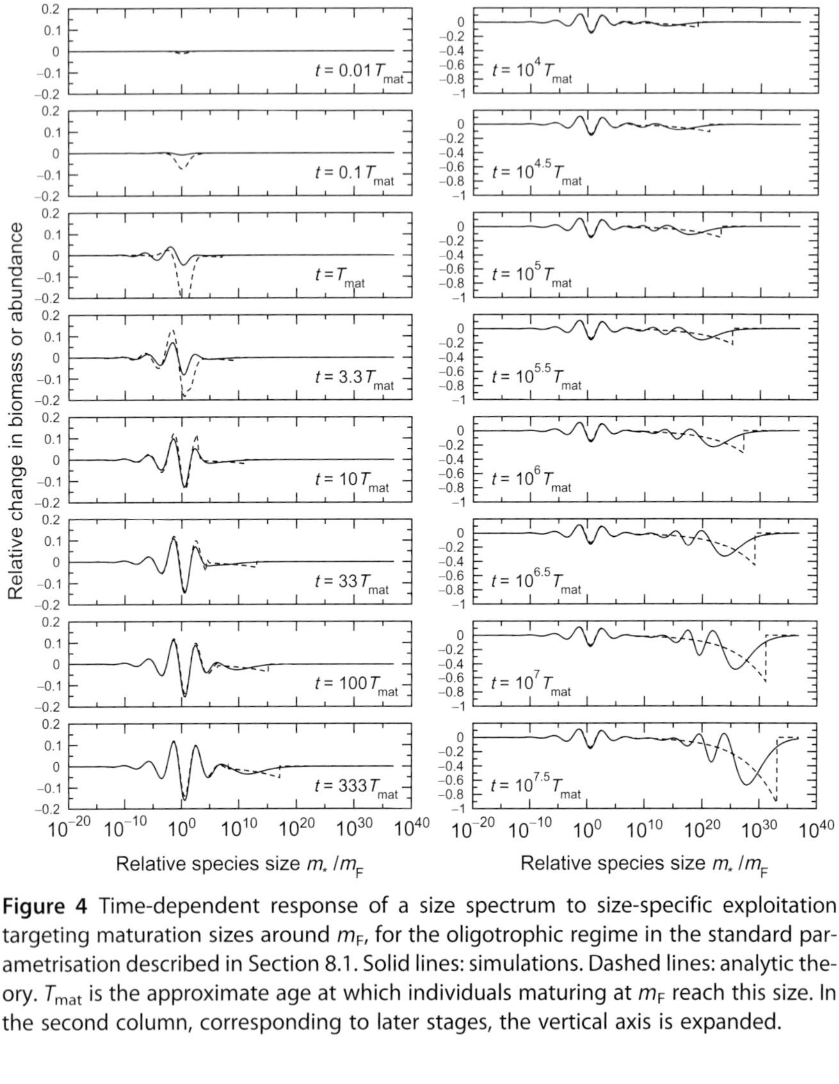

Figure 4 Time-dependent response of a size spectrum to size-specific exploitation targeting maturation sizes around m_F, for the oligotrophic regime in the standard parametrisation described in Section 8.1. Solid lines: simulations. Dashed lines: analytic theory. T_{mat} is the approximate age at which individuals maturing at m_F reach this size. In the second column, corresponding to later stages, the vertical axis is expanded.

The upward cascade evolves much slower than the downward cascade. This is mostly a consequence of the allometric scaling of biological rates, encapsulated in the factor $(Me^{u})^{n-1}$ in Eq. (78). The analytic approximation captures reasonably well the timing of the formation of the first maximum of the upward cascade around 10–33 T_{mat}, the formation of the second minimum around 100 T_{mat}, and the full emergence of the second maximum around 333 T_{mat}. Thereafter (second column in Fig. 4), the two trophic cascades have mostly stabilised.

What follows is the propagation of a downward-bending front to larger species sizes. Continuation of the simulation until $t = 10^{7.5} T_{\mathrm{mat}}$ and over an (unrealistically) large body-size range demonstrates three general facts. Firstly, the analytic theory makes a good prediction of the velocity of the downward-bending front on the u-axis. Secondly, the modulations behind the front travel in its wake, and, after sufficient waiting time, subside (compare, e.g. the simulation results at the point $m/m_{\mathrm{F}} = 10^{10}$ for $t = 10^5 T_{\mathrm{mat}}$ and $t = 10^{7.5} T_{\mathrm{mat}}$). Presumably, these secondary excitations of the size spectrum would be captured when including the neglected sum over j in Eq. (100) as an additional inhomogeneity contributing to $H_k(z, t)$. Thirdly, whereas in the analytic approximation the front quickly becomes sharp as it moves to larger u (as does the front of the upward trophic cascade), it is blurred in simulations. This indicates the presence of additional diffusive effects along the u-axis that were not captured in the present approximation.

The case displayed in Fig. 4 is a typical example for the degree of agreement between simulations and analytic approximations. Despite many differences in detail, the analytic approximation captures the main features and the times of their emergence.

8.4.2 The constrained-domain variant

In Fig. 5, it is shown how the response of the size spectrum changes in simulations when dynamics are constrained to a narrower domain. The dynamic domain was chosen so that with $m_{\mathrm{F}} = 1$ kg it corresponds approximately to the realised range of species sizes in marine communities. The resulting modification of dynamics evident in Fig. 5 can be interpreted as a minimal adaptation (a reflection on the upper boundary?) required to keep $b(u)$ continuous at the domain boundaries. Otherwise, the effect on dynamics is small, at least for the case considered here. Small dynamic effects of the boundaries do not imply that their effects on the steady-state size spectrum will be small.[9] The latter are discussed briefly in Section 9.6.

[9] The correct form of the boundary conditions that represent the upper and lower size limits of real communities in the Species Size-Spectrum Model is unknown. In an approximation consistent with the model, one can expect them to take the form of a set of inhomogeneous linear conditions on $b(u)$ (Cross and Hohenberg, 1993). A particular, stationary solution of the model compatible with these inhomogeneous boundary conditions represents the effect of "press perturbations by the boundaries" discussed in Sec. IX.F below. These perturbations will be stronger the larger the inhomogeneities in the boundary conditions. After subtracting this particular solution from $b(u)$, the reminder is constrained by a set of homogeneous boundary conditions. Of these, the condition applied here ($b(u) = 0$ for u outside the domain) is one example. As it is independent of the magnitude of the inhomogeneities, its effect on the dynamics inside the domain can be small even when boundary perturbations are strong.

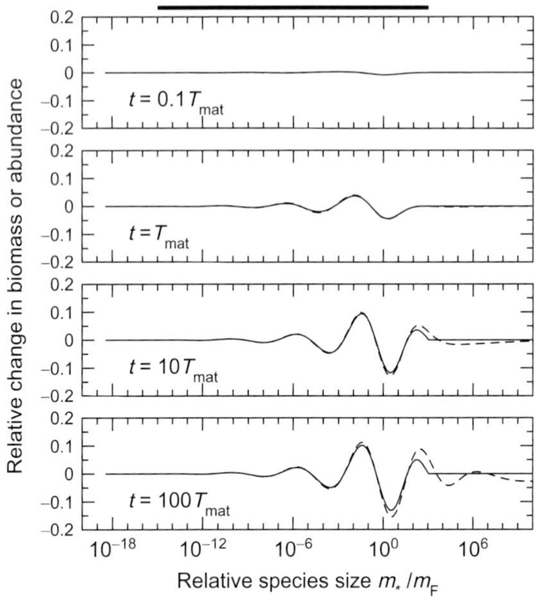

Figure 5 Effect of constraining the domain of the dynamic response of the size spectrum to the size range indicated by the bar above. Simulations with standard parameters in the oligotrophic regime as in Fig. 4, at four points in time *t*. Solid lines: constrained-domain variant. Dashed line: unconstrained variant for comparison.

8.4.3 Characterisation of size-spectrum dynamics by the zeros of $\hat{K}(\xi)$

The solution of Eq. (78) is controlled by the Fourier transform $\hat{K}(\xi)$ of the interaction kernel $\tilde{K}(u)$ entering Eq. (78), specifically by its complex zeros and its first derivatives at these points. A good qualitative understanding of the solution of Eq. (78) can therefore be gained already by studying the geometry of these zeros in the complex plane. For the standard parameter set used here, the zeros of $\hat{K}(\xi)$ and its analytic continuation are marked as open circles in Fig. 6. The arrow attached to each open circle indicates the argument of the first derivative of $\hat{K}(\xi)$ at this point (an arrow pointing straight right would mean a real, positive first derivative). The filled circles represent poles of $\hat{K}(\xi)$. Figure 6B offers a broader view of the same configuration.

Results derived in Section 7.4 translate into the following geometric picture: zeros of $\hat{K}(\xi)$ located below the real axis ($\mathrm{Im}\{\xi\}=0$) correspond to modulations or bending of $b(u)$ that increases towards larger u (larger body sizes), and *vice versa* for zeros above the real axis. A non-modulated (bending) contribution corresponds to a zero on the imaginary axis ($\mathrm{Re}\{\xi\}=0$).

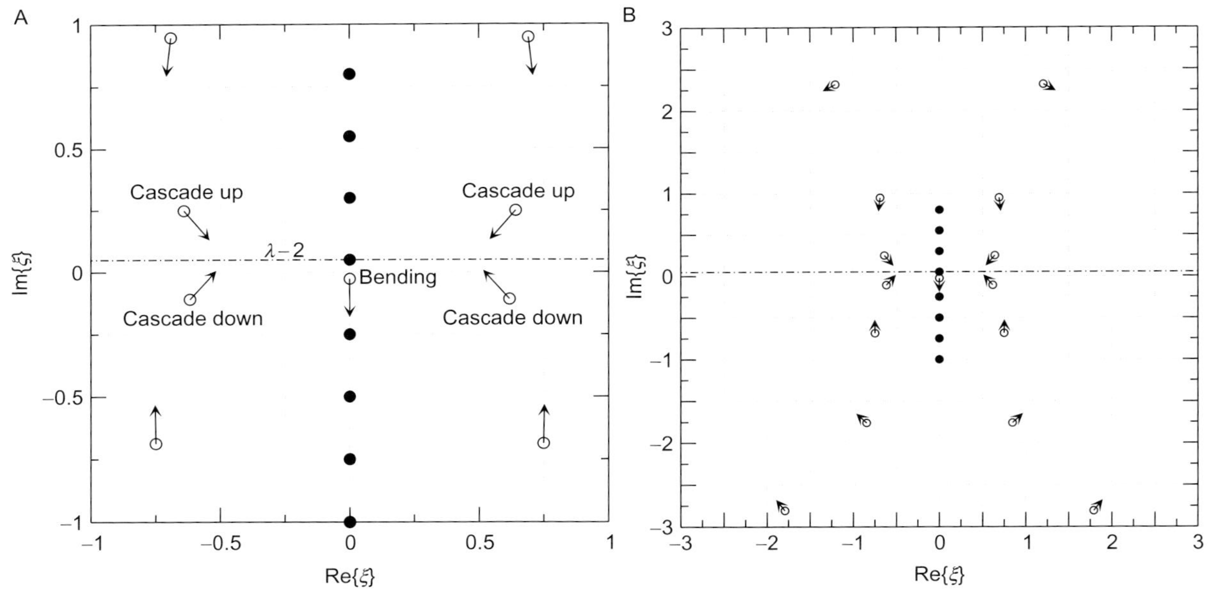

Figure 6 (A) Locations of zero (open circles) and poles (closed circles) of $\hat{K}(\xi)$ in the complex plane as well as the argument of $\hat{K}'(\xi)$ at the zeros (arrows), annotated with the corresponding perturbation responses. Standard parameters. (B) gives a wider perspective.

All other zeros always come in pairs ($\xi = \pm\xi_r + i\xi_i$), guaranteeing that the sum of all contributions approximating $b(u)$ by Eq. (93) is real-valued. When expressing deviations from the unperturbed equilibrium state in terms of relative changes in biomass or abundance (rather than by absolute differences in biomass density), modulations and bending increase towards positive u if the corresponding zero is located below the line $\mathrm{Im}\{\xi\} = \lambda - 2$, and *vice versa*.

The direction into which the front corresponding to each of these contributions propagates is determined by the vertical component of the orientation of the attached arrows. Arrows pointing downwards correspond to fronts propagating into the direction of positive u (upward cascades), that is, towards larger body size. Arrows pointing upwards correspond to fronts propagating in the direction of smaller body size (downward cascades). Combining the positions of zeros with the orientations of arrows, it follows that for zeros with arrows pointing towards the line $\mathrm{Im}\{\xi\} = \lambda - 2$ the corresponding modulations of relative abundance (or non-modulated bending) become smaller away from the size class targeted by fishing, or any other press perturbation. Effects on relative abundance corresponding to zeros with arrows pointing away from the line $\mathrm{Im}\{\xi\} = \lambda - 2$ increase as they propagate along the size axis. Such contributions correspond to the amplifying contributions mentioned at the end of Section 7.4. For the standard parameter set, the only contribution of this type is the upward-moving front of size-spectrum bent.[10]

Among zeros with attached arrows pointing towards the line $\mathrm{Im}\{\xi\} = \lambda - 2$, their contributions decay the faster along the u-axis the farther they are from the line $\mathrm{Im}\{\xi\} = \lambda - 2$. Further, zeros at some distance from $\mathrm{Im}\{\xi\} = \lambda - 2$ generally appear to be of the convectively stable type (see Figs. 6B, 6.7B, and 8B). Often, good qualitative images of the system response to press perturbations can be obtained from considering just a few zeros close to the line $\mathrm{Im}\{\xi\} = \lambda - 2$. In Fig. 6A, the five zeros of $\hat{K}(\xi)$ closest to this line have been annotated.

[10] The fact that in the oligotrophic regime there is always a purely imaginary zero corresponding to an amplifying, upward-moving front follows from the observation that $\hat{K}(\xi)$ is real along the imaginary axis, the existence of the two poles of $\hat{K}(\xi)$ at $\xi = -(1-n)i$ and $\xi = (q-n)i$, their respective sign structures, and continuity considerations. The sign structure follows heuristically from the fact that $\tilde{K}(u)$ describes feeding on smaller species and predation by larger species. The corresponding analytic argument makes use of the observation that the denominator in Eq. (46) is positive, so $\alpha I_1 - I_2 > kh^{-1} I_1$, a comparison of Eqs. (83) and (84), and the observation that Eq. (83) has a zero coinciding with the pole at $\xi = (q-n)i$.

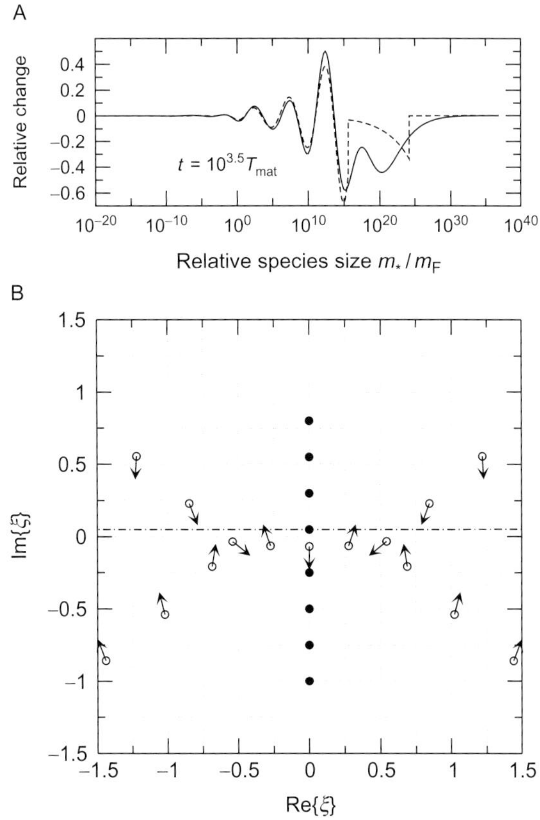

Figure 7 Perturbation response of the size spectrum (A) and graphic characterisation of $\hat{K}(\xi)$ (B) for the case discussed in Section 8.4.4, which leads to an amplifying upward trophic cascade. As in Fig. 4, solid and dashed lines in panel (A) correspond to simulation and analytic approximation, respectively. As in Fig. 6, open and closed circles in (B) correspond to zeros and poles of $\hat{K}(\xi)$ respectively, and arrows to the argument of $\hat{K}'(\xi)$ at the zeros.

8.4.4 An amplifying upward trophic cascade

When changing model parameters, other types of dynamics can be found. Figure 7A displays simulation and analytic results when, compared with the previous case, the preferred predator–prey mass ratio is increased from $\beta = 10^2$ to $\beta = 10^5$. The upward cascade now becomes amplifying, which means that the modulated system response becomes stronger towards species sizes larger than m_F (Fig. 7). Figure 7B displays the configuration of zeros and poles of $\hat{K}(\xi)$ in the complex plane. The amplifying upward cascade is generated by the pair of zeros near ± 0.5 with attached arrows pointing away from the line $\mathrm{Im}\{\xi\} = \lambda - 2$.

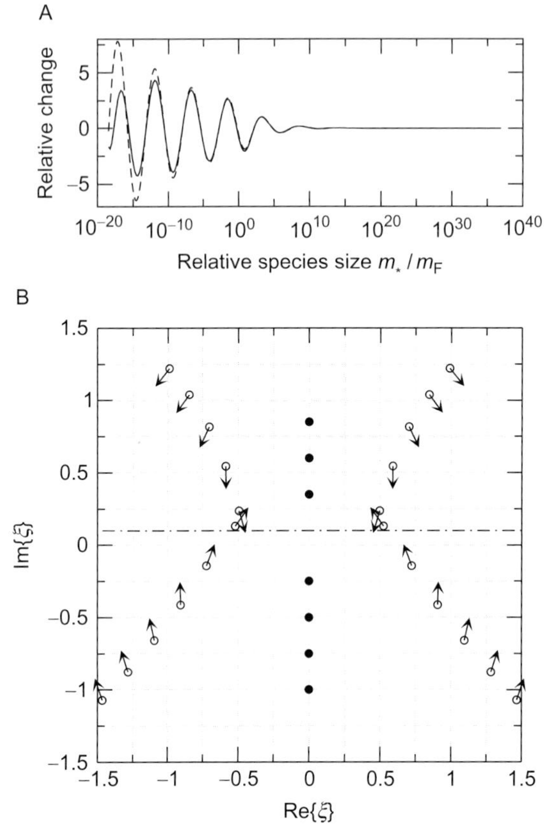

Figure 8 Perturbation response of the size spectrum (A) and graphic characterisation of $\hat{K}(\xi)$ (B) for the case discussed in Section 8.4.5, which leads to an amplifying downward trophic cascade. Symbols as in Figs. 4 and 6. As in Fig. 6, solid and dashed lines in (A) correspond to simulation and analytic approximation, respectively. As in Fig. 6, open and closed circles in (B) correspond to zeros and poles of $\hat{K}(\xi)$, respectively, and arrows to the argument of $\hat{K}'(\xi)$ at the zeros.

8.4.5 An amplifying downward trophic cascade

Finally, an example for the eutrophic regime is considered. In addition to setting $\beta = 10^5$ as in the previous example, standard parameters are modified by increasing the width of the predator–prey size-ratio window (Eq. 109) from $\sigma = 1$ to $\sigma = 3$. The coefficient of the allometric scaling law for respiration is increased from $k = 10g^{1-n}\text{year}^{-1}$ to $k = 49.09g^{1-n}\text{year}^{-1}$, giving a size-spectrum slope of $\lambda = 2.1$ via the constraint Eq. (39). For ecological consistency, all biomasses and abundances must be sufficiently large to fully satiate all consumers. For the system response to the fishing regime used here, this is clearly not the case. As can be seen in Fig. 8A, the predicted linear response grossly exceeds changes by 100% in biomass. Realistically, one should therefore apply less fishing pressure. Reducing fishing pressure by, say, one-tenth would yield exactly the same response as shown in Fig. 8, only with a ten times lower amplitude. Fishing pressure was retained at the standard value in Fig. 8A to allow easy comparison with the previous examples.

Now, the downward cascade becomes amplifying: abundance modulations increase towards lower species sizes. As the downward cascade propagates at constant speed on the m_*^{1-n}-axis, it reaches the lower end of the simulated range after finite time (about 100 T_{mat}). The simulation results differ from the analytic predictions for very low species sizes because of the numerical stabilisation applied near the lower boundary (see Section 8.2). The occurrence of a convectively unstable downward cascade can be predicted from the fact that in Fig. 8B there is one pair of zeros above the line $\mathrm{Im}\{\xi\}=\lambda-2$ with attached arrows pointing away from it. The observation that the amplitude of the system response here is considerably larger than in the previous cases might be related to the fact that these two zeros are both close to other zeros, which could lead to small derivatives of $\hat{K}(\xi)$ at these zeros (when $\hat{K}(\xi)$ has a double zero, $\hat{K}'(\xi)$ vanishes), small values of $|\nu_k|$, and hence large modulation amplitudes by Eq. (108).

Comparing Figs. 7A and 8A, one can see that in the eutrophic regime the pole of $\hat{\beta}(-\xi-\nu i)$ at $\xi=(\lambda-2)i$ has disappeared. It has been cancelled in $\hat{K}(\xi)$ by the zero of $\hat{\kappa}_0(\xi+\nu i)$, Eq. (83), at the same position. This zero results from the interplay between predation mortality (first term in Eq. (83)) and release from predation (second term in Eq. (83)) under *ad libitum* feeding. The pole of $\hat{\beta}(-\xi-\nu i)$ is a result of taking the limit $m_0, x_0 \to 0$ in Section 6.3. When regularising $\hat{\beta}(\xi)$, for example, by fixing x_0 at some small but positive value, $\hat{\beta}(-\xi-\nu i)$ would instead attain a large, but finite, value at $\xi=(\lambda-2)i$. Then $\hat{K}(\xi)$ at this point would be zero and $\hat{K}'(\xi)$ large, corresponding to perturbation responses in the form of small amplitude, rapidly upward-moving fronts, by Eqs. (99) and (108). These fronts represent the mechanism by which the overall abundances of larger species are adjusted to match those of smaller species in the eutrophic regime. In the limit $m_0, x_0 \to 0$, this mechanism degenerates to an immediate, rigid regulation, concealing the regulating dynamics.

9. IMPLICATIONS, DISCUSSION, AND OUTLOOK

9.1. The final steady state of a perturbed species size spectrum

The theory of size-spectrum dynamics developed in Section 8 now allows the problem of computing the steady-state solution reached long after initiating a size-specific press perturbation to be revisited. As argued in Section 7.2, this solution must be of the form

$$b(u) = b_0(u) + \sum_j A_j e^{i\xi_j u}, \qquad [117]$$

where $b_0(u)$ is given by its Fourier transform in Eq. (90) and A_j are constants to be determined. By comparing with the related decomposition Eq. (93) of the time-dependent solution and taking the limit $t \to +\infty$, one obtains

$$\lim_{t\to+\infty} b_c(u,t) = b_0(u) + \sum_j \left[A_j - \lim_{t\to+\infty} a_j(u,t) \right] e^{i\xi_j u}. \qquad [118]$$

The fact that $b_0(u)$ has a Fourier transform and the conjecture that $\lim_{t\to+\infty} b_c(u,t)$ is strongly localised (Section 7.5) and can now be used to compute the coefficients A_j. Consider first those cases where $|\exp(i\xi_j u)|$ diverges as $u \to +\infty$ (i.e. where $\mathrm{Im}\{\xi_j\} < 0$). The conditions on $b_c(u)$ and $b_0(u)$ require that

$$A_j = \lim_{t,\,u\to+\infty} a_j(u,t). \qquad [119]$$

For fronts propagating to negative u, that is, $\mathrm{Re}\{v_j\} < 0$, this implies $A_j = 0$. In the opposite case, $\mathrm{Re}\{v_j\} > 0$, application of Eqs. (107) and (99) gives

$$\begin{aligned} A_j &= -\frac{M^{2-\lambda}(1-n)}{v_j} \hat{\Phi}(\xi_j + vi) \\ &= \frac{iM^{v} \hat{\Phi}(\xi_j + vi)}{\hat{K}'(\xi_j)}. \end{aligned} \qquad [120]$$

The corresponding results in the case $\mathrm{Im}\{\xi\} > 0$ has the opposite sign for A_j. The close similarity between this expression for A_j and Eq. (90)

for $\hat{b}_0(\xi)$ has the interesting consequence that the final steady state $b(u) = \lim_{t\to+\infty} b(u,t)$ can be computed by formally evaluating the inverse Fourier transformation, Eq. (1), of $\hat{b}_0(\xi)$,

$$b(u) = \frac{1}{2\pi}\int_C \hat{b}_0(\xi)\mathrm{e}^{i\xi u}\mathrm{d}\xi. \qquad [121]$$

However, the path of integration C in the complex plane must be chosen such that, in the geometric picture introduced in Fig. 6A, it remains above all zeros ξ_j (open circles) with attached arrows pointing upwards, and below all zeros with attached arrows pointing downwards. Using Eq. (120) and the residue theorem of functional analysis, it is easily verified that the additional terms resulting form deviations of this path C from the real axis are exactly the non-zero contributions $A_j\mathrm{e}^{i\xi_j u}$ in Eq. (117).

9.2. Distribution of individual sizes in the final steady state

From $b(u)$, the final deviations of the distribution of biomass over logarithmic species sizes from the unperturbed distribution, one can compute the corresponding deviations $c(u)$ of the distribution of biomass over logarithmic individual sizes by a convolution with the intraspecific mass distribution $c(u) = \left[\tilde{\beta}*b\right](u)$. For simplicity, only those situations are considered here where $\mathrm{Im}\{\xi_k\} > -(1-n)$ for all k for which $A_k \neq 0$. Graphically, this means that all zeros (open circles) with attached arrows pointing away form the dash-dotted line must lie above the highest pole (filled circle) in the lower half plane. In all scenarios considered above (Figs. 6–8), this is the case. Using Eq. (117) and again invoking the residue theorem, one can then evaluate

$$\begin{aligned} c(u) &= \left[\tilde{\beta}*b_0\right](u) + \sum_j \hat{\beta}(\xi_j)A_j\mathrm{e}^{i\xi_j u} \\ &= \frac{1}{2\pi}\int_C \hat{c}_0(\xi)\mathrm{e}^{i\xi u}\mathrm{d}\xi, \end{aligned} \qquad [122]$$

with $\hat{c}_0(\xi) = \hat{\beta}(\xi)\hat{b}_0(\xi)$ and the path C taken identical to that in Eq. (121). Interestingly, when considering the case of fishing specific to the size of individuals (rather than species) and ignoring food-web effects, Eqs. (90), (79), and (92) evaluate $\hat{c}_0(\xi)$ to the simple expression

$$\hat{c}_0(\xi) = \frac{\hat{\beta}(\xi) M^\nu \hat{\Phi}(\xi + \nu i)}{\hat{K}(\xi)}$$
$$= \frac{\hat{\beta}(\xi) M^\nu \tilde{B}_{\text{tot}} F \hat{W}_{\sigma_F}(\xi + \nu i) \hat{\beta}(-\xi - \nu i) (M/m_F)^{i\xi - \nu}}{\tilde{B}_{\text{tot}} \hat{\kappa}_0(\xi + \nu i) \hat{\beta}(-\xi - \nu i) \hat{\beta}(\xi)} \quad [123]$$
$$= \frac{F M^\nu \hat{W}_{\sigma_F}(\xi + \nu i)}{\hat{\kappa}_0(\xi + \nu i)} \left(\frac{M}{m_F} \right)^{i\xi - \nu}.$$

This result depends on ecological detail only through the feeding interactions described through $\hat{\kappa}_0$, not through the life-history characteristics of species. These enter $c(u)$ only through path C determined by $\hat{K}(\xi)$. Formally, $c(u)$ is a steady-state solution of the balance equation

$$\frac{\partial c(u)}{\partial t} = (Me^u)^{n-1} \left[\tilde{K}_1 * c \right](u) - \tilde{B}_{\text{tot}} F (Me^u)^{2-\lambda} W_{\sigma_F}\left(u - \ln \frac{m_F}{M} \right), \quad [124]$$

with $\tilde{K}_1(u) = \tilde{B}_{\text{tot}} e^{\nu u} \kappa_0(u)$ (so that $\hat{K}_1(\xi) = \tilde{B}_{\text{tot}} \hat{\kappa}_0(\xi + \nu i)$). Making use of the definition of $\kappa_0(u)$, Eq. (70), $\tilde{K}_1(u)$ can be expressed as

$$\tilde{K}_1\left(\ln \frac{m}{m'} \right) = \tilde{B}_{\text{tot}} m^{2-\lambda} \left[\alpha h \frac{\delta f(m)}{\delta \mathcal{B}(m')} - m^{1-n} \frac{\delta \mu_{\text{p}}(m)}{\delta \mathcal{B}(m')} \right], \quad [125]$$

were $\mathcal{B}(m) = m\mathcal{N}(m)$ denotes the density of biomass on the (linear) body-size scale. The two terms in brackets describe effects on individuals of size m by trophic interactions with individual of size m'. The leading factor $\tilde{B}_{\text{tot}} m^{2-\lambda}$ scales as the size-dependence of the biomass of affected individuals.

Oddly, Eq. (124) therefore describes the mass balance in a size-structured community of individuals that feed upon each other but, akin to microorganisms, essentially only to multiply their numbers rather than to grow in size—despite the fact that the equation was obtained from a model for interacting highly size-structured populations. The fact that the true steady-state $c(u)$ solves Eq. (124) justifies the use of simplifying pictures where "large species eat small species" in verbal arguments. However, over-interpretation of such pictures or corresponding numerical models can be fatal: the dynamic response of $c(u)$ to fishing pressure predicted by the Eq. (124) will generally be very different from that predicted via the Species Size-Spectrum Model, Eq. (78), by setting $c(u) = \left[\tilde{\beta} * b \right](u)$. There can be differences in the timing of events, as demonstrated in Section 9.4, when the propagation velocities $\text{Re}\{v_k\}$ of perturbation responses (see Eq. (105)) have different magnitudes but the same signs in both cases, and the dynamic selection of entirely different steady states when some velocities $\mathrm{Re}\{v_k\}$ have different signs.

9.3. The population-level predator–prey size-ratio window

Models of interacting size-structured populations often invoke an individual-level predator–prey size-ratio window such as the function $s(w)$ entering the model of Hartvig et al. (2011). On the other hand, models of unstructured populations differentiating species by size (e.g. Brännström et al., 2010) often depend on a population-level predator–prey size-ratio window $S(w)$ that describes the size-dependence of the amount consumed of a species of size m_* by another species of size $m'_* = e^w m_*$. The QNA in the form developed in Section 6 allows the derivation of $S(w)$ from $s(w)$.

The computation is analogous to the derivation of the interaction kernel $\tilde{K}(u)$ from individual-level interactions, but now contributions related to active and passive feeding are handled separately. Consider the special case of a linear functional response ($h \to \infty$) in the oligotrophic regime. The contribution of predation mortality to $\kappa_0(w)$, Eq. (70), then simplifies to $-\gamma s(w)$. Verifying that $K(m_*, m'_*)$ specifies, according to Eqs. (57) and (62), a rate of change (dimension 1/Time) normalised to the biomass density of the affecting species, and replacing $\kappa_0(w)$ by $-\gamma s(w)$ in the formula for $K(m_*, m'_*)$, Eq. (72), one obtains—in the mean-field approximation underlying this calculation—the rate of population decay of a species of size m_* resulting from consumption by a species of size m'_* with biomass density B' as $\gamma m'^{\,q-1}_* S(\ln(m'_*/m_*))B'$, where

$$S(w) = \left[s * \tilde{\beta} * \tilde{\beta}_\nu^-\right](w), \quad \text{with}\, \tilde{\beta}_\nu^- \overset{\text{def}}{=} \mathrm{e}^{\nu w}\tilde{\beta}(-w) = \mathrm{e}^{(1-q)w}\tilde{\beta}(-w). \qquad [126]$$

The quantity $\gamma m'^{\,q-1}_*$ was factored out to make this result comparable with the corresponding formula for $s(w)$, that is, Eq. (13) with $\mathcal{N}(m)$ expressed in terms of biomass density $\mathcal{B}(m) = m\mathcal{N}(m)$.

Conversely, the rate of population growth of a species of size m'_* resulting from consumption of a species of size m_* with biomass density B evaluates to $\alpha\gamma m'^{\,q-1}_* S(\ln(m'_*/m_*))B$. Therefore, surprisingly, the population-level assimilation efficiency is here identical to the individual-level assimilation efficiency α.

The convolutions in Eq. (126) generally need to be evaluated numerically, for example using Fourier techniques. The resulting form of $S(w)$ for standard parameters is shown in Fig. 9. However, some general characteristics of $S(w)$ can be derived directly from Eq. (126). Recalling the

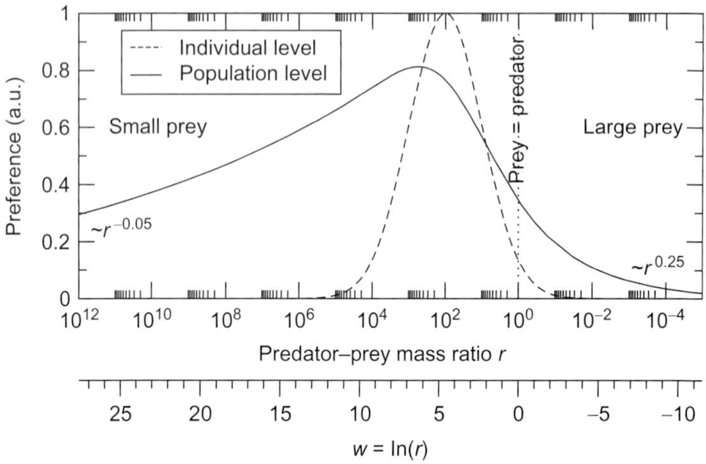

Figure 9 Comparison between the predator–prey size-ratio windows at individual level $s(w)$ and at population level $S(w)$. The mass ratio $r = e^w$ is the body-size ratio m'/m for $s(w)$ and the species-size m'_*/m_* for $S(w)$, with the prime denoting the predator. To display the curves from the predator's perspective, the horizontal axis is inverted. Curves correspond to standard parameters with $x_0 \to 0$. For finite parent-offspring size ratios, the tails of $S(w)$ will be truncated.

assumption that $s(w)$ decays faster than exponentially for $w \to \pm\infty$ (Section 4) and the observation that $\tilde{\beta}(w) \propto e^{(1-n)w}$ as $w \to -\infty$ (see Eqs. (34), (37), and (71)), it follows immediately that (i) $S(\ln(m'_*/m_*)) \propto (m'_*/m_*)^{1-n}$ for predator species smaller than their prey, (ii) $S(\ln(m'_*/m_*)) \propto (m_*/m'_*)^{q-n}$ for prey species much smaller than their predators, and (iii) typical $s(w)$, localised much sharper than intraspecific size structures $\tilde{\beta}(w)$, have little influence on the structure of $S(w)$ apart from determining a preferred predator–prey size ratio.

The conclusion that the population-level predator–prey size-ratio window is wide appears at odds with the conclusion reached analytically by Nakazawa et al. (2011), consistent with empirical analyses by Woodward and Warren (2007) (see also Gilljam et al., 2011; Woodward et al., 2010b), that the mean predator–prey size ratio at population level must be smaller than that at individual level. However, the paradox is easily resolved by noting that Nakazawa et al. (2011) and Woodward and Warren (2007) weighted all individuals equally when averaging over populations, whereas here each individual is weighted by its body mass as a proxy for reproductive value. With this modification, the analysis by Nakazawa et al. (2011) does indeed lead to the opposite conclusions: the mean population-level predator–prey size ratio is larger than that at individual level. This is the more adequate choice. As demonstrated by Rossberg and Farnsworth (2011) analytically and through simulations, weighting individuals within populations by reproductive value leads to better predictions of community dynamics than equal weighting.

9.4. Slow and fast responses to size-selective fishing

The timescale for the long-term community responses to fishing is determined by the constants ν_k entering Eq. (107), which are defined as $\nu_k = i(1-n)M^{n-1}\hat{K}'(\xi_k)$. These depend on ecological details through the factor $\hat{K}'(\xi_k)$, given by Eq. (92).

If intraspecific biomass distributions over the logarithmic mass scale $\tilde{\beta}(w)$ are wide and smooth, then their Fourier transforms $\hat{\beta}(\xi)$ are narrow on the $\mathrm{Re}\{\xi\}$-axis, as exemplified by Eq. (115). As a result, all ξ_k (except for the purely imaginary zero) are likely to lie outside the region where $\hat{\beta}(\xi)$ is large. Typically, this is the case when the logarithmic predator–prey mass ratio ln β is small compared with the width of $\tilde{\beta}(w)$, evidenced by cannibalistic feeding. This leads, by Eq. (92), to small $\hat{K}'(\xi_k)$ and a sluggish formation of trophic cascades compared, for example, with expectations from the naive picture of Eq. (124)—at least as long as $\hat{K}'(\xi_k)$ is not dominated by food-web effects through the second term in Eq. (92). Expressed in ecological terms, trophic cascades form slowly because population-level predator–prey size-ratio windows are broad: even species differing in maturation sizes by exactly the predator–prey mass ratio β do not form a simple predator–prey pair, rather they interact through a mixture of predator–prey, prey–predator, and competitive relationships resulting from the tremendous ontogenetic growth of fish and other aquatic organisms.

For the purely imaginary zero corresponding to size-spectrum bending, the situation is opposite. As shown in Figs. 6 and 7, this zero of $\hat{K}(\xi)$ is typically located near the pole of $\hat{K}(\xi)$ that corresponds to the first pole of $\hat{\beta}(-\xi-\nu i)$. Because of the resulting rapid transition of $\hat{K}(\xi)$ from zero to infinity, $\hat{K}'(\xi_k)$ is generally large for this zero. The downward bending of the size spectrum for large species is therefore fast, at least when compared with the biological rates of these species (despite being slow in absolute terms, see Fig. 4). The underlying ecological mechanism, which follows immediately from back-tracing how $\hat{\beta}(-\xi-\nu i)$ enters this result, is that larger species are being depleted directly through fishing mortality of their juveniles, even if the proportion of these in the catch—which scales as $(m_F/m_*)^{1-n}$—is relatively low. The mechanism is further amplified by a depletion of the food resources of juveniles of large species.

The effects of intraspecific size structure on response times can be demonstrated by a comparison of simulations of the Species Size-Spectrum Model with simulations of Eq. (124), which does not account for intraspecific size structure. To stabilise solutions of Eq. (124), some technical adjustments are required (caption of Fig. 10). As seen in Fig. 10, the initial response of the two systems to exploitation targeting a narrow range of body sizes is very different. Formation of the downward cascade is by a factor 3–10 slower in the full model than in Eq. (124). Depletion of larger organisms, however, proceeds by about a factor 3 faster for the reasons explained above. Yet, the equilibrium states approached by both systems are very similar: indeed, by the arguments of Section 9.2, they are possibly the same.

9.5. Neutrality and the validity of the QNA

Perfect ecological neutrality in the sense of Hubbell (2001) is a situation of perfectly even competition. Apart from drift driven by stochastic processes (which are not modelled here), communities of neutrally competing species are demographically stable for any combination of their relative abundances. In situations of approximate neutrality, populations approach particular equilibrium values, but only slowly in comparison with typical life-history time scales. The QNA, invoked in Section 6, exploits the resulting separation of the timescales of intraspecific and interspecific population dynamics to simplify descriptions of community dynamics (Rossberg and Farnsworth, 2011). Approximate neutrality among species with similar maturation sizes in a size-structured community is a likely phenomenon (O'Dwyer et al., 2009). Indeed, in the Species Size-Spectrum Model, the contribution from food-web effects had to be included in the interaction kernel $\tilde{K}(u)$ exactly to regularise model artefacts arising from perfect neutrality within size classes (Section 6.4).

The QNA as applied in the Species Size-Spectrum Model, however, invokes approximate neutrality not only within size classes, but, to a certain degree, also across size classes. This is justified *a posteriori* by the substantial competitive components in the interactions between species of different size, which arise from broad population-level predator–prey size-ratio windows, leading to slow dynamics. To quantify the resulting degree of timescale separation, and hence to gauge the reliability of the QNA, one can compare typical intra- and interspecific relaxation times in the model. In the simulation with standard parameters, shown in Fig. 4, population sizes near the targeted size class $m_* = m_F$ relax on a timescale of 4–10 T_{mat}, with

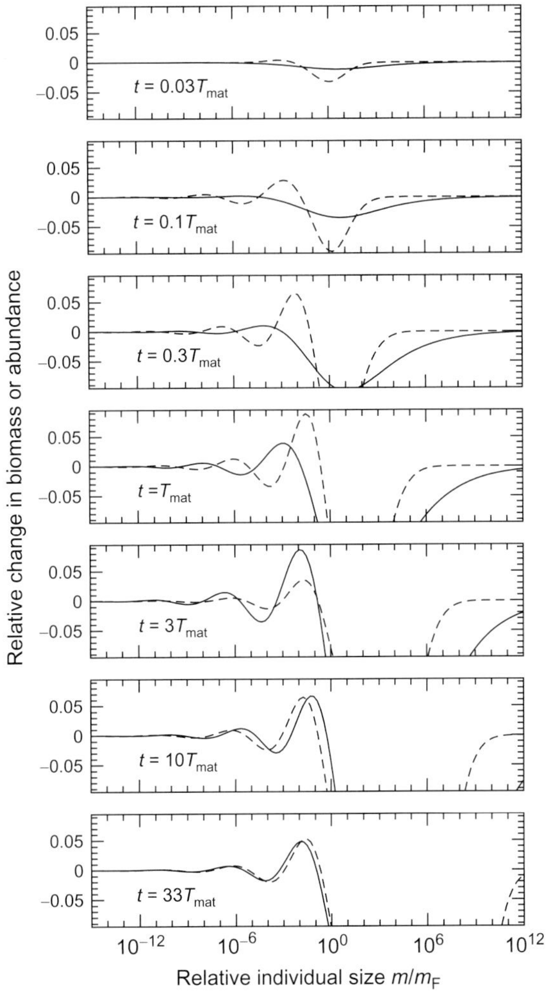

Figure 10 Comparison of the response of individual (i.e. Sheldon-type) size spectra to size-specific feeding as predicted by the Species Size-Spectrum Model (Eqs. (78) and (79), solid lines) and a corresponding model that disregards intraspecific size structure (Eq. (124), dashed lines). Simulations with standard parameters for the oligotrophic regime, however, with 10-fold fishing mortality F, with $\tilde{K}_1(u) = \tilde{B}_{\text{tot}}[e^{\nu}\kappa_0(u) + X(u)]$ in Eq. (124) to stabilise short-wavelength instabilities, $X(w)$ in Eq. (86) replaced by $\left[X * \beta * \tilde{\beta}_{\nu}^{-}\right](w)$ for consistency, and the parameter r in $X(w)$ set to 20γ to suppress amplifying cascades that would otherwise arise.

$T_{\text{mat}} = m_{\text{F}}^{1-n}\tilde{g}_0^{-1}(1-n)^{-1}$. The intraspecific relaxation timescale is given by the inverse size of the spectral gap of the McKendrick–von Foerster operator $\underline{L}_{m_*}$, which is estimated in Appendix as $\tilde{\mu}(m_*/\eta)^{n-1}$. As $\tilde{\mu} = \tilde{g}_0$ by Eqs. (35) and (37), the ratio between the two timescales is 4–10 times $(1-n)^{-1}\eta^{n-1}$, which is ≈ 20 to 60. These are fairly large values, from which at least a semi-quantitative validity of the QNA can be expected. Indeed, Hartvig et al. (2011) find for their closely related model, using simulations that explicitly account for both intra- and interspecific population dynamics in small communities, that dynamical details are generally much more complex than the Species Size-Spectrum Model predicts. Yet, the structures emerging when averaging across time and communities (Hartvig et al., 2011) are similar to those derived here.

9.6. How do power-law size spectra arise?

Just as in the paper of Andersen and Beyer (2006), the current analysis began with the assumption of a power-law size spectrum and then investigated its consequences, for example, constraints on the power-law exponent. This does not immediately answer the question of why power laws emerge in the first place. In addressing this question, the oligotrophic and eutrophic regimes have to be distinguished.

In the oligotrophic regime, where $\lambda = 2 + q - n$, conditions on overall abundance, Eq. (38) for $\tilde{\mathcal{N}}$ or Eq. (116) for $\tilde{B}_{\text{tot}}$, must be satisfied for steady-state power-law size spectra. Press perturbations of the size spectrum lead to modifications of the power law that spread on the m_*-axis away from the point of perturbation. Pulse perturbations can be described as two press perturbations of opposite sign initiated with a short delay. These are therefore predicted to lead to pulses travelling along the m_*-axis away from the point of perturbation. Assuming that any deviation of the size spectrum from the ideal form can be generated by applying appropriate pulse perturbations, one can conclude that any such deviations will eventually travel towards the upper and lower ends of the m_* range realised in a community, and there, depending on the effective boundary conditions, either be reflected back or disappear. When reflection is sufficiently weak and/or all modes are sufficiently damped, all deviations will eventually disappear when no further perturbations are applied. What remains is a power-law size spectrum.

Complications of this scenario can arise only when the boundaries of the ecologically realised m_* range effectively impose press perturbations on the size spectrum, which will then lead to persistent upward or downward cascades or bends originating at the upper or lower end of the realised range. In particular, insufficient biomass at the lower end acts as a press perturbation which will generate an upward-travelling front of reduced abundance corresponding to the purely imaginary zero of $\hat{K}(\xi)$ (Fig. 6A), and eventually to a bent-down steady-state size spectrum. This has, after back-translating result of Section 7 to the linear m_*-axis and disregarding trophic cascades, the form

$$B(m_*) = m_*^{-\lambda+1}\tilde{B}_{\text{tot}}\left(1 + Qm_*^{-\lambda-2-\text{Im}\{\xi_1\}} + \text{h.o.t.}\right), \qquad [127]$$

with a scale parameter $Q<0$, rather than following a power law. Overabundance of biomass at the lower end corresponds to $Q>0$. Higher-order terms in Q, not captured by the linearised QNA, are indicated by h.o.t. However, the exponent $\lambda - 2 - \text{Im}\{\xi_1\}$ can be numerically quite small (Fig. 6A), so the deviations from a perfect power law described by Eq. (127) might remain undetected in empirical data. Possibly, all that can be seen is a change in the best-fitting power-law exponent ("slope").

Besides, mechanisms not considered here could help satisfying Eq. (38) at least approximately, to keep deviations from a power law in Eq. (127) small. For example, temporal modulations of abundances, as observed by Hartvig et al. (2011), can modify the effective value of the interaction coefficient γ in Eq. (38). An effect observed in simulations of size-structured food webs may also be important (Rossberg et al., 2008): evolutionary forces can lead to adjustments of attack rates at community level such that effective food availability becomes of similar magnitude as physiological food demand. In the present case, this would mean that $\gamma\tilde{N}/k$ becomes approximately of the order of magnitude of one, which is indeed what Eq. (38) requires.

In the eutrophic regime, a power-law spectrum can form at any abundance level $\tilde{\mathcal{N}}$, provided $\tilde{\mathcal{N}}$ is sufficiently large to guarantee *ad libitum* feeding. Perturbations of the size spectrum will, over time, subside as in the oligotrophic case, leaving a power-law size spectrum, possibly overlaid with boundary-induced trophic cascades. Therefore, power-law size spectra form naturally in the eutrophic regime.

In size-spectrum theories pre-dating Benoît and Rochet (2004), an individual's rate of food uptake is generally thought to be density independent, as here in the eutrophic regime (e.g. Brown et al., 2004; Platt and Denman, 1977, 1978), leading to perfect power laws over a wide range of conditions. A similar situation arises in the models investigated analytically by Benoît and Rochet (2004) and Law et al. (2009), where all growth and mortality terms scale linearly with abundance, so that abundance can be factored out. Contrary to a statement by Capitán and Delius (2010), the two cases (no density dependence or uniform linear density dependence) are neither ecologically nor mathematically the same. Yet, it is characteristic for both that power-law size spectra can form at any abundance level, and that the size-spectrum slope λ depends on several physiological parameters (in the present case, according to Eq. (39), on α, h, k, n, and the predator–prey size-ratio window $s(\cdot)$).

9.7. Comparison with the theory of Andersen and Beyer (2006)

An important result of the present work is that in size-structured communities of populations covering wide size distributions the boundary condition of the McKendrick–von Foerster Equation (which balances production and outgrowth of newborns) implies that the specific physiological mortality of immature individuals is constrained to $\tilde{a}=1$ over most of their life history. This finding differs from the analytic results of Andersen and Beyer (2006), which Hartvig et al. (2011), similar to Andersen et al. (2008), evaluate to $\tilde{a}\approx 0.42$ for parameters as in Table 3. The theory of (Andersen and Beyer, 2006) does not consider the boundary condition and is therefore not "complete" in the sense used here. As a result, it predicts "equilibrium" population structures that violate the condition that mean lifetime reproductive output per individual (known as R_0) should equal one (Andersen et al., 2008). Resolution of the discrepancy between the two approaches requires understanding how Andersen and Beyer (2006) arrive at their constraint on $\tilde{a}$.

The case considered by Andersen and Beyer (2006) corresponds to the oligotrophic regime, because the feeding level is assumed fixed at an intermediate value $0<f_0<1$. They were the first to observe that this alone determines the size-spectrum slope as $\lambda=2+p-n$ (Capitán and Delius, 2010, make a similar observation). The crucial difference from the model of Hartvig et al. (2011)—investigated here—seems to be that Andersen and Beyer (2006) assume the rate of food intake to follow a different allometric scaling law than metabolic loss so that the loss becomes relevant only as individuals approach their maximal attainable size. For younger individuals, both growth $g(m, m_*)$ and predation mortality $\mu_p(m)$ are proportional to what Andersen and Beyer (2006) call the *metabolic requirement for food*. This leads to a value $\tilde{a}=\lim_{m/m_*\to 0} m\mu_p(m)/g(m, m_*)$ that is fully determined by basic physiological parameters. By contrast, the density-dependent bioenergy available for growth in the model of Hartvig et al. (2011) is always discounted by substantial density-independent metabolic loss so that $\tilde{a}$ (mortality over growth) becomes density dependent. This leaves room for $\tilde{a}$ to be adjusted at community level by the mechanisms explored in Sections 5 and 9.6.

9.8. Outlook

Although providing answers and clarifications to many questions related to community dynamics, the present analysis also led to a number of new questions and challenges. In closing, some of these are discussed briefly.

The distinction between the mechanisms shaping size spectra in oligotrophic and eutrophic regime made throughout the analysis immediately raises the question as to which empirical systems are best described by which regime, or whether either of these exclusively describes the situation in the field. An easy approach to this question would be to measure how close somatic growth and reproduction rates in a given community are to those found in laboratory experiments with *ad libitum* feeding. By the definition of the eutrophic regime, they should be identical in this case, whereas for the oligotrophic regime, growth and reproduction would be slower in the field than in the laboratory.

Based on such identifications of the relevant regimes, further progress may be possible in relating size-spectrum slopes to trophic status (Ahrens and Peters, 1991; Sprules and Munawar, 1986), temperature (Yvon-Durocher et al., 2011), and other environmental conditions. Obviously, changes in the slope resulting from environmental change can have great impacts on the production at higher trophic levels available to human consumption.

Another question that deserves further study is the distinction between damped and amplifying upward and downward cascades in field data. As mentioned in Section 2.5, this will probably require analysis of temporally resolved size-spectrum data. Clearly, the distinction between upward and downward cascades has implications for approaches to management of size-structured communities. Provided that amplifying cascades can be identified, the theoretical problem arises of extending the linear theory for size-spectrum dynamics to include non-linear corrections, because these could limit the amplification. Perturbative approaches developed in other contexts (Newell and Whitehead, 1969) might be employed usefully here.

Important in the context of fisheries management are the three related questions of what determines the upper cut-off of the size spectrum (i.e. the natural size of the largest species), how the boundary condition resulting from this cut-off is best incorporated into size-spectrum models, and how large the expected perturbation of the size spectrum resulting from this boundary conditions is. A way forward on these questions might be to study them first in corresponding species-resolved models of size-structured communities (Hartvig et al., 2011; Rossberg et al., 2008; Shephard et al., 2012), where the upper cut-off naturally emerges.

Finally, the observations of Section 9.4 highlight that care must be taken to incorporate population size structure and trophic interactions at all life stages adequately into management models, because this may drastically modify dynamics. Consistency of model steady states with observed conditions is an insufficient criterion for model validation: this insight might allow improvements of some models currently in use.

As exemplified throughout this work, the language of mathematics has its use in science not only for reasons of numerical precision. Often, the outcome is just precision of thought: the definition of a quantity through a formula allows investigation of logical relations to other quantities; equations stating mechanistic relations between quantities allow tracking down their implications; a description of a system by combinations of such equations (rather than verbal models) allows evaluation of its logical consistency, its descriptive completeness, its mechanical consistency, and at least the qualitative consistency of its predictions with observations. Verbal descriptions and analyses of complex systems are more prone to the risks of being incomplete, overlooking essential feedback loops, or incorporating hidden inconsistencies. Ecology as a field might be underestimating the benefits inherent to the mathematical idiom.

Especially in a context of rapid, global change leading into a world with environmental conditions not experienced before, there is need for good understanding of the general mechanisms controlling high-level ecosystem states (Mintenbeck et al., 2012; Möllmann and Diekmann, 2012). This understanding should ideally derive not solely from syntheses of past observations but be anchored in systematic analyses of the consequences of fundamental principles.

Formal reasoning using mathematical language can help to navigate the complexities of such analyses. The present work hopefully, therefore, does not only provide insights into the workings of size-structured communities but also encourages researchers to make greater use of formal reasoning in order to advance our understanding of complex ecological patterns and processes.

ACKNOWLEDGEMENTS

Discussion and comments on this work by Eric Benoît, Martin Hartvig, Ken Haste Andersen, Linus Carlsson, Åke Brännström, Christian Mulder, Andy Payne, and three anonymous reviewers, as well as editorial input by Julia Reiss, Gabriel Yvon-Durocher, Ute Jacob, and Guy Woodward are gratefully acknowledged. This Beaufort Marine Research Award is carried out under the *Sea Change* Strategy and the Strategy for Science Technology and Innovation (2006–2013), with the support of the Marine Institute, funded under the Marine Research Sub-Programme of the Irish National Development Plan 2007–2013.

APPENDIX. NUMERICAL STUDY OF THE SPECTRUM OF THE MCKENDRICK–VON FOERSTER OPERATOR

This appendix reports results of a numerical study of the spectrum of the linear operator $\underline{L}_{m_*}$, as defined by Eq. (47). Parameters were chosen as in Section 8.1. To avoid numerical difficulties at small body size, the eigenvalue problem $\underline{L}_{m_*} W(m) = lW(m)$ for the density of individuals along the linear mass axis was transformed into the corresponding eigenvalue problem for the density of biomass along the logarithmic mass axis $f(u) = \mathrm{e}^{2u} W(\mathrm{e}^u)$, with $u = \ln(m/m_*)$. The equivalent eigenvalue problem is then $H_{m_*} f(u) = lf(u)$, with

$$H_{m_*} f(u) = \tilde{\mu} m_*^{n-1} \left[\left(\frac{1}{4} \mathrm{e}^{-u/4} - \eta^{1/4} \right) f(u) - \left(\mathrm{e}^{-u/4} - \eta^{1/4} \right) \frac{\mathrm{d}f(u)}{\mathrm{d}u} + \eta^{1/4} \delta(u - u_0) \int_{u_0}^{-\ln \eta} f(v) \mathrm{d}v \right] \qquad [128]$$

conditional to $f(u) = 0$ for $u < u_0$, where $u_0 \overset{\text{def}}{=} \ln(m_0/m_*) = \ln(x_0)$. As in Sec. VIII.A, the maximum attainable relative body mass was set to $\eta^{-1} = 4$. Results reported here are for $x_0 = 10^{-5}$. Results depend only weakly on this particular choice. The last term in Eq. (128) enforces the condition

$$f(u_0) = \eta^{1/4} (x_0^{-1/4} - \eta^{1/4})^{-1} \int_{u_0}^{-\ln \eta} f(v) \mathrm{d}v \qquad [129]$$

on eigenfunctions $f(u)$.

The problem was discretized on a grid with G points, spaced equally on the u-axis from $u_0 + \Delta\mathrm{u}$ to $-\ln \eta$, where $\Delta\mathrm{u} = (-\ln \eta - u_0)/G$. The derivative of $f(u)$ in Eq. (128) was approximated by backward differences. To compute this difference for the first point, $u = u_0 + \Delta\mathrm{u}$, that is, $f'(u_0 + \Delta\mathrm{u}) \approx (\Delta\mathrm{u})^{-1} [f(u_0 + \Delta\mathrm{u}) - f(u_0)]$, Eq. (129) (trivially approximated by a sum) was used for $f(u_0)$. The factor $\tilde{\mu} m_*^{n-1}$ in Eq. (128) can be eliminated by rescaling all eigenvalues. The eigenvalue problem for H_{m_*} is then approximated by a corresponding numerical eigenvalue problem for a $G \times G$ matrix H_{jk}.

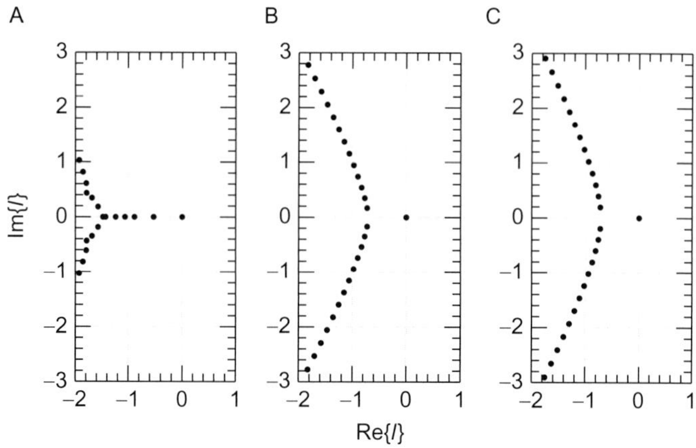

Figure 11 Eigenvalues l of the matrix H_{ij}, a discretization of the McKendrick–von Foerster operator, Eq. (47). (A) Naive implementation, exhibiting discretization artefacts. (B) Regularised version, obtained by removing the last two rows and columns of H_{ij}. (C) After removal of yet another row and column. All eigenvalues are given in units of $\tilde{\mu}m_*^{n-1}$.

Figure 11A shows the eigenvalues of H_{jk} for $G=2000$. All eigenvalues have multiplicity one. The largest eigenvalue found, $l=-0.00018\,\tilde{\mu}m_*^{n-1}$, is close to zero. Further, one can distinguish 5–6 negative real eigenvalues, and a family of complex eigenvalues.[11] The negative real eigenvalues are numerical artefacts. They arise from the fact that, because $g(e^u m_*, m_*)$ approaches 0 linearly as $u\rightarrow -\ln\,\eta$ (the largest adults do not grow), the matrix H_{jk} has for large G in its lower right corner the approximate structure

$$\begin{pmatrix} \ddots & & & \vdots & & \\ & -4A-B & 0 & 0 & 0 & 0 \\ & 3A & -3A-B & 0 & 0 & 0 \\ \cdots & 0 & 2A & -2A-B & 0 & 0 \\ & 0 & 0 & A & -A-B & 0 \\ & 0 & 0 & 0 & 0 & -B \end{pmatrix}, \qquad [130]$$

with positive constants A and B that are independent of G. This leads to one eigenvector with eigenvalue $-B$, localised on the last lattice point with small contributions in all other components, and a series of eigenvectors of the approximate forms $(\ldots, 0, 0, -1, 1, 0)^T$, $(\ldots, 0, 1, -2, 1, 0)^T$, $(\ldots, -1, 3, -3, 1, 0)^T$, ... and corresponding eigenvalues $-(2A+B)$, $-(3A+B)$, $-(4A+B)$, ... The strong localisation of these eigenvectors on the u-axis in the limit $G \to \infty$ and their alternating sign structures exclude them from the set of possible eigenvectors of the exact problem.

These artefacts can be suppressed by removing the last two points of the discretization lattice. The spectrum of the regularised $(G-2)\times(G-2)$ matrix is shown in Fig. 11B. When removing instead the last three lattice point, reducing H_{jk} to a $(G-3)\times(G-3)$ matrix, the structure of the spectrum remains unchanged (Fig. 11C), indicating that the regularisation was successful. The leading eigenvalue $l = -0.00018\tilde{\mu}m_*^{n-1}$ remains essentially unchanged close to zero. The corresponding eigenvector, appropriately normalised, differs by no more than 0.5×10^{-5} at any lattice point from the exact solution of $H_{m_*}f(u)=0$, given for $u_0 < u < -\ln\eta$ by $f(u)=\tilde{\beta}(u)$ (as in Eq. (114)) and $f(u)=0$ otherwise. This result confirms the numerical method used.

All eigenvalues but the leading one are complex. They are separated from the leading eigenvalue by a gap of about $0.7\tilde{\mu}m_*^{n-1}$. The size of the gap approximately equals $\tilde{\mu}(m_*/\eta)^{n-1}$, the mortality of the largest adults, and it was verified to exhibit the corresponding dependence on η. The density of the subdominant eigenvalues along a line in the complex plane (Fig. 11B) was verified to increase approximately as $G^{1/2}$. The results suggest that these subdominant eigenvalues becomes a continuous family of eigenvalues in the limit $G \to \infty$, separated from zero by gap of a size given by the mortality of the largest adults of the m_* size class considered.

REFERENCES

Ahrens, M.A., Peters, R.H., 1991. Patterns and limitations in limnoplankton size spectra. Can. J. Fish. Aquat. Sci. 48 (10), 1967–1978.

Akhiezer, A., Polovin, R., 1971. Criteria for wave growth. Soviet Phys. Uspekhi 14, 278.

Andersen, K.H., Beyer, J.E., 2006. Asymptotic size determines species abundance in the marine size spectrum. Am. Nat. 168 (1), 54–61.

Andersen, K.H., Pedersen, M., 2010. Damped trophic cascades driven by fishing in model marine ecosystems. Proc. R. Soc. B 277 (1682), 795–802.

Andersen, K., Ursin, E., 1977. A multispecies extension to the Beverton and Holt theory of fishing, with accounts of phosphorus circulation and primary production. Meddr. Danm. Fisk.- og Havunders 7, 319–435.

Andersen, K.H., Beyer, J.E., Pedersen, M., Andersen, N.G., Gislason, H., 2008. Life-history constraints on the success of the many small eggs reproductive strategy. Theor. Popul. Biol. 73, 490–497.

Arino, O., Shin, Y.-J., Mullon, C., 2004. A mathematical derivation of size spectra in fish populations. C. R. Biol. 327 (3), 245–254.

Arim, M., Berazategui, M., Barreneche, J.M., Ziegler, L., Zarucki, M., Abades, S.R., 2011. Determinants of density-body size scaling within food webs and tools for their detection. Adv. Ecol. Res. 45, 1–39.

Bays, J.S., Crisman, T.L., 1983. Zooplankton—trophic state relationships in Florida lakes. Can. J. Fish. Aquat. Sci. 40, 1813–1819.

Belgrano, A., Reiss, J., 2011. The Role of Body Size in Multispecies Systems. Advances in Ecological ResearchVol. 45. Academic Press, London.
Bender, E.A., Case, T.J., Gilpin, M.E., 1984. Perturbation experiments in community ecology: theory and practice. Ecology 65, 1–13.
Benoît, E., Rochet, M.-J., 2004. A continuous model of biomass size spectra governed by predation and the effects of fishing on them. J. Theor. Biol. 226 (1), 9–21.
Berlowa, L., Dunne, J.A., Martinez, N.D., Starke, P.B., Williams, R.J., Brose, U., 2009. Simple prediction of interaction strengths in complex food webs. Proc. Natl. Acad. Sci. U.S.A. 106 (1), 187–191.
Beyer, J., 1989. Recruitment stability and survival: simple size-specific theory with examples from the early life dynamic of marine fish. Dana 7, 45–147.
Blanchard, J.L., Jennings, S., Law, R., Castle, M.D., McCloghrie, P., Rochet, M.J., Benoît, E., 2009. How does abundance scale with body size in coupled size-structured food webs? J. Anim. Ecol. 78 (1), 270–280.
Blanco, J., Echevarria, F., Garcia, C., 1994. Dealing with size-spectra: some conceptual and mathematical problems. Sci. Mar. (Barcelona) 58 (1), 17–29.
Boccara, N., 1990. Functional Analysis: An Introduction for Physicists. Academic Press, New York.
Boudreau, P., Dickie, L., 1992. Biomass spectra of aquatic ecosystems in relation to fisheries yield. Can. J. Fish. Aquat. Sci. 49 (8), 1528–1538.
Bourassa, N., Morin, A., 1995. Relationships between size structure of invertebrate assemblages and trophy and substrate composition in streams. J. N. Am. Benthol. Soc. 14 (3), 393–403.
Brännström, Å., Loeuille, N., Loreau, M., Dieckmann, U., 2010. Emergence and maintenance of biodiversity in an evolutionary food-web model. Theor. Ecol. 4 (4), 467–478.
Brown, J.H., Gillooly, J.F., Allen, A.P., Savage, V.M., West, G.B., 2004. Toward a metabolic theory of ecology. Ecology 85 (7), 1771–1789.
Camacho, J., Solé, R., 2001. Scaling in ecological size spectra. Europhys. Lett. 55 (6), 774–780.
Capitán, J.A., Delius, G.W., 2010. Scale-invariant model of marine population dynamics. Phys. Rev. E 81 (6), 061901.
Castle, M.D., Blanchard, J.L., Jennings, S., 2011. Predicted effects of behavioural movement and passive transport on individual growth and community size structure in marine ecosystems. Adv. Ecol. Res. 45, 41–66.
Christensen, V., Pauly, D., 1992. Ecopath II—a software for balancing steady-state ecosystem models and calculating network characteristics. Ecol. Model. 61, 169–185.
Cross, M.C., Hohenberg, P.C., 1993. Pattern formation outside of equilibrium. Rev. Mod. Phys. 65, 851.
Cury, P., Pauly, D., 2000. Patterns and propensities in reproduction and growth of marine fishes. Ecol. Res. 15, 101–106.
Daan, N., Gislason, H., Pope, J.G., Rice, J.C., 2005. Changes in the North Sea fish community: evidence of indirect effects of fishing? ICES J. Mar. Sci. 62, 177–188.
Datta, S., Delius, G.W., Law, R., Plank, M.J., 2010. A stability analysis of the power-law steady state of marine size spectra. J. Math. Biol. 63 (4), 779–799.
Dortch, Q., Packard, T.T., 1989. Differences in biomass structure between oligotrophic and eutrophic marine ecosystems. Deep Sea Res. Part A Oceanogr. Res. Pap. 2, 223–240.
Duplisea, D.E., 2000. Benthic organism biomass size-spectra in the Baltic Sea in relation to the sediment environment. Limnol. Oceanogr. 45 (3), 558–568.
European Commission, 2010. Commission Decision on criteria and methodological standards on good environmental status of marine waters. Official Journal of the European Union 232(14), 2.9.2010.
Fisher, R.A., 1930. The Genetical Theory of Natural Selection. Oxford University Press, Oxford.

Friberg, N., Bonada, N., Bradley, D.C., Dunbar, M.J., Edwards, F.K., Grey, J., Hayes, R.B., Hildrew, A.G., Lamouroux, N., Trimmer, M., Woodward, G., 2011. Biomonitoring of human impacts in freshwater ecosystems: the good, the bad, and the ugly. Adv. Ecol. Res. 44, 1–68.

Fung, T., Farnsworth, K.D., Reid, D.G., Rossberg, A.G., 2012a. Recent data suggest no further recovery in North Sea Large Fish Indicator. ICES J. Mar. Sci. 69 (2), 235–239.

Fung, T., Farnsworth, K.D., Shephard, S., Reid, D.G., Rossberg, A.G., 2012b. Recovery of community size-structure from fishing requires multiple decades, submitted for publication.

Gaedke, U., 1992a. Identifying ecosystem properties: a case study using plankton biomass size distributions. Ecol. Model. 63, 277–298.

Gaedke, U., 1992b. The size distribution of plankton biomass in a large lake and its seasonal variability. Limnol. Oceanogr. 37 (6), 1202–1220.

Gaedke, U., Seifried, A., Adrian, R., 2004. Biomass size spectra and plankton diversity in a shallow eutrophic lake. Int. Rev. Hydrobiol. 89 (1), 1–20.

Garcia, H.E., Locarnini, R.A., Boyer, T.P., Antonov, J.I., 2006. World Ocean Atlas 2005 Volume 4: Nutrients (phosphate, nitrate, silicate). Vol. 64 of NOAA Atlas NESDIS 64. U.S. Government Printing Office, Washington, DC.

Gerlach, S., Hahn, A., Schrage, M., 1985. Size spectra of benthic biomass and metabolism. Mar. Ecol. Prog. Ser. 26 (1–2), 161–173.

Gilljam, D., Thierry, A., Edwards, F.K., Figueroa, D., Ibbotson, A.T., Jones, J.I., Lauridsen, R.B., Petchey, O.L., Woodward, G., Ebenman, B., 2011. Seeing double: size-based and taxonomic views of food web structure. Adv. Ecol. Res. 45, 67–133.

Greenstreet, S.P.R., Rogers, S.I., Rice, J.C., Piet, G.J., Guirey, E.J., Fraser, H.M., Fryer, R.J., 2011. Development of the EcoQO for the North Sea fish community. ICES J. Mar. Sci. J. du Conseil 68 (1), 1–11.

Hall, S.J., Collie, J.S., Duplisea, D.E., Jennings, S., Bravington, M., Link, J., 2006. A length-based multispecies model for evaluating community responses to fishing. Can. J. Fish. Aquat. Sci. 63 (6), 1344–1359.

Hartvig, M., Andersen, K.H., Beyer, J.E., 2011. Food web framework for size-structured populations. J. Theor. Biol. 272 (1), 113–122.

Hindmarsh, A.C., Brown, P.N., Grant, K.E., Lee, S.L., Serban, R., Shumaker, D.E., Woodward, C.S., 2005. SUNDIALS: suite of nonlinear and differential/algebraic equation solvers. ACM Trans. Math. Softw. 31 (3), 363–396.

Houde, E.D., 2009. Recruitment variability. In: Jakobsen, T., Fogarty, M.J., Megrey, B.A., Moksness, E. (Eds.), Fish Reproductive Biology: implications for assessment and management. Wiley-Blackwell, Oxford, pp. 91–171.

Hubbell, S.P., 2001. The Unified Neutral Theory of Biodiversity and Biogeography. Princeton University Press, Princeton, NJ.

Hui, A.K., Armstrong, B.H., Wray, A.A., 1978. Rapid computation of the Voigt and complex error functions. J. Quant. Spectrosc. Radiat. Transf. 19, 509–516.

ICES, 2010. Report of the Working Group on the Ecosystem Effects of Fishing Activities (WGECO). ICES Document CM 2010/ACOM: 23, Copenhagen.

Jackson, J.D., 1962. Classical Electrodynamics. John Wiley & Sons, New York.

Jacob, U., Thierry, A., Brose, U., Arntz, W.E., Berg, S., Brey, T., Fetzer, I., Jonsson, T., Mintenbeck, K., Möllmann, C., Petchey, O.L., Riede, J.O., Dunne, J.A., 2011. The role of body size in complex food webs: a cold case. Adv. Ecol. Res. 45, 181–223.

Jennings, S., Blanchard, J.L., 2004. Fish abundance with no fishing: predictions based on macroecological theory. J. Anim. Ecol. 73 (4), 632–642.

Jeppesen, E., Lauridsen, T.L., Mitchell, S.F., Christoffersen, K., Burns, C.W., 2000. Trophic structure in the pelagial of 25 shallow New Zealand lakes: changes along nutrient and fish gradients. J. Plankton Res. 22 (5), 951.

Johnson, R.S., 1997. A modern introduction to the mathematical theory of water waves. Cambridge University Press, Cambridge.
Keitt, T., 1997. Stability and complexity on a lattice: coexistence of species in an individual-based food web model. Ecol. Model. 102 (2–3), 243–258.
Kerr, S., 1974. Theory of size distribution in ecological communities. J. Fish. Res. Board Can. 31, 1859–1862.
Kerr, S., Dickie, L.M., 2001. The biomass spectrum. Columbia University Press, New York.
Kevorkian, J., Cole, J.D., 1996. Multiple Scale and Singular Perturbation Methods. Springer-Verlag, New York.
Koyama, H., Kira, T., 1956. Intraspecific competition among higher plants. VIII. Frequency distribution of individual plant weight as affected by the interaction between plants. Journal of the Institute of Polytechnics, Osaka City University, Series D 7, 73–94.
Lassen, H., Medley, P., 2001. Virtual population analysis: a practical manual for stock assessment. FAO Fisheries Technical Paper 400. Food and Agriculture Organization of the United Nations, Rome.
Law, R., Plank, M.J., James, A., Blanchard, J.L., 2009. Size-spectra dynamics from stochastic predation and growth of individuals. Ecology 90 (3), 802–811.
Marquet, P.A., Quiñones, R.A., Abades, S., Labra, F., Tognelli, M., 2005. Scaling and power-laws in ecological systems. J. Exp. Biol. 208, 1749–1769.
Maury, O., Faugeras, B., Shin, Y.J., Poggiale, J.C., Ari, T.B., Marsac, F., 2007a. Modeling environmental effects on the size-structured energy flow through marine ecosystems. Part 1: the model. Prog. Oceanogr. 4, 479–499.
Maury, O., Shin, Y.-J., Faugeras, B., Ari, T.B., Marsac, F., 2007b. Modeling environmental effects on the size-structured energy flow through marine ecosystems. Part 2: simulations. Prog. Oceanogr. 4, 500–514.
McKendrick, A.G., 1926. Application of mathematics to medical problems. Proc. Edinburgh Math. Soc. 44, 98–130.
Melian, C.J., Vilas, C., Baldo, F., Gonzalez-Ortegon, E., Drake, P., Williams, R.J., 2011. Eco-evolutionary dynamics of individual-based food webs. Adv. Ecol. Res. 45, 225–268.
Millennium Ecosystem Assessment, 2005. Ecosystems and Human Well-being: Synthesis. Island Press, Washington, DC.
Mintenbeck, K., Barrera-Oro, E.R., Brey, T., Jacob, U., Knust, R., Mark, F.C., Moreira, E., Strobel, A., Arntz, W.E., 2012. Impact of Climate Change on Fishes in Complex Antarctic Ecosystems. Adv. Ecol. Res. 46, 351–426.
Möllmann, C., Diekmann, R., 2012. Marine ecosystem regime shifts induced by 1387 climate and overfishing – a review for the Northern hemisphere. Adv. Ecol. Res. 47. In press.
Mulder, C., Elser, J.J., 2009. Soil acidity, ecological stoichiometry and allometric scaling in grassland food webs. Glob. Chang. Biol. 15 (11), 2730–2738.
Mulder, C., Den Hollander, H.A., Hendriks, A.J., 2008. Aboveground herbivory shapes the biomass distribution and flux of soil invertebrates. PLoS One 3 (10), e3573.
Mulder, C., Hollander, H.A.D., Vonk, J.A., Rossberg, A.G., Jagers op Akkerhuis, G.A.J.M., Yeates, G.W., 2009. Soil resource supply influences faunal size-specific distributions in natural food webs. Naturwissenschaften 96 (7), 813–826.
Mulder, C., Boit, A., Bonkowski, M., Ruiter, P.C.D., Mancinelli, G., van Der Heijden, M.G.A., van Wijnen, H.J., Vonk, J.A., Rutgers, M., 2011. A belowground perspective on dutch agroecosystems: how soil organisms interact to support ecosystem services. Adv. Ecol. Res. 44, 277–358.
Naisbit, R.E., Rohr, R.P., Rossberg, A.G., Kehrli, P., Bersier, L.-F., 2012. Phylogeny vs. body size as determinants of food-web structure. Proc. R. Soc. B 279, 3291–3297.
Nakazawa, T., Ushio, M., Kondoh, M., 2011. Scale dependence of predator–prey mass ratio: determinants and applications. Adv. Ecol. Res. 45, 269–302.

Neubert, M., Blumenshine, S., Duplisea, D., Jonsson, T., Rashlei, B., 2000. Body size and food web structure: testing the equiprobability assumption of the cascade model. Oecologia 123, 241–251.

Newell, A.C., Whitehead, J.A., 1969. Finite bandwidth, finit amplitude convection. J. Fluid Mech. 38, 279.

O'Dwyer, J.P., Lake, J.K., Ostling, A., Savage, V.M., Green, J.L., 2009. An integrative framework for stochastic, size-structured community assembly. Proc. Natl. Acad. Sci. U.S.A. 106 (15), 6170–6175.

O'Gorman, E., Pichler, D.E., Adams, G., Benstead, J.P., Craig, N., Cross, W.F., Demars, B.O.L., Friberg, N., Gislason, G.M., Gudmundsdottier, R., Hawczak, A., Hood, J.M., Hudson, L.N., Johansson, L., Johansson, M., Junker, J.R., Laurila, A., Manson, J.R., Mavromati, E., Nelson, D., Olafsson, J.S., Perkins, D.M., Petchey, O.L., Plebani, M., Reuman, D.C., Rall, B.C., Stewart, R., Thompson, M.S.A., Woodward, G., 2012. Impacts of warming on the structure and function of aquatic communities: individual- to ecosystem-level responses. Adv. Ecol. Res. 47. In press.

OSPAR, 2006. Report on North Sea Pilot Project on Ecological Quality Objectives. Tech. Rep. 2006/239, OSPAR.

Pace, M.L., 1986. An empirical analysis of zooplankton community size structure across lake trophic gradients. Limnol. Oceanogr. 31 (1), 45–55.

Pauly, D., Christensen, V., Walters, C., 2000. Ecopath, Ecosim and Ecospace as tools for evaluating ecosystem impact of fisheries. ICES J. Mar. Sci. 57, 697–706.

Perry, G.L.W., Enright, N.J., 2006. Spatial modelling of vegetation change in dynamic landscapes: a review of methods and applications. Progr. Phys. Geogr. 30 (1), 47.

Peters, R.H., 1983. The ecological implications of body size. Cambridge University Press, Cambridge.

Pigolotti, S., Lopez, C., Hernandez-Garcia, E., 2007. Species clustering in competitive Lotka-Volterra models. Phys. Rev. Lett. 98, 258101.

Platt, T., Denman, K., 1977. Organization in the pelagic ecosystem. Helgol. Wiss. Meeresunters. 30, 575–581.

Platt, T., Denman, K., 1978. The structure of pelagic ecosystems. Rapports et Procès-Verbaux des Réunions du Conseil International pour l'Exploration de la Mer 173, 60–65.

Pope, J.G., Rice, J.C., Daan, N., Jennings, S., Gislason, H., 2006. Modelling an exploited marine fish community with 15 parameters—results from a simple size-based model. ICES J. Mar. Sci. J. du Conseil 63 (6), 1029–1044.

Pratap, A., Govind, Tripathi, R.S., 1999. Magnetic dynamics of bilayer cuprate superconductors. Phys. Rev. B 60, 6775–6780.

Quiñones, R.A., Platt, T., Rodríguez, J., 2003. Patterns of biomass-size spectra from oligotrophic waters of the Northwest Atlantic. Prog. Oceanogr. 57, 405–427.

Reiss, J., Forster, J., Cássio, F., Pascoal, C., Stewart, R., Hirst, A.G., 2010. When microscopic organisms inform general ecological theory. Adv. Ecol. Res. 43 (10), 45–85.

Reuman, D.C., Mulder, C., Raffaelli, D., Cohen, J.E., 2008. Three allometric relations of population density to body mass: theoretical integration and empirical tests in 149 food webs. Ecol. Lett. 11 (11), 1216–1228.

Rice, J., Arvanitidis, C., Borja, A., Frid, C., Hiddink, J.G., Krause, J., Lorance, P., Ragnarsson, S., Sköld, M., Trabucco, B., 2011. Indicators for sea-floor integrity under the European Marine Strategy Framework Directive. Ecological Indicators 12, 174-184.

Riede, J.O., Rall, B.C., Banasek-Richter, C., Navarrete, S.A., Wieters, E.A., Brose, U., 2010. Scaling of food web properties with diversity and complexity across ecosystems. Adv. Ecol. Res. 42, 139–170.

Rodriguez, J., Mullin, M., 1986a. Relation between biomass and body weight of plankton in a steady state oceanic ecosystem. Limnol. Oceanogr. 31, 361–370.
Rodriguez, J., Mullin, M.M., 1986b. Diel and interannual variation of size distribution of oceanic zooplanktonic biomass. Ecology 67 (1), 215–222.
Rodriguez, J., Jiménez, F., Bautista, B., Rodriguez, V., 1987. Planktonic biomass spectra dynamics during a winter production pulse in mediterranean coastal waters. J. Plankton Res. 9 (6), 1183–1194.
Rossberg, A.G., 2007. Some first principles of complex systems theory. Publ. RIMS 1551, 129–136.
Rossberg, A.G., Farnsworth, K.D., 2011. Simplification of structured population dynamics in complex ecological communities. Theor. Ecol. 4 (4), 449–465.
Rossberg, A.G., Ishii, R., Amemiya, T., Itoh, K., 2008. The top-down mechanism for body-mass–abundance scaling. Ecology 89 (2), 567–580.
Rossberg, A.G., Farnsworth, K.D., Satoh, K., Pinnegar, J.K., 2011. Universal power-law diet partitioning by marine fish and squid with surprising stability-diversity implications. Proc. R. Soc. B 278 (1712), 1617–1625.
Schwinghammer, P., 1981. Characteristic size distributions of integral benthic communities. Can. J. Fish. Aquat. Sci. 38, 1255–1263.
Sheldon, R.W., Prakash, A., Sutcliffe Jr., W.H., 1972. The size distribution of particles in the ocean. Limnol. Oceanogr. 17, 327–340.
Sheldon, R., Sutcliffe, W.H.J., Paranjape, M., 1977. Structure of pelagic food chain and relationship between plankton and fish production. J. Fish. Res. Board Can. 34, 2344–2355.
Shephard, S., Fung, T., Houle, J.E., Farnsworth, K.D., Reid, D.G., Rossberg, A.G., 2012. Size-selective fishing drives species composition in the Celtic Sea. ICES J. Mar. Sci. 69 (2), 223–234.
Shin, Y.J., Cury, P., 2004. Using an individual-based model of fish assemblages to study the response of size spectra to changes in fishing. Can. J. Fish. Aquat. Sci. 61 (3), 414–431.
Silvert, W., Platt, T., 1978. Energy flux in the pelagic ecosystem: a time-dependent equation. Limnol. Oceanogr. 23 (4), 813–816.
Silvert, W., Platt, T., 1980. Dynamic energy-flow model of the particle size distribution in pelagic ecosystems. In: Kerfoot, W.C. (Ed.), Evolution and Ecology of Zooplankton Communities. University Press of New England, Hanover, New Hampshire and London, England, pp. 754–763.
Solomon, S., Qin, D., Manning, M., Marquis, M., Averyt, K., Tignor, M.M.B., Miller Jr., H.L., Zhenlin, C. (Eds.), 2007. Climate Change 2007: The Physical Science Basis. Cambridge University Press, Cambridge, UK.
Speirs, D.C., Guirey, E.J., Gurney, W.S.C., Heath, M.R., 2010. A length-structured partial ecosystem model for cod in the North Sea. Fish. Res. 106 (3), 474–494.
Sprules, W.G., Goyke, A.P., 1994. Size-based structure and production in the pelagia of Lakes Ontario and Michigan. Can. J. Fish. Aquat. Sci. 51, 2603–2611.
Sprules, W.G., Munawar, M., 1986. Plankton size spectra in relation to ecosystem productivity, size, and perturbation. Can. J. Fish. Aquat. Sci. 43, 1789–1794.
Sprules, W.G., Casselman, J.M., Shuter, B.J., 1983. Size distributions of pelagic particles in lakes. Can. J. Fish. Aquat. Sci. 40, 1761–1769.
Thibault, K.M., White, E.P., Hurlbert, A.H., Ernest, S., 2011. Multimodality in the individual size distributions of bird communities. Glob. Ecol. Biogeogr. 20 (1), 145–153.
Thiebaux, M.L., Dickie, L.M., 1993. Structure of the body-size spectrum of the biomass in aquatic ecosystems: a consequence of allometry in predator-prey interactions. Can. J. Fish. Aquat. Sci. 50, 1308–1317.

Twomey, M., Jacob, U., Emmerson, M.E., 2012. Perturbing a marine food web-consequences for food web structure and trivariate patterns. Adv. Ecol. Res. 47. In press.

Vasseur, D.A., Gaedke, U., 2007. Spectral analysis unmasks synchronous and compensatory dynamics in plankton communities. Ecology 88 (8), 2058–2071.

von Foerster, H., 1959. Some remarks on changing populations. In: Stohlman, F. (Ed.), The Kinetics of Cellular Proliferation. Grune & Stratton, New York, pp. 382–407.

Wilson, W., Lundberg, P., Vázquez, D., Shurin, J., Smith, M., Langford, W., Gross, K., Mittelbach, G., 2003. Biodiversity and species interactions: extending Lotka–Volterra community theory. Ecol. Lett. 6 (10), 944–952.

Woodward, G., Warren, P.H., 2007. Body size and predatory interactions in freshwaters: scaling from individuals to communities. In: Hildrew, A.G., Raffaelli, D., Edmonds-Brown, R. (Eds.), Body Size: The Structure and Function of Aquatic Ecosystems. Cambridge University Press, Cambridge, pp. 98–117.

Woodward, G., Benstead, J.P., Beveridge, O.S., Blanchard, J., Brey, T., Brown, L.E., Cross, W.F., Friberg, N., Ings, T.C., Jacob, U., Jennings, S., Ledger, M.E., Milner, A.M., Montoya, J.M., O'Gorman, E., Olesen, J.M., Petchey, O.L., Pichler, D.E., Reuman, D.C., Thompson, M.S.A., van Veen, F.J.F., Yvon-Durocher, G., 2010a. Ecological networks in a changing climate. Adv. Ecol. Res. 42, 72–138.

Woodward, G., Blanchard, J., Lauridsen, R.B., Edwards, F.K., Jones, J.I., Figueroa, D., Warren, P.H., Petchey, O.L., 2010b. Individual-based food webs: species identity, body size and sampling effects. Adv. Ecol. Res. 43, 211–266.

Yodzis, P., 1998. Local trophodynamics and the interaction of marine mammals and fisheries in the Benguela ecosystem. J. Anim. Ecol. 67 (4), 635–658.

Yvon-Durocher, G., Montoya, J.M., Trimmer, M., Woodward, G., 2011. Warming alters the size spectrum and shifts the distribution of biomass in freshwater ecosystems. Glob. Chang. Biol. 17 (4), 1681–1694.

Zhou, M., Huntley, M.E., 1997. Population dynamics theory of plankton based on biomass spectra. Mar. Ecol. Prog. Ser. 159, 61–73.

INDEX

Note: Page numbers followed by "*f*" indicate figures, "*t*" indicate tables, and "*b*" indicate boxes.

I

M

Z

ADVANCES IN ECOLOGICAL RESEARCH VOLUME 1–46

CUMULATIVE LIST OF TITLES